DIFFERENTIAL

EQUATIONS

with

Maple

V®

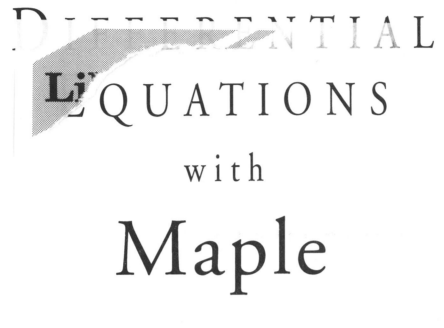

DIFFERENTIAL EQUATIONS

WITH

Maple V®

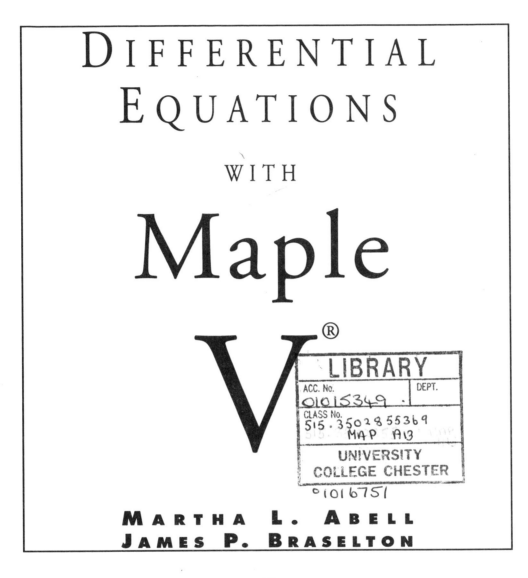

MARTHA L. ABELL
JAMES P. BRASELTON

AP PROFESSIONAL
Boston   San Diego   New York
London   Sydney   Tokyo   Toronto

Maple and Maple V are registered trademarks
of Waterloo Maple Software

AP PROFESSIONAL
955 Massachusetts Avenue, Cambridge, MA 02139

An Imprint of ACADEMIC PRESS, INC.
A Division of HARCOURT BRACE & COMPANY

United Kingdom Edition published by
ACADEMIC PRESS LIMITED
24–28 Oval Road, London NW1 7DX

Library of Congress Cataloging-in-Publication Data

Abell, Martha L., 1962-
  Differential equations with Maple V / Martha L. Abell and James P.
Braselton
      p.   cm.
  Includes bibliographical references and index.
  ISBN 0-12-041548-8
  1. Differential equations--Data processing. 2. Maple (Computer
file) I. Braselton, James P., 1965- . II. Title.
  QA371.5.D37A23 1994
  515'.35'028553--dc20   94-19965
                                                                    CIP

ISBN 0-12- 041548-8

Printed in the United States of America
94 95 96 97 98 IP 9 8 7 6 5 4 3 2 1

Contents

APPENDIX   GETTING HELP FROM MAPLE V   635

Preface

Maple V's diversity makes it particularly well suited to performing many calculations encountered when solving ordinary and partial differential equations. In some cases, Maple's built-in functions can immediately solve a differential equation by providing an explicit, implicit, or numerical solution; in other cases, Maple can be used to perform the calculations encountered when solving a differential equation. Since one goal of differential equations courses is to introduce the student to basic methods and algorithms and for the student to gain proficiency in them, nearly every topic covered in *Differential Equations with Maple V* includes typical examples solved by traditional methods and examples solved using Maple. Consequently, we feel that we have addressed one issue frequently encountered when implementing computer-assisted instruction. In addition, *Differential Equations with Maple V* uses Maple to establish well-known algorithms for solving elementary differential equations.

Taking advantage of the capabilities of Release 2 of Maple V, *Differential Equations with Maple V* introduces the fundamental concepts of differential equations as encountered in typical introductory differential equations courses and uses Maple V to solve typical problems of interest to students, instructors, and scientists. Other features to help make *Differential Equations with Maple V* as easy to use as possible include the following:

1. Getting Started. The Appendix provides a brief introduction to Maple V, including discussions about entering and evaluating commands, loading miscellaneous library functions and packages, and taking advantage of Maple's extensive help facilities.

2. Release 2 Compatibility. All examples illustrated in *Differential Equations with Maple V* were completed using Release 2 of Maple V. Although most computations can continue to be carried out with Release 1 of Maple V, we have taken advantage of the new features in Release 2 as much as possible.

3. Detailed Table of Contents. The table of contents includes all chapter, section, and subsection headings. Along with the comprehensive index, we hope that users will be able to locate information quickly and easily.

4. Comprehensive Index. In the index, mathematical examples are listed by topic, or name, as well as commands along with frequently used options: particular mathematical examples as

well as examples illustrating how to use frequently used commands are easy to locate. In addition, commands listed in the index are cross-referenced with frequently used options. Functions contained in the various available packages are cross-referenced both by package and alphabetically.

5. Glossary. A glossary of the Maple V commands used in the text is included at the end of the book and complements the Quick Reference included inside the front and back covers.

Differential Equations with Maple V may be used as a handbook that addresses some ways to use Maple for computation of explicit or numerical solutions of a variety of familiar differential equations and as a supplement for beginning courses in ordinary and/or partial differential equations.

Of course, we must express our appreciation to those who assisted in this project. Most important, we would like to thank our assistant, Lori Braselton, for typing, running, and verifying a substantial portion of the code that appears in the text in addition to proofreading a large amount of it. We would like to express appreciation to our editor, Charles B. Glaser, and our production editor, Reuben Kantor, at AP PROFESSIONAL for providing a pleasant environment in which to work. In addition, Waterloo Maple Software, particularly Janet Cater, Benton L. Leong, and J. Stan Devitt, have been most helpful in supplying us with information about Maple. Finally, we thank those close to us, especially Imogene Abell and Lori Braselton, for enduring with us the pressures of meeting a deadline and for graciously accepting our demanding work schedules. We certainly could not have completed this task without their care and understanding.

M. L. Abell
J. P. Braselton
Statesboro, Georgia

Introduction to
Differential Equations

1.1 PURPOSE

The purpose of *Differential Equations with Maple V* is twofold. First, we introduce and discuss in a very standard manner all topics typically covered in an undergraduate course in ordinary differential equations as well as some supplementary topics such as Laplace transforms, Fourier series, and partial differential equations that are not. Second, we illustrate how Maple is used to enhance the study of differential equations not only by eliminating the computational difficulties but also by overcoming the visual limitations associated with the solutions of differential equations. In each chapter, we first briefly present the material in a manner similar to most differential equations texts and then illustrate how Maple can be used to solve typical problems. For example, in Chapter 2, we introduce the topic of first-order equations. First, we show how to solve the problems by hand and then show how Maple can be used to perform the same solution procedures. Finally, we illustrate how commands like **dsolve** can be used to solve some equations directly. In Chapter 3 we discuss some applications of first-order equations. Since we are experienced and understand the methods of solution covered in Chapter 2, we make use of **dsolve** and similar commands to obtain solutions. In doing so, we are able to emphasize the applications themselves as opposed to becoming bogged down in calculations.

The advantages of using Maple in the study of differential equations are numerous, but perhaps the most useful is that of being able to produce the graphics associated with solutions of differential equations. This is particularly beneficial in the discussion of applications because many physical situations are modeled with differential equations. For example, we will see that the motion of a pendulum can be modeled by a differential equation. When we solve the problem of the motion of a pendulum, we use technology to actually watch the pendulum move. The same is true for the motion of a mass attached to the end of a spring as well as many other problems. With this ability, the study of differential equations becomes much more meaningful as well as interesting.

If you are a beginning Maple V user and, especially, new to Release 2, the Appendix contains an introduction to Maple V, including discussions about entering and evaluating commands, loading miscellaneous library functions and packages, and taking advantage of Maple's extensive help facility. In addition, the Glossary includes brief descriptions of all Maple V commands used in the text.

Although Chapter 1 is short in length, the vocabulary introduced will be used throughout the text. Consequently, even though, to a large extent, it may be read quickly, subsequent chapters will take advantage of the terminology and techniques discussed here.

1.2 DEFINITIONS AND CONCEPTS

We begin our study of differential equations by explaining what a differential equation is. As the two words **differential** and **equation** suggest, a **differential equation** is an equation containing derivatives of a function.

DEFINITION	*Differential Equation*
	A **differential equation** is an equation which contains the derivative or differentials of one or more dependent variables with respect to one or more independent variables. If the equation contains only ordinary derivatives (of one or more dependent variables) with respect to a single independent variable, then the equation is called an **ordinary differential equation**.

EXAMPLE: Determine which of the following are examples of ordinary differential equations:

(a) $\dfrac{dy}{dx} = \dfrac{x^2}{y^2 \cos(y)}$; (b) $\dfrac{dy}{dx} + \dfrac{du}{dx} = u + x^2 y$; (c) $(y-1)\,dx + x\cos(y)\,dy = 1$; (d) $\dfrac{\partial u}{\partial t} = \dfrac{\partial^2 u}{\partial x^2}$;

(e) $x^2 y'' + xy' + \left(x^2 - n^2\right) y = 0$; and (f) $\dfrac{\partial^2 u}{\partial t^2} = \dfrac{\partial^2 u}{\partial x^2}$.

SOLUTION: The equations in parts (a), (b), (c), and (e) are ordinary differential equations. The equations in parts (d) and (f) are not since they contain partial derivatives.

Our goal in this course is to construct a solution or a numerical approximation of the solution to a given differential equation. In fact, we will see that given an **arbitrary** differential equation, constructing an explicit solution is nearly always impossible. Consequently, although mathematicians were first concerned with finding analytic (or explicit) solutions to differential equations, after realizing that explicit solutions were usually impossible to construct, mathematicians have since (frequently) turned their attention to addressing properties of the solution and finding algorithms to approximate the solution.

If the equation contains partial derivatives of one or more dependent variables, then the equation is called a **partial differential equation**.

DEFINITION | *Partial Differential Equation*

A **partial differential equation** is an equation which contains the partial derivatives or differentials of one or more dependent variables with respect to more than one independent variables.

EXAMPLE: Determine which of the following are examples of partial differential equations:

(a) $u\dfrac{\partial u}{\partial t} = \dfrac{\partial u}{\partial x}$; (b) $uu_x + u = u_{yy}$; (c) $\dfrac{\partial^2 u}{\partial x^2} + \dfrac{\partial^2 u}{\partial y^2} = 0$; and (d) $\dfrac{\partial u}{\partial t} = \dfrac{\partial^2 u}{\partial x^2}$.

SOLUTION: All of these equations are partial differential equations.

Differential equations can be categorized into groups of equations that may be solved in similar ways. The first level of classification, distinguishing ordinary and partial differential equations, was just discussed. We extend this classification system with the following definition.

DEFINITION | *Order*

The highest derivative in the differential equation is called the **order** of the equation.

EXAMPLE: Determine the order of each of the following differential equations:

(a) $\dfrac{dy}{dx} = \dfrac{x^2}{y^2\cos(y)}$; (b) $u_{xx} + u_{yy} = 0$; (c) $\left(\dfrac{dy}{dx}\right)^4 = y + x$; and (d) $y^3 + \dfrac{dy}{dx} = 1$.

SOLUTION: (a) This equation is first order since it includes only one first-order derivative, $\dfrac{dy}{dx}$. (b) This equation is classified as second order since the highest order derivative, both u_{xx}, representing $\dfrac{\partial^2 u}{\partial x^2}$, and u_{yy}, representing $\dfrac{\partial^2 u}{\partial y^2}$, is of the second order. Hence, Laplace's equation is a second-order partial differential equation. (c) This equation is classified as first order since the highest order derivative is the first derivative. Raising that derivative to the fourth power does not affect the order of the equation. The expressions

$$\left(\frac{dy}{dx}\right)^4 \text{ and } \frac{d^4y}{dx^4}$$

do not represent the same quantities: $\left(\frac{dy}{dx}\right)^4$ represents the derivative of y with respect to x, $\frac{dy}{dx}$, raised to the fourth power; $\frac{d^4y}{dx^4}$ represents the fourth derivative of y with respect to x. (d) Again, we have a first-order equation since the highest order derivative is the first derivative.

The next level of classification is based on the following definition.

DEFINITION

> *Linear Differential Equation*
>
> An ordinary differential equation is **linear** (of order n) if it is of the form
>
> $$a_n(x)\frac{d^n y}{dx^n} + a_{n-1}(x)\frac{d^{n-1}y}{dx^{n-1}} + \cdots + a_2(x)\frac{d^2y}{dx^2} + a_1(x)\frac{dy}{dx} + a_0(x)y = f(x),$$
>
> where the functions $a_j(x), j = 0, 1, \ldots, n$; and $f(x)$ are given and at least $a_n(x)$ is not the zero function.
>
> If the equation does not meet the requirements of this definition, then the equation is said to be **nonlinear**. A similar classification is followed for partial differential equations. In this case, the coefficients in a linear partial differential equation are functions of the independent variables.

EXAMPLE: Determine which of the following differential equations are linear: (a) $\frac{dy}{dx} = x^3$; (b) $\frac{d^2u}{dx^2} + u = e^x$; (c) $(y - 1)\,dx + x\cos(y)\,dy = 1$; (d) $\frac{d^3y}{dx^3} + y\frac{dy}{dx} = x$; (e) $\frac{dy}{dx} + x^2y = x$; (f) $\frac{d^2x}{dt^2} + \sin x = 0$; (g) $u_{xx} + yu_y = 0$; and (h) $u_{xx} + uu_y = 0$.

SOLUTION: (a) This equation is linear because the nonlinear term x^3 is the function $f(x)$ in the general formula. (b) This equation is also linear. Using u as the dependent variable name does not affect the linearity. (c) Solving for $\frac{dy}{dx}$ we have $\frac{dy}{dx} = \frac{1-y}{x\cos y}$. Since the right-hand side of this equation is a nonlinear function of y, the equation is nonlinear. (d) The coefficient of the term

$\dfrac{dy}{dx}$ is y and, thus, is not a function of x. Hence, this equation is nonlinear. (e) This equation is linear. The term x^2 is merely the coefficient function. (f) This equation, known as the **pendulum equation** because it models the motion of a pendulum, is nonlinear since it involves a function of x, the dependent variable in this case (t is the independent variable). This function is $\sin x$. (g) This partial differential equation is linear, because the coefficient of u_y is a function of one of the independent variables. (h) In this case, there is a product of u and one of its derivatives. Therefore, the equation is nonlinear.

In the same manner that we consider systems of equations in algebra, we can also consider systems of differential equations. For example, if x and y represent functions of t, we will learn to solve the **system of linear equations**

$$\begin{cases} x' = ax + by \\ y' = cx + dy \end{cases},$$

where a, b, c, and d represent constants, in Chapter 9. We will see that systems of differential equations arise naturally in many physical situations that are modeled with more than one equation and involve more than one dependent variable.

1.3 SOLUTIONS OF DIFFERENTIAL EQUATIONS

When faced with a differential equation, our goal is frequently, but not always, to determine solutions to the equation. Hence, we state the following definition.

DEFINITION

Solution

A **solution** of a differential equation on a given interval is a function that is continuous on the interval and has all the necessary derivatives that are present in the differential equation such that when substituted into the equation yields an identity for all values on the interval.

In later chapters, we will discuss methods for solving differential equations. Here, in order to understand what is meant to be a solution, we give both the equation and a solution, and we verify the solution.

EXAMPLE: Verify that the given function is a solution to the corresponding differential equation: (a) $\dfrac{dy}{dx} = 3y$, $y(x) = e^{3x}$; (b) $\dfrac{d^2u}{dx^2} + 16u = 0$, $u(x) = \cos 4x$; and (c) $y'' + 2y' + y = 0$, $y(x) = xe^{-x}$.

SOLUTION: (a) Differentiating y we have $\dfrac{dy}{dx} = 3e^{3x}$. Since $y = e^{3x}$, $\dfrac{dy}{dx} = 3y$. (b) Two derivatives are required in this case: $\dfrac{du}{dx} = -4\sin 4x$ and $\dfrac{d^2u}{dx^2} = -4\cos 4x$. Therefore,

$$\frac{d^2u}{dx^2} + 4u = -4\cos 4x + 4\cos 4x = 0.$$

(c) In this case, we illustrate how to use Maple. If you are a beginning Maple V user, see the Appendix for help getting started with Maple V. After defining y,

```
> y:=x->x*exp(-x);latex(");
```
$$y := x \longrightarrow xe^{-x}$$

we use **diff** to compute $y' = e^{-x} - xe^{-x}$, naming the resulting output **d1**.

```
> d1:=diff(y(x),x);
```
$$d1 := e^{-x} - xe^{-x}$$

Similarly, we use **diff** together with **$** to compute $y'' = -2e^{-x} + xe^{-x}$, naming the resulting output **d2**.

```
> d2:=diff(y(x),x$2);
```
$$d2 := -2e^{-x} + xe^{-x}$$

Finally, we compute $y'' + 2y' + y = -2e^{-x} + xe^{-x} + 2(e^{-x} - xe^{-x}) + xe^{-x} = 0$. In this case, we see that the result is simplified.

```
> d2+2*d1+y(x);
```
$$0$$

In cases when Maple does not simplify results automatically, try using commands like **simplify** or **combine**. To graph $y(x)$, use **plot**. For example, entering

```
> plot(y(x),x=-1..1);
```

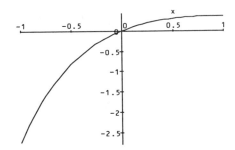

graphs $y(x)$ on the interval $[-1, 1]$.

In each previous example, the solution is given as a function of the independent variable. In these cases, the solution is said to be **explicit**. In solving some differential equations, an explicit solution cannot be determined, but we are able to determine an **implicit** solution as illustrated in the following example.

EXAMPLE: Verify that the given implicit function satisfies the differential equation.

$$\text{Function: } 2x^2 + y^2 - 2xy + 5x = 0$$
$$\text{Differential Equation :} \frac{dy}{dx} = \frac{2y - 4x - 5}{2y - 2x}$$

SOLUTION: We first use implicit differentiation to compute the derivative of $2x^2 + y^2 - 2xy + 5x = 0$.

$$4x + 2y\frac{dy}{dx} - 2x\frac{dy}{dx} - 2y + 5 = 0$$
$$\frac{dy}{dx}(2y - 2x) = 2y - 4x - 5$$
$$\frac{dy}{dx} = \frac{2y - 4x - 5}{2y - 2x}$$

We also illustrate how to use Maple to differentiate $2x^2 + y^2 - 2xy + 5x = 0$. After clearing all prior definitions of x and y, if any, by entering x:='x':y:='y':, we use D to differentiate $2x^2 + y^2 - 2xy + 5x = 0$, naming the result step_1.

```
> x:='x':y:='y':
  step_1:=D(2*x^2+y^2-2*x*y+5*x=0);
```

$step\_1 := 4D(x)x + 2D(y)y - 2D(x)y - 2xD(y) + 5D(x) = 0$

We then replace each occurrence of `D(x)` in `step_1` by 1 with `subs`, naming the result `step_2`. We interpret `step_2` to be equivalent to the equation $4x + 2y\dfrac{dy}{dx} - 2y - 2x\dfrac{dy}{dx} + 5 = 0$.

```
> step_2:=subs(D(x)=1,step_1);
```

$$step\_2 := 4x + 2D(y)y - 2y - 2xD(y) + 5 = 0$$

Finally, we obtain the derivative by solving `step_2` for `D(y)` with `solve`.

```
> solve(step_2,D(y));
```

$$-\frac{4x - 2y + 5}{2y - 2x}$$

Hence, the given implicit solution satisfies the differential equation $\dfrac{dy}{dx} = \dfrac{2y - 4x - 5}{2y - 2x}$. The solution (an ellipse) is graphed using `implicitplot`, which is contained in the `plots` package. First, we load the `plots` package by entering

```
> with(plots):
```

Note that the commands contained in the `plots` package are not displayed because a colon (`:`) is included at the end of the command instead of a semicolon (`;`). If a semicolon had been included, a list of the commands contained in the `plots` package would have been returned.
   Then, entering

```
> implicitplot(2*x^2+y^2-2*x*y+5*x=0,x=-7..2,y=-7..2);
```

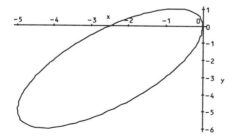

graphs the equation $2x^2 + y^2 - 2xy + 5x = 0$ on the rectangle $[-7, 2] \times [-7, 2]$.

Most differential equations have more than one solution. We illustrate this property in the following example.

EXAMPLE: Verify that the given solution which depends on an arbitrary constant satisfies the differential equation.

$$\text{Solution: } y = C \sin x, C = \text{ any constant}$$
$$\text{Differential Equation: } \frac{d^2y}{dx^2} + y = 0$$

SOLUTION: Differentiating we obtain $\frac{dy}{dx} = C \cos x$ and $\frac{d^2y}{dx^2} = -C \sin x$. Therefore,

$$\frac{d^2y}{dx^2} + y = -C \sin x + C \sin x = 0.$$

Some of the members of the family of solutions $y = C \sin x$ are graphed with **plot**. First, we use **seq** and **subs** to generate a set of eleven functions obtained by replacing C in $y = C \sin x$ by $-2.5, -2, -1.5, -1, -.5, 0, .5, 1, 1.5, 2$, and 2.5, naming the resulting set **to_plot**. Then the set of functions **to_plot** is graphed with **plot**. Notice that these functions are the sine function with various amplitudes.

```
> c_vals:=seq(-2.5+.5*i,i=0..10):
  to_plot:={seq(subs(c=i,c*sin(x)),i=c_vals)};
```

$$to\_plot := \{0, 2.5 \sin(x), 2.0 \sin(x), 1.5 \sin(x), 1.0 \sin(x), 0.5 \sin(x),$$
$$-0.5 \sin(x), -1.0 \sin(x), -1.5 \sin(x), -2.0 \sin(x), -2.5 \sin(x)\}$$

```
> plot(to_plot,x=0..4*Pi);
```

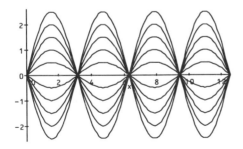

1.4 INITIAL- AND BOUNDARY-VALUE PROBLEMS

In many applications of differential equations, we are given not only a differential equation to solve but also one or more conditions that must be satisfied by the solution(s). The number of the conditions typically equals the order of the equation. For example, consider the first-order equation that models the exponential growth of a population,

$$\frac{dx}{dt} = x,$$

where $x(t)$ represents the population at time t. The solution of this equation is $x(t) = ce^t$, where c is a constant. Since this solution depends on an arbitrary constant, we call this a **general solution**. However, in a problem such as this, we usually know the initial population. Therefore, we must determine the one solution that satisfies the given **initial condition**. Suppose that if $t = 0$, we have $x = 10$, which means that $x(0) = 10$. Substitution into the general solution and solving for c yields $c = 10$. Therefore, the solution to the **initial-value problem**

$$\begin{cases} \dfrac{dx}{dt} = x \\ x(0) = 10 \end{cases}$$

is $x(t) = 10e^t$. Notice that this first-order equation requires one auxiliary condition to eliminate the unknown coefficient in the general solution.

EXAMPLE: Consider the first-order equation that determines the velocity of an object of mass $m = 1$ subjected to air resistance equivalent to the instantaneous velocity of the object:

$$\frac{dv}{dt} = 32 - v,$$

where $v(t)$ represents the object's velocity at time t. A general solution to this equation is $v(t) = 32 + ce^{-t}$. If the initial velocity of the object is $v(0) = 0$, then determine the solution that satisfies this initial condition.

SOLUTION: Substituting into the general solution, we have $v(0) = 32 + c = 0$. Hence, $c = -32$, and the solution to the initial-value problem is $v(t) = 32 - 32e^{-t}$.

Since first-order equations involve a single auxiliary condition, which is usually referred to as an initial condition, we use the following example to distinguish between **initial-** and **boundary-value problems**, which involve higher-order equations.

EXAMPLE: Consider the second-order differential equation $x'' + x = 0$, which models the motion of a mass with $m = 1$ attached to the end of a spring with spring constant $k = 1$, where $x(t)$ represents the distance of the mass from the equilibrium position $x = 0$ at time t. A general solution of this differential equation is $x(t) = A\cos t + B\sin t$. Since this is a second-order equation, we need two auxiliary conditions to determine the unknown constants.

(a) Suppose that the initial position of the mass is $x(0) = 0$ and the initial velocity $x'(0) = 1$. Then this is an **initial-value problem** because we have two auxiliary conditions at the same value of t, namely $t = 0$. Use these initial conditions to determine the solution of this problem. (b) On the other hand, suppose that we know the position at two different values of t such as $x(0) = 0$ and $x\left(\dfrac{\pi}{2}\right) = -4$. Since the conditions are given at different values of t, we call this a **boundary-value problem**. Use the given boundary conditions to determine the solution to this problem.

SOLUTION: (a) Since we need the first derivative of the general solution, we calculate $x'(t) = -A\sin t + B\cos t$. Now, substitution yields $x(0) = A = 0$ and, thus, $x'(0) = B = 1$. Hence, the solution is $x(t) = \sin t$. (b) Substitution into the general solution gives us $x(0) = A = 0$ and $x(\pi/2) = B = -4$. Therefore, the solution to the boundary-value problem is $x(t) = -4\sin t$.

1.5 DIRECTION FIELDS

The geometrical interpretation of solutions to first-order differential equations of the form

$$\frac{dy}{dx} = f(x, y)$$

is important to the basic understanding of problems of this type. Suppose that a solution to this equation is a function $y = \psi(x)$. Hence, the solution is merely the graph of the function ψ. Therefore, if (x, y) is a point on this graph, the slope of the tangent line is given by $f(x, y)$. A set of short line segments representing the tangent lines can be constructed for a large number of points. This collection of line segments (or vectors) is known as the **direction field** of the differential equation and provides a great deal of information concerning the behavior of the family of solutions. This is due to the fact that by determining the slope of the tangent line for a large number of points in the plane, the shape of the solutions can be seen without actually having a formula for the solution. The direction field for a differential equation provides a geometric interpretation about the behavior of the solutions of the equation. Throughout this text, we will frequently display graphs of various solutions to a differential equation along with a graph of the direction field. At this point, you need not worry about how solutions to differential equations are constructed, but you should understand the graphical interpretation of the direction field. Note that the direction field is generated by determining the direction vectors at many points. When possible, we will generate both direction fields and a general solution for a differential equation. The resulting graphs can help us interpret the meaning of the solutions to the differential equation, especially when the differential equation is being used to solve an applied problem.

EXAMPLE: (a) Sketch various solutions of the differential equation $\frac{dy}{dx} = e^{-x} - 2y$. (b) Sketch the direction field associated with the equation.

SOLUTION: (a) To find a general solution of the linear equation $\frac{dy}{dx} = e^{-x} - 2y$, we first define the equation in $Diff\_Eq$ and then use $dsolve$ to find a general solution, naming the resulting output Gen_Sol. In Gen_Sol, $\_C1$ represents the arbitrary constant in the solution.

```
> y:='y':
  Diff_Eq:=diff(y(x),x)=exp(-x)-2*y(x);
  Gen_Sol:=dsolve(Diff_Eq,y(x));
```

$$Diff\_Eq := \frac{d}{dx}y(x) = e^{-x} - 2y(x)$$

$$Gen\_Sol := y(x) = e^{-x} + e^{-2x}\_C1$$

Thus, a general solution of $\frac{dy}{dx} = e^{-x} - 2y$ is given by $y = e^{-x} + Ce^{-2x}$. At this point, we name $y(x)$ the result obtained in Gen_Sol with $assign$ and then use seq and $subs$ to define the set of seven functions, to_plot, obtained by replacing $\_C1$ in $y(x)$ by i for $i = -3, -2, -1, 0, 1, 2,$ and 3. This set of seven functions is then graphed on the interval $[-1/2, 1]$ with $plot$.

```
> assign(Gen_Sol):
  to_plot:={seq(subs(_C1=i,y(x)),i=-3..3)}:
  plot(to_plot,x=-1/2..2,-1..1);
```

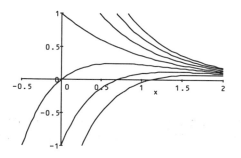

(b) To graph the direction field associated with the equation, we use `DEplot1`, which is contained in the `DEtools` package. After loading the `DEtools` package by entering `with(DEtools)`,

> `with(DEtools);`

   [*DEplot, DEplot1, DEplot2, Dchangevar, PDEplot, dfieldplot, phaseportrait*]

we use `DEplot1` to graph the direction field associated with the equation on the rectangle $[-1, 2] \times [-1, 1]$.

> `y:='y':`
  `DEplot1(Diff_Eq,y(x),x=-1/2..1,y=-1..1);`

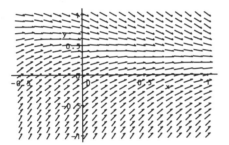

We can also use `DEplot1` to graph solutions without explicitly generating them, which is particularly useful if an explicit formula is either not wanted or impossible to obtain. For example, entering

> `DEplot1(Diff_Eq,y(x),x=-1/2..1,{[0,.75],[0,.5],[0,0],`
  `[0,-.5],[0,-.75]},y=-1..1);`

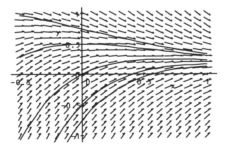

graphs the direction field associated with the equation along with the solutions that satisfy $y(0) = .75$, $y(0) = .5$, $y(0) = 0$, $y(0) = -.5$, and $y(0) = -.75$. If instead the command had been entered as

```
DEplot1(Diff_Eq,y(x),x=-1/2..1,{[0,.75],[0,.5],[0,0],
  [0,-.5],[0,-.75]},y=-1..1,arrows=NONE)
```

the solutions would have been graphed, but the direction field would not have been displayed.

Maple allows us to graph solutions of equations and associated direction fields that would be nearly impossible by traditional methods, as shown in the following example.

EXAMPLE: (a) Sketch various solutions of $\dfrac{dy}{dx} = \dfrac{\cos y - y \cos x}{x \sin y + \sin x - 1}$. (b) Sketch the direction field associated with the equation.

SOLUTION: (a) We begin by finding a general solution of $\dfrac{dy}{dx} = \dfrac{\cos y - y \cos x}{x \sin y + \sin x - 1}$ with **dsolve**, naming the resulting output **Gen_Sol**.

```
> y:='y':
  Gen_Sol:=dsolve(diff(y(x),x)=(cos(y(x))-y(x)*cos(x))/
    (x*sin(y(x))+sin(x)-1),y(x));
```

$Gen\_Sol := -x \cos(y(x)) + y(x) \sin(x) - y(x) = \_C1$

Thus, a general solution of $\dfrac{dy}{dx} = \dfrac{\cos y - y \cos x}{x \sin y + \sin x - 1}$ is $y \sin x - x \cos y - y = C$. Next we graph the solution for various values of C and the direction field associated with $\dfrac{dy}{dx} = \dfrac{\cos y - y \cos x}{x \sin y + \sin x - 1}$. First, we note that the graph of $y \sin x - x \cos y - y = C$ for various values of C is the same as the graph of the level curves of $f(x, y) = y \sin x - x \cos y$. We define **to_graph** to be $y \sin x - x \cos y$ by using **lhs** to extract the left-hand side of the equation **Gen_Sol** and then using **subs** to replace each occurrence of $y(x)$ by y:

```
> to_graph:=subs(y(x)=y,lhs(Gen_Sol));
```

$to\_graph := -x \cos(y) + y \sin(x) - y$

Then, after loading the **plots** package, we use **contourplot**, which is contained in the **plots** package, to graph several level curves of **to_graph** on the rectangle $[0, 4\pi] \times [0, 4\pi]$.

```
> with(plots):
  contourplot(to_graph,x=0..4*Pi,y=0..4*Pi,grid=[40,40],
   axes=NORMAL);
```

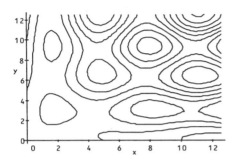

(b) To graph the direction field associated with the equation, we use DEplot1 in the same manner as we did in the previous example.

```
> with(DEtools):
  DEplot1(diff(y(x),x)=(cos(y(x))-y(x)*cos(x))/
   (x*sin(y(x))+sin(x)-1),y(x),x=0..4*Pi,y=0..4*Pi);
```

First-Order Ordinary Differential Equations

In this chapter we introduce frequently used first-order ordinary differential equations and methods to construct their solutions. The equations and methods of solution found in this chapter are standard. Although much of the material in this chapter is briefly discussed, several of the equations found here will be used in other chapters of the text.

2.1 SEPARATION OF VARIABLES

Differential equations are first encountered in beginning integral calculus courses. Although the phrase *differential equation* is not frequently used at that point, the problem of finding a function whose derivative is a given function is a differential equation.

DEFINITION

Separable Differential Equation

A differential equation that can be written in the form

$$g(y)y' = f(x)$$

is called a **separable differential equation**.

Separable differential equations are solved by collecting all the terms involving y on one side of the equation and all the terms involving x on the other side of the equation and integrating. Rewriting $g(y)y' = f(x)$ in the form

$$g(y)\frac{dy}{dx} = f(x)$$

yields $g(y)\,dy = f(x)\,dx$ so that

$$\int g(y)\,dy = \int f(x)\,dx + C,\ C \text{ a constant.}$$

Therefore, in the case of a separable differential equation, we simply separate the variables and integrate both sides of the equation.

EXAMPLE: Show that the equation

$$\frac{dy}{dx} = \frac{2y^{1/2} - 2y}{x}$$

is separable, and solve by separation of variables.

SOLUTION: The equation $\dfrac{dy}{dx} = \dfrac{2y^{1/2} - 2y}{x}$ is separable since it can be written in the form

$$\frac{dy}{2y^{1/2} - 2y} = \frac{dx}{x}.$$

To solve the equation, integrate both sides and simplify. Observe that

$$\int \frac{dy}{2y^{1/2} - 2y} = \int \frac{dx}{x} + C_1$$

is the same as

$$\int \frac{1}{2y^{1/2}} \frac{dy}{(1 - y^{1/2})} = \int \frac{dx}{x} + C_1.$$

To evaluate the integral on the left-hand side, let $u = 1 - y^{1/2}$ so that $du = \dfrac{-dy}{2y^{1/2}}$. We then obtain $\int \dfrac{-du}{u} = \int \dfrac{dx}{x} + C_1$ so that $-\ln u = \ln x + C_1$. Recall that $-\ln u = \ln \dfrac{1}{u}$. Then $\dfrac{1}{u} = Cx$, where $C = e^{C_1}$, and resubstituting we find that

$$\frac{1}{1 - y^{1/2}} = Cx$$

is a general solution of the equation $\dfrac{dy}{dx} = \dfrac{2y^{1/2} - 2y}{x}$.

To graph the direction field for

$$\frac{dy}{dx} = \frac{2y^{1/2} - 2y}{x},$$

we first load the **DEtools** package, define **Eq1** to be the differential equation, and then use **DEplot1** to graph the direction field for the equation on the rectangle $[0,1] \times [0,1]$.

```
> with(DEtools):
  Eq1:=diff(y(x),x)=(2*y^(1/2)-2*y)/x;
```

$$Eq1 := \frac{\partial}{\partial x} y(x) = \frac{2\sqrt{y} - 2y}{x}$$

The symbol **[x,y]** in the following command instructs Maple that in the solution, y is to be a function of x. Replacing **[x,y]** by **y(x)** returns the same graph.

```
> DEplot1(Eq1,[x,y],x=0..1,y=0..1);
```

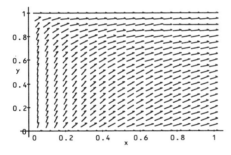

Since any differential equation can be accompanied by one or more auxiliary conditions, a separable equation can be stated along with an initial condition. Therefore, the equation is solved through the following steps:

1. Find a general solution of the differential equation using separation of variables.
2. Use the initial condition to determine the unknown constant in the general solution.

EXAMPLE: Solve the following initial-value problem:

$$y \cos(x)\, dx - (1 + y^2)\, dy = 0, \, y(0) = -1.$$

SOLUTION: Unlike the previous example, this equation is in differential form because dx and dy appear as multiples in the equation instead of the form $\dfrac{dy}{dx}$. The separation of variables is relatively simple, however, with

$$\cos(x)\, dx = \frac{1 + y^2}{y}\, dy.$$

Notice that the right-hand side can be written as $\dfrac{1}{y} + \dfrac{y^2}{y} = \dfrac{1}{y} + y$. Therefore, integration gives us

$$\sin x + C_1 = \ln|y| + \frac{1}{2}y^2.$$

By substituting $y(0) = -1$ into this equation, we find that $C_1 = \dfrac{1}{2}$, so the implicit solution is given by

$$\sin x + \frac{1}{2} = \ln|y| + \frac{1}{2}y^2.$$

As in the previous example, we can take advantage of the **DEplot1** command contained in the **DEtools** package to graph the direction field of the equation. We first rewrite the equation $\cos(x)\, dx = \dfrac{1 + y^2}{y}\, dy$ in the form $\dfrac{dy}{dx} = \dfrac{y \cos x}{1 + y^2}$ and then define **Eq** to be this equation.

```
> Eq:=diff(y(x),x)=y*cos(x)/(1+y^2);
```

$$Eq := \frac{d}{dx} y(x) = \frac{y \cos(x)}{1 + y^2}$$

Next, we load the **DEtools** package and use **DEplot1** to graph the direction field for **Eq** along with the solution satisfying $y(0) = -1$. Note that if you have already loaded the **DEtools** package during your current Maple session, you do not need to re-execute the command **with(DEtools)**.

```
> with(DEtools):
  DEplot1(Eq,[x,y],x=0..3*Pi,[0,-1],y=-Pi..Pi);
```

Maple can be used, in many cases, to perform the integration and algebraic simplification associated with a separable differential equation and sometimes even solve them with the `dsolve` command. Commands like `DEplot1` contained in the `DEtools` package can be used to graph various solutions of equations without actually solving a particular equation. In cases when an explicit solution is known, commands like `plot`, `contourplot`, or `implicitplot` can be used to graph solutions. Note that both `contourplot` and `implicitplot` are contained in the `plots` package.

EXAMPLE: Solve the equation $\dfrac{dy}{dx} = \dfrac{x^2}{\sqrt{9 - x^2}e^y\cos(y)}$ subject to $y(0) = 0$.

SOLUTION: Proceeding as in the previous examples, we first rewrite the equation as $e^y \cos y\, dy = \dfrac{x^2}{\sqrt{9 - x^2}}\, dx$. The integral $\displaystyle\int e^y \cos\ y\, dy$ can be computed by using integration by parts twice, while $\displaystyle\int \dfrac{x^2}{\sqrt{9 - x^2}}\, dx$ can be evaluated by using a trigonometric substitution. Results obtained using `int` are as follows:

```
> left_side:=int(exp(y)*cos(y),y);
  right_side:=int(x^2/sqrt(9-x^2),x);
```

$$left\_side := \frac{1}{2}e^y \cos(y) + \frac{1}{2}e^y \sin(y)$$

$$right\_side := -\frac{1}{2}x\sqrt{9 - x^2} + \frac{9}{2}\arcsin\left(\frac{1}{3}x\right)$$

Thus, a general solution is obtained by setting `left_side` equal to the sum of `right_side` and an arbitrary constant, denoted by C in `gen_sol`:

```
> gen_sol:=left_side=right_side+C;
```

$$gen\_sol := \frac{1}{2}e^y\cos(y) + \frac{1}{2}e^y\sin(y) = -\frac{1}{2}x\sqrt{9-x^2} + \frac{9}{2}\arcsin\left(\frac{1}{3}x\right) + C$$

To find the value of C that satisfies $y(0) = 0$, we use **subs** to substitute these conditions into **gen_sol**, naming the result **find_C**, and use **eval** to evaluate **find_C**. In this case, we see that $C = \frac{1}{2}$. In some cases, we might have to use **solve** to solve the resulting equation for C.

```
> find_C:=subs({x=0,y=0},gen_sol);
```

$$find\_C := \frac{1}{2}e^0\cos(0) + \frac{1}{2}e^0\sin(0) = \frac{9}{2}\arcsin(0) + C$$

```
> eval(find_C);
```

$$\frac{1}{2} = C$$

The desired solution is given by substituting $C = \frac{1}{2}$ into **gen_sol**.

```
> to_plot:=subs(C=1/2,gen_sol);
```

$$to\_plot := \frac{1}{2}e^y\cos(y) + \frac{1}{2}e^y\sin(y) = -\frac{1}{2}x\sqrt{9-x^2} + \frac{9}{2}\arcsin\left(\frac{1}{3}x\right) + \frac{1}{2}$$

We can graph the equation **to_plot** with the **implicitplot** command contained in the **plots** package. First, we load the **plots** package

```
> with(plots):
```

and then use **implicitplot** to graph the equation **to_plot** on the rectangle $[-3,3] \times [-2\pi, 2\pi]$.

```
> implicitplot(to_plot,x=-3..3,y=-2*Pi..2*Pi);
```

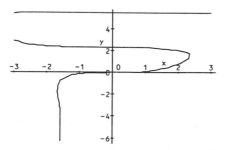

Of course, as in previous examples, we can also take advantage of the `DEplot1` command to graph the direction field for the equation:

```
> with(DEtools):
  DEplot1(diff(y(x),x)=x^2/(exp(y)*cos(y)*sqrt(9-x^2)),[x,y],
    x=-3..3,y=-2*Pi..Pi);
```

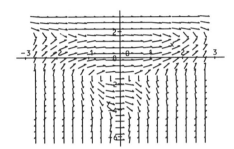

Our last example illustrates how to use **dsolve** to find a general solution of a separable equation.

EXAMPLE: Solve $y' = \dfrac{dy}{dx} = \dfrac{x^2 + 8}{(x^2 - 5x + 6)y^2 \cos(y)}$.

SOLUTION: For this problem, we use **dsolve** to find a general solution, naming the resulting output **Sol**.

```
> Sol:=dsolve(diff(y(x),x)=(x^2+8)/((x^2-5*x+6)*
  y(x)^2*cos(y(x))),y(x));
```

$Sol := y(x)^2 \sin(y(x)) - 2\sin(y(x)) + 2y(x)\cos(y(x)) - x + 12\ln(x - 2)$
$-17\ln(x - 3) = \_C1$

2.2 HOMOGENEOUS EQUATIONS

DEFINITION | *Homogeneous Differential Equation*

A differential equation that can be written in the form $M(x, y)\,dx + N(x, y)\,dy = 0$, where $M(tx, ty) = t^n M(x, y)$ and $N(tx, ty) = t^n N(x, y)$, is called a **homogeneous differential equation (of degree n)**.

It is a good exercise to show that an equation is homogeneous if we can write it in either of the forms $\dfrac{dy}{dx} = F\left(\dfrac{y}{x}\right)$ or $\dfrac{dy}{dx} = G\left(\dfrac{x}{y}\right)$.

EXAMPLE: Show that the equation $(x^2 + yx)\,dx - y^2\,dy = 0$ is homogeneous.

SOLUTION: Let $M(x, y) = x^2 + yx$ and $N(x, y) = -y^2$. Since

$$M(tx, ty) = (tx)^2 + (ty)(tx) = t^2(x^2 + yx) = t^2 M(x, y) \quad \text{and} \quad N(tx, ty) = -t^2 y^2 = t^2 N(x, y),$$

the equation $(x^2 + yx)\,dx - y^2\,dy = 0$ is homogenous of degree 2.

Homogeneous equations can be reduced to separable equations by the substitution of either $y = ux$ or $x = vy$. Note that either substitution will always reduce the equation to a separable equation. In most cases, the integration that results after separation of variables is of equivalent difficulty no matter which substitution is chosen. In general, however, use $y = ux$ if $M(x, y)$ is less complicated than $N(x, y)$, and use $x = vy$ if $N(x, y)$ is less complicated than $M(x, y)$. If a difficult integration problem is encountered after a substitution is made, then try the other substitution to see if it yields an easier problem.

EXAMPLE: Solve the equation $(x^2 - y^2)\,dx + xy\,dy = 0$.

SOLUTION: In this case, let $M(x, y) = x^2 - y^2$ and $N(x, y) = xy$. Then $M(tx, ty) = t^2 M(x, y)$ and $N(tx, ty) = t^2 N(x, y)$, which means that $(x^2 - y^2)\,dx + xy\,dy = 0$ is a homogeneous equation of degree 2. Assume $x = vy$. Then, $dx = v\,dy + y\,dv$, and directly substituting into the equation and simplifying yields

$$
\begin{aligned}
0 &= (x^2 - y^2)\,dx + xy\,dy \\
&= (v^2 y^2 - y^2)(v\,dy + y\,dv) + vy\,y\,dy \\
&= v^3 y^2\,dy - y^2 v\,dy + v^2 y^3\,dv - y^3\,dv + vy^2\,dy \\
&= (v^3 y^2 - y^2 v + vy^2)\,dy + (v^2 y^3 - y^3)\,dv \\
&= y^2 v^3\,dy + y^3(v^2 - 1)\,dv.
\end{aligned}
$$

Dividing this equation by $y^3 v^3$ yields the separable differential equation

$$\frac{dy}{y} + \frac{(v^2 - 1)\,dv}{v^3} = 0.$$

We solve this equation by rewriting it in the form

$$\frac{dy}{y} = \frac{(1 - v^2)\,dv}{v^3} = \left(\frac{1}{v^3} - \frac{1}{v}\right) dv$$

and integrating. This yields

$$\ln y = \frac{-2}{v^2} - \ln v + C_1,$$

which can be simplified as $\ln(vy) = \dfrac{-2}{v^2} + C_1$, so $vy = Ce^{-2/v^2}$, where $C = e^{C_1}$. Since $x = vy$, $v = \dfrac{x}{y}$ and resubstituting into the above equation yields

$$x = Ce^{-2y^2/x^2}$$

as a general solution of the equation $(x^2 - y^2)\,dx + xy\,dy = 0$. We graph the equation $x = Ce^{-2y^2/x^2}$ for various values of C by using **seq** to define **c_vals** to be the sequence of numbers consisting of $\dfrac{1}{2}, 1, \ldots, \dfrac{9}{2}, 5$ and then defining **to_plot** to be the set of equations obtained by replacing C in the equation $x = Ce^{-2y^2/x^2}$ by each value in **c_vals**.

> ```
> c_vals:=seq(n/2,n=1..10);
> ```

$$c\_vals := \frac{1}{2}, 1, \frac{3}{2}, 2, \frac{5}{2}, 3, \frac{7}{2}, 4, \frac{9}{2}, 5$$

> ```
> to_plot:={seq(x=c*exp(-2*y^2/x^2),c=c_vals)};
> ```

$$to\_plot := \left\{ x = \frac{1}{2}e^{-2\frac{y^2}{x^2}}, x = 5e^{-2\frac{y^2}{x^2}}, x = e^{-2\frac{y^2}{x^2}}, \right.$$
$$x = \frac{3}{2}e^{-2\frac{y^2}{x^2}}, x = 2e^{-2\frac{y^2}{x^2}}, x = \frac{5}{2}e^{-2\frac{y^2}{x^2}}, x = 3e^{-2\frac{y^2}{x^2}},$$
$$\left. x = \frac{7}{2}e^{-2\frac{y^2}{x^2}}, x = 4e^{-2\frac{y^2}{x^2}} \right\}$$

Each equation in **to_plot** is then graphed on the rectangle $[0, 6] \times [-3, 3]$ and all 10 graphs are shown together with **implicitplot**.

```
> with(plots):
  implicitplot(to_plot,x=0..6,y=-3..3);
```

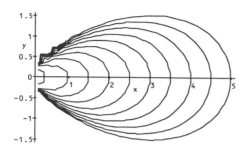

The next example illustrates how Maple can be used to help solve homogeneous equations.

EXAMPLE: Solve the equation $(x^{1/3}y^{2/3} + x)\,dx + (x^{2/3}y^{1/3} + y)\,dy = 0$.

SOLUTION: We begin by identifying $M(x, y) = x^{1/3}y^{2/3} + x$ and $N(x, y) = x^{2/3}y^{1/3} + y$ and then defining M and N.

```
> M:=(x,y)->x^(1/3)*y^(2/3)+x:
  N:=(x,y)->x^(2/3)*y^(1/3)+y:
```

Next, we verify that the equation is homogeneous of degree 1 by computing and factoring $M(tx, ty)$ and $N(tx, ty)$ with **factor**.

```
> factor(M(t*x,t*y));
  factor(N(t*x,t*y));
```

$$t\left(y^{2/3} + x^{1/3} + x\right)$$
$$t\left(x^{1/3}x^{2/3} + y\right)$$

In this case, we see that **dsolve** computes a general solution of the equation, as follows:

```
> dsolve(M(x,y(x))+N(x,y(x))*diff(y(x),x)=0,y(x));
```

$$x = \frac{\_C1x\left(x^{8/3} - y(x)^{4/3}x^{4/3} + y(x)^{8/3}\right)^{1/4}}{\sqrt{y(x)^{4/3} + x^{4/3}}(x^4 + y(x)^4)^{1/4}}$$

We next illustrate how Maple can be used to implement the steps we use to solve homogeneous equations. First, we let $x = vy$.

```
>  x:=v*y;
```

$$x := vy$$

We see that $D(x)$ represents dx. Similarly, $D(v)$ represents dv and $D(y)$ represents dy.

```
>  D(x);
```

$$D(v)y + vD(y)$$

Next, we evaluate the equation with the substitution $x = vy$, naming the result step_1.

```
>  step_1:=M(x,y)*D(x)+N(x,y)*D(y)=0;
```

$$step\_1 := \left(v^{1/3}y + vy\right)(D(v)y + vD(y)) + \left(v^{2/3}y + y\right)D(y) = 0$$

To see that the equation in step_1 is separable, we begin by using collect to collect together those terms containing $D(v)$, representing dv, and $D(y)$, representing dy.

```
>  step_2:=collect(step_1,{D(v),D(y)});
```

$$step\_2 := \left(v^{1/2}y + vy\right)yD(v) + \left(\left(v^{1/3}y + vy\right)v + v^{2/3}y + y\right)D(y) = 0$$

and then divide both sides of the resulting equation by $y^2\left[v\left(v^{1/3} + v\right) + v^{2/3} + 1\right]$.

```
>  step_3:=simplify(step_2/(y^2*(v*(v^(1/3)+v)+v^(2/3)+1)));
```

$$step\_3 := \frac{v^{4/3}D(y) + v^{1/3}yD(v) + D(y)}{y\left(v^{4/3} + 1\right)} = 0$$

```
>  step_4:=collect(step_3,{D(v),D(y)});
```

$$step\_4 := \frac{v^{1/3}D(v)}{v^{4/3} + 1} + \frac{D(y)}{y} = 0$$

The output in step_4 is equivalent to the equation $\dfrac{1}{y}dy = -\dfrac{v^{1/3}}{v^{4/3} + 1}dv$. We use int to evaluate

$\displaystyle\int \frac{1}{y}dy$ and $\displaystyle\int -\frac{v^{1/3}}{v^{4/3} + 1}dv$, naming the results left_side and right_side, respectively.

```
>  left_side:=int(1/y,y);
   right_side:=-int(v^(1/3)/(v^(4/3)+1),v);
```

$$left\_side := \ln(y)$$

$$right\_side := -\frac{1}{2}\ln(v^{4/3} + 1) + \frac{1}{4}\ln(1 - v^{4/3} + v^{8/3}) - \frac{1}{4}\ln(1 + v^4)$$

Finally, we assemble a general solution to the equation by first constructing a general solution of the separable equation in `step_5` and then substituting $v = \dfrac{x}{y}$ into `step_5`.

> `step_5:=left_side=right_side+C;`

$$step\_5 := \ln(y) = -\frac{1}{2}\ln(v^{4/3} + 1) + \frac{1}{4}\ln(1 - v^{4/3} + v^{8/3}) - \frac{1}{4}\ln(1 + v^4) + C$$

> `x:='x':v:='v':y:='y':`
> `gen_sol:=subs(v=x/y,step_5);`

$$gen\_sol := \ln(y) = -\frac{1}{2}\ln\left(\frac{x^{4/3}}{y^{4/3}} + 1\right) + \frac{1}{4}\ln\left(1 - \frac{x^{4/3}}{y^{4/3}} + \frac{x^{8/3}}{y^{8/3}}\right) - \frac{1}{4}\ln\left(1 + \frac{x^4}{y^4}\right) + C$$

> `simplify(gen_sol);`

$$\ln(y) = -\frac{1}{2}\ln(x^{4/3} + y^{4/3}) + \ln(y) + \frac{1}{4}\ln(y^{8/3} - x^{4/3}y^{4/3} + x^{8/3}) - \frac{1}{4}\ln(y^4 + x^4) + C$$

> `combine(gen_sol,ln);`

$$\ln(y) = C + \ln\left(\frac{y(y^{8/3} - x^{4/3}y^{4/3} + x^{8/3})^{1/4}}{\sqrt{x^{4/3} + y^{4/3}}(y^4 + x^4)^{1/4}}\right)$$

Thus, we see that a general solution of the equation is

$$y = \frac{Cy\left(y^{8/3} - x^{4/3}y^{4/3} + x^{8/3}\right)^{1/4}}{\sqrt{x^{4/3} + y^{4/3}}(y^4 + x^4)^{1/4}},$$

where C represents an arbitrary positive constant. We leave it to the reader to verify that this solution is equivalent to the one obtained with `dsolve` earlier. After writing the equation as $\dfrac{dy}{dx} = -\dfrac{x^{1/3}y^{2/3} + x}{x^{2/3}y^{1/3} + y}$, we can use `DEplot1` to graph the direction field for the equation along with various solutions. For example, entering

```
> with(DEtools):
  DEplot1(diff(y(x),x)=-(x^(1/3)*y^(2/3)+x)/(x^(2/3)*
  y^(1/3)+y),[x,y],x=-3..3,{[0,2],[0,1],[0,-1],[0,-2]},
  y=-3..3,stepsize=0.1);
```

graphs the direction field for the equation along with the solutions satisfying $y(0) = 2$, $y(0) = 1$, $y(0) = -1$, and $y(0) = -2$.

In the following example, we solve an initial-value problem.

EXAMPLE: Solve $xy^3\, dx - (x^4 + y^4)\, dy = 0$ subject to the initial condition $y(1) = 1$.

SOLUTION: We begin by clearing all prior definitions of x and y, if any, and then use $\mathtt{dsolve}$ to find a general solution of the equation, naming the resulting output $\mathtt{gen\_sol}$.

```
> x:='x':y:='y':
  gen_sol:=dsolve(x*y(x)^3-(x^4+y(x)^4)*diff(y(x),x)=0,y(x));
```

$$gen\_sol := x = \frac{\_C1xe^{\frac{1}{3}\sqrt{3}\arctan\left(\frac{1}{3}\frac{(-2y(x)^2+x^2)\sqrt{3}}{x^2}\right)}}{y(x)}$$

To find the particular solution which satisfies the condition $y(1) = 1$, we substitute these values into $\mathtt{gen\_sol}$, naming the resulting output $\mathtt{step\_1}$, evaluating $\mathtt{step\_1}$, and then solving the result for $\mathtt{\_C1}$, which represents the arbitrary constant in the general solution given before.

```
> step_1:=subs({x=1,y(x)=1},gen_sol);
```

$$step\_1 := 1 = \_C1e^{-\frac{1}{3}\sqrt{3}\arctan\left(-\frac{1}{3}\sqrt{3}\right)}$$

```
> step_2:=eval(step_1);
```

$$step\_2 := 1 = \_C1e^{-\frac{1}{18}\sqrt{3}\pi}$$

```
> step_3:=simplify(solve(step_2,_C1));
```

$$step\_3 := e^{\frac{1}{18}\sqrt{3}\pi}$$

We then substitute the value obtained for the constant and replace all occurrences of $y(x)$ by y in `gen_sol`, naming the solution `to_plot`. The result, an equation in the variables x and y, can be graphed with `implicitplot`.

```
> to_plot:=subs({_C1=step_3,y(x)=y},gen_sol);
```

$$to\_plot := x = \frac{e^{\frac{1}{18}\sqrt{3}\pi}\, xe^{\frac{1}{3}\sqrt{3}\arctan\left(\frac{1}{3}\frac{(-2y^2+x^2)\sqrt{3}}{x^2}\right)}}{y}$$

To graph `to_plot`, we first load the **plots** package and then use `implicitplot` to graph `to_plot` on the rectangle $[0.1, 2] \times [0.1, 2]$. The option `grid=[40,40]`, in the `implicitplot` command, instructs Maple to sample 40 points in each of the x- and y-directions, resulting in a smoother graph than if the default, `grid=[25,25]`, had been used.

```
> with(plots):
  implicitplot(to_plot,x=0.1..2,y=0.1..2,grid=[40,40]);
```

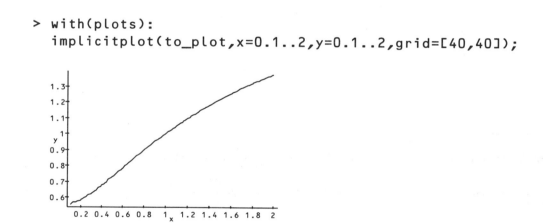

2.3 EXACT EQUATIONS

<table>
<tr><td>**DEFINITION**</td><td>*Exact Differential Equation*

A differential equation that can be written in the form $M(x, y)\, dx + N(x, y)\, dy = 0$, where $\dfrac{\partial N}{\partial x} = \dfrac{\partial M}{\partial y}$ is called an **exact differential equation**.</td></tr>
</table>

EXAMPLE: Show that the equation $2xy^3\, dx + (1 + 3x^2y^2)\, dy = 0$ is exact and that the equation $x^2y\, dx + 5xy^2\, dy = 0$ is not exact.

SOLUTION: Since $\dfrac{\partial}{\partial y}(2xy^3) = 6xy^2 = \dfrac{\partial}{\partial x}(1 + 3x^2y^2)$, the equation $2xy^3\, dx + (1 + 3x^2y^2)\, dy = 0$ is an exact equation. On the other hand, the equation $x^2y\, dx + 5xy^2\, dy = 0$ is not exact because $\dfrac{\partial}{\partial y}(x^2y) = x^2 \neq 5y^2 = \dfrac{\partial}{\partial x}(5xy^2)$. Note, however, that $x^2y\, dx + 5xy^2\, dy = 0$ is separable.

Remember from calculus that the **total differential** of the function $f(x, y)$ is

$$df = \frac{\partial f}{\partial x}(x, y)\, dx + \frac{\partial f}{\partial y}(x, y)\, dy.$$

Therefore, the equation $M(x, y)\, dx + N(x, y)\, dy = 0$ is exact if there exists a function $f(x, y)$ such that $M(x, y) = \dfrac{\partial f}{\partial x}(x, y)$ and $N(x, y) = \dfrac{\partial f}{\partial y}(x, y)$. We use these relationships to solve an exact differential equation. If we can find the function f such that $M(x, y) = \dfrac{\partial f}{\partial x}(x, y)$ and $N(x, y) = \dfrac{\partial f}{\partial y}(x, y)$, then the differential equation becomes

$$M(x, y)\, dx + N(x, y)\, dy = 0$$
$$df = 0.$$

Hence, a general solution of the exact equation is $f(x, y) = C$, where C is a constant.

In multivariable calculus, we learned that $\dfrac{\partial^2 f}{\partial x \partial y} = \dfrac{\partial^2 f}{\partial y \partial x}$ on an open region R if f, $\dfrac{\partial f}{\partial x}$, $\dfrac{\partial f}{\partial y}$, $\dfrac{\partial^2 f}{\partial x \partial y}$, and $\dfrac{\partial^2 f}{\partial x \partial y}$ are continuous on R. Hence, the test for exactness, $\dfrac{\partial N}{\partial x} = \dfrac{\partial M}{\partial y}$, really states that if f exists, then

$$\frac{\partial N}{\partial x} = \frac{\partial}{\partial x}\left(\frac{\partial f}{\partial y}\right) = \frac{\partial}{\partial y}\left(\frac{\partial f}{\partial x}\right) = \frac{\partial M}{\partial y},$$

which is required by the theorem from calculus.

≡ Solving the Exact Differential Equation $M(x, y)\, dx + N(x, y)\, dy = 0$

1. Assume that $M(x, y) = \dfrac{\partial f}{\partial x}(x, y)$ and $N(x, y) = \dfrac{\partial f}{\partial y}(x, y)$.
2. Integrate $M(x, y)$ with respect to x. (Add an arbitrary function of y, $g(y)$.)

3. Differentiate the result in Step 2 with respect to y, and set the result equal to $N(x, y)$. Solve for $g'(y)$.

4. Integrate $g'(y)$ with respect to y to obtain an expression for $g(y)$. (There is no need to include an arbitrary constant.)

5. Substitute $g(y)$ into the result obtained in Step 2 for $f(x, y)$.

6. A general solution is $f(x, y) = c$, where c is a constant.

7. Apply the initial condition if given.

Note: A similar algorithm can be stated so that in Step 2 $N(x, y)$ is integrated with respect to y.

EXAMPLE: Solve $(2x - y^2 \sin(xy)) \, dx + (\cos(xy) - xy \sin(xy)) \, dy = 0$.

SOLUTION: Because

$$\frac{\partial}{\partial y}(2x - y^2 \sin(xy)) = -2y \sin(xy) - y^2 \cos(xy) \cdot x$$
$$= -2y \sin(xy) - xy^2 \cos(xy)$$

and

$$\frac{\partial}{\partial x}(\cos(xy) - xy \sin(xy)) = -\sin(xy) \cdot y - y \sin(xy) - xy \cos(xy) \cdot y$$
$$= -2y \sin(xy) - xy^2 \cos(xy)$$

are equal, we know that the equation is exact. Let $f(x, y)$ be a function with $\frac{\partial f}{\partial x} = 2x - y^2 \sin(xy)$ and $\frac{\partial f}{\partial y} = \cos(xy) - xy \sin(xy)$. Integrating $\frac{\partial f}{\partial x} = 2x - y^2 \sin(xy)$ with respect to x results in

$$f(x, y) = \int (2x - y^2 \sin(xy)) \, dx = x^2 + y \cos(xy) + g(y),$$

where $g(y)$ represents a function of y. Differentiating with respect to y gives us

$$\frac{\partial f}{\partial y} = \frac{\partial}{\partial y}(x^2 + y \cos(xy) + g(y)) = \cos(xy) - xy \sin(xy) + g'(y).$$

Because we must also have that $\frac{\partial f}{\partial y} = \cos(xy) - xy \sin(xy)$, we conclude that $g'(y) = 0$ so that $g(y) = C_1$, where C_1 represents a constant. Thus, $f(x, y) = x^2 + y \cos(xy) + C_1$, so a general solution of the exact equation is given by

$$x^2 + y \cos(xy) + C_1 = C,$$

where C represents a constant. We simplify by combining the constant $C - C_1$ into a constant k so that a general solution of the equation is given by

$$x^2 + y\cos(xy) = k.$$

We can graph this general solution for various values of k by observing that the level curves of the function $g(x, y) = x^2 + y\cos(xy)$ correspond to the graphs of the equation $x^2 + y\cos(xy) = k$ for various values of k. We now use `contourplot` to graph several level curves of $g(x, y) = x^2 + y\cos(xy)$ on the rectangle $[0, 3\pi] \times [0, 3\pi]$. In this case, the option `axes=BOXED` instructs Maple to place a box around the resulting graph, while the option `grid=[70,70]` instructs Maple to sample 70 points in each of the x- and y-directions, helping assure that the resulting graph appears smooth.

```
> with(plots):
  g:=(x,y)->x^2+y*cos(x*y):
  contourplot(g(x,y),x=0..3*Pi,y=0..3*Pi,axes=BOXED,
   grid=[70,70]);
```

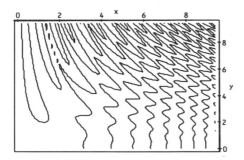

In the next two examples, we consider initial-value problems.

EXAMPLE: Solve $\left(2xe^{x^2}y + 2xe^{-y}\right) dx + \left(e^{x^2} - x^2e^{-y} + 1\right) dy = 0$ subject to the initial condition $y(0) = 0$.

SOLUTION: We begin by identifying $M(x, y) = 2xe^{x^2}y + 2xe^{-y}$ and $N(x, y) = e^{x^2} - x^2e^{-y} + 1$ and then defining M and N.

```
> M:=(x,y)->2*x*exp(x^2)*y+2*x*exp(-y):
  N:=(x,y)->exp(x^2)-x^2*exp(-y)+1:
```

We see that the equation is exact because $\dfrac{\partial M}{\partial y} = \dfrac{\partial N}{\partial x}$ as shown here with **diff** and **simplify**.

```
> simplify(diff(M(x,y),y));
  simplify(diff(N(x,y),x));
```
$$2xe^{x^2} - 2xe^{-y}$$
$$2xe^{x^2} - 2xe^{-y}$$

In this case, we see that **dsolve** can be used to find a general solution of the equation, as follows:

```
> dsolve(M(x,y(x))+N(x,y(x))*diff(y(x),x)=0,y(x));
```
$$e^{x^2} y(x) + x^2 e^{-y(x)} + y(x) = \_C1$$

In fact, **dsolve** can be used to find the solution which satisfies the initial condition $y(0) = 0$, which is computed as follows and named **Sol**. The function **W** appearing in the result represents the **Omega function**, $\Omega(x)$, which satisfies $\Omega(x) \cdot e^{\Omega(x)} = x$. In Maple, **W(x)** represents the unique branch of the Omega function that is analytic at 0.

```
> Sol:=dsolve({M(x,y(x))+N(x,y(x))*diff(y(x),x)=0,
  y(0)=0},y(x));
```
$$Sol := y(x) = W\left(-\frac{e^0 x^2}{e^{x^2} + 1}\right)$$

The explicit solution given in **Sol** is then assigned the name $y(x)$ with **assign** and graphed on the interval $[0, 4]$ with **plot**.

```
> assign(Sol):
  plot(y(x),x=0..4);
```

As with previous techniques we have discussed, Maple can also be used to implement the steps necessary to solve exact equations. Compute the integral of $M(x, y)$ with respect to x and name the result **f**. The result means that the function $f(x, y)$ which satisfies $\dfrac{\partial f}{\partial x} = 2xe^{x^2}y + 2xe^{-y}$ and $\dfrac{\partial f}{\partial y} = e^{x^2} - x^2e^{-y} + 1$ has the form $f(x, y) = e^{x^2}y + x^2e^{-y} + g(y)$ for some function of y, $g(y)$.

```
> f:=int(M(x,y),x)+g(y);
```
$$f := e^{x^2}y + x^2e^{-y} + g(y)$$

Differentiating **f** with respect to y and comparing the result to $N(x, y)$, defined before, we see that $g'(y) = 1$.

```
> diff(f,y);
```
$$e^{x^2}y - x^2e^{-y} + \left(\frac{\partial}{\partial y}g(y)\right)$$

```
> N(x,y);
```
$$e^{x^2}y - x^2e^{-y} + 1$$

Thus, $g(y) = y + C_1$

```
> g(y):=int(1,y);
```
$$g(y) := y$$

and $e^{x^2}y + x^2e^{-y} + y = C$ is a general solution of the equation.

```
> f;
```
$$e^{x^2}y + x^2e^{-y} + y$$

To calculate the value of C that satisfies the initial condition $y(0) = 0$, we substitute the value 0 for both x and y in **f**. Keeping in mind that **f** represents the left-hand side of the general solution, we interpret the result to mean that $C = 0$. Thus, the solution to the equation which satisfies the initial condition $y(0) = 0$ is $e^{x^2}y + x^2e^{-y} + y = 0$.

```
> subs({x=0,y=0},f);
```
$$0$$

We now use **implicitplot** to graph the equation $e^{x^2}y + x^2e^{-y} + y = 0$ on the rectangle $[0, 4] \times [-5, 0]$.

```
> with(plots):
  implicitplot(f=0,x=0..4,y=-5..0);
```

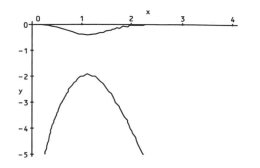

EXAMPLE: Solve $(1 + 5x - y) dx - (x + 2y) dy = 0$ subject to the initial condition $y(0) = 1$.

SOLUTION: We rewrite $(1 + 5x - y) dx - (x + 2y) dy = 0$ in the form $\dfrac{dy}{dx} = \dfrac{1 + 5x - y}{x + 2y}$ and then define **Eq1**.

```
> Eq1:=diff(y(x),x)=(1+5*x-y)/(x+2*y);
```

$$Eq1 := \frac{d}{dx}y(x) = \frac{1 + 5x - y}{x + 2y}$$

As in the previous example, **dsolve** can be used to find a general solution of the equation

```
> gen_sol:=dsolve(Eq1,y(x));
```

$$gen\_sol := -x - \frac{5}{2}x^2 + xy(x) + y(x)^2 = \_C1$$

and also can be used to solve the initial-value problem.

```
> Sol:=dsolve({Eq1,y(0)=1},y(x));
```

$$Sol := y(x) = -\frac{1}{2}x - \frac{1}{2}\sqrt{11x^2 + 4x + 4}, \, y(x) = -\frac{1}{2}x + \frac{1}{2}\sqrt{11x^2 + 4x + 4}$$

The result given in **gen_sol** means that a general solution of the equation is $f(x, y) = C$, where $f(x, y) = -x - \dfrac{5}{2}x^2 + xy + y^2$. To verify this solution, we first observe that the equation

$(1 + 5x - y)\,dx - (x + 2y)\,dy = 0$ is equivalent to the equation $-(1 + 5x - y)\,dx + (x + 2y)\,dy = 0$.
We then define f and compute $\dfrac{\partial f}{\partial x}$ and $\dfrac{\partial f}{\partial y}$ with **diff**. We conclude that the result given in
gen_sol is correct because $\dfrac{\partial f}{\partial x} = -(1 + 5x - y)$ and $\dfrac{\partial f}{\partial y} = x + 2y$.

```
> f:=(x,y)->-x-5/2*x^2+x*y+y^2;
  diff(f(x,y),x);
  diff(f(x,y),y);
```

$$f := (x, y) \longrightarrow -x - \frac{5}{2}x^2 + xy + y^2$$

$$-1 - 5x + y$$

$$x + 2y$$

As in the previous example, we can graph various solutions by graphing level curves of the function $f(x, y) = -x - \dfrac{5}{2}x^2 + xy + y^2$ with **contourplot**. In this case, the option **axes=NORMAL** is used to instruct Maple to include axes in the resulting graph; otherwise, no axes are displayed.

```
> with(plots):
  contourplot(f(x,y),x=-4..4,y=-4..4,axes=NORMAL);
```

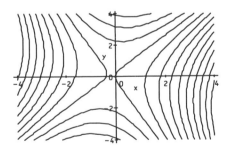

In addition, we can use commands like **DEplot1** to graph the direction field for the equation along with various solutions. We use **DEplot1** to graph the direction field for the equation along with a graph of the solution satisfying $y(0) = 1$.

```
> with(DEtools):
  DEplot1(Eq1,[x,y],x=-4..4,{[0,1]},y=-4..4);
```

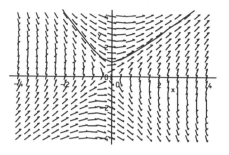

2.4 LINEAR EQUATIONS

In the previous sections, we have seen that some first-order equations may be classified as separable equations, others as homogeneous, and others as exact equations. Of course, most differential equations are neither separable, homogeneous, nor exact. Calculating explicit or implicit closed-form solutions of most first-order equations may be a formidable task, at best. However, equations of the form $\dfrac{dy}{dx} + p(x)y = q(x)$ can always be solved (although we will see that the resulting expressions might include integrals which are difficult or impossible to evaluate exactly), so we discuss their method of solution in the section.

DEFINITION

> *First-Order Linear Differential Equation*
>
> A differential equation which can be written in the form
>
> $$\frac{dy}{dx} + p(x)y = q(x)$$
>
> is called a **first-order linear differential equation**.

First-order linear equations are particularly important because as long as the necessary integrations can be carried out, an explicit solution can be produced. We show how linear first-order equations are solved in the following. Consider the equation

$$\frac{dy}{dx} + p(x)y = q(x),$$

Then multiplying through by the $e^{\int p(x)\,dx}$ yields

$$e^{\int p(x)\,dx}\frac{dy}{dx} + e^{\int p(x)\,dx}p(x)y = e^{\int p(x)\,dx}q(x).$$

By the product rule and the Fundamental Theorem of Calculus,

$$\frac{d}{dx}\left(e^{\int p(x)\,dx}y\right) = e^{\int p(x)\,dx}\frac{dy}{dx} + e^{\int p(x)\,dx}p(x)y$$

so

$$\frac{d}{dx}\left(e^{\int p(x)\,dx}y\right) = e^{\int p(x)\,dx}q(x).$$

Integrating, we obtain

$$e^{\int p(x)\,dx}y = \int e^{\int p(x)\,dx}q(x)\,dx,$$

and dividing by $e^{\int p(x)\,dx}$ yields a general solution of $\dfrac{dy}{dx} + p(x)y = q(x)$,

$$y = \frac{\int e^{\int p(x)\,dx}q(x)\,dx}{e^{\int p(x)\,dx}} = e^{-\int p(x)\,dx}\int e^{\int p(x)\,dx}q(x)\,dx.$$

The term $e^{\int p(x)\,dx}$ is called an **integrating factor** for the linear equation $\dfrac{dy}{dx} + p(x)y = q(x)$.

EXAMPLE: Find a general solution of $x\dfrac{dy}{dx} + y = x\cos x.$

SOLUTION: Dividing the equation by x yields

$$\frac{dy}{dx} + \frac{1}{x}y = \cos x,$$

where $p(x) = \dfrac{1}{x}$ and $q(x) = \cos x$. Then an integrating factor is

$$e^{\int \frac{1}{x}\,dx} = e^{\ln|x|} = x, \text{ for } x > 0,$$

and

$$\frac{d}{dx}(xy) = x\frac{dy}{dx} + y = x\cos x$$

so

$$xy = \int x\cos x\,dx.$$

Using the integration by parts formula, $\int u\,dv = uv - \int v\,du$, with $u = x$ and $dv = \cos x$, we obtain $du = dx$ and $v = \sin x$, so

$$xy = \int x\cos xx\,dx = x\sin x - \int \sin x\,dx = x\sin x + \cos x + C.$$

Therefore, a general solution of the equation $x\dfrac{dy}{dx} + 3y = x\cos x$ for $x > 0$ is $y = \dfrac{x\sin x + \cos x + C}{x}$.

We graph this equation for various values of C, such as $C = -3, -2, -1, 0, 1, 2, 3$, by first defining y:

```
> y:=(x*sin(x)+cos(x)+C)/x:
```

and then using **seq** to create a sequence of C-values, named **c_vals**.

```
> c_vals:=seq(i,i=-3..3):
```

We then define **to_plot** to be the set of functions obtained by replacing C in y by each number in **c_vals**.

```
> to_plot:={seq(y,C=c_vals)};
```

$$to\_plot := \left\{ \frac{x\sin(x) + \cos(x) - 1}{x}, \frac{x\sin(x) + \cos(x)}{x}, \right.$$
$$\frac{x\sin(x) + \cos(x) + 1}{x}, \frac{x\sin(x) + \cos(x) + 2}{x}, \frac{x\sin(x) + \cos(x) + 3}{x},$$
$$\left. \frac{x\sin(x) + \cos(x) - 3}{x}, \frac{x\sin(x) + \cos(x) - 2}{x} \right\}$$

The set of functions in **to_plot** is graphed on the interval $[0, 2\pi]$ with **plot**. The option **-10..10** instructs Maple that the set of y-values displayed, corresponding to the vertical axis, corresponds to the interval $[-10, 10]$.

```
> plot(to_plot,x=0..2*Pi,-10..10);
```

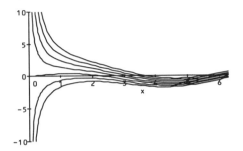

As with other types of equations, we solve initial-value problems by first finding a general solution of the equation and then applying the initial condition to determine the value of the constant.

EXAMPLE: Solve the initial-value problem $\dfrac{dy}{dx} + 5x^4y = x^4$, $y(0) = -7$.

SOLUTION: As we have seen in many previous examples, dsolve can be used to find a general solution of the equation and the solution to the initial-value problem, as done below in gen_sol and Sol.

```
> x:='x':y:='y':
  Eq:=diff(y(x),x)+5*x^4*y(x)=x^4;
  gen_sol:=dsolve(Eq,y(x));
```

$$Eq := \left(\frac{d}{dx}y(x)\right) + 5x^4y(x) = x^4$$

$$gen\_sol := y(x) = \frac{1}{5} + e^{-x^5}\_C1$$

```
> Sol:=dsolve({Eq,y(0)=-7},y(x));
```

$$Sol := y(x) = \frac{1}{5} - \frac{36}{5}e^{-x^5}$$

The result given in Sol is graphed by first using assign to assign $y(x)$ the solution obtained in Sol and then using plot to graph $y(x)$.

```
> assign(Sol):
  plot(y(x),x=-1..2);
```

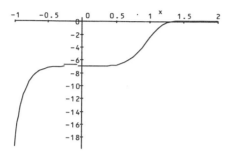

We can also use Maple to carry out the steps necessary to solve linear equations. We begin by identifying the integrating factor $e^{\int 5x^4\,dx} = e^{x^5}$, computed as follows with `int`.

```
> y:='y':
  int_fac:=exp(int(5*x^4,x));
```

$$int\_fac := e^{x^5}$$

Therefore, the equation can be written as

$$\frac{d}{dx}(e^{x^5}y) = x^4 e^{x^5}$$

so that the integration of both sides of the equation yields

$$e^{x^5}y = \frac{1}{5}e^{x^5} + C.$$

```
> right_side:=int(int_fac*x^4,x);
```

$$right\_side := \frac{1}{5}e^{x^5}$$

Hence, a general solution is

$$y = \frac{1}{5} + Ce^{-x^5}.$$

Note that we compute y by using the miscellaneous library function `isolate` to solve the equation $e^{x^5}y = \frac{1}{5}e^{x^5} + C$ for y.

```
>  readlib(isolate):
   step_1:=isolate(exp(x^5)*y=right_side+C,y);
```

$$step\_1 := y = \frac{\frac{1}{5}e^{x^5} + C}{e^{x^5}}$$

```
>  gen_sol:=expand(step_1);
```

$$gen\_sol := y = \frac{1}{5} + \frac{C}{e^{x^5}}$$

We find the unknown constant C by substituting the initial condition $y(0) = -7$. This gives us $-7 = \frac{1}{5} + C$, so $C = -\frac{36}{5}$.

```
>  find_C:=eval(subs({x=0,y=-7},gen_sol));
```

$$find\_C := -7 = \frac{1}{5} + C$$

```
>  C:=solve(find_C,C);
```

$$C := -\frac{36}{5}$$

Therefore, the solution to the initial-value problem is $y = \frac{1}{5} - \frac{36}{5}e^{-x^5}$.

```
>  gen_sol;
```

$$y = \frac{1}{5} - \frac{36}{5}\frac{1}{e^{x^5}}$$

If the integration cannot be carried out, the solution can often be approximated numerically by taking advantage of numerical integration techniques, as illustrated in the following example.

EXAMPLE: Graph the solution to the initial-value problem $y' - \sin(2\pi x)y = 1$, $y(0) = 1$.

SOLUTION: After defining **Eq**, we attempt to use **dsolve** to find a general solution of the equation. However, we see that **dsolve** is unsuccessful since the necessary integrations cannot be carried out.

```
> Eq:=diff(y(x),x)-sin(3*x)*y(x)=1;
```

$$Eq := \left(\frac{\partial}{\partial x}y(x)\right) - \sin(3x)y(x) = 1$$

```
> dsolve(Eq,y(x));
```

$$y(x) = e^{-\frac{1}{3}\cos(x)(4\cos(x)^2-3)} \int e^{\frac{1}{3}\cos(x)(4\cos(x)^2-3)} \, dx$$
$$+ e^{-\frac{1}{3}\cos(x)(4\cos(x)^2-3)} \_C1$$

However, in this case we are able to generate a numerical solution to the equation by using **dsolve** together with the **numeric** option.

```
> num_sol:=dsolve({Eq,y(0)=1},y(x),numeric);
```

$$\text{num\_sol} := \text{proc}(x) \text{ 'dsolve/numeric/result2'}(x, 3018872, [1]) \text{ end}$$

The result of using **dsolve** together with the **numeric** option is a procedure which can be evaluated for particular values of x. For example, entering

```
> num_sol(0.5);
```

$$\text{h}x = 0.5000000000, y(x) = 1.975336564\text{j}$$

returns an ordered pair corresponding to x, which in this case is 0.5, and $y(x)$, which in this case is 1.975336564. The procedure which results from using **dsolve** together with the **numeric** option is graphed using the command **odeplot** contained in the **plots** package. We load the **plots** package and then graph the result given in **num_sol** for $0 \le x \le 2\pi$.

```
> with(plots):
  odeplot(num_sol,[x,y(x)],0..2*Pi);
```

If, instead, we needed only to graph the solution and not generate a numerical solution which can be evaluated for particular numbers, we could use a command like **DEplot1** to graph the direction field for the equation along with the solution satisfying $y(0) = 1$, as follows.

```
> with(DEtools):
  DEplot1(Eq,[x,y],x=0..2*Pi,{[0,1]});
```

▼ A P P L I C A T I O N

Kidney Dialysis

The primary purpose of the kidney is to remove waste products, such as urea, creatinine, and excess fluid, from blood. When kidneys are not working properly, wastes accumulate in the blood; when toxic levels are reached, death is certain. The leading causes of chronic kidney failure in the United States are hypertension (high blood pressure) and diabetes mellitus. In fact, one-quarter of all patients requiring **kidney dialysis** have diabetes. Fortunately, **kidney dialysis** removes waste products from the blood of patients with improperly working kidneys. During the hemodialysis process, the patient's blood is pumped through a **dialyzer**, usually at a rate of 1 to 3 deciliters per minute. The patient's blood is separated from the cleaning fluid by a semipermeable membrane, which permits wastes (but not blood cells) to diffuse to the cleaning fluid; the cleaning fluid contains some substances beneficial to the body that diffuse to the blood. The cleaning fluid, called the **dialysate**, is flowing in the **opposite** direction as the blood, usually at a rate of 2 to 6 deciliters per minute. Waste products from the blood diffuse to the dialysate through the membrane at a rate proportional to the difference in concentration of the waste products in the blood and dialysate. If we let $u(x)$ represent the concentration of wastes in the blood, $v(x)$ the concentration of wastes in the dialysate, where x is the distance along the dialyzer, Q_D the flow rate of the dialysate through the machine, and Q_B the flow rate of the blood through the machine, then

$$\begin{cases} Q_B u' = -k(u - v) \\ -Q_D v' = k(u - v) \end{cases}'$$

where k is the proportionality constant.

If we let L denote the length of the dialyzer and the initial concentration of wastes in the blood is $u(0) = u_0$, while the initial concentration of wastes in the dialysate is $v(L) = 0$, then we must solve the initial-value problem

$$\begin{cases} Q_B u' = -k(u - v) \\ -Q_D v' = k(u - v) \\ u(0) = u_0, v(L) = 0 \end{cases}.$$

Solving the first equation for u' and the second equation for v', we obtain the equivalent system

$$\begin{cases} u' = -\dfrac{k}{Q_B}(u - v) \\ -v' = \dfrac{k}{Q_D}(u - v) \\ u(0) = u_0, v(L) = 0 \end{cases}.$$

Adding these two equations results in the linear equation in $u - v$,

$$u' - v' = -\frac{k}{Q_B}(u - v) + \frac{k}{Q_B}(u - v)$$

$$(u - v)' = -\left(\frac{k}{Q_B} - \frac{k}{Q_B}\right)(u - v).$$

Let $\alpha = \dfrac{k}{Q_B} - \dfrac{k}{Q_D}$ and $y = u - v$. Then we must solve the equation $y' = -\alpha y$, which is done with **dsolve**, naming the resulting output **step_1**.

```
> y:='y':
  step_1:=dsolve(diff(y(x),x)=-alpha*y(x),y(x));
```

$$step\_1 := y(x) = e^{-\alpha x}\_C1$$

Next, we use **assign** to name $y(x)$ the result obtained in **step_1**. Using the facts that $u' = -\dfrac{k}{Q_B}(u - v) = -\dfrac{k}{Q_B}y$ and $u(0) = u_0$, we are able to use **dsolve** to find $u(x)$.

```
> assign(step_1):
  step_2:=dsolve({diff(u(x),x)=-k/Qb*y(x),u(0)=u0},u(x));
```

$$step\_2 := u(x) = \frac{e^{-\alpha x}k - C1}{\alpha Qb} - \frac{k - C1}{\alpha Qb} + u0$$

In the same manner as before, we use `assign` to name $u(x)$ the result obtained in `step_2`.

> `assign(step_2):`

Because $y = u - v$, $v = u - y$. Consequently, because $v(L) = 0$, at this point we are able to compute `_C1` and determine u and v. First, we determine u.

> `left_side:=subs(x=L,u(x)-y(x));`

$$left\_side := \frac{e^{-\alpha L}k\_C1}{\alpha Qb} - \frac{k\_C1}{\alpha Qb} + u0 - e^{-\alpha L}\_C1$$

> `_C1:=solve(left_side=0,_C1);`

$$\_C1 := -\frac{u0}{\frac{e^{-\alpha L}k}{\alpha Qb} - \frac{k}{\alpha Qb} - e^{-\alpha L}}$$

> `eval(u(x));`

$$-\frac{e^{-\alpha x}ku0}{\alpha Qb\left(\frac{e^{-\alpha L}k}{\alpha Qb} - \frac{k}{\alpha Qb} - e^{-\alpha L}\right)} + \frac{ku0}{\alpha Qb\left(\frac{e^{-\alpha L}k}{\alpha Qb} - \frac{k}{\alpha Qb} - e^{-\alpha L}\right)} + u0$$

> `u:=simplify(eval(u(x)));`

$$u := \frac{u0\left(e^{-\alpha x}k - e^{-\alpha L}k + e^{-\alpha L}\alpha Qb\right)}{-e^{-\alpha L}k + k + e^{-\alpha L}\alpha Qb}$$

Finally, using $v = u - y$, we compute v.

> `v:=simplify(eval(u-y(x)));`

$$v := \frac{u0\left(-e^{-\alpha x}k + e^{-\alpha L}k - e^{-\alpha L}\alpha Qb + e^{-\alpha x}\alpha Qb\right)}{-e^{-\alpha L}k + k + e^{-\alpha L}\alpha Qb}$$

For example, in healthy adults, typical urea nitrogen levels are 11 to 23 milligrams per deciliter (1 deciliter = 100 milliliters), while serum creatinine levels range from 0.6 to 1.2 milligrams per deciliter and the total volume of blood is 4 to 5 liters (1 liter = 1000 milliliters). Suppose that hemodialysis is performed on a patient with a urea nitrogen level of 34 mg/dl and serum creatinine level of 1.8 using a dialyzer with $k = 2.25$ and $L = 1$. If the flow rate of blood, Q_B, is 2 dl/min while the flow rate of the dialysate, Q_D, is 4 dl/min, will the level of wastes in the patient's blood reach normal levels after dialysis is performed?

After defining the appropriate constants, we evaluate u and v

```
> alpha:=k/Qb-k/Qd:
  k:=2.25:
  L:=1:
  Qb:=2:
  Qd:=4:
  u0:=(34+1.8):
  u,v;
```

$50.06232707e^{-.5625000000x} - 14.26232707,$
$25.03116353e^{-.5625000000x} - 14.26232707$

and then graph u and v on the interval $[0, 1]$ with **plot**. Remember that the dialysate is moving in the direction **opposite** the blood. Thus, we see from the graphs that as levels of waste in the blood decrease, levels of waste in the dialysate increase and at the end of the dialysis procedure, levels of waste in the blood are within normal ranges.

```
> plot(u,x=0..1);
  plot(v,x=0..1);
```

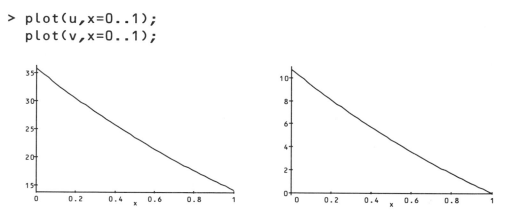

Typically, hemodialysis is performed 3 to 4 hours at a time 3 or 4 times per week. In some cases, a kidney transplant can free patients from the restrictions of dialysis. Of course, transplants have other risks not necessarily faced by those on dialysis; the number of available kidneys also affects the number of transplants performed. For example, in 1991 over 130,000 patients were on dialysis, while only 7000 kidney transplants had been performed. [P]

[P]**Sources:** D. N. Burghes and M. S. Borrie, *Modelling with Differential Equations*, Ellis Horwood Limited (1981), Halsted Press, NY, pp. 41–45. Joyce M. Black and Esther Matassarin-Jacobs, *Luckman and Sorensen's Medical-Surgical Nursing: A Psychophysiological Approach*, 4th ed. Philadelphia: W. B. Saunders Company (1993), pp. 1509–1519, 1775–1808.

2.5 SOME SPECIAL DIFFERENTIAL EQUATIONS

We take this opportunity to discuss several important differential equations. Several of the differential equations discussed in this section can be reduced to first-order equations by an appropriate substitution; another first-order equation can be transformed into a second-order equation which can sometimes be solved. These equations are important because of their historical importance along with applications in other areas of mathematics, physics, and engineering.

≡ Bernoulli Equations

A **Bernoulli equation** is a nonlinear equation of the form

$$y' + p(x)y = q(x)y^n.$$

It is named for the Swiss mathematician Jacques Bernoulli. Equations of this form can be expressed as a first-order linear equation if an appropriate substitution is made. Notice that if $n = 0$, then the equation is first-order linear. Also, if $n = 1$, then the equation can be written as $y' + [p(x) - q(x)]y = 0$, which is also a first-order linear equation. Therefore, for $n = 0$ or $n = 1$, we use the techniques of the previous section or separation of variables to find a general solution.

When solving the equation $y' + p(x)y = q(x)y^n$ for $n \neq 0, 1$, we make the substitution

$$w = y^{1-n}$$

to transform the nonlinear equation into a linear equation.

Differentiating $w = y^{1-n}$ with respect to x results in

$$\frac{dw}{dx} = (1-n)y^{-n}\frac{dy}{dx},$$

and substituting into the equation $y' + p(x)y = q(x)y^n$ yields

$$\frac{y^n}{1-n}\frac{dw}{dx} + p(x)y^n w = q(x)y^n.$$

Multiplying by $\dfrac{1-n}{y^n}$ gives us the first-order linear equation

$$\frac{dw}{dx} + (1-n)p(x)w = (1-n)q(x),$$

which can be solved for $w(x)$ using the techniques of the previous section. Once this solution is found, $y(x)$ is found by using the relationship $w = y^{1-n}$ or $y = w^{1/(1-n)}$.

EXAMPLE: Solve the Bernoulli equation $\dfrac{dy}{dx} + \dfrac{y}{x} = \dfrac{1}{xy^2}$, $x > 0$.

SOLUTION: In this case, $p(x) = \dfrac{1}{x}$, $q(x) = \dfrac{1}{x}$, and $n = -2$. Therefore, we make the substitution

$$w = y^{1-(-2)} = y^3,$$

which means that

$$\frac{dw}{dx} = 3y^2 \frac{dy}{dx}.$$

Substitution of this derivative into the differential equation gives us

$$\frac{1}{3y^2} \frac{dw}{dx} + \frac{y}{x} = \frac{1}{xy^2}.$$

Multiplying by $3y^2$ then gives us

$$\frac{dw}{dx} + \frac{3y^3}{x} = \frac{3}{x}.$$

Of course, we must also make the substitution $w = y^3$ to obtain the first-order linear equation in w,

$$\frac{dw}{dx} + \frac{3w}{x} = \frac{3}{x}.$$

The solution of this equation is found with the integrating factor $e^{\int \frac{3}{x} dx} = e^{3\ln(x)} = x^3$. This means that $\dfrac{dw}{dx} + \dfrac{3w}{x} = \dfrac{3}{x}$ is equivalent to $\dfrac{d}{dx}(x^3 w) = 3x^2$. Integrating both sides of this equation yields $x^3 w = x^3 + C$, so $w = 1 + Cx^{-3}$. Since $w = y^3$, we have $y^3 = 1 + Cx^{-3}$, which implies that $y = (1 + Cx^{-3})^{1/3}$ is a general solution of the equation.

As with many other examples we have seen, $dsolve$ can be used to find a general solution of the equation.

```
> gen_sol:=dsolve(diff(y(x),x)+y(x)/x=1/(x*y(x)^2),y(x));
```

$$gen\_sol := y(x)^3 = \frac{x^3 + \_C1}{x^3}$$

We can graph the general solution given in `gen_sol` for various values of the arbitrary constant, `_C1`, by solving the equation given in `gen_sol` for `_C1`, naming the resulting output `to_graph`,

```
> to_graph:=solve(subs(y(x)=y,gen_sol),_C1);
```

$$to\_graph := (y^3 - 1)x^3$$

and then using `contourplot` to graph various level curves of the function of x and y given in `to_graph`.

```
> with(plots):
  contourplot(to_graph,x=-1..1,y=-1..1,grid=[50,50],
    axes=NORMAL);
```

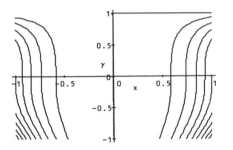

We can also take advantage of the `DEplot1` command to graph the direction field for the equation.

```
> with(DEtools):
  DEplot1(diff(y(x),x)+y(x)/x=1/(x*y(x)^2),[x,y],
    x=-1..1,y=-1..1);
```

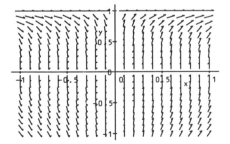

▼   **A P P L I C A T I O N**

Modeling the Spread of a Disease

Suppose that a disease is speading among a population of size N. In some diseases, like chickenpox, once an individual has had the disease, the individual becomes immune to the disease. In other diseases, like most venereal diseases, once an individual has had the disease and recovers from it, the individual does not become immune to the disease; subsequent encounters can lead to recurrences of the infection.

Let $S(t)$ denote the percent of the population susceptible to a disease at time t, $I(t)$ the percent of the population infected with the disease, and $R(t)$ the percent of the population unable to contract the disease. For example, $R(t)$ could represent the percent of persons who have had a particular disease, recovered, and subsequently become immune to the disease.

In order to model the spread of various diseases, we begin by making several assumptions and introducing some notation.

1. Susceptible and infected individuals die at a rate proportional to the number of susceptible and infected individuals with proportionality constant μ called the **daily death removal rate**; the number $\dfrac{1}{\mu}$ is the **average lifetime** or **life expectancy**.

2. The constant λ represents the **daily contact rate**: on average, an infected person will spread the disease to λ people per day.

3. Individuals recover from the disease at a rate proportional to the number infected with the disease with proportionality constant γ. The constant γ is called the **daily recovery removal rate**; the **average period of infectivity** is $\dfrac{1}{\gamma}$.

4. The **contact number** $\sigma = \dfrac{\lambda}{\gamma + \mu}$ represents the average number of contacts an infected person has with both susceptible and infected persons.

If a person becomes susceptible to a disease after recovering from it (like gonorrhea, meningitis, and streptococcal sore throat), then the percent of persons susceptible to becoming infected with the disease, $S(t)$, and the percent of people in the population infected with the disease, I(t), can be modeled by the system

$$\begin{cases} S'(t) = -\lambda IS + \gamma I + \mu - \mu S \\ I'(t) = \lambda IS - \gamma I - \mu I \\ S(0) = S_0, \ I(0) = I_0, \ S(t) + I(t) = 1 \end{cases}$$

This model is called an **SIS** (susceptible-infected-susceptible) model since once an individual has recovered from the disease, the individual again becomes susceptible to the disease.

Since $S(t) = 1 - I(t)$, we can write $I'(t) = \lambda IS - \gamma I - \mu I$ as

$$I'(t) = \lambda I(1 - I) - \gamma I - \mu I,$$

and thus we need to solve the initial-value problem

$$\begin{cases} I'(t) = [\lambda - (\gamma + \mu)]I - \lambda I^2 \\ I(0) = I_0 \end{cases}.$$

We see that the equation $I'(t) = \lambda I(1 - I) - \gamma I - \mu I$ is a Bernoulli equation because rewriting it gives us

$$\begin{aligned} I'(t) &= \lambda I(1 - I) - \gamma I - \mu I \\ &= \lambda I - \lambda I^2 - \gamma I - \mu I \\ I' + (\gamma + \mu - \lambda) &= -\lambda I^2. \end{aligned}$$

In the following, we use i to represent I, thus avoiding a conflict with the built-in constant $I = \sqrt{-1}$, and **Gamma**, **Mu**, and **Lambda** to represent the constants γ, μ, and λ, respectively, thus avoiding a conflict with the built-in constant **gamma** $= \gamma \approx 0.5772156649\ldots$. After defining **Eq**, we use **dsolve** to find the solution to the initial-value problem, naming the result **Sol**.

```
> i:='i':
  Eq:=diff(i(t),t)+(Gamma+Mu-Lambda)*i(t)=-Lambda*i(t)^2;
```

$$Eq := \left(\frac{\partial}{\partial t} i(t) \right) + (\Gamma + M - \Lambda)i(t) = -\Lambda\, i(t)^2$$

```
> Sol:=dsolve({Eq,i(0)=i0},i(t));
```

$$Sol := i(t) = -(-\Gamma - M + \Lambda)\Bigg(-\Lambda$$
$$-\frac{(-\Gamma - M + \Lambda - i0\,\Lambda)e^{-(-\Gamma-M+\Lambda)t}\Gamma}{i0\Gamma + i0M - i0\Lambda}$$
$$-\frac{(-\Gamma - M + \Lambda - i0\,\Lambda)e^{-(-\Gamma-M+\Lambda)t}M}{i0\,\Gamma + i0\,M - i0\,\Lambda}$$
$$-\frac{(-\Gamma - M + \Lambda - i0\,\Lambda)e^{-(-\Gamma-M+\Lambda)t}\Lambda}{i0\,\Gamma + i0\,M - i0\,\Lambda} \Bigg)$$

We can use this result to see how a disease might spread through a population. First, we use **assign** to name **i(t)** the result given in **Sol**. For example, we compute **i(t)** in the case where $\lambda = 0.50$, $\gamma = 0.75$, and $\mu = 0.65$. In this case, we see that the contact number is

$$\sigma = \frac{\lambda}{\gamma + \mu} \approx 0.3571428571.$$

```
> assign(Sol):
  Lambda:=0.5:
  Gamma:=0.75:
  Mu:=0.65:
  Lambda/(Gamma+Mu);
  eval(i(t));
```

0.3571428571

$$0.90\frac{1}{-0.5 - 1.000000000\frac{(-.90-.5i0)e^{.90t}}{i0}}$$

Next, we use **seq** and **subs** to substitute various initial conditions into **i(t)**, naming the resulting set of nine functions **to_plot1**.

```
> to_plot1:={seq(subs(i0=.1*j,eval(i(t))),j=1..9)}:
```

We then graph the functions in **to_plot1** for $0 \leq t \leq 5$. Apparently, regardless of the initial percent of the population infected, under these conditions, the disease is eventually removed from the population. This makes sense because the contact number is less than one.

```
> plot(to_plot1,t=0..5);
```

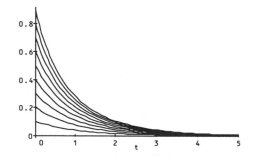

On the other hand, in the case where $\lambda = 1.50$, $\gamma = 0.75$, and $\mu = 0.65$, we see that the contact number is $\sigma = \dfrac{\lambda}{\gamma + \mu} \approx 1.071428571$.

```
> Lambda:=1.5:
  Gamma:=0.75:
  Mu:=0.65:
  Lambda/(Gamma+Mu);
  eval(i(t));
```

1.071428571

$$-0.10\dfrac{1}{-1.5 - 1.000000000\dfrac{(.10 - 1.5i0)e^{-.10t}}{i0}}$$

Proceeding as before, we graph the solution using different initial conditions. In this case, we see that no matter what percent of the population is initially infected, a certain percent of the population is always infected. This makes sense because the contact number is greater than one. In fact, it is a theorem that[*]

$$\lim_{t \to \infty} I(t) = \begin{cases} 1 - \frac{1}{\sigma}, & \text{if } \sigma > 1 \\ 0, & \text{if } \sigma \leq 1 \end{cases}.$$

```
> to_plot2:={seq(subs(i0=.1*j,eval(i(t))),j=1..9)}:
  plot(to_plot2,t=0..5);
  to_plot3:={seq(subs(i0=.01*j,eval(i(t))),j=1..9)}:
  plot(to_plot3,t=0..20);
```

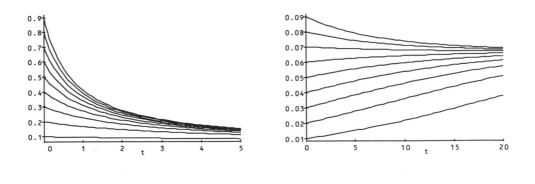

[*]**Source:** Herbert W. Hethcote, "Three Basic Epidemiological Models," in *Applied Mathematical Ecology*, edited by Simon A. Levin, Thomas G. Hallan, and Louis J. Gross. New York: Springer-Verlag (1989), pp. 119–143.

☰ *Clairaut Equations*

Equations of the form

$$f(xy' - y) = g(y')$$

are called **Clairaut equations** after the French mathematician Alexis Clairaut (1713–1765) who studied these equations in 1734. Solutions to this equation are determined by differentiating each side of the equation with respect to x. In doing this, we must make use of the chain rule. The derivative of $f(xy' - y)$ is

$$f'(xy' - y)(xy'' + y' - y') = f'(xy' - y)(xy''),$$

where $'$ denotes differentiation with respect to the argument of the function, x. Therefore, the derivative of the left-hand side of the equation is $g'(y')y''$. This gives us

$$f'(xy' - y)(xy'') = g'(y')y'',$$

which is equivalent to

$$[f'(xy' - y)x - g'(y')] y'' = 0.$$

Therefore, $y'' = 0$ or $f'(xy' - y)x - g'(y') = 0$. If $y'' = 0$, then $y' = c$, where c is a constant. Substituting this into the differential equation $f(xy' - y) = g(y')$ gives us $f(xc - y) = g(c)$. Therefore, a general solution to the equation $f(xy' - y) = g(y')$ is $f(xc - y) = g(c)$.

In constrast, if $f'(xy' - y)x - g'(y') = 0$, then this equation can be used along with $f(xy' - y) = g(y')$ to determine another solution by eliminating y'. This is called the **singular solution** of the Clairaut equation.

EXAMPLE: Solve the equation $xy' - (y')^3 = y$.

SOLUTION: We begin by placing this equation in the appropriate form $xy' - y = (y')^3$. This implies that $f(xy' - y) = xy' - y$ and $g(y') = (y')^3$. Hence, $f(x) = x$ and $g(x) = x^3$. Therefore, a general solution is

$$xc - y = c^3 \text{ or } y = xc - c^3.$$

The singular solution is obtained by differentiating $xy' - (y')^3 = y$ with respect to x. This gives us $xy'' + y' - 3(y')^2y'' = y'$, which can be simplified to $[x - 3(y')^2]y'' = 0$. Of course $y'' = 0$ yields the general solution obtained earlier, so we use $x - 3(y')^2 = 0$ and $xy' - y = (y')^3$ to determine the singular solution. Solving $x - 3(y')^2 = 0$ for y' yields $y' = \left(\dfrac{x}{3}\right)^{1/2}$. Then substituting into

$xy' - y = (y')^3$ and simplifying leads to the singular solution $x^3 = \dfrac{27}{4}y^2$. We see next that $dsolve$ is able to compute both the general solution and the singular solutions that we computed earlier.

```
> gen_sol:=dsolve(x*diff(y(x),x)-diff(y(x),x)^3=y(x),y(x));
```

$$gen\_sol := y(x) = -\frac{2}{9}\sqrt{3}x^{3/2}, \, y(x) = \frac{2}{9}\sqrt{3}x^{3/2}, \, y(x) = -\_C1^3 + x\_C1$$

The resulting three functions are extracted from gen_sol with $[\dots]$ as follows:

```
> gen_sol[1],gen_sol[2],gen_sol[3];
```

$$y(x) := -\frac{2}{9}\sqrt{3}x^{3/2}, \; y(x) = \frac{2}{9}\sqrt{3}x^{3/2}, \; y(x) = -\_C1^3x\_C1$$

Thus, to graph the singular solution, we use rhs to extract the right-hand side of the first two equations in gen_sol, naming the resulting set of two functions of x $singular\_sol$, and then graph the two functions in $singular\_sol$ with $plot$.

```
> singular_sol:=rhs(gen_sol[1]),rhs(gen_sol[2]);
```

$$singular\_sol := \left\{ -\frac{2}{9}\sqrt{3}x^{3/2}, \, \frac{2}{9}\sqrt{3}x^{3/2} \right\}$$

```
> plot(singular_sol,x=0..9);
```

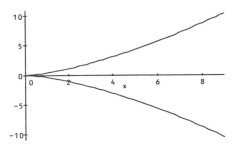

Similarly, to graph the general solution for various values of the arbitrary constant, $\_C1$, use seq and $subs$ to replace the occurrences of $\_C1$ in the right-hand side of the general solution by various numbers and then use $plot$ to graph the resulting set of functions.

```
> to_plot:={seq(subs(_C1=i,rhs(gen_sol[3])),i=-5..5)}:
```

```
> plot(to_plot,x=-10..10);
```

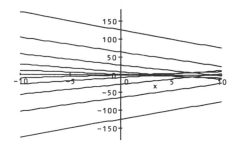

≡ *Lagrange Equations*

Equations of the form

$$y = x f(y') + g(y')$$

are called **Lagrange equations**. These equations are solved in a manner similar to Clairaut equations in that an appropriate substitution must be made and differentiation carried out. In this case, let $p = y'(x)$. Then differentiating $y = x f(y') + g(y')$ with respect to x yields

$$y' = x f'(y')y'' + f(y') + g'(y')y'',$$

and substituting p into the equation yields

$$p = x f'(p)\frac{dp}{dx} + f(p) + g'(p)\frac{dp}{dx} = f(p) + \frac{dp}{dx}[x f'(p) + g'(p)].$$

Solving for $\dfrac{dx}{dp}$ yields the linear equation

$$\frac{dx}{dp} = \frac{x f'(p) + g'(p)}{p - f(p)},$$

which is equivalent to

$$\frac{dx}{dp} + \frac{f'(p)}{f(p) - p}x = \frac{g'(p)}{p - f(p)}.$$

This linear first-order equation can be solved for x in terms of p. Then $x(p)$ can be used with $y = xf(p) + g(p)$ to obtain an equation for y.

EXAMPLE: Solve the equation $y = x\left(\dfrac{dy}{dx}\right)^2 + 3\left(\dfrac{dy}{dx}\right)^2 - 2\left(\dfrac{dy}{dx}\right)^3$.

SOLUTION: We begin by observing that the equation is a Lagrange equation with $f(x) = x^2$ and $g(x) = 3x^2 - 2x^3$. Let $p = \dfrac{dy}{dx}$. Then differentiating $y = x\left(\dfrac{dy}{dx}\right)^2 + 3\left(\dfrac{dy}{dx}\right)^2 - 2\left(\dfrac{dy}{dx}\right)^3$ with respect to x, substituting $\dfrac{dy}{dx}$ by p, and solving for $\dfrac{dx}{dp}$ yields the equation

$$\frac{dx}{dp} - x\left(\frac{f'(p)}{p - f(p)}\right) = \frac{g'(p)}{p - f(p)}.$$

Since $f(x) = x^2$, $\dfrac{-f'(p)}{p - f(p)} = \dfrac{-2p}{p - p^2}$, and since $g(x) = 3x^2 - 2x^3$, $\dfrac{g'(p)}{p - f(p)} = 6$. Thus,

$$\frac{dx}{dp} - x\left(\frac{f'(p)}{p - f(p)}\right) = \frac{g'(p)}{p - f(p)}$$

becomes

$$\frac{dx}{dp} + \frac{2p}{p - p^2}x = 6.$$

We solve this linear equation by first computing the integrating factor

$$e^{\int \frac{2p}{p - p^2}\,dp} = e^{-2\ln|1 - p|} = \frac{1}{(1 - p)^2}$$

and multiplying the equation $\dfrac{dx}{dp} + \dfrac{2p}{p - p^2}x = 6$ by $\dfrac{1}{(1 - p)^2}$. To solve this equation, we must compute

$$\int \frac{6}{(1 - p)^2}\,dp = \frac{-6}{p - 1} + C.$$

Therefore, $\dfrac{1}{(1 - p)^2}x = \dfrac{-6}{p - 1} + C$, so a general solution of $\dfrac{dx}{dp} + \dfrac{2p}{p - p^2}x = 6$ is

$$x(p) = (1 - p)^2\left(\frac{-6}{p - 1} + C\right) = 6(1 - p) + C(1 - p)^2.$$

Because $y = x \left(\dfrac{dy}{dx} \right)^2 + 3 \left(\dfrac{dy}{dx} \right)^2 - 2 \left(\dfrac{dy}{dx} \right)^3$ and $p = \dfrac{dy}{dx}$,

$$
\begin{aligned}
y(p) &= x(p)\, p^2 + 3p^2 - 2p^3 \\
&= \left[6(1 - p) + C(1 - p)^2 \right] p^2 + 3p^2 - 2p^3 \\
&= C p^2 (1 - p)^2 + 9p^2 - 8p^3.
\end{aligned}
$$

Therefore, a general solution of $y = x \left(\dfrac{dy}{dx} \right)^2 + 3 \left(\dfrac{dy}{dx} \right)^2 - 2 \left(\dfrac{dy}{dx} \right)^3$ is given by the parametric equations

$$
\begin{cases}
x(p) = 6(1 - p) + C(1 - p)^2 \\
y(p) = C p^2 (1 - p)^2 + 9p^2 - 8p^3
\end{cases}.
$$

Although **dsolve** can be used to find a general solution, due to the complexity of the result, it is probably easier to graph various solutions using the parametric representation we just obtained. We begin by defining **x** and **y**.

```
> x:=p->6*(1-p)+C*(1-p)^2:
  y:=p->C*p^2*(1-p)^2+9*p^2-8*p^3:
```

For example, to graph the solution corresponding to $C = 1$, we define **pair_1** with **subs**. Note that when graphing parametric equations, the values of the argument for which the parametric equations are to be graphed are **enclosed within the brackets**. The parametric equations given in **pair_1** are then graphed with **plot**.

```
> pair_1:=subs(C=1,[x(p),y(p),p=-2..2]);
```

$pair\_1 := [6 - 6\,p + (1 - p)^2, p^2(1 - p)^2 + 9\,p^2 - 8\,p^3, p = -2 \ldots 2]$

```
> plot(pair_1);
```

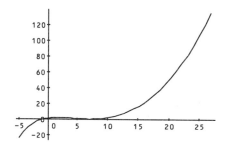

In `pairs`, we use `seq` and `subs` to generate a set of parametric equations that are then graphed with `plot`. In this case, the option `view=[-25..25,25..25]` is used to instruct Maple that both the horizontal and vertical axes displayed correspond to the interval $[-25, 25]$.

> `pairs:={seq(subs(C=i,[x(p),y(p),p=-4..4]),i=-5..5)}:`

> `plot(pairs,view=[-25..25,-25..25]);`

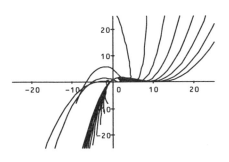

≡ Ricatti Equations

A **Ricatti equation**, named for the Italian mathematician Jacopo Francesco Ricatti (1676–1754), is a nonlinear equation of the form

$$y' + a(x)y^2 + b(x)y + c(x) = 0.$$

As with the Bernoulli and Clairaut equations, we make a substitution in order to solve Ricatti equations. In this case, let

$$y(x) = \frac{w'(x)}{w(x)} \frac{1}{a(x)}.$$

Then, by the quotient rule, we have

$$y'(x) = \frac{[w(x)a(x)]w''(x) - [w(x)a'(x) + w'(x)a(x)]w'(x)}{[w(x)a(x)]^2}$$

$$= \frac{w''(x)}{a(x)\,w(x)} - \frac{(w'(x))^2}{a(x)\,(w(x))^2} - \frac{a'(x)\,w'(x)}{(a(x))^2 w(x)},$$

and substituting into the equation $y' + a(x)y^2 + b(x)y + c(x) = 0$ yields the second-order equation

$$\frac{w''(x)}{a(x)\,w(x)} - \frac{(w'(x))^2}{a(x)\,(w(x))^2} - \frac{a'(x)\,w'(x)}{(a(x))^2 w(x)} + \frac{(w'(x))^2}{a(x)\,(w(x))^2} + \frac{b(x)\,w'(x)}{a(x)\,w(x)} + c(x) = 0.$$

Multiplying the equation by $a(x)w(x)$, we obtain $w''(x) - \dfrac{a'(x)\,w'(x)}{a(x)} + b(x)\,w'(x) + a(x)\,c(x)\,w(x) = 0$

and simplifying yields the second-order equation $w'' - \left(\dfrac{a'(x)}{a(x)} - b(x)\right)w' + a(x)\,c(x)\,w = 0.$

EXAMPLE: Convert the Ricatti equation

$$y' + (x^4 + x^2 + 1)y^2 + \frac{2(1 - x + x^2 - 2x^3 + x^4)}{1 + x^2 + x^4}y + \frac{1}{x^4 + x^2 + 1} = 0$$

to a second-order equation.

SOLUTION: Unlike the previous three examples, **dsolve** is unable to find a general solution to the equation.

```
> gen_sol:=
    dsolve(diff(y(x),x)+(x^4+x^2+1)*y(x)^2+
    2*(1-x+x^2-2*x^3+x^4)/(1+x^2+x^4)*y(x)+
    1/(x^4+x^2+1)=0,y(x));
```

  gen_sol :=

However, proceeding by traditional methods, we begin by identifying $a(x) = x^4 + x^2 + 1$, $b(x) = \dfrac{2(1 - x + x^2 - 2x^3 + x^4)}{1 + x^2 + x^4}$, and $c(x) = \dfrac{1}{x^4 + x^2 + 1}$. Letting $y(x) = \dfrac{w'(x)}{w(x)}\dfrac{1}{a(x)}$ yields the second-order equation

$$w'' - \left(\frac{4x^3 + 2x}{x^4 + x^2 + 1} - \frac{2(1 - x + x^2 - 2x^3 + x^4)}{1 + x^2 + x^4}\right)w' + (x^4 + x^2 + 1)\frac{1}{x^4 + x^2 + 1}w = 0,$$

which simplifies to $w'' - 2w' + w = 0$. In Chapter 4, we will learn how to find a general solution of $w'' - 2w' + w = 0$. For now, we compute a general solution with **dsolve**, naming the result **Sol**.

```
> Sol:=dsolve(diff(w(x),x$2)-2*diff(w(x),x)+w(x)=0,w(x));
```

  Sol := $w(x) = \_C1\,e^x + \_C2\,e^x x$

Since $y(x) = \dfrac{w'(x)}{w(x)} \dfrac{1}{a(x)}$, we have that

$$y(x) = \frac{C_1 e^x + C_2(e^x + xe^x)}{(x^4 + x^2 + 1)(C_1 e^x + C_2 xe^x)}$$

is a solution of $y' + (x^4 + x^2 + 1)y^2 + \dfrac{2(1 - x + x^2 - 2x^3 + x^4)}{1 + x^2 + x^4}y + \dfrac{1}{x^4 + x^2 + 1} = 0.$

We use the solution obtained in **Sol** to define y.

```
> assign(Sol):
  a:=x->x^4+x^2+1;
  y:=diff(w(x),x)/(w(x)*a(x));
```

$$a := x \rightarrow x^4 + x^2 + 1$$

$$y := \frac{\_C1\,e^x + \_C2\,e^x x + \_C2\,e^x}{(\_C1\,e^x + \_C2\,e^x)(x^4 + x^2 + 1)}$$

To graph the solution, y, for various values of the arbitrary constants, **_C1** and **_C2**, we first define a set of ordered pairs that we will use for **_C1** and **_C2**. Note that **minus** is used to remove the ordered pair corresponding to $(0, 0)$. If **_C1** and **_C2** are both replaced by 0, division by 0 occurs and an error results.

```
> pairs:={seq(seq([i/4,j/4],i=-1..1),j=-1..1)} minus {[0,0]}:
```

Next, we use **proc** and **subs** to define the function g. Given an ordered pair, **pair**, **g(pair)** replaces **_C1** in y by the first element of **pair** and **_C2** by the second and returns the result. We illustrate **g** using the ordered pair corresponding to $(2, 3)$.

```
> g:=proc(pair)
    subs({_C1=pair[1],_C2=pair[2]},y);
    end:
```

```
> g([2,3]);
```

$$\frac{5e^x + 3e^x x}{(2e^x + 3e^x x)(x^4 + x^2 + 1)}$$

map is then used to apply **g** to each ordered pair in the set **pairs**. The result is not displayed because a colon is included at the end of the command. However, we use **nops** to see that seven elements are in **to_graph** and **op** to display the first two elements of **to_graph**.

```
> to_graph:=map(g,pairs):
  nops(to_graph);
```

7

```
> op(1..2,to_graph);
```

$$\frac{1}{x^4 + x^2 + 1}, \quad -4\frac{-\frac{1}{4}e^x - \frac{1}{4}e^x x}{e^x x(x^4 + x^2 + 1)}$$

Finally, `plot` is used to graph the set of functions `to_graph` on the interval $[-2, 2]$.

```
> plot(to_graph,x=-2..2,-8..8);
```

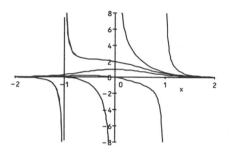

2.6 THEORY OF FIRST-ORDER EQUATIONS

In order to understand the types of initial-value problems that yield a unique solution, the following theorem is stated.

THEOREM   *Existence and Uniqueness*

Consider the initial-value problem

$$y' = f(x, y), \quad y(x_0) = y_0.$$

If f and $\dfrac{\partial f}{\partial y}$ are continuous functions on the rectangular region R, $R = \{(x, y) : a < x < b, \, c < y < d\}$ containing the point (x_0, y_0), then there exists an interval $|x - x_0| < h$ centered at x_0 on which there exists one and only one solution to the differential equation that satisfies the initial condition.

EXAMPLE: Solve the initial-value problem

$$\frac{dy}{dx} = \frac{x}{y},$$
$$y(0) = 0.$$

Does this result contradict the Existence and Uniqueness Theorem?

SOLUTION: This equation can be solved by the separation of variables that yields the equation $y\,dy = x\,dx$ that is easily solved to determine the family of solutions $y^2 - x^2 = C$, as done here with `solve`.

> `Sol:=dsolve(diff(y(x),x)=x/y(x),y(x));`

$Sol := y(x)^2 = x^2 + \_C1$

We note that the graph of $y^2 = x^2 + C$ for various values of C is the same as the graph of the level curves of $f(x,y) = y^2 - x^2$. After defining `to_graph` to be $y^2 - x^2$ and loading the `plots` package, several members of this family of hyperbolas (for $C \neq 0$) are graphed with `contourplot`, which is contained in the `plots` package.

> `to_graph:=solve(subs(y(x)=y,Sol),_C1);`

$to\_graph := y^2 - x^2$

> `with(plots):`
`contourplot(to_graph,x=-6..6,y=-6..6,`
`  axes=NORMAL);`

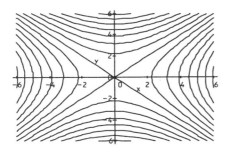

Hence, if $x = y = 0$, then $C = 0$. Therefore, solutions that pass through $(0,0)$, satisfy $y^2 - x^2 = 0$. Hence, two solutions $y = x$ and $y = -x$ satisfy the differential equation and the initial condition. Although more than one solution satisfies this initial-value problem, the Existence and Uniqueness Theorem is **not** contradicted because the function $\dfrac{x}{y}$ is not continuous at the point $(0,0)$.

EXAMPLE: Verify that the initial-value problem $\frac{dy}{dx} = y$, $y(0) = 1$ has a unique solution.

SOLUTION: Notice that in this case, $f(x, y) = y$, $x_0 = 0$, and $y_0 = 1$. Hence, both f and $\frac{\partial f}{\partial y}$ are continuous on all rectangular regions containing the point $(x_0, y_0) = (0, 1)$. Therefore, by the Existence and Uniqueness Theorem, there exists a unique solution to the differential equation that also satisfies the initial condition $y(0) = 1$. We can verify this by solving the initial-value problem. This equation is separable and equivalent to $\frac{dy}{y} = dx$. A general solution is given by $y = Ce^x$. Since we must have that $y(0) = 1$, we find that $C = 1$. Therefore, the unique solution is by given $y = e^x$, which is computed with **dsolve**.

> `Sol:=dsolve({diff(y(x),x)=y(x),y(0)=1},y(x));`

$Sol := y(x) = e^x$

> `assign(Sol):`
> `plot(y(x),x=-1..1);`

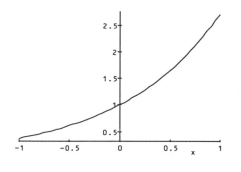

2.7 NUMERICAL APPROXIMATION OF FIRST-ORDER EQUATIONS
••

☰ Built-In Methods

Numerical approximations of solutions to differential equations can be obtained with **dsolve** together with the **numeric** option. This command is particularly useful when working with nonlinear equations for which **dsolve** alone is unable to find an explicit solution. This command is entered in the form **dsolve({deq,ics},fun,numeric)**, where **deq** is solved for **fun**.

Note that the number of initial conditions in **ics** must equal the order of the differential equation indicated in **deq**. In order to illustrate the command **dsolve** together with the **numeric** option, we consider the nonlinear equations in the following example.

EXAMPLE: Graph the solution of $\dfrac{dy}{dx} = \sin(2x - y)$ subject to the initial condition $y(0) = 0.5$ on the interval $[0, 15]$.

SOLUTION: First, we define **Eq** to be the equation $\dfrac{dy}{dx} = \sin(2x - y)$ and then use **dsolve** to approximate the solution of **Eq** subject to the initial condition $y(0) = 0.5$, naming the resulting output **Sol**. The resulting output is a procedure that represents an approximate function obtained through interpolation.

```
> x:='x':y:='y':
  Eq:=D(y)(x)=sin(2*x-y(x)):
```

```
> Sol:=dsolve({Eq,y(0)=.5},y(x),numeric);
```

   Sol := proc (x)'dsolve/numeric/result2' $(x, 2986564, [1])$ end

Sol can be evaluated for particular values of x as follows. Entering

```
> Sol(1);
```

   $\{x = 1, y(x) = .8758947813\}$

returns an ordered pair corresponding to x and $y(x)$ if $x = 1$. Entering

```
> Sol(1)[2];
```

   $y(x) = .8758947813$

returns the second part of **Sol**. Entering

```
> rhs(Sol(1)[2]);
```

   .8758947813

returns the value of the solution, $y(x)$, if $x = 1$.

   We then graph the solution by using the command **odeplot**, which is contained in the **plots** package. Generally, we will graph numerical solutions of differential equations obtained with **dsolve** together with the **numeric** option with **odeplot**:

```
> with(plots):
  odeplot(Sol,[x,y(x)],0..15);
```

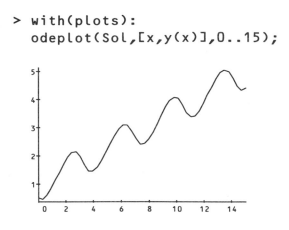

We can also use **DEplot1** to graph solutions. However, in this case, a numerical solution which can be evaluated, like that obtained with **dsolve**, is not generated. For example, entering

```
> with(DEtools):
  DEplot1(diff(y(x),x)=sin(2*x-y),y(x),
    x=0..15,{[0,1],[0,-1]});
```

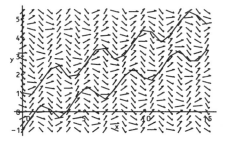

graphs the solutions of the equation satisfying $y(0) = 1$ and $y(0) = -1$ along with the direction field for the equation on the interval $[0, 15]$. The direction field is not displayed if the option **arrows=NONE** is included in the **DEplot1** command.

We can also use **DEplot1** to graph solutions to a differential equation under changing initial conditions.

EXAMPLE: Graph the solution of $y' = \sin(xy)$ subject to the initial condition $y(0) = i$ on the interval $[0, 7]$ for $i = 0.5, 1.0, 1.5, 2.0$, and 2.5.

SOLUTION: We begin by using `seq` to define the set of ordered pairs $(i, i/2)$ for $i = 1, 2, \ldots, 5$. These correspond to the initial conditions $y(0) = i/2$ for $i = 1, 2, \ldots, 5$.

```
> inits:={seq([0,i/2],i=1..5)}:
```

Next, we use `DEplot1` to graph the solutions to $y' = \sin(xy)$ for the initial conditions specified in `inits`. The option `arrows=NONE` is included so that the direction fields for the equation are not included in the graph. The option `stepsize=0.1` instructs Maple to use a smaller step size, helping to assure that the resulting graphs appear smooth.

```
> with(DEtools):
  DEplot1(diff(y(x),x)=sin(x*y),y(x),x=0..7,inits,
    arrows=NONE,stepsize=0.1);
```

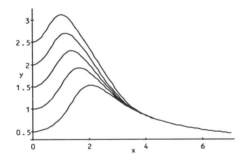

APPLICATION

The SIS Model

The initial-value problem for an **SIS model** is

$$\begin{cases} i'(t) = [a - (c + b)]i - a\,i^2 \\ i(0) = i_0 \end{cases},$$

where $a, b, c,$ and d represent positive constants corresponding to the **contact rate**, **death rate**, and **removal rate**, respectively, and $i(t)$ represents the percentage of the population infected by the disease. The incidence of some diseases, such as measles, rubella, and gonorrhea, oscillates seasonally. To model these diseases, we may wish to replace the constant contact rate, a, by a periodic function $a(t)$.

EXAMPLE: Sketch the graph of the solution to the SIS model for various initial conditions if (a) $a(t) = 3 - 2.5 \sin(6t)$, $b = 2$, and $c = 1$ and (b) $a(t) = 3 - 2.5 \sin(6t)$, $b = 1$, and $c = 1$.

SOLUTION: For (a), we begin by defining a, b, c, and Eq.

```
> a:=t->3-2.5*sin(6*t):
  b:=2:
  c:=1:

> i:='i':
  Eq:=diff(i(t),t)=(a(t)-(c+b))*i-a(t)*i^2:
```

We will graph the solutions satisfying the initial conditions $i(0) = 0.1, i(0) = 0.2, \ldots, i(0) = 0.9$, so we use seq to define inits to be the set of nine ordered pairs corresponding to the initial conditions. Note that the first coordinate of each pair corresponds to the t-value, while the second coordinate corresponds to the value of i for that t-value.

```
> inits:={seq([0,i/10],i=1..9)};
```
$$inits := \left\{ \left[0, \frac{1}{10}\right], \left[0, \frac{1}{5}\right], \left[0, \frac{2}{5}\right], \left[0, \frac{3}{10}\right], \left[0, \frac{1}{2}\right], \left[0, \frac{3}{5}\right], \left[0, \frac{4}{5}\right], \left[0, \frac{7}{10}\right], \left[0, \frac{9}{10}\right] \right\}$$

Last, we use DEplot1 to graph Eq for $0 \le t \le 10$ subject to the initial conditions specified in inits. The option stepsize=0.05 reduces the step size in the DEplot1 calculations, helping to assure that the resulting graphs are smooth. In this case, the direction field for Eq is not displayed because we include the option arrows=NONE.

```
> DEplot1(Eq,i(t),0..10,inits,stepsize=0.05,arrows=NONE);
```

For (b), we proceed in the same manner as in (a). In this case, however, we see that the direction field for the equation is also displayed.

```
> a:=t->3-2.5*sin(6*t):
  b:=1:
  c:=1:
  i:='i':
  Eq:=diff(i(t),t)=(a(t)-(c+b))*i-a(t)*i^2;
```

$$Eq := \frac{\partial}{\partial t} i(t) = (1 - 2.5 \sin(6\,t))i - (3 - 2.5 \sin(6\,t))i^2$$

```
> DEplot1(Eq,i(t),0..10,inits,stepsize=0.05);
```

In other cases, you may wish to implement your own numerical algorithms to approximate solutions of differential equations. We briefly discuss three familiar methods (Euler's method, the improved Euler's method, and the Runge-Kutta method) and illustrate how to implement these algorithms using Maple. Details regarding these and other algorithms, including discussions of the error involved in implementing them, can be found in most numerical analysis texts or other references like the *Handbook of Differential Equations* by Daniel Zwillinger, which is published by Academic Press (Cambridge, MA).

▲

☰ *Euler's Method*

In many cases, we cannot obtain a formula for the solution of an initial-value problem of the form

$$y' = f(x, y), y(x_0) = y_0.$$

Therefore, we approximate the solution using a numerical method. One of these methods is called **Euler's method** and is based on tangent line approximations. Let h represent a small change, or **step size**, in the independent variable x. Then we approximate the value of y at the sequence of x-values, $x_1, x_2, x_3, \ldots$, where

$$x_1 = x_0 + h$$
$$x_2 = x_1 + h = x_0 + 2h$$
$$x_3 = x_2 + h = x_0 + 3h$$
$$\vdots$$
$$x_n = x_{n-1} + h = x_0 + nh.$$

The slope of the tangent line to the graph of y at these values of x is given by the differential equation $y' = f(x, y)$. For example, at $x = x_0$, the slope of the tangent line is $f(x_0, y(x_0)) = f(x_0, y_0)$. Therefore, the tangent line to the graph of y is

$$y - y_0 = f(x_0, y_0)(x - x_0) \text{ or } y = f(x_0, y_0)(x - x_0) + y_0.$$

Using this line to find the value of y at $x = x_1$ (which we call y_1) then yields

$$y_1 = f(x_0, y_0)(x_1 - x_0) + y_0 = hf(x_0, y_0) + y_0.$$

Therefore, we obtain the approximate value of y at $x = x_1$.
Next, we use the point (x_1, y_1) to estimate the value of y when $x = x_2$. Using a similar procedure, we approximate the tangent line at $x = x_1$ with

$$y - y_1 = f(x_1, y_1)(x - x_1) \text{ or } y = f(x_1, y_1)(x - x_1) + y_1.$$

Then, at $x = x_2$,

$$y_2 = f(x_1, y_1)(x_2 - x_1) + y_1 = hf(x_1, y_1) + y_1.$$

Continuing with this procedure, we see that at $x = x_n$,

$$y_n = hf(x_{n-1}, y_{n-1}) + y_{n-1}.$$

Using this formula, we obtain a sequence of points of the form (x_n, y_n) $(n = 1, 2, \ldots)$, where y_n is the approximate value of $y(x_n)$.

EXAMPLE: Use Euler's method with $h = 0.1$ and 0.05 to approximate the solution of $y' = xy$, $y(0) = 1$ on $0 \le x \le 1$. Also, determine the exact solution and compare the results.

SOLUTION: First, we note that $f(x, y) = xy$, $x_0 = 0$, and $y_0 = 1$. Then, with $h = 0.1$, we have the formula

$$y_n = hf(x_{n-1}, y_{n-1}) + y_{n-1} = 0.1x_{n-1}y_{n-1} + y_{n-1}.$$

Then, for $x_1 = x_0 + h = 0.1$, we have

$$y_1 = 0.1x_0y_0 + y_0 = 0.1(0)(1) + 1 = 1.$$

Similarly, for $x_2 = x_0 + 2h = 0.2$,

$$y_2 = 0.1x_1y_1 + y_1 = 0.1(0.1)(1) + 1 = 1.01.$$

In the following, we define f, h, x, and y to calculate

$$y_n = hf(x_{n-1}, y_{n-1}) + y_{n-1}.$$

Note that in defining the recursively defined function y, we take advantage of the option `remember`. This instructs Maple to remember the values of y computed, and thus, when computing $y(n)$, Maple need not recompute $y(n - 1)$.

```
> f:=(x,y)->x*y:
  h:=0.1:
  x:=n->n*h:
  y:=proc(n) option remember;
    y(n-1)+h*f(x(n-1),y(n-1))
    end:
  y(0):=1:
```

We then use `seq` to calculate $y_0, y_1, \ldots, y_9, y_{10}$.

```
> seq(y(i),i=0..10);
```

1, 1, 1.01, 1.0302, 1.061106, 1.10355024, 1.158727752, 1.228251417, 1.314229016,
1.419367337, 1.547110397

To compare these results to the exact solution, we use **dsolve** to find the exact solution, naming
the resulting output **Sol**, and then we use **assign** to name $Y(X)$ the result obtained in **Sol**.
Note that we use capital letters to avoid conflict with the definition of y entered before.

```
> Sol:=dsolve({diff(Y(X),X)=X*Y(X),Y(0)=1},Y(X));
  assign(Sol):
```

$$Sol := Y(X) = e^{\frac{1}{2}X^2}$$

We then use **array**, **seq**, and **evalf** to compare the approximate and exact values. The
numbers in the second column correspond to the approximation, while the numbers in the third
column correspond to the exact value.

```
> array([seq([n,y(n),evalf(subs(X=n/10,Y(X)))],n=0..10)]);
```

$$
\begin{bmatrix}
0 & 1 & 1. \\
1 & 1 & 1.005012521 \\
2 & 1.01 & 1.020201340 \\
3 & 1.0302 & 1.046027860 \\
4 & 1.061106 & 1.083287068 \\
5 & 1.10355024 & 1.133148453 \\
6 & 1.158727752 & 1.197217363 \\
7 & 1.228251417 & 1.277621313 \\
8 & 1.314229016 & 1.377127764 \\
9 & 1.419367337 & 1.499302500 \\
10 & 1.547110397 & 1.648721271
\end{bmatrix}
$$

Then, for $h = 0.05$, we use

$$y_n = hf(x_{n-1}, y_{n-1}) + y_{n-1} = 0.05x_{n-1}y_{n-1} + y_{n-1}$$

to obtain an approximation. We first redefine x, h, and y

```
> x:='x':y:='y':
  h:=0.05:
  x:=n->n*h:
  y:=proc(n) option remember;
    y(n-1)+h*f(x(n-1),y(n-1))
    end:
  y(0):=1:
```

and then, as previously, compare the approximation to the exact value. Notice that the approximation obtained with $h = 0.05$ is better than the approximation obtained with $h = 0.1$.

```
> array([seq([n,y(n),evalf(subs(X=n/20,Y(X)))],n=0..20)]);
```

$$
\begin{bmatrix}
0 & 1 & 1. \\
1 & 1 & 1.001250782 \\
2 & 1.0025 & 1.005012521 \\
3 & 1.00751250 & 1.011313519 \\
4 & 1.015068844 & 1.020201340 \\
5 & 1.025219532 & 1.031743407 \\
6 & 1.038034776 & 1.046027860 \\
7 & 1.053605298 & 1.063164672 \\
8 & 1.072043391 & 1.083287068 \\
9 & 1.093484259 & 1.106553245 \\
10 & 1.118087655 & 1.133148453 \\
11 & 1.146039846 & 1.163287444 \\
12 & 1.177555942 & 1.197217363 \\
13 & 1.212882620 & 1.235221122 \\
14 & 1.252301305 & 1.277621313 \\
15 & 1.296131851 & 1.324784759 \\
16 & 1.344736795 & 1.377127764 \\
17 & 1.398526267 & 1.435122197 \\
18 & 1.457963633 & 1.499302500 \\
19 & 1.523571997 & 1.570273801 \\
20 & 1.595941667 & 1.648721271
\end{bmatrix}
$$

≡ *Improved Euler's Method*

Euler's method can be improved by using an average slope over each interval. Using the tangent line approximation of the curve through (x_0, y_0), $y = f(x_0, y_0)(x - x_0) + y_0$, we find the approximate value of y at $x = x_1$, which we now call y_1^*. Therefore,

$$y_1^* = hf(x_0, y_0) + y_0.$$

Then, with the differential equation $y' = f(x, y)$, we find that the approximate slope of the tangent line at $x = x_1$ is $f(x_1, y_1^*)$. Then the average of the two slopes, $f(x_0, y_0)$ and $f(x_1, y_1^*)$, is $\dfrac{f(x_0, y_0) + f(x_1, y_1^*)}{2}$, and the equation of the line through (x_0, y_0) with slope $\dfrac{f(x_0, y_0) + f(x_1, y_1^*)}{2}$ is

$$y = \frac{f(x_0, y_0) + f(x_1, y_1^*)}{2}(x - x_0) + y_0.$$

Therefore, at $x = x_1$, we find the approximate value of f given by

$$y_1 = \frac{f(x_0, y_0) + f(x_1, y_1^*)}{2}(x_1 - x_0) + y_0 = \frac{f(x_0, y_0) + f(x_1, y_1^*)}{2}h + y_0.$$

Continuing in this manner, the approximation in each step in the improved Euler's method depends on the following two calculations:

$$y_n^* = hf(x_{n-1}, y_{n-1}) + y_{n-1}$$
$$y_n = \frac{f(x_{n-1}, y_{n-1}) + f(x_n, y_n^*)}{2}h + y_{n-1}$$

EXAMPLE: Use the improved Euler's method to approximate the solution of $y' = xy$, $y(0) = 1$ on $0 \le x \le 1$ for $h = 0.1$. Also, compare the results to the exact solution.

SOLUTION: In this case, $f(x, y) = xy$, $x_0 = 0$, and $y_0 = 1$. Therefore, we use the equations

$$y_n^* = hx_{n-1}y_{n-1} + y_{n-1}$$

and

$$y_n = \frac{x_{n-1}y_{n-1} + x_n y_n^*}{2}h + y_{n-1}$$

for $n = 1, 2, \ldots, 10$. For example, if $n = 1$, we have

$$y_1^* = hx_0 y_0 + y_0 = (0.1)(0)(1) + 1 = 1$$

and

$$y_1 = \frac{x_0 y_0 + x_1 y_1^*}{2} h + y_0 = \frac{(0)(1) + (0.1)(1)}{2}(0.1) + 1 = 1.005.$$

Then

$$y_2^* = hx_1 y_1 + y_1 = (0.1)(0.1)(1.005) + 1.005 = 1.01505.$$

and

$$y_2 = \frac{x_1 y_1 + x_2 y_2^*}{2} h + y_1 = \frac{(0.1)(1.005) + (0.2)(1.01505)}{2}(0.1) + 1.005 = 1.0201755.$$

As in the previous example, we define x, f, h, and y:

```
> x:='x':y:='y':
  f:=(x,y)->x*y:
  h:=0.1:
  x:=n->n*h:
  y:=proc(n) option remember;
    y(n-1)+h/2*(f(x(n-1),y(n-1))+
    f(x(n-1),y(n-1)+h*f(x(n-1),y(n-1))))
    end:
  y(0):=1:
```

We find the exact solution of the equation with **dsolve**:

```
> Sol:=dsolve({diff(Y(X),X)=X*Y(X),Y(0)=1},Y(X));
  assign(Sol):
```

$$Sol := Y(X) = e^{\frac{1}{2}X^2}$$

and then compare the approximate solution to the exact solution. From the results, we see that the approximation using the improved Euler's method results in a slight improvement from that obtained in the first example.

```
> array([seq([n,y(n),evalf(subs(X=n/10,Y(X)))],n=0..10)]);
```

$$
\begin{bmatrix}
0 & 1 & 1. \\
1 & 1 & 1.005012521 \\
2 & 1.010050000 & 1.020201340 \\
3 & 1.030453010 & 1.046027860 \\
4 & 1.061830304 & 1.083287068 \\
5 & 1.105152980 & 1.133148453 \\
6 & 1.161792070 & 1.197217363 \\
7 & 1.233590820 & 1.277621313 \\
8 & 1.322964475 & 1.377127764 \\
9 & 1.433035119 & 1.499302500 \\
10 & 1.567812072 & 1.648721271
\end{bmatrix}
$$

≡ *The Runge-Kutta Method*

In an attempt to improve on the approximation obtained with Euler's method as well as avoid the analytic differentiation of the function $f(x, y)$ to obtain y'', y''', ..., the Runge-Kutta method, which involves many more computations at each step, is introduced. Let us begin with the Runge-Kutta method of order 2. Suppose that we know the value of y at x_n. We now use the point (x_n, y_n) to approximate the value of y at a nearby value $x = x_n + h$ by assuming that

$$
y_{n+1} = y_n + Ak_1 + Bk_2,
$$

where

$$
k_1 = hf(x_n, y_n) \text{ and } k_2 = hf(x_n + ah, y_n + bk_1).
$$

We can also use the Taylor series expansion of y to obtain another representation of $y_{n+1} = y(x_n + h)$, as follows:

$$
y(x_n + h) = y(x_n) + hy'(x_n) + h^2 \frac{y''(x_n)}{2!} + \cdots = y_n + hy'(x_n) + h^2 \frac{y''(x_n)}{2!} + \cdots
$$

Now, since

$$
y_{n+1} = y_n + Ak_1 + Bk_2 = y_n + Ahf(x_n, y_n) + Bhf(x_n + ah, y_n + bhf(x_n, y_n)),
$$

we wish to determine values of A, B, a, and b such that these two representations of y_{n+1} agree. Notice that if we let $A = 1$ and $B = 0$, then the relationships match up to order h. However, we can choose these parameters more wisely so that agreement occurs up through terms of order h^2. This is accomplished by considering the Taylor expansion of a function F of two variables about (x_0, y_0) that is given by

$$F(x, y) = F(x_0, y_0) + \frac{\partial F}{\partial x}(x_0, y_0)(x - x_0) + \frac{\partial F}{\partial y}(x_0, y_0)(y - y_0) + \cdots.$$

In our case, we have

$$f(x_n + ah, y_n + bhf(x_n, y_n)) = f(x_n, y_n) + ah\frac{\partial f}{\partial x}(x_n, y_n) + bhf(x_n, y_n)\frac{\partial f}{\partial x}(x_n, y_n) + O(h^2).$$

The power series is then substituted into the following expression and simplified to yield

$$y_{n+1} = y_n + Ahf(x_n, y_n) + bhf(x_n + ah, y_n + bhf(x_n, y_n))$$
$$= y_n + (A + B)hf(x_n, y_n) + aBh^2\frac{\partial f}{\partial x}(x_n, y_n) + bBh^2 f(x_n, y_n)\frac{\partial f}{\partial x}(x_n, y_n) + O(h^3).$$

Comparing this expression to the following power series obtained directly from the Taylor series of y,

$$y(x_n + h) = y(x_n) + hf(x_n, y_n) + \frac{h^2}{2}\frac{\partial f}{\partial x}(x_n, y_n) + \frac{h^2}{2}f(x_n, y_n)\frac{\partial f}{\partial y}(x_n, y_n) + O(h^3)$$

or

$$y_{n+1} = y_n + hf(x_n, y_n) + \frac{h^2}{2}\frac{\partial f}{\partial x}(x_n, y_n) + \frac{h^2}{2}f(x_n, y_n)\frac{\partial f}{\partial y}(x_n, y_n) + O(h^3),$$

we see that A, B, a, and b must satisfy the following system of nonlinear equations:

$$A + B = 1, \quad aA = \frac{1}{2}, \quad \text{and } bB = \frac{1}{2}.$$

Therefore, choosing $a = b = 1$, the Runge-Kutta method of order two uses the equations

$$y_{n+1} = y(x_n + h) = y_n + \frac{1}{2}hf(x_n, y_n) + \frac{1}{2}hf(x_n + h, y_n + hf(x_n, y_n))$$

$$= y_n + \frac{1}{2}(k_1 + k_2),$$

where $k_1 = hf(x_n, y_n)$ and $k_2 = hf(x_n + h, y_n + k_1)$. Notice that this method is equivalent to the improved Euler method.

EXAMPLE: Use the Runge-Kutta method with $h = 0.1$ to approximate the solution of the initial value problem $y' = xy$, $y(0) = 1$ on $0 \le x \le 1$.

SOLUTION: In this case, $f(x, y) = xy$, $x_0 = 0$, and $y_0 = 1$. Therefore, on each step we use the three equations

$$k_1 = hf(x_n, y_n) = 0.1x_ny_n,$$
$$k_2 = hf(x_n + h, y_n + k_1) = 0.1(x_n + 0.1)(y_n + k_1),$$

and

$$y_{n+1} = y_n + \frac{1}{2}(k_1 + k_2).$$

For example, if $n = 0$, then

$$k_1 = 0.1x_0y_0 = 0.1(0)(1) = 0,$$
$$k_2 = 0.1(x_0 + 0.1)(y_0 + k_1) = 0.1(0.1)(1) = 0.01,$$

and

$$y_1 = y_0 + \frac{1}{2}(k_1 + k_2) = 1 + \frac{1}{2}(0.01) = 1.005.$$

Therefore, the Runge-Kutta method of order 2 approximates that the value of y at $x = 0.1$ is 1.005. In the same manner as in the previous examples, we first define f, h, c, and yrk:

```
> yrk:='yrk':
  f:=(x,y)->x*y:
  h:=0.1:
  x:=n->n*h:
  yrk:=proc(n)
    local k1,k2;
    option remember;
    k1:=h*f(x(n-1),yrk(n-1));
    k2:=h*f(x(n-1)+h,yrk(n-1)+k1);
    yrk(n-1)+1/2*(k1+k2)
    end:
  yrk(0):=1:
```

Then we compute the exact solution of the equation

```
> Sol:=dsolve({diff(Y(X),X)=X*Y(X),Y(0)=1},Y(X));latex(");
assign(Sol):
```

$$Sol := Y(X) = e^{\frac{1}{2}X^2}$$

and then display the results obtained for the other values on $0 \leq x \leq 1$ using the Runge-Kutta method of order 2. Note that the Runge-Kutta method provides a much better approximation than either the Euler's method or the improved Euler's method.

```
> array([seq([n,yrk(n),evalf(subs(X=n/10,Y(X)))],n=0..10)]);
```

0	1	1.
1	1.005000000	1.005012521
2	1.020175500	1.020201340
3	1.045985940	1.046027860
4	1.083223039	1.083287068
5	1.133051299	1.133148453
6	1.197068697	1.197217363
7	1.277392007	1.277621313
8	1.376773105	1.377127764
9	1.498755202	1.499302500
10	1.647881345	1.648721271

The terms of the power series expansions used in the derivation of the Runge-Kutta method of order 2 can be made to match up to order 4. The approximation at each step is found to be made with

$$y_{n+1} = y_n + \frac{h}{6} [k_1 + 2k_2 + 2k_3 + k_4], n = 0, 1, 2, \ldots,$$

where

$$k_1 = f(x_n, y_n), k_2 = f\left(x_n + \frac{h}{2}, y_n + \frac{hk_1}{2}\right), k_3 = f\left(x_n + \frac{h}{2}, y_n + \frac{hk_2}{2}\right),$$

and

$$k_4 = f(x_{n+1}, y_n + hk_3).$$

EXAMPLE: Use the fourth-order Runge-Kutta method to approximate the solution of the problem $y' = xy$, $y(0) = 1$ on $0 \le x \le 1$.

SOLUTION: With $f(x, y) = xy$, $x_0 = 0$, and $y_0 = 1$, the formulas are

$$k_1 = f(x_n, y_n) = x_n y_n,$$

$$k_2 = f\left(x_n + \frac{h}{2}, y_n + \frac{hk_1}{2}\right) = \left(x_n + \frac{0.1}{2}\right)\left(y_n + \frac{0.1k_1}{2}\right),$$

$$k_3 = f\left(x_n + \frac{h}{2}, y_n + \frac{hk_2}{2}\right) = \left(x_n + \frac{0.1}{2}\right)\left(y_n + \frac{0.1k_2}{2}\right),$$

$$k_4 = f(x_{n+1}, y_n + hk_3) = x_{n+1}(y_n + 0.1k_3),$$

and

$$y_{n+1} = y_n + \frac{h}{6}[k_1 + 2k_2 + 2k_3 + k_4] = y_n + \frac{0.1}{6}[k_1 + 2k_2 + 2k_3 + k_4].$$

For $n = 0$, we have $k_1 = x_0 y_0 = (0)(1) = 0$, $k_2 = \left(x_0 + \frac{0.1}{2}\right)\left(y_0 + \frac{0.1k_1}{2}\right) = (0.05)(1) = 0.05$, $k_3 = \left(x_0 + \frac{0.1}{2}\right)\left(y_0 + \frac{0.1k_2}{2}\right) = (0.05)(1 + 0.0025) = 0.050125$, and $k_4 = x_1(y_0 + 0.1k_3) = (0.1)(1 + 0.0050125) = 0.10050125$. Therefore,

$$y_1 = y_0 + \frac{0.1}{6}[k_1 + 2k_2 + 2k_3 + k_4] = 1 + \frac{0.1}{6}[0 + 0.05 + 0.050125 + 0.10050125] = 1.005012521.$$

We list the results for the Runge-Kutta method of order 4 and compare these results to those obtained in the previous example. Notice that this method yields the most accurate approximation of the methods used to this point.

```
> yrk4:='yrk4':
  f:=(x,y)->x*y:
  h:=0.1:
  x:=n->n*h:
  yrk4:=proc(n)
    local k1,k2,k3,k4;
    option remember;
    k1:=f(x(n-1),yrk4(n-1));
    k2:=f(x(n-1)+h/2,yrk4(n-1)+h*k1/2);
    k3:=f(x(n-1)+h/2,yrk4(n-1)+h*k2/2);
    k4:=f(x(n),yrk4(n-1)+h*k3);
    yrk4(n-1)+h/6*(k1+2*k2+2*k3+k4)
    end:
  yrk4(0):=1:

> array([seq([n,yrk(n),yrk4(n),
  evalf(subs(X=n/10,Y(X)))],n=0..10)]);
```

$$
\begin{bmatrix}
0 & 1 & 1 & 1. \\
1 & 1.005000000 & 1.005012521 & 1.005012521 \\
2 & 1.020175500 & 1.020201340 & 1.020201340 \\
3 & 1.045985940 & 1.046027859 & 1.046027860 \\
4 & 1.083223039 & 1.083287065 & 1.083287068 \\
5 & 1.133051299 & 1.133148446 & 1.133148453 \\
6 & 1.197068697 & 1.197217347 & 1.197217363 \\
7 & 1.277392007 & 1.277621279 & 1.277621313 \\
8 & 1.376773105 & 1.377127694 & 1.377127764 \\
9 & 1.498755202 & 1.499302362 & 1.499302500 \\
10 & 1.647881345 & 1.648721007 & 1.648721271
\end{bmatrix}
$$

3
Applications of First-Order Ordinary Differential Equations

When the space shuttle is launched from the Kennedy Space Center, its velocity and position at a later time can be determined by solving a first-order ordinary differential equation. The same can be said for finding the flow of electromagnetic forces, the temperature of a cup of coffee, the population of a species, and numerous other applications. In this chapter, we show how these problems can be expressed as first-order equations. Since the techniques for solving these problems were introduced in Chapter 2, we focus our attention on setting up the problems and explaining the meaning of the subsequent solutions.

3.1 ORTHOGONAL TRAJECTORIES

We begin our discussion with a topic that is encountered in the study of electromagnetic fields and heat flow. Before we can give any specific applications, however, we must state the following definition.

DEFINITION

Orthogonal Curves

Two lines, L_1 and L_2, with slopes m_1 and m_2, respectively, are **orthogonal** (or perpendicular) if their slopes satisfy the relationship $m_1 = \dfrac{-1}{m_2}$. Hence, two curves, C_1 and C_2, are **orthogonal** (or perpendicular) at a point if their respective tangent lines to the curves at that point are perpendicular.

EXAMPLE: Use the definition of orthogonality to verify that the curves given by $y = x$ and $y = \sqrt{1 - x^2}$ are orthogonal at the point $\left(\dfrac{\sqrt{2}}{2}, \dfrac{\sqrt{2}}{2} \right)$.

SOLUTION: First note that the point $\left(\dfrac{\sqrt{2}}{2}, \dfrac{\sqrt{2}}{2} \right)$ lies on the graph of both $y = x$ and $y = \sqrt{1 - x^2}$. The derivatives of the functions are given by $y' = 1$ and $y' = \dfrac{-x}{\sqrt{1 - x^2}}$, respectively, as shown using **D**.

```
> y[1]:=x->x:
  y[2]:=x->sqrt(1-x^2):
  D(y[1])(x);
  D(y[2])(x);
```

$$1$$

$$-\frac{x}{\sqrt{1 - x^2}}$$

Hence, the slope of the tangent line to $y = x$ at $x = \dfrac{\sqrt{2}}{2}$ is 1. Substitution of $x = \dfrac{\sqrt{2}}{2}$ into $y' = \dfrac{-x}{\sqrt{1 - x^2}}$ yields -1 as the slope of the tangent line at $x = \dfrac{\sqrt{2}}{2}$.

```
> D(y[2])(sqrt(2)/2);
```

$$-1$$

Since the slopes of the lines tangent to the graphs of $y = x$ and $y = \sqrt{1 - x^2}$ at the point $\left(\dfrac{\sqrt{2}}{2}, \dfrac{\sqrt{2}}{2} \right)$ are negative reciprocals, the curves are orthogonal at the point $\left(\dfrac{\sqrt{2}}{2}, \dfrac{\sqrt{2}}{2} \right)$. We graph these two curves along with the tangent line to $y = \sqrt{1 - x^2}$ at $\left(\dfrac{\sqrt{2}}{2}, \dfrac{\sqrt{2}}{2} \right)$ with **plot** to illustrate that the two are orthogonal.

```
> plot({y[1](x),y[2](x),-x+sqrt(2)},x=-1..1,-.5..1.5);
```

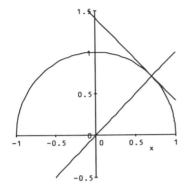

The next step in our discussion of orthogonal curves is to determine the set of orthogonal curves to a given family of curves. Typically, we refer to this set of orthogonal curves as the **family of orthogonal trajectories**. In other words, if we are given a family of curves F, then we would like to find a set of curves that are orthogonal to each curve in F. This is not a difficult procedure, because we know the relationship between orthogonal curves.

Suppose that the family of curves is defined as $F(x, y) = c$ and that the slope of the tangent line at any point on these curves is

$$\frac{dy}{dx} = f(x, y)$$

(which can be obtained by differentiating $F(x, y) = c$ with respect to x and solving for $\frac{dy}{dx}$). Then the slope of the tangent line on the orthogonal trajectory is $\frac{dy}{dx} = -\frac{1}{f(x, y)}$.

Hence, the family of orthogonal trajectories is found by solving the first-order equation

$$\frac{dy}{dx} = -\frac{1}{f(x, y)}.$$

We illustrate how to find the family of orthogonal trajectories in the example following.

EXAMPLE: Determine the family of orthogonal trajectories to the family of curves $y = cx^2$ (parabolas).

SOLUTION: First, we must find the slope of the tangent line at any point on the parabola $y = cx^2$. Therefore, we differentiate with respect to x. This gives us

$$\frac{dy}{dx} = 2cx.$$

However, from $y = cx^2$, we have that

$$c = \frac{y}{x^2}.$$

Substitution into $\frac{dy}{dx} = 2cx$ then yields $\frac{dy}{dx} = 2cx = 2\left(\frac{y}{x^2}\right)x = \frac{2y}{x}$ on the parabolas. Hence, we must solve

$$\frac{dy}{dx} = -\frac{x}{2y}$$

to determine the orthogonal trajectories. Of course, this equation is separable, so we write it as $2y\,dy = -x\,dx$, and then integrating both sides gives us

$$y^2 = -\frac{x^2}{2} + k,$$

where k is a constant. Alternatively, we can use **dsolve** to find a general solution as follows:

```
> sol:=dsolve(diff(y(x),x)=-x/(2*y(x)),y(x));
```

$$Sol := y(x)^2 = -\frac{1}{2}x^2 + \_C1$$

This implicit solution can be expressed as

$$y^2 + \frac{x^2}{2} = k,$$

which we recognize as a family of ellipses.

To graph the family of parabolas $y = cx^2$, the family of ellipses $y^2 + \frac{x^2}{2} = k$, and the two families of curves together, we use the commands **implicitplot** and **display**, both of which are contained in the **plots** package. First, we load the **plots** package. Remember that if a colon had been included at the end of the command instead of a semicolon, the commands would have been loaded but not displayed.

```
> with(plots);
```

> [*animate, animate3d, conformal, contourplot, cylinderplot, densityplot,
> display, display3d, fieldplot, fieldplot3d, gradplot, gradplot3d,
> implicitplot, implicitplot3d, loglogplot, logplot, matrixplot, odeplot,
> pointplot, polarplot, polygonplot, polygonplot3d, polyhedraplot,
> replot, setoptions, setoptions3d, spacecurve, sparsematrixplot,
> sphereplot, surfdata, textplot, textplot3d, tubeplot*]

After defining the set of numbers `c_vals`, we use `seq` to define `parabs` to be the set of nine equations consisting of $y = cx^2$, where c has been replaced by each of the numbers in `c_vals`.

```
> c_vals:=seq(-1+i/4,i=0..8):
  parabs:={seq(y=c*x^2,c=c_vals)};
```

$$parabs := \left\{ y = \frac{3}{4}x^2, y = x^2, y = \frac{1}{4}x^2, y = \frac{1}{2}x^2, y = -\frac{1}{4}x^2, \right.$$
$$\left. y = -\frac{1}{2}x^2, y = 0, y = -\frac{3}{4}x^2, y = -x^2 \right\}$$

We then use `implicitplot` to graph these nine equations on the rectangle $[-3, 3] \times [-3, 3]$, naming the resulting graph `IP_1`. `IP_1` is then displayed with `display`.

```
> IP_1:=implicitplot(parabs,x=-3..3,y=-3..3):
  display({IP_1});
```

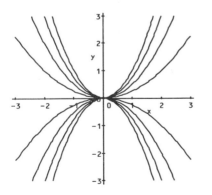

Similarly, we define `ellipses` to be the set of nine equations consisting of $y^2 + \dfrac{x^2}{2} = k$, where k is replaced by 1, 2, 3, ..., 8. These eight equations are also graphed on the rectangle $[-3, 3] \times [-3, 3]$. Then we use `display` to show the eight ellipses and again to show the parabolas and ellipses together.

```
> ellipses:={seq(y^2+x^2/2=k,k=1..8)}:
  IP_2:=implicitplot(ellipses,x=-3..3,y=-3..3):
  display({IP_2});
  display({IP_1,IP_2});
```

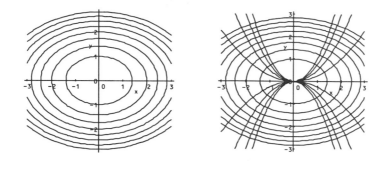

EXAMPLE: Let $T(x, y)$ represent the temperature at the point (x, y). The curves given by $T(x, y) = c$ (where c is constant) are called **isotherms**. The orthogonal trajectories are curves along which heat will flow. Determine the isotherms if the curves of heat flow are given by $y^2 + 2xy - x^2 = c$.

SOLUTION: We begin by finding the slope of the tangent line at each point on the heat flow curves $y^2 + 2xy - x^2 = c$ by using implicit differentiation.

We use **D** to compute the derivative of $y^2 + 2xy - x^2 = c$, naming the resulting output **step_1**.

```
> Eq_1:=y^2+2*x*y-x^2=c:
  step_1:=D(Eq_1);
```

$$step\_1 := 2D(y)y + 2D(x)y + 2xD(y) - 2D(x)x = D(c)$$

We then replace each occurrence of **D(x)** in **step_1** by 1 and each occurrence of **D(c)** by 0 (because c is a constant) with **subs** and name the resulting output **step_2**. We interpret **step_2** to be equivalent to the equation $2y\dfrac{dy}{dx} + 2y + 2x\dfrac{dy}{dx} - 2x = 0$.

```
> step_2:=subs({D(x)=1,D(c)=0},step_1);
```

$$step\_2 := 2D(y)y + 2y + 2xD(y) - 2x = 0$$

We calculate $\dfrac{dy}{dx}$ by solving **step_2** for **D(y)** with **solve** and name the result **im_deriv**.

```
> im_deriv:=solve(step_2,D(y));
```

$$im\_deriv := -\frac{2y - 2x}{2y + 2x}$$

Thus,

$$\frac{dy}{dx} = \frac{x - y}{x + y},$$

so the orthogonal trajectories satisfy the differential equation

$$\frac{dy}{dx} = -\frac{x + y}{x - y}.$$

```
> step_3:=simplify(-1/im_deriv);
```

$$step\_3 := \frac{y + x}{y - x}$$

Writing this equation in differential form as

$$(x + y)\,dx + (x - y)\,dy = 0,$$

we see that this equation is exact because $\frac{\partial}{\partial y}(x + y) = 1$ and $\frac{\partial}{\partial x}(x - y) = 1$. Hence, we solve the equation by integrating $x + y$ with respect to x to yield

$$f(x, y) = \frac{x^2}{2} + xy + g(y),$$

and then differentiating f with respect to y gives us

$$f_y(x, y) = x + g'(y).$$

Then $x + g'(y) = x - y$, so $g'(y) = -y$, which implies that $g(y) = -\frac{y^2}{2}$. This means that the family of orthogonal trajectories (isotherms) is given by $\frac{x^2}{2} + xy - \frac{y^2}{2} = k$.

We obtain the same results with **dsolve** as follows:

```
> Diff_Eq:=diff(y(x),x)=subs(y=y(x),step_3);
```

$$Diff\_Eq := \frac{\partial}{\partial x}y(x) = \frac{y(x) + x}{y(x) - x}$$

```
> Sol:=dsolve(Diff_Eq,y(x));
```

$$Sol := xy(x) + \frac{1}{2}x^2 - \frac{1}{2}y(x)^2 = \_C1$$

To graph $y^2 + 2xy - x^2 = c$ and $\frac{x^2}{2} + xy - \frac{y^2}{2} = k$ for various values of c and k and to see that the curves are orthogonal, we first define **Eq_2** to be the equation $\frac{x^2}{2} + xy - \frac{y^2}{2} = k$.

```
> Eq_2:=subs({y(x)=y,_C1=k},Sol);
```

$$Eq\_2 := xy + \frac{1}{2}x^2 - \frac{1}{2}y^2 = k$$

Then, in the same manner as in the previous example, we define **set_1** to be the set of seven equations obtained by replacing c in $y^2 + 2xy - x^2 = c$ by $-3, -2, -1, 0, 1, 2$, and 3 and **set_2** to be the set of seven equations obtained by replacing k in $\frac{x^2}{2} + xy - \frac{y^2}{2} = k$ by $-3, -2, -1, 0, 1,$ 2, and 3. Both sets of equations are graphed with **implicitplot** and then displayed together with **display**.

```
> set_1:={seq(subs(c=i,Eq_1),i=-3..3)}:
  set_2:={seq(subs(k=i,Eq_2),i=-3..3)}:
  with(plots):
  IP_1:=implicitplot(set_1,x=-4..4,y=-4..4):
  display({IP_1});
  IP_2:=implicitplot(set_2,x=-4..4,y=-4..4):
  display({IP_2});
  display({IP_1,IP_2});
```

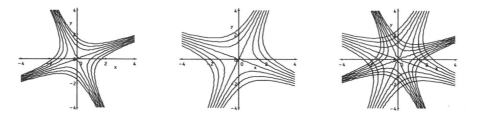

3.2 POPULATION GROWTH AND DECAY

Many interesting problems involving population can be solved with first-order differential equations. These include the determination of the number of cells in a bacteria culture, the number of citizens in a country, and the amount of radioactive substance remaining in a fossil. We begin our discussion by solving a population problem.

☰ The Malthus Model

Suppose that the rate at which a population $y(t)$ changes is proportional to the amount present. Mathematically, this statement is represented as the first-order initial-value problem

$$\frac{dy}{dt} = ky, \; y(0) = y_0,$$

where y_0 is the initial population. If $k > 0$, then the population increases (growth) while the population decreases (decay) if $k < 0$. Problems of this nature arise in fields such as cell population growth in biology as well as radioactive decay in physics. This model is known as the **Malthus model** because it was identified by the English clergyman and economist Thomas R. Malthus.

☰ Solution of the Malthus Model

We solve the Malthus model for all values of k and y_0. This enables us to refer to this solution in other problems without solving the differential equation again. Rewriting $\frac{dy}{dt} = ky$ in the form $\frac{dy}{y} = k\,dt$, we see that this is a separable differential equation. Integrating and simplifying results in

$$\int \frac{dy}{y} = \int k\,dt$$
$$\ln|y| = kt + C_1$$
$$y = Ce^{kt} \; (C = e^{C_1})$$

To find C, we apply the initial condition to obtain $y_0 = y(0) = Ce^{k\cdot 0} = C$. Thus, the solution to the initial-value problem $\frac{dy}{dt} = ky$, $y(0) = y_0$ is

$$y = y_0 e^{kt}.$$

Note that we obtain the same result with `dsolve`:

```
> dsolve({diff(y(t),t)=k*y(t),y(0)=y0},y(t));latex(");
```

$$y(t) = e^{kt}y0$$

This formula can be used for population growth and decay problems as shown in the following example.

EXAMPLE: Forms of a given element with different numbers of neutrons are called **nuclides**. Some nuclides are not stable. For example, potassium-40 (^{40}K) naturally decays to reach argon-40 (^{40}Ar). This decay was first observed, but not understood, by Henri Becquerel (1852–1908) in 1896. Marie Curie, however, began studying this decay in 1898, named it **radioactivity**, and discovered the radioactive substances polonium and radium. Marie Curie (1867–1934), along with her husband, Pierre Curie (1859–1906), and Henri Becquerel, received the Nobel Prize in physics in 1903 for their work on radioactivity. Marie Curie subsequently received the Nobel Prize in chemistry in 1910 for discovering polonium and radium. Given a sample of ^{40}K of sufficient size, after 1.2×10^9 years approximately half of the sample will have decayed to ^{40}Ar. The **half-life** of a nuclide is the time for half the nuclei in a given sample to decay. Since the half-life of a given nuclide is constant and independent of the sample size, we see that the rate of decay is proportional to the amount present. If the half-life of Polonium ^{209}Po is 100 years, then determine the percentage of the original amount of ^{209}Po that remains after 50 years.

SOLUTION: Let y_0 represent the original amount of ^{209}Po that is present. Then the amount present after t years is $y(t) = y_0e^{kt}$. Since $y(100) = \frac{1}{2}y_0$ and $y(100) = y_0e^{100k}$, we solve $y_0e^{100k} = \frac{1}{2}y_0$ for k with **solve**.

```
> k:=solve(y0*exp(100*k)=1/2*y0,k);
```

$$k := -\frac{1}{100}\ln(2)$$

Hence, $y(t) = y_0\left(\frac{1}{2}\right)^{t/100}$.

```
> y:=t->y0*exp(k*t);
  simplify(y(t));
```

$$y := t \rightarrow y0\, e^{kt}$$
$$y0\, 2^{-\left(\frac{1}{100}t\right)}$$

In order to determine the percentage of y_0 that remains, we evaluate $y(50)$ and obtain an approximation with **evalf**.

```
> y(50);
  evalf(y(50));
```

$$\frac{1}{2}y0\sqrt{2}$$

.7071067810y0

Therefore, approximately 70.71% of the original amount of ^{209}Po remains after 50 years.

In the previous example, we see that we can determine the amount of y_0 that remains even though we do not know the value of y_0. Hence, instead of letting $y(t)$ represent the amount of the substance present after time t, we can let it represent the fraction of y_0 that remains after time t. In doing this, we use the initial condition $y(0) = 1$ to indicate that 100% of y_0 is present at $t = 0$. We illustrate this idea in the following example.

EXAMPLE: The wood of an Egyptian sarcophagus (burial case) is found to contain 63% of the carbon-14 found in a present-day sample. What is the age of the sarcophagus?

SOLUTION: The half-life of carbon-14 is 5730 years. Let $y(t)$ be the percentage of carbon-14 in the sample after t years. Then $y(0) = 1$. Since $y(t) = y_0 e^{kt}$, $y(5730) = e^{5730k} = \frac{1}{2}$. Solving for k with **solve** yields $k = \dfrac{-\ln 2}{5730}$.

```
> k:='k':
  k:=solve(exp(5730*k)=1/2,k);
```

$$k := -\frac{1}{5730}\ln(2)$$

Thus, $y(t) = e^{kt} = e^{\frac{-\ln 2}{5730}t} = 2^{-t/5730}$.

```
> y:=t->exp(k*t):
  simplify(y(t));
```

$$2^{\left(-\frac{1}{5730}t\right)}$$

We must find the value of t for which $y(t) = .63$. Solving this equation with **solve** results in $t \approx 3819.48$. We conclude that the sarcophagus is approximately 3819 years old.

```
> solve(y(t)=.63);
```

3819.482007

Alternatively, we can graph $y(t)$ with `plot`

```
> plot({.63,y(t)},t=0..6000);
```

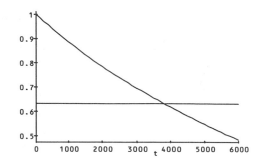

and then use `fsolve` to approximate the solution of $y(t) = .63$ as follows:

```
> fsolve(y(t)=.63,t,3000..4000);
```

3819.482007

To observe some of the limitations of the Malthus model, we consider a population problem in which the rate of growth of the population does not exclusively depend on the population present.

EXAMPLE: The population of the United States was recorded as 5.3 million in 1800. Use the Malthus model to approximate the population for years after 1800 if k was experimentally determined to be 0.03. Compare these results to the actual population. Is this a good approximation for years after 1800?

SOLUTION: In this example, $k = 0.03$ and $y_0 = 5.3$, and our model for the population of the United States at time t (where t is the number of years from 1800) is $y(t) = 5.3e^{0.03t}$.

```
> peq:=dsolve({diff(y(t),t)=k*y(t),y(0)=y0},y(t));
```

$peq := y(t) = e^{kt}y0$

```
> pop:=proc(t0,k0,y00) subs({t=t0,k=k0,y0=y00},op(2,peq)) end:
  pop(t,0.03,5.3);
```

$5.3e^{.03t}$

In order to compare this model with the actual population of the United States, census figures for the population of the United States for various years are listed along with the corresponding value of $y(t)$.

Year (t)	Actual Population (in millions)	Value of $y(t) = 5.3e^{0.03t}$	Year (t)	Actual Population (in millions)	Value of $y(t) = 5.3e^{0.03t}$
1800 (0)	5.30	5.30	1870 (70)	38.56	43.28
1810 (10)	7.24	7.15	1880 (80)	50.19	58.42
1820 (20)	9.64	9.66	1890 (90)	62.98	78.86
1830 (30)	12.68	13.04	1900 (100)	76.21	106.45
1840 (40)	17.06	17.60	1910 (110)	92.23	143.70
1850 (50)	23.19	23.75	1920 (120)	106.02	193.97
1860 (60)	31.44	32.06	1930 (130)	123.20	261.83

Although the model appears to approximate the data for several years after 1800, the accuracy of the approximation diminishes over time. This is because the population of the United States does not exclusively increase at a rate proportional to the population. Hence, another model that better approximates the population by taking other factors into account is needed. The graph of $y(t) = 5.3e^{0.03t}$ is shown along with the data points to show how the approximation becomes less accurate as t increases.

```
> popplot:=plot(pop(t,0.03,5.3),t=0..100):
  pdata:=[0,5.3,10,7.2,20,9.6,30,12.9,40,17,
    50,23.2,60,31.4,70,38.6,80,50.2,90,63,100,76.2]:
  dataplot:=plot(pdata,style=POINT):
  with(plots):
  display({popplot,dataplot});
```

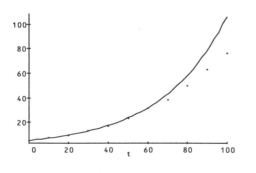

☰ *The Logistic Equation*

Since the approximation obtained with the Malthus model is less than desirable in the previous example, we see that another model is needed. The **logistic equation** (or **Verhulst equation**) is

$$y'(t) = (r - a\,y(t))y(t),$$

where r and a are constants, subject to the condition $y(0) = y_0$. This equation was first introduced by the Belgian mathematician Pierre Verhulst to study population growth. This equation differs from the Malthus model in that the term $(r - a\,y(t))$ is not constant. This equation can be written as $\dfrac{dy}{dt} = (r - ay)y = ry - ay^2$, where the term $(-y^2)$ represents an inhibitive factor or "death rate."

Hence, the population under these assumptions is not allowed to grow out of control as it was in the Malthus model. Also, the population does not grow or decay constantly.

☰ *Solution of the Logistic Equation*

The logistic equation is separable and thus can be solved by separation of variables. For convenience, we remove the independent variable t from the equation to obtain

$$y' = (r - a\,y)y \text{ or } \frac{dy}{dt} = (r - ay)y.$$

Separating variables and using partial fractions to integrate with respect to y, we have

$$\frac{dy}{(r - ay)y} = dt$$

$$\left(\frac{a/r}{r - ay} + \frac{1/r}{y}\right)dy = dt$$

$$\frac{1}{r}\left(\frac{a}{r - ay} + \frac{1}{y}\right)dy = dt$$

$$\left(\frac{a}{r - ay} + \frac{1}{y}\right)dy = r\,dt$$

$$-\ln|r - ay| + \ln|y| = rt + c.$$

Of course, we would like to solve this expression for y. Using the properties of logarithms yields

$$\ln\left|\frac{y}{r-ay}\right| = rt + c$$

$$\frac{y}{r-ay} = e^{rt+c} = Ke^{rt} \ (K = e^c)$$

$$\frac{r-ay}{y} = \frac{1}{K}e^{-rt}$$

$$\frac{r}{y} = \frac{1}{K}e^{-rt} + a$$

$$y = r\left(\frac{1}{K}e^{-rt} + a\right)^{-1}.$$

Applying the initial condition $y(0) = y_0$ to solve for K, we find that

$$\frac{y_0}{r - ay_0} = K.$$

After substitution of this constant into the general solution and simplification, the solution can be written as

$$y = \frac{ry_0}{ay_0 + (r - ay_0)}e^{-rt}.$$

We are also able to use **dsolve** to solve this initial-value problem, naming the resulting output **Sol**.

```
> y:='y':
  Sol:=dsolve({diff(y(t),t)=(r-a*y(t))*y(t),
  y(0)=y0},y(t));
```

$$Sol := y(t) = -\frac{r}{-a - \frac{e^{(-rt)}(r-y0a)}{y0}}$$

Then we use **assign** to name $y(t)$ the result obtained in **Sol** and **simplify** to simplify $y(t)$.

```
> assign(Sol):
  simplify(y(t));
```

$$-\frac{r\,y0}{-y0\,a - e^{-rt}r + e^{-rt}y0a}$$

Notice that $\lim_{t \to \infty} y(t) = \frac{r}{a}$. This makes the solution to the logistic equation different from that of the Malthus model in that the solution to the logistic equation approaches a limit as $t \to \infty$, while that of the Malthus model approaches infinity or zero as $t \to \infty$.

EXAMPLE: Use the logistic equation to approximate the population of the United States using $r = 0.03$, $a = 0.0001$, and $y_0 = 5.3$. Compare this result with the actual census values given in the following table. Use the model obtained to predict the population of the United States in the year 2000.

SOLUTION: We substitute the indicated values of r, a, and y_0 into $y = \dfrac{ry_0}{ay_0 + (r - ay_0)e^{-rt}}$ to obtain the approximation of the population of the United States at time t, where t represents the number of years since 1800,

$$y(t) = \frac{0.03 \cdot 5.3}{0.0001 \cdot 5.3 + (0.03 - 0.0001 \cdot 5.3)e^{-.03t}} = \frac{0.159}{0.00053 + 0.02947e^{-0.03t}}.$$

```
> y:='y':
  y:=t->0.159/(0.00053+0.02947*exp(-0.03*t)):
```

We compare the approximation of the population of the United States given by the approximation $y(t)$ with the actual population obtained from census figures. Note that this model appears to approximate more closely the population over a longer period of time than did the Malthus model.

Year (t)	Actual Population (in millions)	Value of $y(t)$	Year (t)	Actual Population (in millions)	Value of $y(t)$
1800 (0)	5.30	5.30	1900 (100)	76.21	79.61
1810 (10)	7.24	7.11	1910 (110)	92.23	98.33
1820 (20)	9.64	9.52	1920 (120)	106.02	119.08
1830 (30)	12.68	12.71	1930 (130)	123.20	141.14
1840 (40)	17.06	16.90	1940 (140)	132.16	163.59
1850 (50)	23.19	22.38	1950 (150)	151.33	185.45
1860 (60)	31.44	29.44	1960 (160)	179.32	205.82
1870 (70)	38.56	38.42	1970 (170)	203.30	224.05
1880 (80)	50.19	49.63	1980 (180)	226.54	239.78
1890 (90)	62.98	63.33	1990 (190)	248.71	252.94

```
> with(plots):
  pdata:=[0,5.3,10,7.2,20,9.6,30,12.9,40,17,
   50,23.2,60,31.4,70,38.6,80,50.2,90,63,100,76.2,
   110,92.23,120,106.02,130,123.2,140,132.16,
   150,151.33,160,179.32,170,203.3,180,226.54,
   190,248.71]:
  dataplot:=plot(pdata,style=POINT):
  plot_y:=plot(y(t),t=0..200):
  display({plot_y,dataplot});
```

To predict the population of the United States in the year 2000 with this model, we evaluate $y(200)$.

```
> y(200);
```

263.6602427

Thus, we predict that the population will be approximately 263.66 million in the year 2000. Note that projections of the population of the United States in the year 2000 made by the Bureau of the Census range from 259.57 million to 278.23 million.

3.3 NEWTON'S LAW OF COOLING

First-order linear differential equations can be used to solve a variety of problems that involve temperature. For example, a medical examiner can find the time of death in a homicide case, a chemist can determine the time required for a plastic mixture to cool to a hardening temperature, and an engineer can design the cooling and heating system of a manufacturing facility. Although distinct, each of these problems depends on a basic principle that is used to develop the associated differential equation. We discuss this important law now.

☰ Newton's Law of Cooling

Newton's law of cooling states that the rate at which the temperature $T(t)$ changes in a cooling body is proportional to the difference between the temperature of the body and the constant temperature T_s of the surrounding medium. This situation is represented as the first-order initial-value problem

$$\frac{dT}{dt} = k(T - T_s), \quad T(0) = T_0,$$

where T_0 is the initial temperature of the body and k is the constant of proportionality. We investigate problems involving Newton's law of cooling in the following examples.

☰ Solution of the Equation

Notice that the equation

$$\frac{dT}{dt} = k(T - T_s)$$

is separable. Separating variables gives us

$$\frac{dT}{T - T_s} = kdt.$$

Hence, $\ln|T - T_s| = kt + C$. Using the properties of the natural logarithm and simplifying yields

$$T = C_1 e^{kt} + T_s,$$

where $C_1 = e^C$. Applying the initial condition implies that $T_0 = C_1 + T_s$, so $C_1 = T_0 - T_s$. Therefore, the solution of the equation is

$$T = (T_0 - T_s)e^{kt} + T_s.$$

Recall that if $k < 0$, then $\lim_{t \to \infty} e^{kt} = 0$. Therefore, $\lim_{t \to \infty} T(t) = T_s$, so the temperature of the body approaches that of its surroundings.

We see that an equivalent solution is obtained with **dsolve**, which we name **DE1** for later use.

```
> DE1:=dsolve({diff(tp(t),t)=k*(tp(t)-temps),
   tp(0)=temp0},tp(t));
```

$DE1 := tp(t) = temps + e^{kt}(-temps + temp0)$

EXAMPLE: A pie is removed from a 350° oven and placed to cool in a room with temperature 75°. In 15 minutes, the pie has a temperature of 150°. Determine the time required to cool the pie to a temperature of 80° so that it may be eaten.

SOLUTION: In this example, $T_0 = 350$ and $T_s = 75$. Substituting these values using **subs** into $T = (T_0 - T_s)e^{kt} + T_s$, which we extract from **DE1** with **rhs**, we obtain $T(t) = (350 - 75)e^{kt} + 75 = 275e^{kt} + 75$.

> `step_1:=subs({temp0=350,temps=75},rhs(DE1));`

$$step\_1 := 75 + 275\, e^{kt}$$

To solve the problem we must find k. We also know that $T(15) = 150$ so $T(15) = 275e^{15k} + 75 = 150$. Solving this equation for k with **solve** gives us $k = \dfrac{1}{15}\ln\left(\dfrac{3}{11}\right)$.

> `k:=solve(subs(t=15,step_1)=150);`

$$k := \frac{1}{15}\ln\left(\frac{3}{11}\right)$$

Thus, $T(t) = 275\left(\dfrac{3}{11}\right)^{t/15} + 75$, as shown in the following using **simplify** together with the **exp** option.

> `simplify(step_1,exp);`

$$75 + 275\left(\frac{3}{11}\right)^{\left(\frac{1}{15}t\right)}$$

To find the value of t for which $T(t) = 80$, we solve the equation $275\left(\dfrac{11}{3}\right)^{-t/15} + 75 = 80$ for t with **solve**. Thus, the pie will be ready to eat after approximately 46 minutes.

> `t00:=solve(step_1=80);`
> `evalf(t00);`

$$t00 := -15\frac{\ln(55)}{\ln\left(\frac{3}{11}\right)}$$

$$46.26397676$$

Alternatively, we can graph $T(t) = 275\left(\dfrac{3}{11}\right)^{t/15} + 75$ with **plot**.

```
> plot({80,step_1},t=0..90);
```

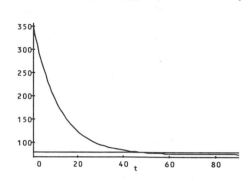

From the graph, we see that the temperature is 80° between $t = 40$ and $t = 50$. To approximate the value of t for which the temperature is 80°, we use fsolve as follows:

```
> fsolve(step_1=80,t,40..50);
```
46.26397676

An interesting question associated with cooling problems is to determine if the pie reaches room temperature. From the formula, $T(t) = 275\left(\dfrac{11}{3}\right)^{-t/15} + 75$, we see that the component $275\left(\dfrac{11}{3}\right)^{-t/15} > 0$, so $T(t) = 275\left(\dfrac{11}{3}\right)^{-t/15} + 75 > 75$. Therefore, the pie never actually reaches room temperature according to our model. However, we see from the graph that its temperature approaches 75° as t increases.

In the previous example, the temperature of the surroundings was assumed to be constant. However, this does not have to be the case. For example, consider the problem of heating and cooling a building. Over the span of a twenty-four hour day, the outside temperature varies. The problem of determining the temperature inside the building, therefore, becomes more complicated. For the meantime, let's assume that the building has no heating or air conditioning system. Hence, the differential equation that should be solved to find the temperature $u(t)$ at time t inside the building is

$$\frac{du}{dt} = k(C(t) - u(t)),$$

where $C(t)$ is a function that describes the outside temperature and $k > 0$ is a constant that depends on the insulation of the building. According to this equation, if $C(t) > u(t)$, then $\dfrac{du}{dt} > 0$, which implies that u increases. Conversely, if $C(t) < u(t)$, then $\dfrac{du}{dt} < 0$, which means that u decreases.

EXAMPLE: (a) Suppose that during the month of April in Atlanta, Georgia, the outside temperature is given by $C(t) = 70 - 10\cos\left(\dfrac{\pi t}{12}\right)$, $0 \le t \le 24$. (Note: This implies that the average value of $C(t)$ is 70°.) Determine the temperature in a building that has an initial temperature of 60° if $k = \frac{1}{4}$. (b) Compare this to the temperature in June when the outside temperature is $C(t) = 80 - 10\cos\left(\dfrac{\pi t}{12}\right)$ and the initial temperature is 70°.

SOLUTION: (a) The initial-value problem that we must solve is

$$\left\{\frac{du}{dt} = \frac{1}{4}\left[70 - 10\cos\left(\frac{\pi t}{12}\right) - u\right], u(0) = 60\right\},$$

which we solve as follows with **dsolve**. Note that the result of using **dsolve** is generally not in simplified form. Thus, **Sol** is not displayed because a colon is included at the end of the command instead of a semicolon because the unsimplified result is rather lengthy.

```
> Sol:=dsolve({diff(u(t),t)=1/4*
  (70-10*cos(Pi*t/12)-u(t)),u(0)=60},u(t));
```

After naming $u(t)$ the solution obtained in **Sol** with **assign**, **simplify** is used to simplify the solution.

```
> assign(Sol):
  simplify(u(t));
```

$$-10\frac{-7\pi^2 + e^{-\frac{1}{4}t}\pi^2 + 3\pi\sin\left(\frac{1}{12}\pi t\right) + 9\cos\left(\frac{1}{12}\pi t\right) - 63}{9 + \pi^2}$$

We then use **plot** to graph the solution for $0 \le t \le 24$.

```
> plot(u(t),t=0..24);
```

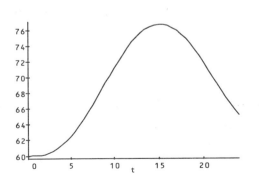

From the graph we see that the temperature reaches its maximum near $t = 15$. A more accurate estimate is obtained with **fsolve** by setting the first derivative of u equal to zero and solving for t.

```
> fsolve(diff(u(t),t)=0,t,14..16);
```

15.15061632

(b) This problem is solved in the same manner as the previous case. Be sure to clear the definition of u by entering **u:='u'** before entering the **dsolve** command; otherwise, error messages will result because u was assigned a definition previously.

```
> u:='u':
  Sol:=dsolve({diff(u(t),t)=1/4*
  (80-10*cos(Pi*t/12)-u(t)),u(0)=70},u(t)):
```

Again, we name $u(t)$ the solution obtained in **Sol** and simplify the result.

```
> assign(Sol):
  simplify(u(t));latex(");
```

$$-10\frac{-8\pi^2 + e^{-\frac{1}{4}t}\pi^2 + 3\pi\sin\left(\frac{1}{12}\pi t\right) + 9\cos\left(\frac{1}{12}\pi t\right) - 72}{9 + \pi^2}$$

This solution is also graphed with **plot**. From the graph, we see that the maximum temperature appears to occur near $t = 15$ hours.

```
> plot(u(t),t=0..24);
```

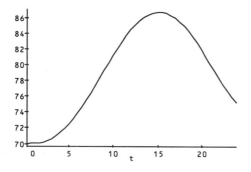

Again, the more accurate value is obtained with **fsolve** by setting the first derivative of u equal to zero and solving for t. This calculation yields approximately 15.15 hours, the same as that in (a).

```
>  fsolve(diff(u(t),t)=0,t,14..16);
```

   15.15061632

3.4 FREE-FALLING BODIES

The motion of some objects can be determined through the solution of a first-order equation. We begin by explaining some of the theory that is needed to set up the differential equation that models the situation.

☰ *Newton's Second Law of Motion*

The rate at which the momentum of a body changes with respect to time is equal to the resultant force acting on the body. Since the body's momentum is defined as the product of its mass and velocity, this statement is modeled as

$$\frac{d}{dt}(mv) = F,$$

where m and v represent the body's mass and velocity, respectively, and F is the sum of the forces acting on the body. Since m is constant, differentiation leads to the well-known equation

$$m\frac{dv}{dt} = F.$$

If the body is subjected to the force due to gravity, then its velocity is determined by solving the differential equation

$$m\frac{dv}{dt} = mg \quad \text{or} \quad \frac{dv}{dt} = g,$$

where $g \cong 32$ ft/s^2 (English system) and 9.8m/s^2 (metric system).

This differential equation is applicable only when the resistive force due to the medium (such as air resistance) is ignored. If this offsetting resistance is considered, we must discuss all of the forces acting on the object. Mathematically, we write the equation as

$$m\frac{dv}{dt} = \sum(\text{forces acting on the object}),$$

where the direction of motion is taken to be the positive direction.

We use a force diagram to set up the differential equation that models the situation. Since air resistance acts against the object as it falls and g acts in the same direction of the motion, we state the initial-value problem in the form

$$m\frac{dv}{dt} = mg + (-F_R) \text{ or } m\frac{dv}{dt} = mg - F_R,$$

where F_R represents this resistive force. Note that down is assumed to be the positive direction. The resistive force is typically proportional to the body's velocity (v) or the square of its velocity (v^2). Hence, the differential equation is linear or nonlinear based on whether or not the resistance of the medium is taken into account.

EXAMPLE: Determine the velocity and position functions of an object with $m = 1$ that is thrown downward with an initial velocity of 2 ft/sec from a height of 1000 ft. Assume that the object is subjected to air resistance that is equivalent to the instantaneous velocity of the object. Also, determine the time at which the object strikes the ground and its velocity when it strikes the ground.

SOLUTION: First, we set up the initial-value problem to determine the velocity of the object. Since the air resistance is equivalent to the instantaneous velocity, we have

$$F_R = v.$$

The formula $m\frac{dv}{dt} = mg - F_R$ then gives us

$$\frac{dv}{dt} = 32 - v.$$

Of course, we must impose the initial velocity $v(0) = 2$. Therefore, the initial-value problem is

$$\left\{ \frac{dv}{dt} = 32 - v, v(0) = 2 \right\},$$

which we solve now with **dsolve**.

```
> v:='v':s:='s':
  step_1:=dsolve({diff(v(t),t)=32-v(t),v(0)=2},v(t));
```
$$step\_1 := v(t) = 32 - 30\,e^{-t}$$

Thus, the velocity of the object is

$$v = 32 - 30e^{-t}.$$

To determine the position $s(t)$, we simply solve the first-order equation

$$\frac{ds}{dt} = v = 32 - 30e^{-t}$$

with initial position $s(0) = 0$. Notice that we use the initial position as a reference and let s represent the distance traveled from this reference point.

```
> assign(step_1):
  step_2:=dsolve({diff(s(t),t)=v(t),s(0)=0},s(t));
```

$$step\_2 := s(t) = 32t + 30e^{-t} - 30$$

The position of the object is, therefore, given by $s = 32t + 30e^{-t} - 30$.

Since we are taking $s(0) = 0$ as our starting point, then the object strikes the ground when $s(t) = 1000$. Therefore, we must solve $s = 32t + 30e^{-t} - 30 = 1000$. The roots of this equation can be approximated with **fsolve**. We begin by graphing the function s with **plot**.

```
> assign(step_2):
  plot({1000,s(t)},t=0..70);
```

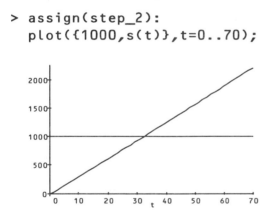

From the graph we see that the time at which $s(t) = 1000$ occurs is between $t = 30$ and $t = 40$. To obtain a better approximation, we use **fsolve**.

```
> t00:=fsolve(s(t)=1000,t,30..40);
```

$$t00 := 32.18750000$$

The velocity at the point of impact is found to be 32.0 ft/sec by evaluating $v(t)$ at the time at which the object strikes the ground.

```
> evalf(subs(t=t00,v(t)));
```

32.00000000

EXAMPLE: Determine a general solution (for the velocity and the position) of the differential equation that models the motion of an object of mass m when directed upward with an initial velocity of v_0 from an initial position y_0 assuming that the air resistance equals cv (c is constant).

SOLUTION: Since the motion of the object is upward, g and F_R act against the upward motion of the object. By drawing a force diagram, we see that g and F_R are in the negative direction. Therefore, the differential equation that must be solved in this case is the linear equation $\frac{dv}{dt} = -g - \frac{c}{m}v$. We solve the initial-value problem

$$\left\{ \frac{dv}{dt} = -g - \frac{c}{m}v, \; v(0) = v_0 \right\}$$

with **dsolve**, naming the resulting output **Sol**.

```
> v:='v':
  Sol:=dsolve({diff(v(t),t)=-g-c/m*v(t),v(0)=v0},v(t));
```

$$Sol := v(t) = -\frac{mg}{c} + e^{-\frac{ct}{m}}\left(\frac{mg}{c} + v0\right)$$

Therefore, the solution to the initial-value problem

$$\left\{ \frac{dv}{dt} = -g - \frac{c}{m}v, \; v(0) = v_0 \right\}$$

is

$$v(t) = -\frac{gm}{c} + \frac{c v_0 + gm}{c}e^{-ct/m}.$$

Next, we use **Sol** to define **velocity**. This function can be employed to investigate numerous situations without solving the differential equation each time.

```
> velocity:=proc(m0,c0,g0,v00,t0)
    subs({m=m0,c=c0,g=g0,v0=v00,t=t0},rhs(Sol))
    end:
  velocity(m,c,g,vo,t);
```

$$-\frac{mg}{c} + e^{-\frac{ct}{m}}\left(\frac{mg}{c} + vo\right)$$

For example, the velocity function for the case with $m = \frac{1}{128}$, $c = \frac{1}{160}$, $g = 32$, and $v_0 = 48$ is $v(t) = 88e^{-4t/5} - 40$.

```
> velocity(1/128,1/160,32,48,t);latex(");
```
$$-40 + 88e^{-\frac{4}{5}t}$$

The position function $s(t)$ that represents the distance above the ground at time t is determined by integrating the velocity function. This is accomplished here with **dsolve** using the initial position y_0. As with the previous case, the output is named **Pos** so that the position formula may be extracted from the result for later use.

```
> y:='y':
  Pos:=dsolve({diff(y(t),t)=velocity(m,c,g,v0,t),
  y(0)=y0},y(t));
```
$$Pos := y(t) = -\frac{gmt}{c} - \frac{m^2 e^{-\frac{ct}{m}}g}{c^2} - \frac{me^{-\frac{ct}{m}}v0}{c} + \frac{m^2 g}{c^2} + \frac{m\,v0}{c} + y0$$

```
> position:=proc(m0,c0,g0,v00,y00,t0)
    subs({m=m0,c=c0,g=g0,v0=v00,y0=y00,t=t0},rhs(Pos))
    end:
  position(m,c,g,v0,y0,t);
```
$$-\frac{gmt}{c} - \frac{m^2 e^{-\frac{ct}{m}}g}{c^2} - \frac{me^{-\frac{ct}{m}}v0}{c} + \frac{m^2 g}{c^2} + \frac{m\,v0}{c} + y0$$

The position and velocity functions are plotted in the following using the parameters $m = \dfrac{1}{128}$, $c = \dfrac{1}{160}$, $g = 32$, and $v_0 = 48$ listed in the previous example as well as $y_0 = 0$.

The time at which the object reaches its maximum height occurs when the derivative of the position is equal to zero. From the graph we see that $s'(t) = v(t) = 0$ when $t \approx 1$. Application of Newton's method, or another appropriate numerical algorithm, yields the better approximation $t \approx 0.985572$.

```
> plot({velocity(1/128,1/160,32,48,t),
    position(1/128,1/160,32,48,0,t)},t=0..2);
```

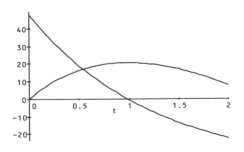

`solve` is then used to determine the time at which the object reaches its maximum height. This time occurs when the derivative of the position is equal to zero.

> `root:=solve(diff(position(1/128,1/160,32,48,0,t),t)=0);`

$$root := -\frac{5}{4} \ln\left(\frac{5}{11}\right)$$

> `evalf(root);`

0.9855717006

We now compare the effect that varying the initial velocity and position has on the position function. Suppose that we use the same values used earlier for m, c, and g. However, we let $v_0 = 48$ in one function and $v_0 = 36$ in the other. We also let $y_0 = 0$ and $y_0 = 6$ in these two functions, respectively.

> `plot({position(1/128,1/160,32,48,0,t),`
>   `position(1/128,1/160,32,36,6,t)},t=0..2);`

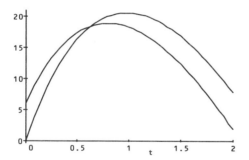

The following plot demonstrates the effect that varying the initial velocity only has on the position function. The values of v_0 used are 48, 64, and 80. Notice that as the initial velocity is increased, the maximum height attained by the object is increased as well.

```
> plot({position(1/128,1/160,32,48,0,t),
    position(1/128,1/160,32,64,0,t),
    position(1/128,1/160,32,80,0,t)},t=0..2);
```

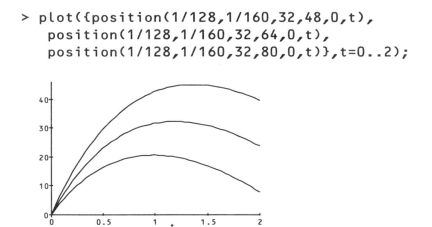

The following graph indicates the effect that varying the initial position and holding all other values constant has on the position function. We use values of 0, 10, and 20 for y_0. Notice that the value of the initial position vertically translates the position function.

```
> plot({position(1/128,1/160,32,48,0,t),
    position(1/128,1/160,32,48,10,t),
    position(1/128,1/160,32,48,20,t)},t=0..2);lprint(");
```

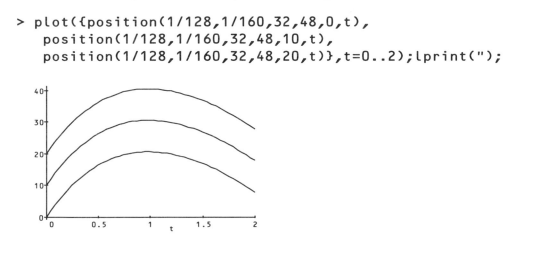

We now combine several of the topics discussed in this section to solve the following problem.

EXAMPLE: An object of mass $m = 1$ is dropped from a height of 50 ft above the surface of a small pond. While the object is in the air, the force due to air resistance is v. However, when the object is in the pond, it is subjected to a buoyancy force equivalent to $6v$. Determine how much time is required for the object to reach a depth of 25 ft in the pond.

SOLUTION: This problem must be broken into two parts: an initial-value problem for the object above the pond, and an initial-value problem for the object below the surface of the pond. Using techniques discussed in previous examples, the initial-value problem above the pond's surface is found to be

$$\frac{dv}{dt} = 32 - v, \ v(0) = 0.$$

However, to define the initial-value problem to find the velocity of the object beneath the pond's surface, the velocity of the object when it reaches the surface must be known. Hence, the velocity of the object above the surface must be determined by solving the initial-value problem above.

```
> s:='s':v:='v':
  step_1:=dsolve({diff(v(t),t)=32-v(t),v(0)=0},v(t));
```
$step\_1 := v(t) = 32 - 32\,e^{-t}$

In order to find the velocity when the object hits the pond's surface, we must know the time at which the position of the object is 0. Thus, we must find the position function. We do this by integrating the velocity function, obtaining $s(t) = 32e^{-t} + 32t - 32$.

```
> assign(step_1):
  step_2:=dsolve({diff(s(t),t)=v(t),s(0)=0},s(t));
```
$step\_2 := s(t) = 32\,t + 32\,e^{-t} - 32$

The position function is graphed with **plot**. The value of t at which the object has traveled 50 ft is needed. This time appears to be approximately 2.5 sec.

```
> assign(step_2):
  plot({50,s(t)},t=0..5);
```

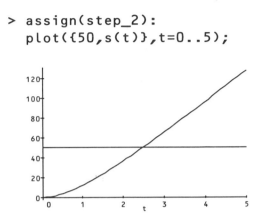

A more accurate value of the time at which the object hits the surface is found now using fsolve. In this case, we obtain $t \approx 2.47864$. The velocity at this time is then determined by substitution into the velocity function resulting in $v(2.47864) \approx 29.3166$. Note that this value is the initial velocity of the object when it hits the surface of the pond.

```
> t1:=fsolve(s(t)=50,t);
```

$$t1 := 2.478643063$$

```
> v1:=evalf(subs(t=t1,v(t)));
```

$$v1 := 29.31657802$$

Thus, the initial-value problem that determines the velocity of the object beneath the surface of the pond is given by $\dfrac{dv}{dt} = 32 - 6v$, $v(0) = 29.3166$. The solution of this initial-value problem is $v(t) = \dfrac{16}{3} + 23.9833\,e^{-t}$:

```
> s:='s':v:='v':
  step_3:=dsolve({diff(v(t),t)=32-6*v(t),v(0)=v1},v(t));
```

$$step\_3 := v(t) = \frac{16}{3} + 23.98324469e^{-6t}$$

and integrating to obtain the position function (the initial position is 0), we obtain $s(t) = 3.99722 - 3.99722e^{-6t} + \dfrac{16}{3}t$.

```
> assign(step_3):
  step_4:=dsolve({diff(s(t),t)=v(t),s(0)=0},s(t));
```

$$step\_4 := s(t) = 5.333333333t - 3.997207448e^{-6t} + 3.997207448$$

This position function is then plotted to determine when the object is 25 ft beneath the surface of the pond. This time appears to be near 4 sec.

```
> assign(step_4):
  plot({25,s(t)},t=0..5);
```

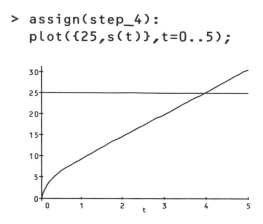

A more accurate approximation of the time at which the object is 25 ft beneath the pond's surface is obtained with **fsolve**. In this case, we obtain $t \approx 3.93802$. Finally, the time required for the object to reach the pond's surface is added to the time needed for it to travel 25 ft beneath the surface to see that approximately 6.41667 sec are required for the object to travel from a height of 50 ft above the pond to a depth of 25 ft below the surface.

```
> t2:=fsolve(s(t)=25,t);
```

$t2 := 3.938023603$

```
> t1+t2;
```

6.416666666

4

Higher-Order Differential Equations

In Chapters 2 and 3 we saw that first-order differential equations can be used to model a variety of physical situations. However, many physical situations exist that need to be modeled by higher-order differential equations. In this chapter, we discuss several methods for solving higher-order differential equations with constant coefficients.

4.1 PRELIMINARY DEFINITIONS AND NOTATION

☰ The nth-Order Ordinary Linear Differential Equation

In order to develop the methods needed to solve higher-order differential equations, we must state several important definitions and theorems. We begin by introducing the types of higher-order equations that we will be solving in this chapter.

DEFINITION

> *nth-Order Ordinary Linear Differential Equation*
>
> An ordinary differential equation of the form
>
> $$a_n(x)\, y^{(n)}(x) + a_{n-1}(x)\, y^{(n-1)}(x) + \cdots + a_1(x)\, y'(x) + a_0(x)\, y(x) = f(x)$$
>
> is called an ***nth-order ordinary linear differential equation***. If $f(x)$ is identically the zero function, the equation is said to be **homogeneous**; if $f(x)$ is not the zero function, the equation is said to be **nonhomogeneous**; and if the functions $a_i(x)$, $i = 0, 1, 2, \ldots, n$ are constants, the equation is said to have **constant coefficients**.

We will see that solutions to *n*th-order ordinary linear differential equations require *n* solutions with the property defined as follows.

DEFINITION | *Linearly Dependent and Linearly Independent*

Let $f_1(x), f_2(x), f_3(x), \ldots, f_{n-1}(x)$, and $f_n(x)$ be a set of n functions at least $n - 1$ times differentiable. (This means that all derivatives at least through the $(n - 1)$st derivative of each of these functions exist.) Let $S = \{f_1(x), f_2(x), f_3(x), \ldots, f_{n-1}(x), f_n(x)\}$. S is **linearly dependent** on an interval I means that there are constants $c_1, c_2, \ldots, c_n$, not all zero, so that

$$\sum_{k=1}^{n} c_k f_k(x) = c_1 f_1(x) + c_2 f_2(x) + \cdots + c_{n-1} f_{n-1}(x) + c_n f_n(x) = 0$$

for every value of x in the interval I.

S is **linearly independent** means that S is not linearly dependent.

DEFINITION | *Wronskian*

Let $S = \{f_1(x), f_2(x), f_3(x), \ldots, f_{n-1}(x), f_n(x)\}$ be a set of n functions for which each is differentiable at least $n - 1$ times. The **Wronskian** of S, denoted by

$$W(S) = W(f_1(x), f_2(x), f_3(x), \ldots, f_{n-1}(x), f_n(x)),$$

is the determinant

$$W(S) = \begin{vmatrix} f_1(x) & f_2(x) & \cdots & f_n(x) \\ f_1'(x) & f_2'(x) & \cdots & f_n'(x) \\ \vdots & \vdots & \vdots & \vdots \\ f_1^{(n-1)}(x) & f_2^{(n-1)}(x) & \cdots & f_n^{(n-1)}(x) \end{vmatrix}.$$

EXAMPLE: Compute the Wronskian for each of the following sets of functions: (a) $S = \{\sin x, \cos x\}$ and (b) $S = \{\cos 2x, \sin 2x, \sin x \cos x\}$.

SOLUTION: For (a), the Wronskian is obtained by evaluating

$$\begin{vmatrix} \sin x & \cos x \\ \dfrac{d}{dx}(\sin x) & \dfrac{d}{dx}(\cos x) \end{vmatrix} = \begin{vmatrix} \sin x & \cos x \\ \cos x & -\sin x \end{vmatrix} = -\sin^2 x - \cos^2 x = -1.$$

We can use Maple to compute the Wronskian by taking advantage of the **Wronskian** command contained in the **linalg** package. We first load the **linalg** package by entering **with(linalg)** and then define S to be the list of functions $S = \{\sin x, \cos x\}$. **Wronskian** is then used to compute the Wronskian matrix, naming the resulting output **ws**. The determinant of **ws** is found with **det**, also contained in the **linalg** package, and named **step_1**. Note that **step_1** is not simplified.

```
>  S:=[sin(x),cos(x)]:
   ws:=Wronskian(S,x);
   step_1:=det(ws);
```

$$ws := \begin{bmatrix} \sin(x) & \cos(x) \\ \cos(x) & -\sin(x) \end{bmatrix}$$

$$step\_1 := -\sin(x)^2 - \cos(x)^2$$

We simplify **step_1** using **simplify**, naming the result **step_2**. In this case, we see that the Wronskian for the set of functions S is -1, agreeing with the result obtained earlier.

```
>  step_2:=simplify(step_1);
```

$$step\_2 := -1$$

For (b), we must compute

$$\begin{vmatrix} \cos 2x & \sin 2x & \sin x \cos x \\ \dfrac{d}{dx}(\cos 2x) & \dfrac{d}{dx}(\sin 2x) & \dfrac{d}{dx}(\sin x \cos) \\ \dfrac{d^2}{dx^2}(\cos 2x) & \dfrac{d^2}{dx^2}(\sin 2x) & \dfrac{d^2}{dx^2}(\sin x \cos x) \end{vmatrix}.$$

We first define S to be the list of functions corresponding to $S = \{\cos 2x, \sin 2x, \sin x \cos x\}$ and then use **Wronskian**, in the same manner as for (a), to compute the Wronskian matrix for S, naming the resulting output **ws**.

```
>  S:=[cos(2*x),sin(2*x),sin(x)*cos(x)]:
   ws:=Wronskian(S,x);
```

$$ws := \begin{bmatrix} \cos(2x) & \sin(2x) & \sin(x)\cos(x) \\ -2\sin(2x) & 2\cos(2x) & \cos(x)^2 - \sin(x)^2 \\ -4\cos(2x) & -4\sin(2x) & -4\sin(x)\cos(x) \end{bmatrix}$$

Similarly, we use **det** to compute the determinant of **ws**. In this case, we see that the Wronskian is 0. You should use the identity $\sin 2x = 2 \sin x \cos x$ to show that the set of functions $S = \{\cos 2x, \sin 2x, \sin x \cos x\}$ is linearly independent.

```
> det(ws);
        0
```

In the previous example, we see that in (a) the Wronskian was not 0, while in (b) the Wronskian was 0. Moreover, we know that the set of functions in (a) is linearly independent, while the set of functions in (b) is linearly dependent. In fact, we can use the Wronskian to determine if a set of functions is linearly dependent or linearly independent.

• •

THEOREM   Let $S = \{f_1(x), f_2(x), f_3(x), \ldots, f_{n-1}(x), f_n(x)\}$ be a set of n functions each differentiable at least $n - 1$ times on an interval I. If $W(S) \neq 0$ for at least one value of x in the interval I, S is linearly independent.

• •

EXAMPLE: Use the Wronskian to classify each of the following sets as linearly independent or linearly dependent: (a) $S = \{1 - 2\sin^2 x, \cos 2x\}$ and (b) $S = \{e^x, xe^x, x^2 e^x\}$.

SOLUTION: For (a), the Wronskian of S is given by

$$\begin{vmatrix} 1 - 2\sin^2 x & \cos 2x \\ \dfrac{d}{dx}(1 - 2\sin^2 x) & \dfrac{d}{dx}(\cos 2x) \end{vmatrix} = \begin{vmatrix} 1 - 2\sin^2 x & \cos 2x \\ -4\sin x \cos x & -2\sin 2x \end{vmatrix}$$

$$= (1 - 2\sin^2 x) \cdot -2\sin 2x - \cos 2x \cdot -4\sin x \cos x$$
$$= -2\sin 2x - 4\sin 2x \sin^2 x + 4\sin x \cos x \cos 2x.$$

Applying the identity $\sin 2x = 2\sin x \cos x$ followed by the identity $\cos 2x = 1 - 2\sin^2 x$ gives us

$$-2\sin 2x - 4\sin 2x \sin^2 x + 4\sin x \cos x \cos 2x = -2\sin 2x - 4\sin 2x \sin^2 x + 2\sin 2x \cos 2x$$
$$= -2\sin 2x(1 - 2\sin^2 x + \cos 2x)$$
$$= -2\sin 2x \cdot 0 = 0.$$

Because the Wronskian is 0, we conclude that the set of functions $S = \{1 - 2\sin^2 x, \cos 2x\}$ is linearly independent. We obtain the same results with Maple using the **Wronskian** and `det` commands, both contained in the `linalg` package, followed by using the `simplify` command.

```
> with(linalg):
  S:=[1-2*sin(x)^2,cos(2*x)];
  ws:=Wronskian(S,x);
  step_1:=det(ws);
  step_2:=simplify(step_1);
```

$$S := [1 - 2\sin(x)^2, \cos(2x)]$$

$$ws := \begin{bmatrix} 1 - 2\sin(x)^2 & \cos(2x) \\ -4\sin(x)\cos(x) & -2\sin(2x) \end{bmatrix}$$

$$step\_1 := -2\sin(2x) + 4\sin(2x)\sin(x)^2 + 4\cos(2x)\sin(x)\cos(x)$$

$$step\_2 := 0$$

For (b), we compute the Wronskian by first defining S to be the list of functions corresponding to the set of functions $S = \{e^x, xe^x, x^2e^x\}$, and then using **Wronskian** to compute the Wronskian matrix for S, naming the resulting matrix **ws**, and finally using **det** to compute the determinant of **ws**. The result is not the zero function, so we conclude that the set of functions is linearly independent.

```
> S:=[exp(x),x*exp(x),x^2*exp(x)];
  ws:=Wronskian(S,x);
  det(ws);
```

$$S := [e^x, xe^x, x^2e^x]$$

$$ws := \begin{bmatrix} e^x & xe^x & x^2e^x \\ e^x & e^x + xe^x & 2xe^x + x^2e^x \\ e^x & 2e^x + xe^x & 2e^x + 4xe^x + x^2e^x \end{bmatrix}$$

$$2(e^x)^3$$

☰ Fundamental Set of Solutions

Obtaining a collection of n linearly independent solutions to an nth-order linear differential equation is of great importance in solving equations of this type. Therefore, we state the following definition. (Note that a nontrivial solution is one that is not identically zero.)

DEFINITION | *Fundamental Set of Solutions*

A set $S = \{f_1(x), f_2(x), f_3(x), \ldots, f_{n-1}(x), f_n(x)\}$ of n linearly independent nontrivial solutions of the nth-order linear homogeneous equation

$$a_n(x)\, y^{(n)}(x) + a_{n-1}(x)\, y^{(n-1)}(x) + \cdots + a_1(x)\, y^{(1)}(x) + a_0(x) = 0$$

is called a **fundamental set of solutions** of the equation.

EXAMPLE: Show that $S = \{e^{-5x}, e^{-x}\}$ is a fundamental set of solutions of the equation $y''(x) + 6y'(x) + 5y(x) = 0$.

SOLUTION: S is linearly independent because

$$W(S) = \begin{vmatrix} e^{-5x} & e^{-x} \\ -5e^{-5x} & -e^{-x} \end{vmatrix} = -e^{-6x} + 5e^{-6x} = 5e^{-6x} \neq 0.$$

Also, we must verify that each function is a solution of the differential equation. Substituting each function into the left-hand side of the equation and simplifying gives us

$$\frac{d^2}{dx^2}(e^{-5x}) + 6\frac{d}{dx}(e^{-5x}) + 5\,e^{-5x} = 25e^{-5x} - 30e^{-5x} + 5e^{-5x} = 0$$

and

$$\frac{d^2}{dx^2}(e^{-x}) + 6\frac{d}{dx}(e^{-x}) + 5e^{-x} = e^{-x} - 6e^{-x} + 5e^{-x} = 0.$$

We conclude that S is a fundamental set of solutions of the equation $y''(x) + 6y'(x) + 5y(x) = 0$.

Of course, we can also use Maple to implement the steps carried out here. We first load the `linalg` package, and then define S to be the list of functions corresponding to $S = \{e^{-5x}, e^{-x}\}$. Then, `simplify`, `det`, and `Wronskian` are used to compute the Wronskian of the set of functions S.

```
> with(linalg):
  S:=[exp(-5*x),exp(-x)]:
  simplify(det(Wronskian(S,x)));
```
$$4e^{-6x}$$

Note that the first element of S is extracted from S with `S[1]`; similarly, the second is extracted with `S[2]`.

```
> S[1];
```

$$e^{-5x}$$

We verify that each element of S is a solution to the equation with **diff**.

```
> diff(S[1],x$2)+6*diff(S[1],x)+5*S[1];
```

0

```
> diff(S[2],x$2)+6*diff(S[2],x)+5*S[2];
```

0

••••••••••••••••••••••••••

THEOREM   *Principle of Superposition*

If $S = \{f_1(x), f_2(x), \ldots, f_{n-1}(x), f_n(x)\} = \{f_i(x)\}_{i=1}^n$ is a fundamental set of solutions of the equation

$$a_n(x)\, y^{(n)}(x) + \cdots + a_1(x)\, y'(x) + a_0(x)\, y(x) = \sum_{i=0}^{n} a_i(x)\, y^{(i)}(x) = 0$$

and $\{c_1, c_2, \ldots, c_{n-1}, c_n\} = \{c_i\}_{i=1}^n$ is a set of n constants, then

$$f(x) = c_1 f_1(x) + c_2 f_2(x) + \cdots + c_{n-1} f_{n-1}(x) + c_n f_n(x) = \sum_{i=1}^{n} c_i f_i(x)$$

is also a solution of

$$a_n(x)\, y^{(n)}(x) + \cdots + a_1(x)\, y'(x) + a_0(x)\, y(x) = \sum_{i=0}^{n} a_i(x)\, y^{(i)}(x) = 0.$$

••••••••••••••••••••••••••

Therefore, the linear combination of the functions in a fundamental set of solutions of an nth-order homogeneous linear differential equation is also a solution of the differential equation.

EXAMPLE: Show that $S = \{\cos 2x, \sin 2x\}$ is a fundamental set of solutions of the second-order ordinary linear differential equation with constant coefficients $y'' + 4y = 0$. Then show that the linear combination of these two solutions is also a solution of the homogeneous equation.

SOLUTION: We first verify that both functions are solutions of $y'' + 4y = 0$. For $y = \cos 2x$, we have the derivatives $y' = -2\sin 2x$ and $y'' = -4\cos 2x$. Substitution then gives us $y'' + 4y = -4\cos 2x + 4\cos 2x = 0$. Hence, $y = \cos 2x$ is a solution. Similar calculations follow to verify that $y = \sin 2x$. Next, we compute the Wronskian

$$W(S) = \begin{vmatrix} \cos 2x & \sin 2x \\ -2\sin 2x & 2\cos 2x \end{vmatrix} = 2\cos^2 2x + 2\sin^2 2x = 2 \neq 0$$

to show that the solutions are linearly independent. This means that $S = \{\cos 2x, \sin 2x\}$ is a fundamental set of solutions. According to the Principle of Superposition, the linear combination of the functions in the fundamental set of solutions is also a solution. In this case, we must consider $y = c_1 \cos 2x + c_2 \sin 2x$ with derivatives $y' = -2c_1 \sin 2x + 2c_2 \cos 2x$ and $y'' = -4c_1 \cos 2x - 4c_2 \sin 2x$. Substitution then yields

$$y'' + 4y = -4c_1 \cos 2x - 4c_2 \sin 2x + 4(c_1 \cos 2x + c_2 \sin 2x) = 0.$$

Therefore, $y = c_1 \cos 2x + c_2 \sin 2x$ is also a solution of $y'' + 4y = 0$.

As with previous examples, we show how Maple is used to carry out the same steps. After loading the `linalg` package and defining S to be the list of functions corresponding to $S = \{\cos 2x, \sin 2x\}$, we use `simplify`, `det`, and `Wronskian` to compute the Wronskian of $S = \{\cos 2x, \sin 2x\}$.

```
> with(linalg):
  S:=[cos(2*x),sin(2*x)]:
  simplify(det(Wronskian(S,x)));

  2
```

Next, we use `diff` to verify that each element of S is a solution of the equation.

```
> diff(S[1],x$2)+4*S[1];

  0
```

```
> diff(S[2],x$2)+4*S[2];

  0
```

To verify that any linear combination of these functions is also a solution of the equation, we begin by defining y. Note that `c[1]` and `c[2]` represent arbitrary constants.

```
> y:=x->c[1]*cos(2*x)+c[2]*sin(2*x);
```

$$y := x \rightarrow c_{[1]} \cos(2x) + c_{[2]} \sin(2x)$$

Then we verify that $y(x)$ is a solution of the differential equation.

```
> diff(y(x),x$2)+4*y(x);
```

0

We can graph $y(x)$ for various values of the constants `c[1]` and `c[2]`. For example, here we use `seq` and `subs` to define the set of functions `to_plot` obtained by replacing `c[1]` by i and `c[2]` by j for $i = -1, 0,$ and 1 and $j = -1, 0,$ and 1.

```
> to_plot:={seq(seq(subs({c[1]=i,c[2]=j},
    y(x)),i=-1..1),j=-1..1)};
```

$$to\_plot := \{0, \cos(2x) + \sin(2x), -\cos(2x) + \sin(2x), -\cos(2x) - \sin(2x),$$
$$- \cos(2x), - \sin(2x), \cos(2x), \sin(2x), \cos(2x) - \sin(2x)\}$$

The set of functions $S = \{\cos 2x, \sin 2x\}$ and the set of functions `to_plot` are then each graphed with `plot` on the interval $[-\pi, 2\pi]$.

```
> plot({cos(2*x),sin(2*x)},x=-Pi..2*Pi);
  plot(to_plot,x=-Pi..2*Pi);
```

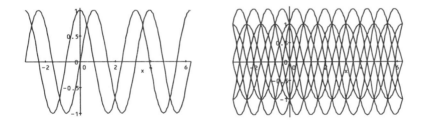

≡ Existence of a Fundamental Set of Solutions

The following two theorems tell us that under reasonable conditions, the nth-order linear homogeneous equation

$$a_n(x) y^{(n)}(x) + a_{n-1}(x) y^{(n-1)}(x) + \cdots + a_1(x) y^{(1)}(x) + a_0(x) = 0$$

has a fundamental set of n solutions.

THEOREM   If $a_i(x)$ is continuous on an open interval I for $i = 0, 1, \ldots, n$, then the nth-order linear homogeneous equation

$$a_n(x)\, y^{(n)}(x) + a_{n-1}(x)\, y^{(n-1)}(x) + \cdots + a_1(x)\, y^{(1)}(x) + a_0(x) = 0$$

has a fundamental set of n solutions.

THEOREM   Any set of $n + 1$ solutions of the nth-order linear homogeneous equation

$$a_n(x)\, y^{(n)}(x) + a_{n-1}(x)\, y^{(n-1)}(x) + \cdots + a_1(x)\, y^{(1)}(x) + a_0(x) = 0$$

is linearly dependent.

We can summarize the results of these theorems by saying that in order to solve an nth-order ordinary linear differential equation, we must find a set S of n functions that satisfy the differential equation such that $W(S) \neq 0$.

4.2 SOLUTIONS OF HOMOGENEOUS EQUATIONS WITH CONSTANT COEFFICIENTS

We now turn our attention to solving linear homogeneous equations with constant coefficients. Nonhomogeneous equations are considered in the following sections.

☰ General Solution

The Principle of Superposition tells us that if

$$S = \{f_1(x), f_2(x), \ldots, f_{n-1}(x), f_n(x)\} = \{f_i(x)\}_{i=1}^{n}$$

is a fundamental set of solutions of the nth-order linear homogeneous equation

$$a_n(x)\, y^{(n)}(x) + \cdots + a_1(x)\, y'(x) + a_0(x)\, y(x) = \sum_{i=0}^{n} a_i(x)\, y^{(i)}(x) = 0,$$

then

$$f(x) = c_1 f_1(x) + c_2 f_2(x) + \cdots + c_{n-1} f_{n-1}(x) + c_n f_n(x) = \sum_{i=1}^{n} c_i f_i(x)$$

is also a solution of the equation for any set of numbers $\{c_1, c_2, \ldots, c_{n-1}, c_n\} = \{c_i\}_{i=1}^{n}$.

DEFINITION | *General Solution*

If $S = \{f_1(x), f_2(x), \ldots, f_{n-1}(x), f_n(x)\} = \{f_i(x)\}_{i=1}^{n}$ is a fundamental set of solutions of the nth-order linear homogeneous equation

$$a_n(x) y^{(n)}(x) + \cdots + a_1(x) y'(x) + a_0(x) y(x) = \sum_{i=0}^{n} a_i(x) y^{(i)}(x) = 0,$$

then a **general solution** of the equation is

$$f(x) = c_1 f_1(x) + c_2 f_2(x) + \cdots + c_{n-1} f_{n-1}(x) + c_n f_n(x) = \sum_{i=1}^{n} c_i f_i(x),$$

where $\{c_1, c_2, \ldots, c_{n-1}, c_n\} = \{c_i\}_{i=1}^{n}$ is a set of n arbitrary constants.

EXAMPLE: Show that $y = c_1 e^{-x} + c_2 e^{-2x}$ is a general solution of $y'' + 3y' + 2y = 0$.

SOLUTION: We begin by differentiating the function $y = c_1 e^{-x} + c_2 e^{-2x}$ to yield $y' = -c_1 e^{-x} - 2c_2 e^{-2x}$ and $y'' = c_1 e^{-x} + 4c_2 e^{-2x}$. Substitution then gives us

$$y'' + 3y' + 2y = c_1 e^{-x} + 4c_2 e^{-2x} + 3(-c_1 e^{-x} - 2c_2 e^{-2x}) + 2(c_1 e^{-x} + c_2 e^{-2x})$$
$$= (c_1 - 3c_1 + 2c_1)e^{-x} + (4c_2 - 6c_2 + 2c_2)e^{-2x} = 0.$$

Therefore, $y = c_1 e^{-x} + c_2 e^{-2x}$ is a general solution of $y'' + 3y' + 2y = 0$. To graph this solution for various values of the constants, we begin by defining $y = c_1 e^{-x} + c_2 e^{-2x}$.

```
> y:=c1*exp(-x)+c2*exp(-2*x);
```
$$y := c1 e^{-x} + c2 e^{-2x}$$

We will graph the four solutions obtained by replacing **c1** and **c2** by 1 and -1, -1 and 1, 2 and 1, and 1 and -2. We define **vals** to be the set of four ordered pairs $(1, -1), (-1, 1), (2, 1)$, and $(1, -2)$.

```
> vals:={[1,-1],[-1,1],[2,1],[1,-2]};
```

$$vals := \{[2,1],[1,-2],[1,-1],[-1,1]\}$$

and then define **f**. Given an ordered pair **pair**, **f** returns the function obtained by replacing each occurrence of **c1** in **y** by the first coordinate of **pair** and each occurrence of **c2** in **y** by the second coordinate of **pair**.

```
> f:=proc(pair)
    subs(c1=pair[1],c2=pair[2],y)
  end:
```

We then use **map** to compute **f** for each pair in the set of ordered pairs **vals**, naming the resulting set of functions **to_plot**.

```
> to_plot:=map(f,vals);
```

$$to\_plot := \{e^{-x} - e^{-2x}, 2e^{-x} + e^{-2x}, e^{-x} - 2e^{-2x}, -e^{-x} + e^{-2x}\}$$

We then use **plot** to graph the functions in **to_plot** on the interval $[-1, 2]$.

```
> plot(to_plot,x=-1..2);
```

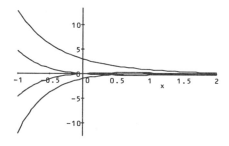

EXAMPLE: Show that $y = e^{-x}(c_1 \cos 4x + c_2 \sin 4x)$ is a general solution of $y'' + 2y' + 17y = 0$.

SOLUTION: We illustrate how Maple can be used to carry out the steps illustrated in the previous example. After defining y, we use **diff** to compute the first and second derivatives (with respect to x) of y.

```
> y:=exp(-x)*(c1*cos(4*x)+c2*sin(4*x));
```

$$y := e^{-x}(c1 \cos(4x) + c2 \sin(4x))$$

```
> diff(y,x);
  diff(y,x$2);
```

$$-e^{-x}(c1\cos(4x) + c\sin(4x)) + e^{-x}(-4c1\sin(4x) + 4c2\cos(4x))$$

$$e^{-x}(c1\cos(4x) + c2\sin(4x)) - 2e^{-x}(-4c1\sin(4x) + 4c2\cos(4x))$$
$$+ e^{-x}(-16c1\cos(4x) - 16c2\sin(4x))$$

We then compute the value of $y'' + 2y' + 17y$, naming the resulting output **step_1**

```
> step_1:=diff(y,x$2)+2*diff(y,x)+17*y;
```

$$step\_1 := 16e^{-x}(c1\cos(4x) + c2\sin(4x))$$
$$+ e^{-x}(-16c1\cos(4x) - 16c2\sin(4x))$$

which we simplify with **simplify**. Because the result is 0 and the set of functions $\{e^{-x}\cos 4x, e^{-x}\sin 4x\}$ is linearly independent, $y = e^{-x}(c_1 \cos 4x + c_2 \sin 4x)$ is a general solution of the equation.

```
> simplify(step_1);
```

0

We can graph the solution for various values of the constant in the same manner as in the previous example. We define **vals** to be the set of ordered pairs consisting of the ordered pairs $(0, 1)$, $(1, 0)$, $(2, 1)$, and $(1, -2)$.

```
> vals:={[0,1],[1,0],[2,1],[1,-2]};
```

$$vals := \{[0, 1], [1, 0], [2, 1], [1, -2]\}$$

The function **f** is defined in exactly the same way as in the previous example.

```
> f:=proc(pair)
    subs({c1=pair[1],c2=pair[2]},y)
    end:
```

Similarly, we use **map** to compute the value of **f** for each ordered pair in **vals**, naming the resulting set of functions **to_plot** and then graphing them on the interval $[-1, 2]$ with **plot**.

```
> to_plot:=map(f,vals);
```

$$to\_plot := \{e^{-x}\sin(4x), e^{-x}\cos(4x), e^{-x}(2\cos(4x) + \sin(4x)), e^{-x}(\cos(4x) - 2\sin(4x))\}$$

```
> plot(to_plot,x=-1..2);
```

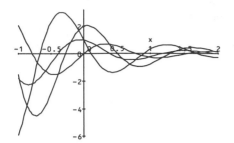

≡ *Finding a General Solution for a Homogeneous Equation with Constant Coefficients*

Solutions of any nth-order homogeneous linear differential equation with constant coefficients are determined by the solutions of the **characteristic equation**.

DEFINITION

Characteristic Equation

The equation

$$a_n m^n + a_{n-1} m^{n-1} + \cdots + a_1 m + a_0 = 0$$

is called the **characteristic equation** of the nth-order homogeneous linear differential equation with constant coefficients

$$a_n y^{(n)}(x) + a_{n-1} y^{(n-1)}(x) + \cdots + a_1 y'(x) + a_0 y(x) = 0.$$

Let us begin our investigation by considering the second-order equation with constant coefficients

$$ay'' + by' + cy = 0$$

with characteristic equation

$$am^2 + bm + c = 0.$$

. . . .

THEOREM   *Solving Second-Order Equations with Constant Coefficients*

Let $ay'' + by' + cy = 0$ be a homogeneous second-order equation with constant coefficients, and let m_1 and m_2 be the solutions of the equation $am^2 + bm + c = 0$.

1. If $m_1 \neq m_2$ and both m_1 and m_2 are real, a general solution of $ay'' + by' + cy = 0$ is

$$y = c_1 e^{m_1 x} + c_2 e^{m_2 x}.$$

2. If $m_1 = m_2$, a general solution of $ay'' + by' + cy = 0$ is $y = c_1 e^{m_1 x} + c_2 x e^{m_1 x}$.

3. If $m_1 = \alpha + i\beta$, $\beta \neq 0$, and $m_2 = \overline{m_1} = \alpha - i\beta$, a general solution of $ay'' + by' + cy = 0$ is

$$y = c_1 e^{\alpha x} \cos \beta x + c_2 e^{\alpha x} \sin \beta x = e^{\alpha x}(c_1 \cos \beta x + c_2 \sin \beta x).$$

. . . .

In (3) above, $\overline{m_1}$ is the **complex conjugate** of m_1: $\overline{m_1} = \overline{\alpha + i\beta} = \alpha - i\beta$. The following three examples illustrate each of these situations.

EXAMPLE: Solve $y'' + 3y' - 4y = 0$ subject to $y(0) = 1$ and $y'(0) = -1$.

SOLUTION: The characteristic equation of $y'' + 3y' - 4y = 0$ is $m^2 + 3m - 4 = (m + 4)(m - 1) = 0$. We see that because the solutions of the characteristic equation are $m = -4$ and $m = 1$, a general solution of $y'' + 3y' - 4y = 0$ is

$$y(x) = c_1 e^{-4x} + c_2 e^x.$$

To find the particular solution of $y'' + 3y' - 4y = 0$ for which $y(0) = 1$ and $y'(0) = -1$, we compute $y'(x) = -4c_1 e^{-4x} + c_2 e^x$. Applying the initial conditions $y(0) = 1$ and $y'(0) = -1$ results in the system of equations

$$\begin{cases} c_1 + c_2 = 1 \\ -4c_1 + c_2 = -1 \end{cases}.$$

Subtracting the second equation from the first results in $5c_1 = 2$ so that $c_1 = \frac{2}{5}$. Then substituting this value of c_1 into $c_1 + c_2 = 1$ gives us the equation $\frac{2}{5} + c_2 = 1$ so that $c_2 = \frac{3}{5}$. Consequently, the desired solution is $y(x) = \frac{2}{5} e^{-4x} + \frac{3}{5} e^x$.

Of course, we can use Maple to graph various solutions and the particular solution obtained here. After we define y, we define **to_plot** to be the set of nine functions obtained by replacing each occurrence of **c1** in y by $-1, 0$, and 1 and each occurrence of **c2** in y by $-1, 0$, and 1. Then we use **plot** to graph the particular solution obtained by replacing **c1** by 2/5 and **c2** by 3/5 on the interval $[-1, 1]$ and the set of functions **to_plot** on the interval $[-1, 1]$. Note that in the second **plot** command, the option **-10..10** instructs Maple that the y-values displayed (the vertical axis) correspond to the interval $[-10, 10]$.

```
> y:=c1*exp(-4*x)+c2*exp(x):
```

```
> to_plot:={seq(seq(subs({c1=i,c2=j},y),i=-1..1),j=-1..1)}:
  plot(subs({c1=2/5,c2=3/5},y),x=-1..1);
  plot(to_plot,x=-1..1,-10..10);
```

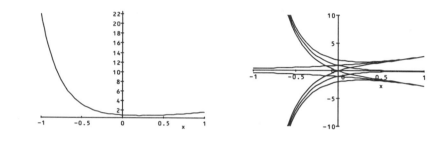

EXAMPLE: Solve $y'' + 2y' + y = 0$.

SOLUTION: The characteristic equation of $y'' + 2y' + y = 0$ is $m^2 + 2m + 1 = (m + 1)^2 = 0$. Because the solution of the characteristic equation is $m = -1$ with multiplicity 2, a general solution of $y'' + 2y' + y = 0$ is $y(x) = c_1 e^{-x} + c_2 x e^{-x}$.

We can graph the solution for various values of the constant in the exact same manner as in the previous example. For example, entering

```
> y:=c1*exp(-x)+c2*x*exp(-x):
  to_plot:={seq(seq(subs({c1=i,c2=j},y),i=-1..1),j=-1..1)}:
  plot(to_plot,x=-1..1);
```

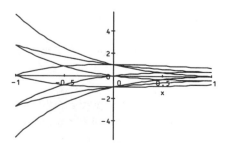

defines y, defines `to_plot` to be the set of nine functions obtained by replacing each occurrence of `c1` in y by $-1, 0$, and 1 and each occurrence of `c2` in y by $-1, 0$, and 1, and then graphs the set of functions `to_plot` on the interval $[-1, 1]$.

EXAMPLE: Solve $y'' + 4y' + 20y = 0$ subject to $y(0) = 3$ and $y'(0) = -1$.

SOLUTION: The characteristic equation of $y'' + 4y' + 20y = 0$ is $m^2 + 4m + 20 = 0$. Completing the square yields $m^2 + 4m + 20 = (m + 2)^2 + 16 = 0$ so the solutions of the characteristic equation are $m = -2 \pm 4i$. Of course, using the quadratic formula to solve $m^2 + 4m + 20 = 0$ produces the same solutions. Since the solutions of the characteristic equation are complex conjugates, a general solution of $y'' + 4y' + 20y = 0$ is

$$y(x) = e^{-2x}(c_1 \cos 4x + c_2 \sin 4x).$$

To find the solution for which $y(0) = 3$ and $y'(0) = -1$, we first calculate

$$y' = 2e^{-2x}(-c_1 \cos 4x + 2c_2 \cos 4x - 2c_1 \sin 4x - c_2 \sin 4x)$$

and then evaluate both $y(0) = c_1$ and $y'(0) = 2(2c_2 - c_1)$, obtaining the system of equations

$$\begin{cases} c_1 = 3 \\ 2(2c_2 - c_1) = -1 \end{cases}.$$

Substituting $c_1 = 3$ into the second equation results in

$$2(2c_2 - 3) = -1$$
$$4c_2 - 6 = -1$$
$$c_2 = \frac{5}{4}.$$

Thus, the solution of $y'' + 4y' + 20y = 0$ for which $y(0) = 3$ and $y'(0) = -1$ is

$$y(x) = e^{-2x}\left(3 \cos 4x + \frac{5}{4} \sin 4x\right).$$

We can graph the solution for various values of the constant in the exact same manner as in the previous example. For example, entering

```
> y:=exp(-2*x)*(c1*cos(4*x)+c2*sin(4*x)):
  to_plot:={seq(seq(subs({c1=i,c2=j},y),i=-1..1),j=-1..1)}:
  plot(subs({c1=3,c2=5/4},y),x=-1..1);
  plot(to_plot,x=-1..1,-10..10);
```

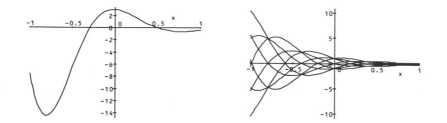

defines y, defines `to_plot` to be the set of nine functions obtained by replacing each occurrence of `c1` in y by $-1, 0$, and 1 and each occurrence of `c2` in y by $-1, 0$, and 1, graphs the particular solution obtained by replacing `c1` by 3 and `c2` by 5/4 in y on the interval $[-1, 1]$, and then graphs the set of functions `to_plot` on the interval $[-1, 1]$.

In each of these cases, we can also use Maple's `dsolve` command to solve each equation in the same manner that we use `dsolve` to solve some first-order equations. For example, entering

```
> y:='y':
  dsolve(diff(y(x),x$2)+3*diff(y(x),x)-4*y(x)=0,y(x));
```

$$y(x) = \_C1\,e^x + \_C2\,e^{-4x}$$

finds a general solution of $y'' + 3y' - 4y = 0$; entering

```
> dsolve({diff(y(x),x$2)+3*diff(y(x),x)-4*y(x)=0,
  y(0)=1,D(y)(0)=-1},y(x));
```

$$y(x) = \frac{3}{5}e^x + \frac{2}{5}e^{-4x}$$

finds the solution of $y'' + 3y' - 4y = 0$ satisfying $y(0) = 1$ and $y'(0) = -1$; entering

```
> dsolve(diff(y(x),x$2)+2*diff(y(x),x)+y(x)=0,y(x));
```

$$y(x) = \_C1\,e^{-x} + \_C2\,e^{-x}x$$

finds a general solution of $y'' + 2y' + y = 0$; entering

```
> dsolve(diff(y(x),x$2)+4*diff(y(x),x)+20*y(x)=0,y(x));
```

$$y(x) = \_C1\,e^{-2x}\sin(4x) + \_C2\,e^{-2x}\cos(4x)$$

finds a general solution of $y'' + 4y' + 20y = 0$; and entering

```
> dsolve({diff(y(x),x$2)+4*diff(y(x),x)+20*y(x)=0,
    y(0)=3,D(y)(0)=-1},y(x));
```

$$y(x) = \frac{5}{4}e^{-2x}\sin(4x) + 3e^{-2x}\cos(4x)$$

finds the solution of $y'' + 4y' + 20y = 0$ satisfying $y(0) = 3$ and $y'(0) = -1$.

After mastering the techniques of solving second-order linear homogeneous differential equations, we are able to move on to higher-order equations. In order to explain the process of finding a general solution of a higher-order equation, we must state the following.

DEFINITION | *Multiplicity*

Suppose that the characteristic equation $a_n m^n + a_{n-1}m^{n-1} + \cdots + a_1 m + a_0 = 0$ can be written in factored form as $(m - m_1)^{k_1}(m - m_2)^{k_2} \cdots (m - m_r)^{k_r} = 0$. Then the roots of the equation are $m = m_1, m = m_2, \ldots, m = m_r$, where the roots have **multiplicity** $k_1, k_2, \ldots, k_r$, respectively.

In the same manner as in the case for a second-order homogeneous equation with real constant coefficients, a general solution of an nth-order ordinary differential equation is also determined by the solutions of the characteristic equation. Hence, we state the following rules for finding a general solution of an nth-order equation for the numerous situations that may be encountered.

☰ Rules for Determining the General Solution of a Higher-Order Equation

1. Let m be a real root of the characteristic equation

$$a_n m^n + a_{n-1}m^{n-1} + \cdots + a_1 m + a_0 = 0$$

of an nth-order homogeneous linear differential equation with real coefficients. Then e^{mx} is the solution associated with the root m.

If m is a real root of multiplicity k, where $k \geq 2$ of the characteristic equation, then the k solutions associated with m are

$$e^{mx}, xe^{mx}, x^2 e^{mx}, \ldots, x^{k-1}e^{mx}.$$

2. Suppose that m and $\bar{m}$ represent the complex conjugate pair $\alpha \pm \beta i$. Then the two solutions associated with these two roots are

$$e^{\alpha x} \cos \beta x \qquad \text{and} \qquad e^{\alpha x} \sin \beta x.$$

If the values $\alpha \pm \beta i$ are each a root of multiplicity k of the characteristic equation, then the other solutions associated with this pair are

$$xe^{\alpha x} \cos \beta x, xe^{\alpha x} \sin \beta x, x^2 e^{\alpha x} \cos \beta x, x^2 e^{\alpha x} \sin \beta x, \ldots, x^{k-1} e^{\alpha x} \cos \beta x, x^{k-1} e^{\alpha x} \sin \beta x.$$

A general solution to the nth-order differential equation is the linear combination of the solutions obtained for all values of m. Note that if $m_1, m_2, \ldots, m_r$ are the roots of the equation of multiplicity $k_1, k_2, \ldots, k_r$, respectively, then $k_1 + k_2 + \cdots + k_r = n$.

We now show how to use these rules to find a general solution. Notice that the key to the process is identifying each root of the characteristic equation and the associated solution(s).

EXAMPLE: Find a general solution of each of the following higher-order equations: (a) $24y''' + 2y'' - 5y' - y = 0$, (b) $y''' + 3y'' + 3y' + y = 0$, (c) $4y^{(4)} + 12y''' + 49y'' + 42y' + 10y = 0$, and (d) $y^{(4)} + 4y''' + 24y'' + 40y' + 100y = 0$.

SOLUTION: (a) In this case, the characteristic equation is $24m^3 + 2m^2 - 5m - 1 = 0$. Factoring, we have

$$24m^3 + 2m^2 - 5m - 1 = (3m + 1)(2m - 1)(4m + 1) = 0,$$

so the roots are

$$m_1 = -\frac{1}{3}, m_2 = \frac{1}{2}, \text{ and } m_3 = -\frac{1}{4}.$$

Therefore, the corresponding solutions are

$$y_1(x) = e^{-\frac{1}{3}x}, y_2(x) = e^{\frac{1}{2}x}, \text{ and } y_3(x) = e^{-\frac{1}{4}x},$$

respectively.

Roots	Multiplicity	Corresponding Solution
$m_1 = -\dfrac{1}{3}$	$k = 1$	$y_1(x) = e^{-\frac{1}{3}x}$
$m_2 = \dfrac{1}{2}$	$k = 1$	$y_2(x) = e^{\frac{1}{2}x}$
$m_3 = -\dfrac{1}{4}$	$k = 1$	$y_3(x) = e^{-\frac{1}{4}x}$

Thus, a general solution is given by

$$y = c_1 e^{-\frac{1}{3}x} + c_2 e^{\frac{1}{2}x} + c_3 e^{-\frac{1}{4}x}.$$

We can graph this general solution for various values of the arbitrary constants in the same way we graph solutions of second-order equations. For example, entering

```
> y:=c1*exp(-1/3*x)+c2*exp(1/2*x)+c3*exp(-1/4*x):
  vals:={[0,-1,1],[1,-1,1],[-1,1,1],[1,0,1]}:
  f:=proc(triple)
     subs({c1=triple[1],c2=triple[2],c3=triple[3]},y)
  end:
  to_plot:=map(f,vals);
```

$$to\_plot := \left\{ -e^{-\frac{1}{3}x} + e^{\frac{1}{2}x} + e^{-\frac{1}{4}x}, -e^{\frac{1}{2}x} + e^{-\frac{1}{4}x}, e^{-\frac{1}{3}x} + e^{-\frac{1}{4}x}, e^{-\frac{1}{3}x} - e^{\frac{1}{2}x} + e^{-\frac{1}{4}x} \right\}$$

defines y; defines $vals$ to be the set of ordered triples $(0, -1, 1)$, $(1, -1, 1)$, $(-1, 1, 1)$, and $(1, 0, 1)$; defines f, which given an ordered triple $triple$ returns the function obtained by replacing $c1$ in y by the first coordinate of $triple$, $c2$ in y by the second coordinate of $triple$, and $c3$ in y by the third coordinate of $triple$; and then defines to_plot to be the list of functions obtained by applying f to the set of ordered triples $vals$. The set of functions to_plot is then graphed with $plot$.

```
> plot(to_plot,x=-2..3);
```

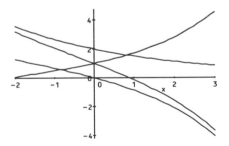

(b) The characteristic equation of $y''' + 3y'' + 3y' + y = 0$ is $m^3 + 3m^2 + 3m + 1 = 0$, and factoring results in

$$m^3 + 3m^2 + 3m + 1 = (m + 1)^3 = 0,$$

so the root is $m = -1$ with multiplicity 3. Therefore, the corresponding solutions are $y_1(x) = e^{-x}$, $y_2(x) = xe^{-x}$, and $y_3(x) = x^2 e^{-x}$. This tells us that a general solution is given by

$$y = c_1 e^{-x} + c_2 x e^{-x} + c_3 x^2 e^{-x}.$$

We can graph this general solution for various values of the constants in the same manner as in (a).

```
> y:=c1*exp(-x)+c2*x*exp(-x)+c3*x^2*exp(-x):
  vals:={[0,-1,1],[1,-1,1],[-1,1,1],[1,0,1],[1,1,1]}:
  f:=proc(triple)
    subs({c1=triple[1],c2=triple[2],c3=triple[3]},y)
    end:
  to_plot:=map(f,vals):
  plot(to_plot,x=-1/2..3/2);
```

(c) The characteristic equation of $4y^{(4)} + 12y''' + 49y'' + 42y' + 10y = 0$ is $4m^4 + 12m^3 + 49m^2 + 42m + 10 = 0$. We use **factor** to try to factor the polynomial $4m^4 + 12m^3 + 49m^2 + 42m + 10$, but see that Maple does not completely factor the polynomial.

```
> factor(4*m^4+12*m^3+49*m^2+42*m+10);
```

$$(m^2 + 2m + 10)(1 + 2m)^2$$

When we include **I** in the command, representing the imaginary number $i = \sqrt{-1}$, Maple is able to factor the polynomial completely.

```
> factor(4*m^4+12*m^3+49*m^2+42*m+10,I);
```

$$(m + 1 - 3I)(m + 1 + 3I)(1 + 2m)^2$$

From the results we see that the roots of the characteristic equation are $m_1 = -1 + 3i$, $m_2 = -1 - 3i$, and $m_3 = -\frac{1}{2}$ with multiplicity 2. As you may suspect, we obtain the same results with **solve**:

```
> solve(4*m^4+12*m^3+49*m^2+42*m+10=0);
```

$$-1 + 3I, -1 - 3I, -\frac{1}{2}, -\frac{1}{2}$$

The corresponding solutions are given by $y_1(x) = e^{-x}\cos 3x$, $y_2(x) = e^{-x}\sin 3x$, $y_3(x) = e^{-\frac{1}{2}x}$, and $y_4(x) = xe^{-\frac{1}{2}x}$. This tells us that a general solution is given by

$$y = e^{-x}(c_1\cos 3x + c_2\sin 3x) + c_3 e^{-\frac{1}{2}x} + c_4 xe^{-\frac{1}{2}x}.$$

As with second-order homogeneous equations, we can obtain the same result with `dsolve`. For example, entering

```
> Sol:=dsolve(4*diff(y(x),x$4)+12*diff(y(x),x$3)+
    49*diff(y(x),x$2)+42*diff(y(x),x)+10*y(x)=0,y(x));
```

$Sol := y(x) = \_C1\,e^{-x}\sin(3x) + \_C2\,e^{-x}\cos(3x) + \_C3\,e^{-\frac{1}{2}x} + \_C4\,e^{-\frac{1}{2}x}x$

finds a general solution of the equation. To name $y(x)$ the result given in `Sol`, we use `assign`. We then graph the general solution for various values of the constants in the same manner as in (a) and (b).

```
> assign(Sol):
  vals:={[1,0,1,0],[0,1,0,1],[1,1,0,1],[1,-1,1,2],
    [0,2,1,-2],[1,-2,1,2]}:
  f:=proc(quad)
    subs({_C1=quad[1],_C2=quad[2],_C3=quad[3],
      _C4=quad[4]},y(x))
    end:
  to_plot:=map(f,vals):
  plot(to_plot,x=-1..2);
```

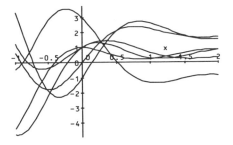

(d) The characteristic equation of $y^{(4)} + 4y''' + 24y'' + 40y' + 100y = 0$ is $m^4 + 4m^3 + 24m^2 + 40m + 100 = 0$. We solve it here with `solve`

```
> solve(m^4+4*m^3+24*m^2+40*m+100=0);
```

$-1 + 3I, -1 - 3I, -1 + 3I, -1 - 3I$

and see that the characteristic roots are $m_1 = -1 + 3i$ and $m_2 = -1 - 3i$, each with multiplicity 2, so the corresponding solutions are $y_1 = e^{-x} \cos 3x$, $y_2 = xe^{-x} \cos 3x$, $y_3 = e^{-x} \sin 3x$, and $y_4 = xe^{-x} \sin 3x$. This tells us that a general solution is given by

$$y = e^{-x}(c_1 \cos 3x + c_2 \sin 3x + c_3 x \cos 3x + c_4 x \sin 3x).$$

We obtain the same general solutions using `dsolve`.

```
> y:='y':
  Sol:=dsolve(diff(y(x),x$4)+4*diff(y(x),x$3)+
  24*diff(y(x),x$2)+40*diff(y(x),x)+100*y(x)=0,y(x));
```

$$Sol := y(x) = \_C1\,e^{-x} \sin(3x) + \_C2\,e^{-x} \cos(3x) + \_C3\,e^{-x} \sin(3x)x + \_C4\,e^{-x} \cos(3x)x$$

Instead of retyping $y(x)$, we use `assign` to name $y(x)$ the formula for the general solution given in `Sol`.

```
> assign(Sol):
```

To graph the solution for various values of the constants, we first define a list of ordered quadruples, `vals`.

```
> vals:=[[5,0,1,0],[0,1,0,-3],[1,3,0,1],
    [1,-1,1,2],[0,2,1,-2],[1,-2,5,2],[0,-3,0,2],
    [3,0,0,2],[1,1,1,1]]:
```

Then we use a `for` loop to graph the solution obtained by replacing $\_Ci$ by the ith coordinate of each ordered quadruple in `vals` on the interval $\left[-\dfrac{1}{2}, \dfrac{3}{2}\right]$.

```
> i:='i':
  for i from 1 to 9 do
    plot(eval(subs({_C1=vals[i][1],_C2=vals[i][2],
      _C3=vals[i][3],_C4=vals[i][4]},y(x))),x=-1/2..3/2)
  od;
```

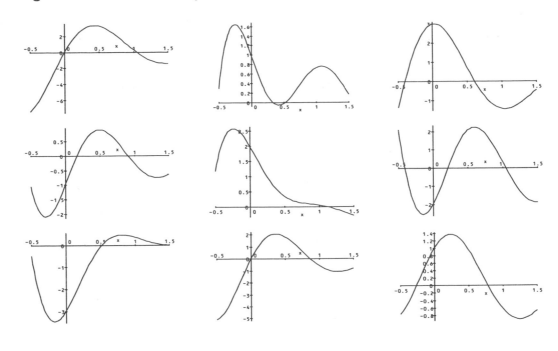

In our previous examples, we have been interested in finding a general solution of a higher-order equation or a solution that satisfies particular conditions. Generally, a problem of the form

$$\begin{cases} a_n(x)y^{(n)} + a_{n-1}(x)y^{(n-1)} + \cdots + a_1(x)y' + a_0(x)y = f(x) \\ y^{(n-1)}(x_0) = y_{n-1}, y^{(n-2)}(x_0) = y_{n-2}, \ldots, y'(x_0) = y_1, y(x_0) = y_0 \end{cases}$$

is called an **initial-value problem** because it includes n initial conditions

$$\left\{ y^{(n-1)}(x_0) = y_{n-1}, y^{(n-2)}(x_0) = y_{n-2}, \ldots, y'(x_0) = y_1, y(x_0) = y_0 \right\}$$

that must be satisfied by a general solution of the differential equation. Note that we call the restrictions $\{y^{(n-1)}(x_0) = y_{n-1}, y^{(n-2)}(x_0) = y_{n-2}, \ldots, y'(x_0) = y_1, y(x_0) = y_0\}$ *initial conditions* because $y(x)$ and its $(n-1)$ derivatives are evaluated at the same value of x. The n initial conditions allow us to solve for the constants that appear in a general solution because a general solution of an nth-order equation involves n arbitrary constants.

Since one of our main concerns is finding the solution of initial-value problems, we state the following theorem, which insures that a solution can be found.

• •

THEOREM   The initial-value problem

$$\begin{cases} a_n(x)y^{(n)} + a_{n-1}(x)y^{(n-1)} + \cdots + a_1(x)y' + a_0(x)y = f(x) \\ y^{(n-1)}(x_0) = y_{n-1}, y^{(n-2)}(x_0) = y_{n-2}, \ldots, y'(x_0) = y_1, y(x_0) = y_0 \end{cases}$$

has a unique solution if the functions $a_n(x), a_{n-1}(x), \ldots, a_1(x), a_0(x)$ are continuous on an interval containing x_0, where the functions $a_k(x) \neq 0$ for all values of x on this interval.

• •

EXAMPLE: (a) Find a general solution of $4y''' + 33y' - 37y = 0$. (b) Solve the initial-value problem

$$\begin{cases} 4y''' + 33y' - 37y = 0 \\ y(0) = 0, y'(0) = -1, y''(0) = 3 \end{cases}.$$

SOLUTION: For (a), we use **dsolve** as in the previous examples, naming the result **Sol**.

```
> y:='y':
  Sol:=dsolve(4*diff(y(x),x$3)+33*diff(y(x),
  x)-37*y(x)=0,y(x));
```

$Sol := y(x) = \_C1\, e^x + \_C2\, e^{-\frac{1}{2}x} \sin(3x) + \_C3\, e^{-\frac{1}{2}x} \cos(3x)$

To graph the solution for various values of the arbitrary constants, we use **assign** to name $y(x)$ the formula given in **Sol**. We then use **seq** and **subs** to to define the set of 12 functions, **to_plot**, obtained by replacing **_C1** by 0 and 1, **_C2** by $-1, 0$, and 1, and **_C3** by -1 and 0 in $y(x)$. The set of functions **to_plot** is then graphed on the interval $[0, 2]$ with **plot**.

```
> assign(Sol);
  to_plot:={seq(seq(seq(
    subs({_C1=i,_C2=j,_C3=k},y(x),i=0..1),j=-1..1),k=-1..0)};
  plot(to_plot,x=0..2);
```

$to\_plot := \{0, e^x, e^x + \%2 - \%1, -\%2 - \%1, e^x - \%2 - \%1, e^x - \%1,$
$\qquad \%2 - \%1, e^x - \%2, e^x + \%2, \%2, -\%2, -\%1\}$

$\%1 := e^{-\frac{1}{2}x} \cos(3x)$

$\%2 := e^{-\frac{1}{2}x} \sin(3x)$

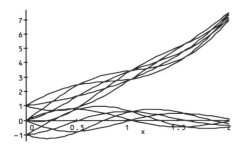

For (b), we again use `dsolve` to solve the initial-value problem. Note that `D(y)(0)=-1` corresponds to the initial condition $y'(0) = -1$, while the repeated composition operator `@@` in `(D@@2)(y)(0)=3` is used to represent the initial condition $y''(0) = 3$.

```
> y:='y':
  Part_Sol:=dsolve({4*diff(y(x),x$3)+33*diff(y(x),
    x)-37*y(x)=0,y(0)=0,D(y)(0)=-1,(D@@2)(y)(0)=3},y(x));
```

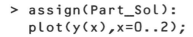

$$Part\_Sol := y(x) = \frac{8}{45}e^x - \frac{19}{45}e^{-\frac{1}{2}x}\sin 3x - \frac{8}{45}e^{-\frac{1}{2}x}\cos(3x)$$

Again, we use `assign` to name $y(x)$ the result obtained in `Part_Sol` and then graph $y(x)$ on the interval $[0, 2]$ with `plot`.

```
> assign(Part_Sol):
  plot(y(x),x=0..2);
```

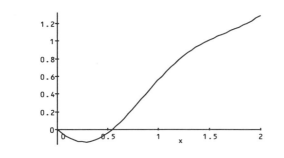

▼ **A P P L I C A T I O N**

Testing for Diabetes

Diabetes mellitus affects approximately 12 million Americans; approximately one-half of these people are unaware that they have diabetes. Diabetes is a serious disease: it is the leading cause of blindness in adults, the leading cause of renal failure, and is responsible for approximately one-half of all nontraumatic amputations in the United States. In addition, people with diabetes have an increased rate of coronary artery disease and strokes. People at risk for developing diabetes include those who are obese; those suffering from excessive thirst, hunger, urination, and weight loss; women who have given birth to a baby with weight greater than nine pounds; those with a family history of diabetes; those who are over 40 years of age. People with diabetes cannot metabolize glucose because their pancreas produces an inadequate or ineffective supply of insulin. Subsequently, glucose levels rise. The body attempts to remove the excess glucose through the kidneys: the glucose acts as a diuretic, resulting in increased water consumption. Since some cells require energy, which is not being provided by glucose, fat and protein are broken down and ketone levels rise. Although there is no cure for diabetes at this time, many cases can be effectively managed by a balanced diet and insulin therapy in addition to maintaining an optimal weight. Diabetes can be diagnosed by several tests. In the **fasting blood sugar** test, a patient fasts for at least 4 hrs, and then the glucose level is measured. In a fasting state, the glucose level in normal adults ranges from 70 to 110 mg/ml. An adult in a fasting state with consistent readings of over 150 mg probably has diabetes. However, since individuals vary greatly, people with mild cases of diabetes might have fasting state glucose levels within the normal range. In these cases, a highly accurate test that is frequently used to diagnose mild diabetes is the **glucose tolerance test** (GTT), which was developed by Drs. Rosevear and Molnar of the Mayo Clinic and Drs. Ackerman and Gatewood of the University of Minnesota. During the GTT, blood and urine samples are taken from a patient in a fasting state to measure the glucose, G_0, and hormone, H_0, and glycosuria levels, respectively. We assume that these values are equilibrium values. The patient is then given 100 g of glucose. Blood and urine samples are then taken at 1-, 2-, 3-, and 4-hr intervals. In a person without diabetes, glucose levels return to normal after 2 hr; in diabetics their blood sugar levels either take longer or never return to normal levels. Let G denote the cumulative level of glucose in the blood, $g = G - G_0$, H the cumulative level of hormones that affects insulin production (like glucagon, epinephrine, cortisone, and thyroxin), and $h = H - H_0$. Notice that g and h represent the fluctuation of the cumulative levels of glucose and hormones from their equilibrium values. The relationship between the rate of change of glucose in the blood and the rate of change of the cumulative levels of the hormones in the blood that affect insulin production is

$$\begin{cases} g' = f_1(g, h) + J(t) \\ h' = f_2(g, h) \end{cases},$$

where $J(t)$ represents the **external** rate at which the blood glucose concentration is being increased. If we assume that f_1 and f_2 are linear functions, then this system of equations becomes

$$\begin{cases} g' = -ag - bh + J(t) \\ h' = -ch + dg \end{cases},$$

where $a, b, c,$ and d represent positive numbers. We define these equations as

```
> EQ1:=diff(g(t),t)=-a*g(t)-b*h(t)+J(t);
  EQ2:=diff(h(t),t)=-c*h(t)+d*g(t);
```

$$EQ1 := \frac{\partial}{\partial t} g(t) = -ag(t) - bh(t) + J(t)$$

$$EQ2 := \frac{\partial}{\partial t} h(t) = -ch(t) + dg(t)$$

Solving **EQ1** for g' gives us

```
> step_1:=solve(EQ1,h(t));
```

$$step\_1 := -\frac{\left(\frac{\partial}{\partial t} g(t)\right) + ag(t) - J(t)}{b}$$

and differentiating with respect to t results in

```
> step_2:=diff(step_1,t);
```

$$step\_2 := -\frac{\left(\frac{\partial^2}{\partial t^2} g(t)\right) + a\left(\frac{\partial}{\partial t} g(t)\right) - \left(\frac{\partial}{\partial t}\right) J(t)}{b}$$

Next, we substitute the results obtained in **step_1** and **step_2** into **EQ2** and multiply both sides of the equation by b.

```
> step_3:=subs({h(t)=step_1,diff(h(t),t)=step_2},EQ2);
```

$$step\_3 := -\frac{\left(\frac{\partial^2}{\partial t^2} g(t)\right) + a\left(\frac{\partial}{\partial t} g(t)\right) - \left(\frac{\partial}{\partial t} J(t)\right)}{b} = \frac{c\left(\left(\frac{\partial}{\partial t} g(t)\right) + ag(t) - J(t)\right)}{b} + dg(t)$$

```
> step_4:=expand(b*step_3);
```

$$step\_4 := -\left(\frac{\partial^2}{\partial t^2} g(t)\right) - a\left(\frac{\partial}{\partial t} g(t)\right) + \left(\frac{\partial}{\partial t} J(t)\right) = c\left(\frac{\partial}{\partial t} g(t)\right) + cag(t) - cJ(t) + bdg(t)$$

Since the glucose solution is consumed at $t = 0$, for $t > 0$ we have that $J(t) = 0$ and $J'(t) = 0$, so for $t > 0$ we can rewrite the equation as

> `step_5:=subs({diff(J(t),t)=0,J(t)=0},step_4);`

$$step\_5 := -\left(\frac{\partial^2}{\partial t^2}g(t)\right) - a\left(\frac{\partial}{\partial t}g(t)\right) = c\left(\frac{\partial}{\partial t}g(t)\right) + cag(t) + bdg(t)$$

It might be reasonable to assume that glucose levels fluctuate in a periodic fashion so that the solutions to the equation are periodic functions. In order to have periodic solutions, we must have that $(a - c)^2 - 4bd < 0$. We use **assume** to assume that $(a - c)^2 - 4bd < 0$ and then use **dsolve** to find a general solution of the equation.

> `assume((a-c)^2-4*b*d<0):`
> `Sol:=dsolve(step_5,g(t));`

$$Sol := g(t) = \_C1\, e^{-\frac{1}{2}(a\sim+c\sim)t}\sin\left(\frac{1}{2}\sqrt{-a\sim^2+2c\sim a\sim-c\sim^2+4b\sim d\sim}\,t\right)$$
$$+ \_C2\, e^{-\frac{1}{2}(a\sim+c\sim)t}\cos\left(\frac{1}{2}\sqrt{-a\sim^2+2c\sim a\sim-c\sim^2+4b\sim d\sim}\,t\right)$$

Letting $\alpha = \frac{1}{2}(a + c)$ and $\omega = \frac{1}{2}\sqrt{4bd - (a - c)^2}$, we can rewrite the general solution we just obtained as

$$G(t) = G_0 + e^{-\alpha t}[A\cos(\omega t) + B\sin(\omega t)].$$

Research has shown that lab results of $\dfrac{2\pi}{\omega} > 4$ indicate a mild case of diabetes.*

▲

*Sources: D. N. Burghes and M. S. Borrie, *Modelling with Differential Equations*, Ellis Horwood Limited (1981), Halsted Press, NY, pp. 113–116. Joyce M. Black and Esther Matassarin-Jacobs, *Luckman and Sorensen's Medical-Surgical Nursing: A Psychophysiologic Approach*, 4th ed. Philadelphia: W. B. Saunders Company (1993), pp. 1775–1808.

4.3 NONHOMOGENEOUS EQUATIONS WITH CONSTANT COEFFICIENTS: THE ANNIHILATOR METHOD

≡ *General Solution of a Nonhomogeneous Equation*

In Section 4.2, we learned how to solve nth-order linear homogeneous equations with constant coefficients. These techniques are also useful in solving some nonhomogeneous equations of the form

$$a_n y^{(n)}(x) + a_{n-1} y^{(n-1)}(x) + \cdots + a_1 y'(x) + a_0 y(x) = \sum_{k=0}^{n} a_k y^{(k)}(x) = g(x).$$

Before describing how to obtain solutions of some nonhomogeneous equations, we first describe what is meant by a "general solution of a nonhomogeneous equation."

DEFINITION | *Particular Solution*

A **particular solution**, $y_p(x)$, of the nonhomogeneous differential equation $a_n y^{(n)}(x) + a_{n-1} y^{(n-1)}(x) + \cdots + a_1 y'(x) + a_0 y(x) = g(x)$ is a specific function (which contains no arbitrary constants) that satisfies the equation.

EXAMPLE: Show that $y_p(x) = -\dfrac{3}{2} \sin x$ is a particular solution of $y'' - 2y' + y = 3\cos x$.

SOLUTION: After defining $y_p(x) = -\dfrac{3}{2} \sin x$,

> `yp:=x->-3/2*sin(x);`

$$yp := x \longrightarrow -\frac{3}{2} \sin x$$

we compute $y_p'' - 2y_p' + y_p$

> `(D@@2)(yp)(x)-2*D(yp)(x)+yp(x);`

$3\cos(x)$

and see that the result is $3\cos x$.

DEFINITION

> *General Solution of a Nonhomogeneous Equation*
>
> A **general solution to the nonhomogeneous equation**
>
> $$a_n y^{(n)}(x) + a_{n-1} y^{(n-1)}(x) + \cdots + a_1 y'(x) + a_0 y(x) = g(x)$$
>
> is
>
> $$y(x) = y_h(x) + y_p(x),$$
>
> where $y_h(x)$ is a general solution of the corresponding homogeneous equation
>
> $$a_n y^{(n)}(x) + a_{n-1} y^{(n-1)}(x) + \cdots + a_1 y'(x) + a_0 y(x) = 0,$$
>
> and $y_p(x)$ is a particular solution to the nonhomogeneous equation.

EXAMPLE: Show that $y(x) = y_h(x) + y_p(x)$ is a general solution of $y'' + 6y' + 13y = 2e^{-2x}\sin x$ if $y_p(x) = -\dfrac{1}{5}e^{-2x}\cos x + \dfrac{2}{5}e^{-2x}\sin x$ and $y_h(x) = e^{-3x}(c_1\cos 2x + c_2\sin 2x)$.

SOLUTION: We first show that $y_p(x) = -\dfrac{1}{5}e^{-2x}\cos x + \dfrac{2}{5}e^{-2x}\sin x$ is a particular solution of $y'' + 6y' + 13y = 2e^{-2x}\sin x$. After defining $y_p(x) = -\dfrac{1}{5}e^{-2x}\cos x + \dfrac{2}{5}e^{-2x}\sin x$, we calculate $y_p'' + 6y_p' + 13y_p$

```
> yp:=x->-1/5*exp(-2*x)*cos(x)+2/5*exp(-2*x)*sin(x):
  diff(yp(x),x$2)+6*diff(yp(x),x)+13*yp(x);
```
$$2e^{-2x}\sin(x)$$

and see that the result is $2e^{-2x}\sin x$. Similarly, we show that $y_h(x) = e^{-3x}(c_1\cos 2x + c_2\sin 2x)$ is a general solution of the corresponding homogeneous equation $y'' + 6y' + 13y = 0$ by first defining $y_h(x) = e^{-3x}(c_1\cos 2x + c_2\sin 2x)$ and then computing $y_h'' + 6y_h' + 13y_h$

```
> yh:=x->exp(-3*x)*(c1*cos(2*x)+c2*sin(2*x)):
  step_1:=diff(yh(x),x$2)+6*diff(yh(x),x)+13*yh(x);
```
$$step\_1 := 4e^{-3x}(c1\cos(2x) + c2\sin(2x)) + e^{-3x}(-4c1\cos(2x) - 4c2\sin(2x))$$

and simplifying the result with **simplify**.

```
> simplify(step_1);
    0
```

We now can graph the particular solution and the general solution for various values of the arbitrary constant.

```
> y:=x->yp(x)+yh(x):
  to_plot:={seq(seq(subs({c1=i,c2=j},y(x)),i=-1..1),j=-1..1)}:
  plot(yp(x),x=0..2);
  plot(to_plot,x=0..2);
```

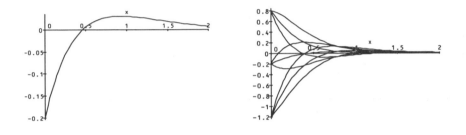

☰ *Operator Notation*

The nth-order derivative of a function y is given in **operator notation** by

$$D^n y = \frac{d^n y}{dx^n}.$$

Thus, the left-hand side of the linear nth-order differential equation with constant coefficients

$$a_n y^{(n)}(x) + a_{n-1} y^{(n-1)}(x) + \cdots + a_1 y'(x) + a_0 y(x) = 0$$

can be expressed in operator notation as

$$
\begin{aligned}
a_n y^{(n)}&(x) + a_{n-1} y^{(n-1)}(x) + \cdots + a_1 y'(x) + a_0 y(x) \\
&= a_n D^n y(x) + a_{n-1} D^{n-1} y(x) + \cdots + a_1 D y(x) + a_0 y(x) \\
&= (a_n D^n + a_{n-1} D^{n-1} + \cdots + a_1 D + a_0) y(x).
\end{aligned}
$$

Hence, the differential equation $a_n y^{(n)}(x) + a_{n-1} y^{(n-1)}(x) + \cdots + a_1 y'(x) + a_0 y(x) = g(x)$ can be written in operator form as

$$(a_n D^n + a_{n-1} D^{n-1} + \cdots + a_1 D + a_0) y(x) = g(x).$$

The expression $p(D) = a_n D^n + a_{n-1} D^{n-1} + \cdots + a_1 D + a_0$ is called an **nth-order differential operator**.

Notice that when we place the differential equation in operator form, we end up with an equation of the form $p(D)y = f(x)$, where $p(D)$ is a function of the differential operator D. Functions of this type will be of great use to us in solving nonhomogeneous equations. Three important properties of differential operators are stated as follows:

··

PROPERTY 1   Let $P = p(D)$ be a differential operator and consider the two functions $f(x)$ and $g(x)$. Then, $P[f(x) + g(x)] = P[f(x)] + P[g(x)]$.

··

PROPERTY 2   The product of two operators P_1 and P_2 is defined by $P_1 P_2[f(x)] = P_1(P_2[f(x)])$. Thus, $(P_1 P_2)[f(x)] = P_1(P_2[f(x)])$, and consequently, $P_1 P_2[f(x)] = P_2 P_1[f(x)]$.

··

PROPERTY 3   Differential operators can be treated as polynomials in D.

··

DEFINITION | *Annihilator*
|
| The differential operator $p(D)$ is said to **annihilate** a function $f(x)$ if $p(D)(f(x)) = 0$ for all x. In this case, $p(D)$ is called an **annihilator** of $f(x)$.

EXAMPLE: Show that the following operators annihilate the corresponding function: (a) $p(D) = D^3$, $f(x) = x^2$; (b) $p(D) = (D + 3)^2$, $f(x) = xe^{-3x}$; and (c) $p(D) = D^2 - 2D + 2$, $f(x) = e^x \cos x$ and $g(x) = e^x \sin x$.

SOLUTION: For (a), $D^3(x^2) = D^2(2x) = D(2) = 0$. For (b) we have:

$$(D + 3)^2(xe^{-3x}) = (D^2 + 6D + 9)(xe^{-3x}) = D^2(xe^{-3x}) + 6D(xe^{-3x}) + 9(xe^{-3x}),$$

which we compute and simplify as follows:

```
> diff(x*exp(-3*x),x$2)+6*diff(x*exp(-3*x),x)+9*x*exp(-3*x);
    0
```

Similarly, for (c) we have $(D^2 - 2D + 2)(e^x \cos x) = D^2(e^x \cos x) - 2D(e^x \cos x) + 2(e^x \cos x)$, which is computed and simplified as follows:

```
> f:=x->exp(x)*cos(x):
  (D@@2)(f)(x)-2*D(f)(x)+2*f(x);
    0
```

In the exact same manner, we see that $(D^2 - 2D + 2)(e^x \sin x) = 0$ by entering

```
> g:=x->exp(x)*sin(x):
  (D@@2)(g)(x)-2*D(g)(x)+2*g(x);

  0
```

The following table lists the annihilator of familiar functions.

Functions	Annihilator
$1, x, x^2, \ldots, x^{n-1}$	D^n
$ee^{kx}, xe^{kx}, \ldots, x^{n-1}e^{kx}$	$(D - k)^n$
$e^{\alpha x}\cos(\beta x), xe^{\alpha x}\cos(\beta x), \ldots, x^{n-1}e^{\alpha x}\cos(\beta x),$ $e^{\alpha x}\sin(\beta x), xe^{\alpha x}\sin(\beta x), \ldots, x^{n-1}e^{\alpha x}\sin(\beta x)$	$[D^2 - 2\alpha D + (\alpha^2 + \beta^2)]^n$

PROPERTY 4 If $P_1 = p_1(D)$ annihilates $f(x)$ and $P_2 = p_2(D)$ annihilates $g(x)$, then $P_1 P_2$ (or $P_2 P_1$) annihilates $af(x) + bg(x)$ where a and b are constants.

We can use this property to determine the operator that annihilates any function that involves functions of the forms discussed earlier. (Note that we can extend this property to include the linear combination of more than two functions.)

EXAMPLE: Determine the differential operator that annihilates the indicated function:
(a) $f(x) = x + \cos x$, (b) $g(x) = 3x^2 - e^{-x}\sin 4x$, (c) $h(x) = x\cos x + 4$, (d) $k(x) = x^2 e^x - x^4 + x$.

SOLUTION: (a) Let $f(x) = f_1(x) + f_2(x)$, where $f_1(x) = x$ and $f_2(x) = \cos x$. The operator D^2 annihilates $f_1(x) = x$ and $(D^2 + 1)$ annihilates $f_2(x) = \cos x$. Therefore, $D^2(D^2 + 1)$ annihilates $f(x) = x + \cos x$.

(b) Let $g(x) = g_1(x) + g_2(x)$, where $g_1(x) = 3x^2$ and $g_2(x) = -e^{-x}\sin 4x$. Then D^3 annihilates $g_1(x) = 3x^2$, and $(D^2 + 2D + 17)$ annihilates $g_2(x) = -e^{-x}\sin 4x$. Hence, $D^3(D^2 + 2D + 17)$ or $(D^2 + 2D + 17)D^3$ annihilates $g(x) = 3x^2 - e^{-x}\sin 4x$.

(c) Let $h(x) = h_1(x) + h_2(x)$, where $h_1(x) = x\cos x$ and $h_2(x) = 4$. Since $(D^2 + 1)^2$ annihilates $h_1(x) = x\cos x$ and D annihilates $h_2(x) = 4$, $(D^2 + 1)^2 D$ or $D(D^2 + 1)^2$ annihilates $h(x) = x\cos x + 4$.

(d) Let $k(x) = k_1(x) + k_2(x) + k_3(x)$, where $k_1(x) = x^2e^x$, $k_2(x) = -x^4$, and $k_3(x) = x$. The operator $(D - 1)^3$ annihilates $k_1(x) = x^2e^x$, and D^5 annihilates $k_2(x) = -x^4$. Since D^5 also annihilates $k_3(x) = x$, the operator $(D - 1)^3D^5$ or $D^5(D - 1)^3$ annihilates $k(x) = x^2e^x - x^4 + x$.

≡ Using the Annihilator Method

The homogeneous linear nth-order differential equation with constant coefficients can be expressed as $p(D)y = g(x)$. When $g(x)$ is a function of one of the forms listed in the following table, another differential operator, $q(D)$, that annihilates $g(x)$ can be determined.

$g(x)$
$1, x, x^2, \ldots, x^{n-1}$
$e^{kx}, xe^{kx}, \ldots, x^{n-1}e^{kx}$
$e^{\alpha x}\cos(\beta x), xe^{\alpha x}\cos(\beta x), \ldots, x^{n-1}e^{\alpha x}\cos(\beta x),$
$e^{\alpha x}\sin(\beta x), xe^{\alpha x}\sin(\beta x), \ldots, x^{n-1}e^{\alpha x}\sin(\beta x)$

Suppose that the differential operator $q(D)$ annihilates $g(x)$. Then applying $q(D)$ to the non-homogeneous equation yields $q(D)p(D)y = q(D)g(x) = 0$. The form of the particular solution is found by solving the homogeneous equation $q(D)p(D)y = 0$.

Procedure to Solve

$p(D)y = g(x)$ if $g(x)$ is a linear combination of the functions $1, x, x^2, \ldots, e^{kx}, xe^{kx}, x^2e^{kx}, \ldots, e^{\alpha x}\cos\beta x,$ $xe^{\alpha x}\cos\beta x, x^2e^{\alpha x}\cos\beta x, \ldots, e^{\alpha x}\sin\beta x, xe^{\alpha x}\sin\beta x, x^2e^{\alpha x}\sin\beta x, \ldots$.

1. Determine the operator $q(D)$ that annihilates $g(x)$ [i.e., $q(D)(q(x)) = 0$].
2. Apply the operator to both sides of the differential equation $q(D)(p(D)y) = q(D)(g(x)) = 0$.
3. Solve the homogeneous equation $q(D)(p(D)y) = 0$.
4. Find the solution $y_h(x)$ of the homogeneous equation corresponding to the original equation, $p(D)y = 0$.
5. Eliminate the terms of the homogeneous solution $y_h(x)$ from the general solution obtained in Step 3. The function that remains is the correct form of the particular solution.
6. Solve for the unknown coefficients in the particular solution.
7. A general solution of the nonhomogeneous equation is $y(x) = y_h(x) + y_p(x)$.

EXAMPLE: Solve the nonhomogeneous equation $y'' + y = x$.

SOLUTION: After writing the differential equation in operator form $(D^2 + 1)y = x$, we note that the operator D^2 annihilates the function $g(x) = x$. Applying this annihilator to both sides of the equation, we obtain the homogeneous equation

$$D^2(D^2 + 1)y = 0,$$

which has characteristic equation $m^2(m^2 + 1) = 0$. Therefore, a general solution of this equation is $y(x) = b_1 + b_2 x + b_3 \cos(x) + b_4 \sin(x)$, where b_1, b_2, b_3, and b_4 are arbitrary constants.

```
> dsolve(diff(y(x),x$4)+diff(y(x),x$2)=0,y(x));
```

$$y(x) = \_C1 + \_C2\,x + \_C3\sin(x) + \_C4\cos(x)$$

When we solve the homogeneous equation $y'' + y = 0$ that corresponds to the original equation, we find that $y_h(x) = c_1 \cos(x) + c_2 \sin(x)$.

```
> Hom_Sol:=dsolve(diff(y(x),x$2)+y(x)=0,y(x));
```

$$Hom\_Sol := y(x) = \_C1\sin(x) + \_C2\cos(x)$$

Eliminating these terms from $y(x) = b_1 + b_2 x + b_3 \cos(x) + b_4 \sin(x)$ indicates that a particular solution has the form $y_p(x) = A + Bx$, where A and B are unknown constants that are determined by substituting $y_p(x)$ into the original equation $y'' + y = x$.

```
> A:='A':B:='B':yp:=x->A+B*x:
  Eq:=(D@@2)(yp)(x)+yp(x)=x;
```

$$Eq := A + Bx = x$$

Since $y_p'(x) = B$ and $y_p''(x) = 0$, we have $y'' + y = 0 + A + Bx = A + Bx = x$. Equating the coefficients of like terms, we find that $A = 0$ and $B = 1$; the same result is obtained as follows with `match`.

```
> match(x=A+B*x,x,'vals');
```

true

```
> vals;
```

$$\{A = 0, B = 1\}$$

Therefore, $y_p(x) = A + Bx = 0 + (1)x = x,$

```
> assign(vals):
  yp(x);
```

x

so a general solution of the nonhomogeneous equation $y'' + y = x$ is $y(x) = y_h(x) + y_p(x) = c_1 \cos(x) + c_2 \sin(x) + x$.

```
> Gen_Sol:=yp(x)+rhs(Hom_Sol);
```

$Gen\_Sol := x + \_C1\sin(x) + \_C2\cos(x)$

```
> to_plot:={seq(seq(subs({_C1=i,_C2=j},Gen_Sol),
    i=-1..1),j=-1..1)}:
  plot(to_plot,x=-Pi..Pi);
```

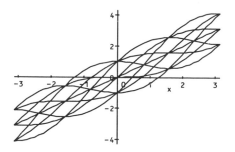

In this case, we see that **dsolve** returns the same general solution as that obtained earlier.

```
> dsolve(diff(y(x),x$2)+y(x)=x,y(x));
```

$y(x) = x + \_C1\sin(x) + \_C2\cos(x)$

EXAMPLE: Find a general solution of $y''' - y'' - 7y' + 15y = x^2e^{-3x} + e^{2x}\cos 3x$.

SOLUTION: As in the previous example, **dsolve** is able to find a general solution of the equation, which we name **Gen_Sol**.

```
> y:='y':
  Gen_Sol:=dsolve(diff(y(x),x$3)-diff(y(x),x$2)-
    7*diff(y(x),x)+15*y(x)=
    x^2*exp(-3*x)+exp(2*x)*cos(3*x),y(x));
```

$$Gen\_Sol := y(x) = -\frac{5}{68}e^{2x}\cos(x)^3 + \_C2\,e^{2x}\sin(x) + \_C3\,e^{2x}\cos(x)$$

$$+ \frac{60}{221}e^{2x}\cos(x) + \frac{12}{221}e^{2x}\sin(x) + \_C1\,e^{-3x} + \frac{5}{338}x^2e^{-3x} + \frac{37}{4394}xe^{-3x}$$

$$+ \frac{3}{68}e^{2x}\sin(x)\cos(x)^2 + \frac{60}{28561}e^{-3x} + \frac{1}{78}e^{-3x}x^3$$

Alternatively, we illustrate how to use Maple to construct a solution via the annihilator method. In operator notation, the equation $y''' - y'' - 7y' + 15y = x^2 e^{-3x} + e^{2x} \cos 3x$ is written as

$$(D^3 - D^2 - 7D + 15)y(x) = x^2 e^{-3x} + e^{2x} \cos 3x.$$

Applying $(D + 3)^3$ and $D^2 - 4D + 13$ to the equation results in

$$(D + 3)^3 (D^2 - 4D + 13)(D^3 - D^2 - 7D + 15)y(x) = 0$$

because $(D + 3)^3$ annihilates $x^2 e^{-3x}$ and $D^2 - 4D + 13$ annihilates $e^{2x} \cos 3x$. The characteristic equation of

$$(D + 3)^3 (D^2 - 4D + 13)(D^3 - D^2 - 7D + 15)y(x) = 0$$

is

$$(m + 3)^3 (m^2 - 4m + 13)(m^3 - m^2 - 7m + 15) = 0,$$

which has solutions -3, with multiplicity 4, $2 \pm 3i$, and $2 \pm i$, as we see here with solve.

```
> solve((m+3)^3*(m^2-4*m+13)*(m^3-m^2-7*m+15)=0);
```
$$-3, -3, -3, 2 + 3I, 2 - 3I, -3, 2 + I, 2 - I$$

Therefore, a general solution of this equation is

$$y(x) = c_1 e^{-3x} + c_2 x e^{-3x} + c_3 x^2 e^{-3x} + c_4 x^3 e^{-3x} + e^{2x}(c_5 \cos 3x + c_6 \sin 3x) + e^{2x}(c_7 \cos x + c_8 \sin x).$$

In contrast, the characteristic equation of the corresponding homogeneous solution is $m^3 - m^2 - 7m + 15 = 0$, which has solutions -3 and $2 \pm 3i$:

```
> solve(m^3-m^2-7*m+15=0);
```
$$-3, 2 + I, 2 - I$$

Thus, $y_h(x) = c_1 e^{-3x} + e^{2x}(c_2 \cos x + c_3 \sin x)$.

```
yh:=x->c[1]*exp(-3*x)+exp(2*x)*(c[2]*cos(x)+c[3]*sin(x));
```
$$yh := x \to c_{[1]} e^{-3x} + e^{2x}(c_{[2]} \cos(x) + c_{[3]} \sin(x))$$

Eliminating these terms from

$$y(x) = c_1 e^{-3x} + c_2 x e^{-3x} + c_3 x^2 e^{-3x} + c_4 x^3 e^{-3x} + e^{2x}(c_5 \cos 3x + c_6 \sin 3x) + e^{2x}(c_7 \cos x + c_8 \sin x)$$

indicates that a particular solution has the form

$$y_p(x) = c_4 x e^{-3x} + c_5 x^2 e^{-3x} + c_6 x^3 e^{-3x} + e^{2x}(c_7 \cos 3x + c_8 \sin 3x).$$

We define **yp** by first defining **funs** to be the list of functions corresponding to the vector

$$(x e^{-3x} \quad x^2 x^{-3x} \quad x^3 e^{-3x} \quad e^{2x} \cos 3x \quad e^{2x} \sin 3x).$$

```
> funs:=[x*exp(-3*x),x^2*exp(-3*x),x^3*exp(-3*x),
  exp(2*x)*cos(3*x),exp(2*x)*sin(3*x)];
```

$$funs := [x e^{-3x}, x^2 e^{-3x}, e^{-3x} x^3, e^{2x} \cos(3x), e^{2x} \sin(3x)]$$

We then define **cs** to be the list of constants corresponding to the vector $(c_4 \quad c_5 \quad c_6 \quad c_7 \quad c_8)$.

```
> cs:=[seq(c[i],i=4..8)];
```

$$cs := [c_{[4]}, c_{[5]}, c_{[6]}, c_{[7]}, c_{[8]}]$$

We then define **yp** by computing the dot product of **funs** and **cs** with **dotprod**, which is contained in the **linalg** package.

```
> with(linalg):
  yp:=x->dotprod(funs,cs):
  yp(x);
```

$$x e^{-3x} c_{[4]} + x^2 e^{-3x} c_{[5]} + e^{-3x} x^3 c_{[6]} + e^{2x} \cos(3x) c_{[7]} + e^{2x} \sin(3x) c_{[8]}$$

```
> LHSeqn:=diff(yp(x),x$3)-diff(yp(x),x$2)-
  7*diff(yp(x),x)+15*yp(x);
```

$$LHSeqn := -40 e^{2x} \sin(3x) c_{[8]} - 40 e^{2x} \cos(3x) c_{[7]} + 52 x e^{-3x} c_{[5]}$$
$$+78 e^{-3x} x^2 c_{[6]} - 60 e^{-3x} x c_{[6]} + 26 e^{-3x} c_{[4]} - 24 e^{2x} \sin(3, x) c_{[7]}$$
$$-24 e^{2x} \cos(3x) c_{[8]} - 20 e^{-3x} c_{[5]} + 6 e^{-3x} c_{[6]}$$

Equating the coefficients of **LHSeqn** and $x^2 e^{-3x} + e^{2x} \cos 3x$ yields the system of equations

$$\begin{cases} 26c_4 - 20c_5 + 6c_6 = 0 \\ 52c_5 - 60c_6 = 0 \\ 78c_6 = 1 \\ -40c_7 - 24c_8 = 1 \\ 24c_7 - 40c_8 = 0 \end{cases}$$

that we solve with **solve** and name the resulting solutions **Sols**.

```
> Sols:=solve({26*c[4]-20*c[5]+6*c[6]=0,
    52*c[5]-60*c[6]=0,78*c[6]=1,
    -40*c[7]-24*c[8]=1,24*c[7]-40*c[8]=0});
```

$$Sols := \left\{ c_{[8]} = \frac{-3}{272}, c_{[7]} = \frac{-5}{272}, c_{[4]} = \frac{37}{4394}, c_{[5]} = \frac{5}{338}, c_{[6]} = \frac{1}{78} \right\}$$

We then use **assign** to name the constants the values obtained in **Sols** and evaluate the particular solution.

```
> assign(Sols);
  yp(x);
```

$$\frac{37}{4394} xe^{-3x} + \frac{5}{338} x^2 e^{-3x} + \frac{1}{78} x^3 e^{-3x} - \frac{5}{272} e^{2x} \cos(3x) \frac{3}{272} e^{2x} \sin(3x)$$

Finally, a general solution is obtained by computing the sum of the homogeneous solution and the particular solution.

```
> y:=x->yh(x)+yp(x):
  y(x);
```

$$c_{[1]} e^{-3x} + e^{2x} \left(c_{[2]} \cos(x) + c_{[3]} \sin(x) \right) + \frac{37}{4394} xe^{-3x} + \frac{5}{338} x^2 e^{-3x} + \frac{1}{78} x^3 e^{-3x}$$

$$- \frac{5}{282} e^{2x} \cos(3x) - \frac{3}{272} e^{2x} \sin(3x)$$

To graph the general solution for various values of the arbitrary constants, we begin by defining **i_vals** to be the set of numbers 0 and 5, **j_vals** to be the set $-1/2$ and $1/2$, and **k_vals** to be the set consisting of $-1/2$ and $1/2$. We then define **to_plot** to be the set of eight functions obtained by replacing **c[1]** by each number in **i_vals**, **c[2]** by each number in **j_vals**, and **c[3]** by each number in **k_vals**. The set of functions **to_plot** is then graphed on the interval $\left[-\frac{1}{2}, \frac{3}{2} \right]$ with **plot**.

```
> i_vals:={0,5}:j_vals:={-1/2,1/2}:k_vals:={-1/2,1/2}:
  to_plot:={seq(seq(seq(subs({c[1]=i,c[2]=j,c[3]=k},y(x)),
  i=i_vals),j=j_vals),k=k_vals)}:
  plot(to_plot,x=-1/2..3/2);
```

As with other differential equations, we can also solve initial-value problems that involve non-homogeneous equations. The steps followed are the same as those followed with homogeneous equations. Namely, we first find a general solution and then use the initial conditions to find the values of the unknown coefficients so that the initial conditions are satisfied.

≡ Solving Initial-Value Problems Involving Nonhomogeneous Equations

1. Determine a general solution to the nonhomogeneous equation.
2. Use the initial conditions to find the unknown coefficients in the general solution obtained in Step 1.

EXAMPLE: Solve $y'' + 2y' - 3y = 4e^x - \sin x$ subject to the conditions $y(0) = 0$ and $y'(0) = 1$.

SOLUTION: First, we notice that $(D - 1)$ annihilates e^x and $(D^2 + 1)$ annihilates $\sin x$. Hence, $(D - 1)(D^2 + 1)$ annihilates $g(x) = 4e^x - \sin x$. Applying this operator to both sides of the equation $y'' + 2y' - 3y = 4e^x - \sin x$ gives us $(D - 1)(D^2 + 1)(D^2 + 2D - 3)y = (D - 1)^2(D^2 + 1)(D + 3)y = 0$ in operator form. The solution of this equation with characteristic equation $(m - 1)^2(m^2 + 1)(m + 3) = 0$ is

$$y(x) = b_1 e^{-3x} + b_2 e^x + b_3 x e^x + b_4 \cos x + b_5 \sin x.$$

The solution to the corresponding homogeneous equation $y'' + 2y' - 3y = 0$ with characteristic equation $m^2 + 2m - 3 = (m + 3)(m - 1) = 0$ is

$$y_h(x) = c_1 e^{-3x} + c_2 e^x.$$

If we eliminate the terms of $y_h(x)$ from $y(x) = b_1 e^{-3x} + b_2 e^x + b_3 x e^x + b_4 \cos x + b_5 \sin x$, we see that a particular solution is $y_p(x) = A x e^x + B \cos x + C \sin x$. Differentiating $y_p(x)$ yields

$$y_p'(x) = A e^x + A x e^x - B \sin x + C \cos x$$

and

$$y_p''(x) = 2 A e^x + A x e^x - B \cos x - C \sin x.$$

Substitution into the equation $y'' + 2y' - 3y = 4e^x - \sin x$ then gives us $4A e^x + (2C - 4B) \cos x + (-4C - 2B) \sin x = 4e^x - \sin x$. Therefore, $4A = 4$, $2C - 4B = 0$, and $-4C - 2B = -1$, which implies that $A = 1$, $B = \frac{1}{10}$, and $C = \frac{1}{5}$. Hence, a particular solution is $y_p(x) = x e^x + \frac{1}{10} \cos(x) + \frac{1}{5} \sin(x)$, so a general solution of $y'' + 2y' - 3y = 4e^x - \sin x$ is

$$y(x) = y_h(x) + y_p(x) = c_1 e^{-3x} + c_2 e^x + x e^x + \frac{1}{10} \cos x + \frac{1}{5} \sin x.$$

Solving $y(0) = c_1 + c_2 + \frac{1}{10} = 0$ and $y'(0) = c_1 - 3c_2 + \frac{6}{5} = 1$ for c_1 and c_2 results in $c_1 = -\frac{1}{8}$ and $c_2 = \frac{1}{40}$ so that the solution of the initial value problem is

$$y(x) = \frac{-e^x}{8} + \frac{e^{-3x}}{40} + x e^x + \frac{\cos x}{10} + \frac{\sin x}{5}.$$

In this case, **dsolve** returns the same result as shown next.

```
> y:='y':
  Sol:=dsolve({diff(y(x),x$2)+2*diff(y(x),x)-3*y(x)=
  4*exp(x)-sin(x),y(0)=0,D(y)(0)=1},y(x));
```

$$Sol := y(x) = x e^x + \frac{1}{10} \cos(x) + \frac{1}{5} \sin(x) - \frac{1}{8} e^x + \frac{1}{40} e^{-3x}$$

We then use **assign** to name $y(x)$ the result obtained in **Sol** and graph $y(x)$ on the interval $[-2, 2]$ with **plot**.

```
> assign(Sol):
  plot(y(x),x=-2..2);
```

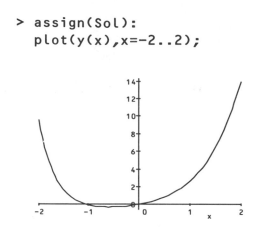

EXAMPLE: Solve $y''' + y' = \cos x + x$ subject to the initial conditions $y(0) = 1$, $y'(0) = 0$, and $y''(0) = -2$.

SOLUTION: We begin by determining the operator that annihilates the function $g(x) = \cos(x) + x$. The operator $D^2(D^2 + 1)$ annihilates $g(x) = \cos(x) + x$ because $(D^2 + 1)$ annihilates $\cos x$ and D^2 annihilates x. Applying this operator to both sides of $y''' + y' = \cos(x) + x$ in operator form yields $D^2(D^2 + 1)(D^3 + D)y = D^2(D^2 + 1)(\cos(x) + x) = 0$. This homogeneous equation has the characteristic equation $m^2(m^2 + 1)(m^3 + m) = m^3(m^2 + 1)^2 = 0$, of multiplicity 3, and $\pm i$, each with multiplicity 2. Therefore, it has a general solution

$$y(x) = b_1 + b_2 x + b_3 x^2 + b_4 \cos(x) + b_5 \sin(x) + b_6 x \cos(x) + b_7 x \sin(x).$$

Now, if we solve the corresponding homogeneous equation $y''' + y' = 0$,

```
> dsolve(diff(y(x),x$3)+diff(y(x),x)=0,y(x));
  y(x) = _C1 + _C2 sin(x) + _C3 cos(x)
```

we see that

$$y_h(x) = c_1 + c_2 \cos x + c_3 \sin x.$$

Comparing the two solutions

$$y_h(x) = c_1 + c_2 \cos(x) + c_3 \sin(x)$$

and

$$y(x) = b_1 + b_2 x + b_3 x^2 + b_4 \cos(x) + b_5 \sin(x) + b_6 x \cos(x) + b_7 x \sin(x),$$

we recognize that $y_p(x) = Ax + Bx^2 + Cx \cos(x) + Dx \sin(x)$ after eliminating the terms of $y_h(x)$ from $y(x)$.

```
> yp:=x->A*x+B*x^2+C*x*cos(x)+D*x*sin(x);
```
$$yp := x \rightarrow Ax + Bx^2 + Cx \cos(x) + Dx \sin(x)$$

Differentiating $y_p(x)$, we have

$$y_p'(x) = A + 2Bx + C \cos(x) - Cx \sin(x) + D \sin(x) + Dx \cos(x),$$
$$y_p''(x)0 = 2B - 2C \sin(x) - Cx \cos(x) + 2D \cos(x) - Dx \sin(x),$$

and

$$y_p'''(x) = -3C \cos(x) + Cx \sin(x) - 3D \sin(x) - Dx \cos(x).$$

Substitution into $y''' + y' = \cos(x) + x$ then gives us $-2C \cos(x) - 2D \sin(x) + 2Bx + A = \cos x + x$.

```
> LHSeqn:=simplify(diff(yp(x),x$3)+diff(yp(x),x));
```
$$LHSeqn := -2C \cos(x) - 2D \sin(x) + A + 2Bx$$

```
> match(x+cos(x)=LHSeqn,x,'Vals'):
  Vals;
```
$$\left\{ B = \frac{1}{2}, D = 0, A = 0, C = -\frac{1}{2} \right\}$$

Hence, A=0, $B = \dfrac{1}{2}$, $C = -\dfrac{1}{2}$, and $D = 0$, so a particular solution of the equation $y''' + y' = \cos(x) + x$ is $y_p(x) = \dfrac{1}{2}x^2 - \dfrac{1}{2}x \cos(x)$. This means that a general solution is

$$y(x) = y_h(x) + y_p(x) = c_1 + c_2 \cos(x) + c_3 \sin(x) + \frac{1}{2}x^2 - \frac{1}{2}x \cos(x).$$

```
> assign(Vals):
  y:=x->c1+c2*cos(x)+c3*sin(x)+yp(x);
```
$$y := x \rightarrow c1 + c2 \cos(x) + c3 \sin(x) + yp(x)$$

To find the solution that satisfies the initial conditions $y(0) = 1$, $y'(0) = 0$, and $y''(0) = -2$, we substitute $x = 0$ into $y(x)$, $y'(x)$, and $y''(x)$, and set the results equal to 1, 0, and -2, respectively.

```
> Eq1:=y(0)=1;
  Eq2:=eval(subs(x=0,diff(y(x),x))=0);
  Eq3:=eval(subs(x=0,diff(y(x),x$2))=-2);
```

$$Eq1 := c1 + c2 = 1$$

$$Eq2 := c3 - \frac{1}{2} = 0$$

$$Eq3 := -c2 + 1 = -2$$

The resulting equations are then solved with **solve**.

```
> Sols:=solve({Eq1,Eq2,Eq3});
  assign(Sols):
  y(x);
```

$$Sols := \left\{ c1 = -2, c3 = \frac{1}{2}, c2 = 3 \right\}$$

$$\frac{1}{2}x^2 + \frac{1}{2}\sin(x) - 2 - \frac{1}{2}\cos(x)x + 3\cos(x)$$

For this problem the same results are obtained with **dsolve**. For example, entering

```
> y:='y':
  Gen_Sol:=dsolve(diff(y(x),x$3)+diff(y(x),x)=x+cos(x),y(x));
```

$$Gen\_Sol := y(x) = \frac{1}{2}x^2 + \frac{1}{2}\sin(x) - 1 - \frac{1}{2}\cos(x)x + \_C1 + \_C2\sin(x) + \_C3\cos(x)$$

finds a general solution of the equation, naming the result **Gen_Sol**, and entering

```
> Part_Sol:=dsolve({diff(y(x),x$3)+diff(y(x),x)=x+cos(x),
  y(0)=1,D(y)(0)=0,(D@@2)(y)(0)=-2},y(x));
```

$$Part\_Sol := y(x) = \frac{1}{2}x^2 + \frac{1}{2}\sin(x) - 2 - \frac{1}{2}\cos(x)\,x + 3\cos(x)$$

solves the initial-value problem, naming the result **Part_Sol**.

As with previous examples, we graph the solution given in **Gen_Sol** for various values of the constants and the solution given in **Part_Sol**.

```
> i_vals:={-2,2}:j_vals:={0,2}:k_vals:={-2,0}:
  to_plot:={seq(seq(seq(subs({_C1=i,_C2=j,_C3=k},
    rhs(Gen_Sol)),i=i_vals),j=j_vals),k=k_vals)}:

> plot(to_plot,x=-Pi..Pi);
  plot(rhs(Part_Sol),x=-Pi..2*Pi);
```

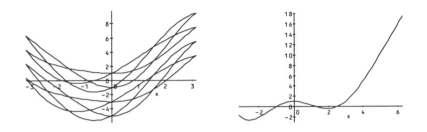

4.4 NONHOMOGENEOUS EQUATIONS WITH CONSTANT COEFFICIENTS: THE METHOD OF UNDETERMINED COEFFICIENTS

We would now like to present another approach (which can be used instead of the annihilator method) that can be used to determine the form of the particular solution of a nonhomogeneous differential equation.

Consider the nonhomogeneous linear nth-order differential equation with constant coefficients

$$a_n y^{(n)}(x) + a_{n-1} y^{(n-1)}(x) + \cdots + a_1 y'(x) + a_0 y(x) = g(x),$$

where $g(x)$ is a linear combination of the functions $1, x, x^2, \ldots, e^{kx}, xe^{kx}, x^2 e^{kx}, \ldots, e^{\alpha x} \cos \beta x, xe^{\alpha x} \cos \beta x,$ $x^2 e^{\alpha x} \cos \beta x, \ldots, e^{\alpha x} \sin \beta x, xe^{\alpha x} \sin \beta x, x^2 e^{\alpha x} \sin \beta x, \ldots$. A general solution of this differential equation is

$$y(x) = y_h(x) + y_p(x),$$

where $y_h(x)$ is a solution of the corresponding homogeneous equation

$$a_n y^{(n)}(x) + a_{n-1} y^{(n-1)}(x) + \cdots + a_1 y'(x) + a_0 y(x) = 0,$$

and $y_p(x)$ is a particular solution involving no arbitrary constants of the nonhomogeneous equation

$$a_n y^{(n)}(x) + a_{n-1} y^{(n-1)}(x) + \cdots + a_1 y'(x) + a_0 y(x) = g(x).$$

Since we learned how to solve homogeneous equations in Section 4.2, we must learn how to find the form of the particular solution to solve nonhomogeneous equations.

☰ Outline of the Method of Undetermined Coefficients

1. Solve the corresponding homogeneous equation for $y_h(x)$.
2. Determine the form of the particular solution $y_p(x)$ (see the next subsection).
3. Determine the unknown coefficients in $y_p(x)$ by substituting $y_p(x)$ into the nonhomogeneous equation and equating the coefficients of like terms.
4. Form a general solution with $y(x) = y_h(x) + y_p(x)$.

☰ Determining the Form of $y_p(x)$ (Step 2):

Suppose that $g(x) = b_1 g_1(x) + b_2 g_2(x) + \cdots + b_j g_j(x)$, where $b_1, b_2, \ldots, b_j$ are constants and each $g_i(x)$, $i = 1, 2, \ldots, j$ is a function of the form x^m, $x^m e^{kx}$, $x^m e^{\alpha x} \cos \beta x$, or $x^m e^{\alpha x} \sin \beta x$.

(1) If $g_i(x) = x^m$, then the associated set of functions is

$$S = \{x^m, x^{m-1}, \ldots, x^2, x, 1\}.$$

(2) If $g_i(x) = x^m e^{kx}$, then the associated set of functions is

$$S = \{x^m e^{kx}, x^{m-1} e^{kx}, \ldots, x^2 e^{kx}, x e^{kx}, e^{kx}\}.$$

(3) If $g_i(x) = x^m e^{\alpha x} \cos \beta x$, or $g_i(x) = x^m e^{\alpha x} \sin \beta x$, then the associated set of functions is

$$S = \{x^m e^{\alpha x} \cos \beta x, x^{m-1} e^{\alpha x} \cos \beta x, \ldots, x^2 e^{\alpha x} \cos \beta x, x e^{\alpha x} \cos \beta x, e^{\alpha x} \cos \beta x,$$
$$x^m e^{\alpha x} \sin \beta x, x^{m-1} e^{\alpha x} \sin \beta x, \ldots, x^2 e^{\alpha x} \sin \beta x, x e^{\alpha x} \sin \beta x, e^{\alpha x} \sin \beta x\}.$$

For each function in S, determine the associated set of functions. If any of the functions in S appears in the homogeneous solution $y_h(x)$, then multiply each function in S by x^r to obtain a new set S'. r is the smallest positive integer so that each function in S' is not a function in $y_h(x)$. The particular solution is obtained by taking the linear combination of all functions in the associated sets where repeated functions should appear only once in the particular solution.

EXAMPLE: Solve the nonhomogeneous equations (a) $y'' + 5y' + 6y = 2e^x$ and (b) $y'' + 5y' + 6y = 3e^{-2x}$.

SOLUTION: (a) We begin by solving the corresponding homogeneous equation $y'' + 5y' + 6y = 0$.

```
> Hom_Sol:=dsolve(diff(y(x),x$2)+5*diff(y(x),x)+6*y(x)=0,
  y(x));
```

$$Hom\_Sol := y(x) = \_C1\,e^{-2x} + \_C2\,e^{-3x}$$

Thus,

$$y_h(x) = c_1e^{-2x} + c_2e^{-3x}$$

is a general solution of the corresponding homogeneous equation $y'' + 5y' + 6y = 0$. Now we determine the form of $y_p(x)$. We choose

$$S = \{e^x\}.$$

because $g(x) = 2e^x$. Since e^x is not a solution to the homogeneous equation, we take $y_p(x)$ to be the linear combination of the functions in S. Therefore,

$$y_p(x) = Ae^x.$$

```
> yp:=x->A*exp(x);
```

$$yp := x \longrightarrow Ae^x$$

Substituting this solution into $y'' + 5y' + 6y = 2e^x$,

```
> LHSeqn:=diff(yp(x),x$2)+5*diff(yp(x),x)+6*yp(x);
```

$$LHSeqn := 12Ae^x$$

we have

$$Ae^x + 5Ae^x + 6Ae^x = 12Ae^x = 2e^x.$$

Equating the coefficients of e^x then gives us $A = \dfrac{1}{6}$.

```
> match(2*exp(x)=12*A*exp(x),x,'Val'):
  Val;
```

$$\left\{A = \frac{1}{6}\right\}$$

Hence, a particular solution is $y_p(x) = \dfrac{1}{6}e^x$,

```
> assign(Val):
  yp(x);
```

$$\frac{1}{6}e^x$$

so a general solution of $y'' + 5y' + 6y = 2e^x$ is

$$y(x) = y_h(x) + y_p(x) = c_1 e^{-2x} + c_2 e^{-3x} + \frac{1}{6}e^x.$$

The same result is obtained with **dsolve**.

```
> Gen_Sol:=dsolve(diff(y(x),x$2)+5*diff(y(x),x)+
  6*y(x)=2*exp(x),y(x));
```

$$Gen\_Sol := y(x) = \frac{1}{6}e^x + \_C1\,e^{-2x} + \_C2\,e^{-3x}$$

As in previous examples, we name $y(x)$ the result given in **Gen_Sol** with **assign** and then use **seq** and **subs** to generate the set of nine functions obtained by replacing each occurrence of **_C1** in $y(x)$ by $-1, 0$, and 1, and each occurrence of **_C2** in $y(x)$ by $-1, 0$, and 1. The set of functions **to_plot** is then graphed on the interval $[-1, 3]$ with **plot**.

```
> assign(Gen_Sol):
  to_plot:={seq(seq(subs({_C1=i,_C2=j},y(x)),
  i=-1..1),j=-1..1)}:
  plot(to_plot,x=-1..3);
```

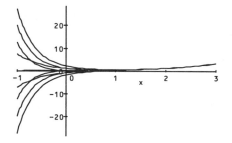

(b) In this case, we see that $g(x) = 3e^{-2x}$. Then the associated set is $S = \{e^{-2x}\}$. However, since e^{-2x} is a solution to the corresponding homogeneous equation, we must multiply this function by x^r so that it is no longer a solution. Since xe^{-2x} is not a solution of $y'' + 5y' + 6y = 0$, we multiply the element of S by x to obtain $S' = \{xe^{-2x}\}$. Hence, $y_p(x) = Axe^{-2x}$.

```
> A:='A':yp:='yp':
  yp:=x->A*x*exp(-2*x);
```

$$yp := x \longrightarrow Axe^{-2x}$$

Differentiating, we have

```
> Diff(yp(x),x)=diff(yp(x),x);
```

$$\frac{\partial}{\partial x} Axe^{-2x} = Ae^{-2x} - 2Axe^{-2x}$$

and

```
> Diff(yp(x),x$2)=diff(yp(x),x$2);
```

$$\frac{\partial^2}{\partial x^2} Axe^{-2x} = -4Ae^{-2x} + 4Axe^{-2x}$$

Substitution then yields

$$y'' + 5y' + 6y = -4Ae^{-2x} + 4Axe^{-2x} + 5(Ae^{-2x} - 2Axe^{-2x}) + 6Axe^{-2x} = Ae^{-2x} = 3e^{-2x}.$$

```
> LHSeqn:=diff(yp(x),x$2)+5*diff(yp(x),x)+6*yp(x);
```

$$LHSeqn := Ae^{-2x}$$

As we see in the following, $A = 3$, so $y_p(x) = 3xe^{-2x}$.

```
> match(3*exp(-2*x)=LHSeqn,x,'Val'):
  Val;
```

$$\{A = 3\}$$

```
> assign(Val):
  yp(x);
```

$$3xe^{-2x}$$

Therefore, a general solution of $y'' + 5y' + 6y = 3e^{-2x}$ is

$$y(x) = y_h(x) + y_p(x) = c_1e^{-2x} + c_2e^{-3x} + 3xe^{-2x}.$$

As in (a), we can use **dsolve** to obtain equivalent results. For example, entering

```
> y:='y':
  Gen_Sol:=dsolve({diff(y(x),x$2)+5*diff(y(x),x)+
  6*y(x)=3*exp(-2*x),y(0)=y0,D(y)(0)=y1},y(x));
```

$$Gen\_Sol := y(x) = 3xe^{-2x} - 3e^{-2x} + (3\,y0 + y1)e^{-2x} + (3 - 2\,y0 - y1)e^{-3x}$$

solves the equation subject to the initial conditions $y(0) = y_0$ and $y'(0) = y_1$ and names the resulting output `Gen_Sol`. Then entering

```
> assign(Gen_Sol):
  to_plot:={seq(seq(subs({y0=i,y1=j},y(x)),
  i=0..2),j=-1..1)}:
  plot(to_plot,x=-1/2..3/2);
```

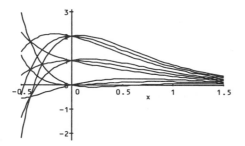

first assigns $y(x)$ the result obtained in `Gen_Sol`; defines `to_plot` to be the set of nine functions corresponding to solutions of $y'' + 5y' + 6y = 3e^{-2x}$ that satisfy the initial conditions $y(0) = i$ and $y'(0) = j$ for $i = 0, 1,$ and 2 and $j = -1, 0,$ and 1; and then graphs the set of functions `to_plot` on the interval $\left[-\dfrac{1}{2}, \dfrac{3}{2} \right]$.

In order to solve an initial-value problem, first determine the general solution and then use the initial conditions to solve for the unknown constants in the general solution.

EXAMPLE: Solve $y''' + 4y'' + 14y' + 20y = 10e^{-2x} - e^{-x} \cos 3x$ subject to the initial conditions $y(0) = 5, y'(0) = 0,$ and $y''(0) = -\dfrac{1}{2}$.

SOLUTION: First, we solve the corresponding homogeneous equation $y''' + 4y'' + 14y' + 20y = 0$, naming the result `Hom_Sol`.

```
> y:='y':
  Hom_Sol:=dsolve(diff(y(x),x$3)+4*diff(y(x),x$2)+
  14*diff(y(x),x)+20*y(x)=0,y(x));
```

$Hom\_Sol := y(x) = \_C1\, e^{-2x} + \_C2\, e^{-x} \sin(3x) + \_C3\, e^{-x} \cos(3x)$

In this case, $g(x) = 10e^{-2x} - e^{-x} \cos 3x$. The set of functions associated with $10e^{-2x}$ is $S_1 = \{e^{-2x}\}$, while the set of functions associated with $e^{-x} \cos 3x$ is $S_2 = \{e^{-x} \cos 3x, e^{-x} \sin 3x\}$. However, we note that both e^{-2x} and $e^{-x} \cos x$ appear in the solution of the corresponding homogeneous equation. Since xe^{-2x} does not appear in the solution of $y''' + 4y'' + 14y' + 20y = 0$, we take $S_1' = \{xe^{-2x}\}$. Similarly, we take $S_2' = \{xe^{-x} \cos 3x, xe^{-x} \sin 3x\}$. Thus, we must find numbers A, B, and C so that $y_p(x) = Axe^{-2x} + xe^{-x}(B \cos 3x + C \sin 3x)$ is a particular solution of the nonhomogeneous equation. After defining yp,

```
> A:='A':B:='B':C:='C':
  yp:=x->A*x*exp(-2*x)+x*exp(-x)*(B*cos(3*x)+C*sin(3*x));
```

$yp := x \rightarrow Axe^{-2x} + xe^{-x}(B \cos(3x) + C \sin(3x))$

we compute and simplify $y''' + 4y'' + 14y' + 20y$.

```
> LHSeqn:=simplify(diff(yp(x),x$3)+4*diff(yp(x),x$2)+
  14*diff(yp(x),x)+20*yp(x));
```

$LHSeqn := 10Ae^{-2x} - 18e^{-x}B \cos(3x) - 18e^{-x}C \sin(3x) - 6e^{-x}B \sin(3x) + 6e^{-x}C \cos(3x)$

LHSeqn must be the same as $10e^{-2x} - e^{-x} \cos 3x$ for all values of x. Equating coefficients results in the system of equations

$$\begin{cases} 10A = 10 \\ -18B + 6C = -1 \\ -18C - 6B = 0 \end{cases}.$$

We now solve this system of equations with solve and name the resulting solutions Vals.

```
> Vals:=solve({10*A=1,-18*B+6*C=-1,-18*C-6*B=0});
```

$Vals := \left\{ A = \dfrac{1}{10}, C = -\dfrac{1}{60}, B = \dfrac{1}{20} \right\}$

We then assign A, B, and C the values obtained in Vals with assign and evaluate yp(x). The result is a particular solution of the equation.

```
> assign(Vals):
  yp(x);
```

$$\frac{1}{10}xe^{-2x} + xe^{-x}\left(\frac{1}{20}\cos(3x) - \frac{1}{60}\sin(3x)\right)$$

Thus, a general solution is given by

$$y(x) = c_1e^{-2x} + e^{-x}(c_2\cos 3x + c_3\sin 3x) + y_p(x)$$

$$= c_1e^{-2x} + e^{-x}(c_2\cos 3x + c_3\sin 3x) + \frac{1}{10}xe^{-2x} + xe^{-x}\left(\frac{1}{20}\cos 3x - \frac{1}{60}\sin 3x\right)$$

To find the values of c_1, c_2, and c_3 so that $y(x)$ satisfies the conditions $y(0) = 5$, $y'(0) = 0$, and $y''(0) = -\frac{1}{2}$, we solve the equations $y(0) = 5$, $y'(0) = 0$, and $y''(0) = -\frac{1}{2}$

```
> y:='y':c1:='c1':c2:='c2':c3:='c3':
  y:=x->c1*exp(-2*x)+exp(-x)*(c2*cos(3*x)+c3*sin(3*x))+
  1/10*x*exp(-2*x)+x*exp(-x)*(1/20*cos(3*x)-1/60*sin(3*x)):
```

```
> Eq1:=y(0)=5;
```

$$Eq1 := c1 + c2 = 5$$

```
> Eq2:=D(y)(0)=0;
```

$$Eq2 := -2c1 - c2 + 3c3 + \frac{3}{20} = 0$$

```
> Eq3:=(D@@2)(y)(0)=-1/2;
```

$$Eq3 := 4c1 - 8c2 - 6c3 - \frac{3}{5} = -\frac{1}{2}$$

for c_1, c_2, and c_3.

```
> Sols:=solve({Eq1,Eq2,Eq3});
```

$$Sols := \left\{c1 = \frac{249}{50}, c3 = \frac{983}{300}, c2 = \frac{1}{50}\right\}$$

Then we substitute these values into $y(x)$.

```
> assign(Sols):
  y(x);
```

$$\frac{249}{50}e^{-2x} + e^{-x}\left(\frac{1}{50}\cos(3x) + \frac{983}{300}\sin(3x)\right) + \frac{1}{10}xe^{-2x} + xe^{-x}\left(\frac{1}{20}\cos(3x) - \frac{1}{60}\sin(3x)\right)$$

We graph this solution on the interval $[0, 2]$.

```
> plot(y(x),x=0..2);
```

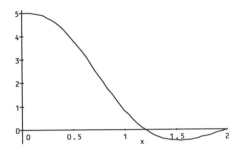

Equivalent results are obtained with **dsolve**. For example, entering

```
> y:='y':
  Gen_Sol:=dsolve(diff(y(x),x$3)+4*diff(y(x),x$2)+
    14*diff(y(x),x)+20*y(x)=
    10*exp(-2*x)-exp(-x)*cos(3*x),y(x));
```

$$Gen\_Sol := y(x) = xe^{-2x} - \frac{1}{25}e^{-x}\cos(x)^3 - \frac{3}{25}e^{-x}\cos(x)^2\sin(x) + \frac{3}{100}e^{-x}\cos(x)$$

$$+ \frac{3}{100}e^{-x}\sin(x) + \frac{1}{5}e^{-2x} - \frac{1}{120}e^{-x}\sin(3x)\cos(6x)$$

$$- \frac{1}{360}e^{-x}\sin(3x)\sin(6x) - \frac{1}{60}e^{-x}\sin(3x)x - \frac{1}{360}e^{-x}\cos(3x)\cos(6x)$$

$$+ \frac{1}{120}e^{-x}\cos(3x)\sin(6x) + \frac{1}{20}e^{-x}\cos(3x)x$$

$$+ \_C1\,e^{-2x} + \_C2\,e^{-x}\sin(3x) + \_C3\,e^{-x}\cos(3x)$$

finds a general solution of the equation, naming the resulting output **Gen_Sol**. In constrast, entering

```
> y:='y':
  Init_Sol:=dsolve(diff(y(x),x$3)+4*diff(y(x),x$2)+
  14*diff(y(x),x)+20*y(x)=10*exp(-2*x)-exp(-x)*cos(3*x),
  y(0)=y0,D(y)(0)=y1,(D@@2)(y)(0)=y2,y(x));
```

$$Init\_Sol := y(x) = xe^{-2x} - \frac{1}{25}e^{-x}\cos(x)^3 - \frac{3}{25}e^{-x}\cos(x)^2\sin(x) + \frac{3}{100}e^{-x}\cos(x)$$

$$+ \frac{3}{100}e^{-x}\sin(x) + \frac{1}{5}e^{-2x} - \frac{1}{120}e^{-x}\sin(3x)\cos(6x) - \frac{1}{360}e^{-x}\sin(3x)\sin(6x)$$

$$- \frac{1}{60}e^{-x}\sin(3x)x - \frac{1}{360}e^{-x}\cos(3x)\cos(6x) + \frac{1}{120}e^{-x}\cos(3x)\sin(6x)$$

$$+ \frac{1}{20}e^{-x}\cos(3x)x + \left(\frac{1}{100} + \frac{1}{5}y1 + \frac{1}{10}y2 + y0\right)e^{-2x}$$

$$+ \left(-\frac{31}{120} + \frac{2}{5}y1 + \frac{1}{30}y2 + \frac{2}{3}y0\right)e^{-x}\sin(3x)$$

$$+ \left(-\frac{71}{360} - \frac{1}{5}y1 - \frac{1}{10}y2\right)e^{-x}\cos(3x)$$

solves the equation $y''' + 4y'' + 14y' + 20y = 10e^{-2x} - e^{-x}\cos 3x$ subject to the initial conditions $y(0) = y_0$, $y'(0) = y_1$, and $y''(0) = y_2$. Thus, entering

```
> assign(Init_Sol):
  i_vals:={-1,2}:j_vals:={-3,0,3}:k_vals:={-2,2}:
  to_plot:={seq(seq(seq(subs({y0=i,y1=j,y2=k},y(x)),
  i=i_vals),j=j_vals),k=k_vals)}:
  plot(to_plot,x=-1/2..3/2);
```

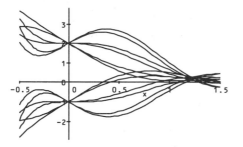

assigns $y(x)$ the result obtained in Init_Sol; defines i_vals to be the set of numbers -1 and 2, j_vals the set of numbers -3, 0, and 3, k_vals the set of numbers -2 and 2, to_plot the set of 12 functions obtained from $y(x)$ by replacing y0 by each number in i_vals, y1 by each number in j_vals, and y2 by each number in k_vals; and then graphs the set of functions to_plot on the interval $\left[-\frac{1}{2}, \frac{3}{2}\right]$.

4.5 NONHOMOGENEOUS EQUATIONS WITH CONSTANT COEFFICIENTS: VARIATION OF PARAMETERS

☰ *Second-Order Equations*

We know that in order to solve the second-order homogeneous linear differential equation

$$a_2(x)\, y'' + a_1(x)\, y' + a_0(x)\, y = 0$$

we need two linearly independent solutions $y_1(x)$ and $y_2(x)$. A general solution is then given by $y(x) = c_1 y_1(x) + c_2 y_2(x)$. Using the Method of Variation of Parameters to solve the nonhomogeneous equation

$$a_2(x)\, y'' + a_1(x)\, y' + a_0(x)\, y = g(x),$$

we assume that the particular solution has a form similar to the general solution by "varying" the parameters c_1 and c_2. Therefore, we let

$$y_p(x) = u_1(x)\, y_1(x) + u_2(x)\, y_2(x),$$

where $y_1(x)$ and $y_2(x)$ are solutions of the corresponding homogeneous equation.

```
> y:='y':y1:='y1':y2:='y2':u1:='u1':u2:='u2':
  p:='p':q:='q':f:='f':yp:='yp':
  yp:=x->y1(x)*u1(x)+y2(x)*u2(x);
```

$$yp := x \longrightarrow y1(x)\, u1(x) + y2(x)\, u2(x)$$

We need two equations to determine the two unknown functions $u_1(x)$ and $u_2(x)$. We obtain one equation by substituting $y_p(x) = u_1(x)\, y_1(x) + u_2(x)\, y_2(x)$ into the nonhomogeneous differential equation $a_2(x)\, y'' + a_1(x)\, y' + a_0(x)\, y = g(x)$. Differentiating this function, we obtain

$$y_p'(x) = u_1(x)\, y_1'(x) + u_1'(x)\, y_1(x) + u_2(x)\, y_2'(x) + u_2'(x)\, y_2(x),$$

which can be simplified to

$$y_p'(x) = u_1(x)\, y_1'(x) + u_2(x)\, y_2'(x)$$

with the assumption that $u_1'(x)\, y_1(x) + u_2'(x)\, y_2(x) = 0$.

```
> first:=diff(yp(x),x);
```

$$first := \left(\frac{\partial}{\partial x} y1(x)\right) u1(x) + y1(x)\left(\frac{\partial}{\partial x} u1(x)\right) + \left(\frac{\partial}{\partial x} y2(x)\right) u2(x) + y2(x)\left(\frac{\partial}{\partial x} u2(x)\right)$$

```
> step_1:=subsop(2=0,3=0,first);
```

$$step\_1 := \left(\frac{\partial}{\partial x}y1(x)\right)u1(x) + y2(x)\left(\frac{\partial}{\partial x}u2(x)\right)$$

This becomes our second equation for $u_1(x)$ and $u_2(x)$. The second derivative is $y_p''(x) = u_1(x)\,y_1''(x) + u_1'(x)\,y_1'(x) + u_2(x)\,y_2''(x) + u_2'(x)\,y_2'(x)$. Substitution into $a_2(x)\,y'' + a_1(x)\,y' + a_0(x)\,y = g(x)$ then yields

$$a_2(x)\,y'' + a_1(x)\,y' + a_0(x)\,y = u_1(x)[a_0(x)\,y_1''(x) + a_1(x)\,y_1'(x) + a_2(x)\,y_1(x)]$$
$$+ u_2(x)[a_0(x)\,y_2''(x) + a_1(x)\,y_2'(x) + a_2(x)\,y_2(x)]$$
$$+ u_1'(x)\,y_1'(x) + u_2'(x)\,y_2'(x)$$
$$= u_1'(x)\,y_1'(x) + u_2'(x)\,y_2'(x)$$

since $y_1(x)$ and $y_2(x)$ are solutions of the corresponding homogeneous equation. Therefore, our second equation for determining $u_1(x)$ and $u_2(x)$ is $u_1'(x)\,y_1'(x) + u_2'(x)\,y_2'(x) = f(x)$.

```
> second:=diff(step_1,x);
```

$$second := \left(\frac{\partial^2}{\partial x^2}y1(x)\right)u1(x) + \left(\frac{\partial}{\partial x}y1(x)\right)\left(\frac{\partial}{\partial x}u1(x)\right) + \left(\frac{\partial}{\partial x}y2(x)\right)\frac{\partial}{\partial x}u2(x)$$

$$+ y2(x)\frac{\partial^2}{\partial x^2}u2(x)$$

```
> step_2:=expand(second+p(x)*step_1+q(x)*yp(x));
```

$$step\_2 := \left(\frac{\partial^2}{\partial x^2}y1(x)\right)u1(x) + \left(\frac{\partial}{\partial x}y1(x)\right)\left(\frac{\partial}{\partial x}u1(x)\right) + \left(\frac{\partial}{\partial x}y2(x)\right)\left(\frac{\partial}{\partial x}u2(x)\right)$$

$$+ y2(x)\left(\frac{\partial^2}{\partial x^2}u2(x)\right) + p(x)\left(\frac{\partial}{\partial x}y1(x)\right)u1(x) + p(x)\,y2(x)\left(\frac{\partial}{\partial x}u2(x)\right)$$

$$+ q(x)\,y1(x)\,u1(x) + q(x)\,y2(x)\,u2(x)$$

```
> step_3:=collect(step_2,[u1(x),u2(x)]);
```

$$step\_3 := \left(\left(\frac{\partial^2}{\partial x^2}y1(x)\right) + p(x)\left(\frac{\partial}{\partial x}y1(x)\right) + q(x)\,y1(x)\right)u1(x) + \left(\frac{\partial}{\partial x}y1(x)\right)\left(\frac{\partial}{\partial x}u1(x)\right)$$

$$+ \left(\frac{\partial}{\partial x}y2(x)\right)\left(\frac{\partial}{\partial x}u2(x)\right) + y2(x)\left(\frac{\partial^2}{\partial x^2}u2(x)\right)$$

$$+ p(x)\,y2(x)\left(\frac{\partial}{\partial x}u2(x)\right) + q(x)\,y2(x)\,u2(x)$$

```
> array([seq([i,op(i,step_3)],i=1..6)]);
```

$$\begin{bmatrix} 1 & \left(\left(\frac{\partial^2}{\partial x^2}y1(x)\right) + p(x)\left(\frac{\partial}{\partial x}y1(x)\right) + q(x)\,y1(x)\right)u1(x) \\ 2 & \left(\frac{\partial}{\partial x}y1(x)\right)\left(\frac{\partial}{\partial x}u1(x)\right) \\ 3 & \left(\frac{\partial}{\partial x}y2(x)\right)\left(\frac{\partial}{\partial x}u2(x)\right) \\ 4 & y2(x)\left(\frac{\partial^2}{\partial x^2}u2(x)\right) \\ 5 & p(x)\,y2(x)\left(\frac{\partial}{\partial x}u2(x)\right) \\ 6 & q(x)\,y2(x)\,u2(x) \end{bmatrix}$$

```
> step_4:=subsop(1=0,4=0,5=0,6=0,step_3);
```

$$step\_4 := \left(\frac{\partial}{\partial x}y1(x)\right)\left(\frac{\partial}{\partial x}u1(x)\right) + \left(\frac{\partial}{\partial x}y2(x)\right)\left(\frac{\partial}{\partial x}u2(x)\right)$$

Hence, we have the system

$$\begin{cases} u_1'(x)\,y_1(x) + u_2'(x)\,y_2(x) = 0 \\ u_1'(x)\,y_1'(x) + u_2'(x)\,y_2'(x) = f(x) \end{cases},$$

and this is written in matrix form as $\begin{pmatrix} y_1(x) & y_2(x) \\ y_1'(x) & y_2'(x) \end{pmatrix}\begin{pmatrix} u_1'(x) \\ u_2'(x) \end{pmatrix} = \begin{pmatrix} 0 \\ f(x) \end{pmatrix}$. In linear algebra, we learn

that this system has a unique solution if and only if $\begin{vmatrix} y_1(x) & y_2(x) \\ y_1'(x) & y_2'(x) \end{vmatrix} \neq 0$. Notice that this determinant

is the Wronskian of the set $S = \{y_1(x), y_2(x)\}$, $W(S)$. We stated in Section 4.1 that $W(S) \neq 0$ if the functions $y_1(x)$ and $y_2(x)$ in the set S are linearly independent. Since $S = \{y_1(x), y_2(x)\}$ represents a fundamental set of solutions of the corresponding homogeneous equation, $W(S) \neq 0$. Hence, this system has a unique solution that can be found with Cramer's Rule to be

$$u_1'(x) = \frac{\begin{vmatrix} 0 & y_2(x) \\ f(x) & y_2'(x) \end{vmatrix}}{W(S)} = \frac{-y_2(x)\,f(x)}{W(S)}$$

and

$$u_2'(x) = \frac{\begin{vmatrix} y_1(x) & 0 \\ y_1'(x) & f(x) \end{vmatrix}}{W(S)} = \frac{y_1(x)\,f(x)}{W(S)},$$

where $S = \{y_1(x), y_2(x)\}$ is the fundamental set of solutions of the corresponding homogeneous equation. Integrating, we then obtain

$$u_1(x) = \int \frac{-y_2(x) f(x)}{W(S)} \, dx \quad \text{and} \quad u_2(x) = \int \frac{y_1(x) f(x)}{W(S)} \, dx,$$

so a general solution of the nonhomogeneous equation $a_2(x) y'' + a_1(x) y' + a_0(x) y = g(x)$ is

$$y(x) = y_h(x) + y_p(x) = c_1 y_1(x) + c_2 y_2(x) + u_1(x) y_1(x) + u_2(x) y_2(x).$$

Note that the problems that were solved in the preceding sections using annihilators or the Method of Undetermined Coefficients can be solved by Variation of Parameters as well.

EXAMPLE: Solve $y'' + \dfrac{1}{4}y = \sec\left(\dfrac{x}{2}\right) + \csc\left(\dfrac{x}{2}\right), 0 < x < \pi.$

SOLUTION: Since the characteristic equation of the corresponding homogeneous equation is $m^2 + \dfrac{1}{4} = 0$ with roots $\pm\dfrac{i}{2}$, $y_h(x) = c_1 \cos\left(\dfrac{x}{2}\right) + c_2 \sin\left(\dfrac{x}{2}\right)$. Hence,

$$S = \{y_1(x), y_2(x)\} = \left\{\cos\left(\frac{x}{2}\right), \sin\left(\frac{x}{2}\right)\right\}$$

and

$$W(S) = \begin{vmatrix} \cos\left(\dfrac{x}{2}\right) & \sin\left(\dfrac{x}{2}\right) \\ -\dfrac{1}{2}\sin\left(\dfrac{x}{2}\right) & \dfrac{1}{2}\cos\left(\dfrac{x}{2}\right) \end{vmatrix} = \frac{1}{2}\cos^2\left(\frac{x}{2}\right) + \frac{1}{2}\sin^2\left(\frac{x}{2}\right) = \frac{1}{2}.$$

We therefore have the following calculations.

$$u_1(x) = \int \frac{-\sin\left(\frac{x}{2}\right)\left(\sec\left(\frac{x}{2}\right) + \csc\left(\frac{x}{2}\right)\right)}{\frac{1}{2}} \, dx = -2\int\left(\frac{\sin\left(\frac{x}{2}\right)}{\cos\left(\frac{x}{2}\right)} + 1\right) dx = -2x + 4\ln\left|\cos\left(\frac{x}{2}\right)\right|$$

and

$$u_2(x) = \int \frac{\cos\left(\frac{x}{2}\right)\left(\sec\left(\frac{x}{2}\right) + \csc\left(\frac{x}{2}\right)\right)}{\frac{1}{2}} = 2\int\left(1 + \frac{\cos\left(\frac{x}{2}\right)}{\sin\left(\frac{x}{2}\right)} + \right) dx = 2x + 4\ln\left|\sin\left(\frac{x}{2}\right)\right|.$$

Then, by Variation of Parameters,

$$y_p(x) = \cos\left(\frac{x}{2}\right)\left[-2x + 4\ln\left|\cos\left(\frac{x}{2}\right)\right|\right] + \sin\left(\frac{x}{2}\right)\left[2x + 4\ln\left|\sin\left(\frac{x}{2}\right)\right|\right]$$

is a particular solution of

$$y'' + \frac{1}{4}y = \sec\left(\frac{x}{2}\right) + \csc\left(\frac{x}{2}\right)$$

and

$$y = c_1 \cos\left(\frac{x}{2}\right) + c_2 \sin\left(\frac{x}{2}\right) + y_p(x)$$

$$= c_1 \cos\left(\frac{x}{2}\right) + c_2 \sin\left(\frac{x}{2}\right) + \cos\left(\frac{x}{2}\right)\left[-2x + 4\ln\left(\cos\left(\frac{x}{2}\right)\right)\right] + \sin\left(\frac{x}{2}\right)\left[2x + 4\ln\left(\sin\left(\frac{1}{2}x\right)\right)\right]$$

is a general solution. Note that $\cos\left(\frac{x}{2}\right) > 0$ and $\sin\left(\frac{x}{2}\right) > 0$ on $0 < x < \pi$, so the absolute value signs can be eliminated.

The **dsolve** command is equally successful in solving the equation. For example, entering

```
> y:='y':
  Sol:=dsolve(diff(y(x),x$2)+1/4*y(x)=sec(x/2)+csc(x/2),y(x));
```

$$Sol := y(x) = 2\sin\left(\frac{1}{2}x\right)x + 4\sin\left(\frac{1}{2}x\right)\ln\left(\sin\left(\frac{1}{2}x\right)\right) + 4\cos\left(\frac{1}{2}x\right)\ln\left(\cos\left(\frac{1}{2}x\right)\right)$$
$$- 2\cos\left(\frac{1}{2}x\right)x + \_C1\sin\left(\frac{1}{2}x\right) + \_C2\cos\left(\frac{1}{2}x\right)$$

finds a general solution of the equation, naming the result **Sol**. Then entering

```
> assign(Sol):
  to_plot:={seq(seq(subs({_C1=i,_C2=j},y(x),i=-1..1),
   j=-1..1)}:
  plot(to_plot,x=0..Pi);
```

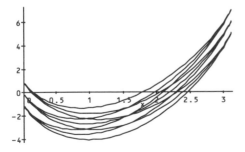

assigns $y(x)$ the solution found in **Sol**, defines **to_plot** to be the set of nine functions obtained by replacing **_C1** in $y(x)$ by -1, 0, and 1 and **_C2** in $y(x)$ by -1, 0, and 1, and then graphs the set of functions **to_plot** on the interval $[0, \pi]$.

EXAMPLE: Find a general solution of $y'' + 4y' + 13y = x\cos^2 3x$.

SOLUTION: We see that **dsolve** finds a general solution of the equation.

```
> y:='y':
  Sol:=dsolve(diff(y(x),x$2)+4*diff(y(x),x)+13*y(x)=
  x*cos(3*x)^2,y(x));
```

$$Sol := y(x) = \frac{1}{39}\sin(3x)\cos(3x)x + \frac{5}{1014}\sin(3x)\cos(3x) + \frac{3}{52}x - \frac{1}{26}x\cos(3x)^2$$
$$- \frac{3}{169} + \frac{2}{169}\cos(3x)^2 + \frac{1}{510}\sin(3x)\cos(9x)x + \frac{77}{86700}\sin(3x)\cos(9x)$$
$$+ \frac{3}{340}\sin(3x)\sin(9x)x - \frac{3}{7225}\sin(3x)\sin(9x) + \frac{3}{340}\cos(3x)\cos(9x)x$$
$$- \frac{3}{7225}\cos(3x)\cos(9x) - \frac{1}{510}\cos(3x)\sin(9x)x - \frac{77}{86700}\cos(3x)\sin(9x)$$
$$+ \_C1\,e^{-2x}\sin(3x) + \_C2\,e^{-2x}\cos(3x)$$

Alternatively, we can use Maple to help us implement the Variation of Parameters procedure. We begin by clearing all prior definitions, if any, of the variables we will use.

```
> f:='f':y:='y':y1:='y1':y2:='y2':yh:='yh':
  yp:='yp':u1:='u1':u2:='u2':
```

The characteristic equation of the associated homogeneous equation is $m^2 + 4m + 13 = 0$, which is solved with **solve**.

```
> solve(m^2+4*m+13=0);
```

$$-2 + 3I, -2 - 3I$$

The results mean that a general solution of $y'' + 4y' + 13y = 0$ is $y_h(x) = e^{-2x}(c_1\cos 3x + c_2\sin 3x)$, so a fundamental set of solutions is $S = \{y_1(x), y_2(x)\} = \{e^{-2x}\cos 3x, e^{-2x}\sin 3x\}$. Next, we define $f(x)$, $y_1(x)$, and $y_2(x)$

```
> f:=x->x*cos(3*x)^2;
```

$$f := x \to x\cos(3x)^2$$

```
> y1:=x->exp(-2*x)*cos(3*x):
  y2:=x->exp(-2*x)*sin(3*x):
```

and then compute and simplify the Wronskian of the set of functions $S = \{e^{-2x}\cos 3x, e^{-2x}\sin 3x\}$ with **Wronskian**, naming the result **wr**.

```
> with(linalg):
  Wr:=Wronskian([y1(x),y2(x)],x);
```

$$Wr := [e^{-2x}\cos(3x) \; e^{-2x}\sin(3x)]$$
$$[-2e^{-2x}\cos(3x) - 3e^{-2x}\sin(3x), -2e^{-2x}\sin(3x) + 3e^{-2x}\cos(3x)]$$

```
> wr:=det(Wr);
```

$$wr := 3(e^{-2x})\cos(3x)^2 + 3(e^{-2x})\sin(3x)^2$$

```
> wr:=simplify(wr);
```

$$wr := 3e^{-4x}$$

To calculate $u_1(x) = \displaystyle\int \frac{-y_2(x)\,f(x)}{3e^{-4x}}\,dx$, we first define **u1prime** to be $\dfrac{-y_2(x)\,f(x)}{3e^{-4x}}$

```
> u1prime:=simplify(-y2(x)*f(x)/wr);
```

$$u1prime := -\frac{1}{3}\sin(3x)\,x\cos(3x)^2 e^{2x}$$

and then use **int** to define $u_1(x) = \displaystyle\int \frac{-y_2(x)\,f(x)}{3e^{-4x}}\,dx$.

```
> u1:=x->int(u1prime,x):
  u1(x);
```

$$-\frac{1}{12}\left(-\frac{9}{85}x + \frac{36}{7225}\right)e^{2x}\cos(9x) - \frac{1}{12}\left(\frac{2}{85}x + \frac{77}{7225}\right)e^{2x}\sin(9x)$$

$$-\frac{1}{12}\left(-\frac{3}{13}x + \frac{12}{169}\right)e^{2x}\cos(3x) - \frac{1}{12}\left(\frac{2}{13}x + \frac{5}{169}\right)e^{2x}\sin(3x)$$

We define $u_2(x) = \displaystyle\int \frac{y_1(x)\,f(x)}{3e^{-4x}}\,dx$ in the exact same way.

```
> u2prime:=simplify(y1(x)*f(x)/wr);
```

$$u2prime := \frac{1}{3}\cos(3x)^3\,xe^{2x}$$

```
> u2:=x->int(u2prime,x):
  u2(x);
```

$$\frac{1}{4}\left(\frac{2}{13}x + \frac{5}{169}\right)e^{2x}\cos(3x) - \frac{1}{4}\left(-\frac{3}{13}x + \frac{12}{169}\right)e^{2x}\sin(3x)$$

$$+ \frac{1}{12}\left(\frac{2}{85}x + \frac{77}{7225}\right)e^{2x}\cos(9x) - \frac{1}{12}\left(-\frac{9}{85}x + \frac{36}{7225}\right)e^{2x}\sin(9x)$$

Then a particular solution of $y'' + 4y' + 13y = x\cos^2 3x$ is given by $y_p(x) = u_1(x)y_1(x) + u_2(x)y_2(x)$,

```
> yp:=x->y1(x)*u1(x)+y2(x)*u2(x):
  yp(x);
```

$$e^{-2x}\cos(3x)\left(-\frac{1}{12}\left(-\frac{9}{85}x + \frac{36}{7225}\right)e^{2x}\cos(9x) - \frac{1}{12}\left(\frac{2}{85}x + \frac{77}{7225}\right)e^{2x}\sin(9x)\right.$$

$$\left.- \frac{1}{12}\left(-\frac{3}{13}x + \frac{12}{169}\right)e^{2x}\cos(3x) - \frac{1}{12}\left(\frac{2}{13}x + \frac{5}{169}\right)e^{2x}\sin(3x)\right)$$

$$+ e^{-2x}\sin(3x)\left(\frac{1}{4}\left(\frac{2}{13}x + \frac{5}{169}\right)e^{2x}\cos(3x) - \frac{1}{4}\left(-\frac{3}{13}x + \frac{12}{169}\right)e^{2x}\sin(3x)\right.$$

$$\left.+ \frac{1}{12}\left(\frac{2}{85}x + \frac{77}{7225}\right)e^{2x}\cos(9x) - \frac{1}{12}\left(-\frac{9}{85}x + \frac{36}{7225}\right)e^{2x}\sin(9x)\right)$$

and a general solution of the corresponding homogeneous equation is given by $y_h(x) = e^{-2x}(c_1\cos 3x + c_2\sin 3x)$, which is defined as follows:

```
> yh:=x->c[1]*y1(x)+c[2]*y2(x):
  yh(x);
```

$$c_{[1]}e^{-2x}\cos(3x) + c_{[2]}e^{-2x}\sin(3x)$$

Thus, a general solution of the equation is given by $y(x) = y_h(x) + y_p(x)$.

```
> y:=x->yp(x)+yh(x):
  y(x);
```

$$e^{-2x}\cos(3x)\left(-\frac{1}{12}\left(-\frac{9}{85}x + \frac{36}{7225}\right)e^{2x}\cos(9x) - \frac{1}{12}\left(\frac{2}{85}x + \frac{77}{7225}\right)e^{2x}\sin(9x)\right.$$

$$\left.- \frac{1}{12}\left(-\frac{3}{13}x + \frac{12}{169}\right)e^{2x}\cos(3x) - \frac{1}{12}\left(\frac{2}{13}x + \frac{5}{169}\right)e^{2x}\sin(3x)\right)$$

$$+ e^{-2x}\sin(3x)\left(\frac{1}{4}\left(\frac{2}{13}x + \frac{5}{169}\right)e^{2x}\cos(3x) - \frac{1}{4}\left(-\frac{3}{13}x + \frac{12}{169}\right)e^{2x}\sin(3x)\right.$$

$$\left.+ \frac{1}{12}\left(\frac{2}{85}x + \frac{77}{7225}\right)e^{2x}\cos(9x) - \frac{1}{12}\left(-\frac{9}{85}x + \frac{36}{7225}\right)e^{2x}\sin(9x)\right)$$

$$+ c_{[1]}e^{-2x}\cos(3x) + c_{[2]}e^{-2x}\sin(3x)$$

In order to graph various solutions corresponding to different values of c_1 and c_2, we first define `i_vals` to be the set of numbers $-5, 0$, and 5 and `j_vals` to be the set of numbers $-4, -2, 0$, 2, and 4. Then the set of 15 functions `to_graph` is created by replacing `c[1]`, corresponding to c_1, by each number in `i_vals` and replacing `c[2]`, corresponding to c_2, by each number in `j_vals`.

```
> i_vals:={-5,0,5}:j_vals:={-4,-2,0,2,4}:
  to_graph:={seq(seq(subs({c[1]=i,c[2]=j},y(x)),
  i=i_vals),j=j_vals)}:
```

The set of functions `to_graph` is then graphed on the interval $[-1, 1]$ with `plot`. Note that the option `view=[-1..1,-20..20]` instructs Maple that the numbers displayed on the x-axis (the horizontal axis) correspond to the interval $[-1, 1]$, while the numbers displayed on the y-axis (the vertical axis) correspond to the interval $[-20, 20]$.

```
> plot(to_graph,x=-1..1,view=[-1..1,-20..20]);
```

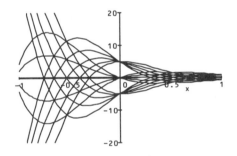

Summary of Variation of Parameters for Second-Order Equations

Given the second-order equation $ay'' + by' + cy = f(x)$,

1. Find a general solution $y_h = c_1 y_1 + c_2 y_2$ of the associated homogeneous equation $ay'' + by' + cy = 0$.

2. Let $W = \begin{vmatrix} y_1 & y_2 \\ y_1' & y_2' \end{vmatrix}$.

3. Let $u_1' = \dfrac{-y_2 f(x)}{W}$ and $u_2' = \dfrac{y_1 f(x)}{W}$.

4. Integrate to obtain u_1 and u_2.

5. A particular solution of $ay'' + by' + cy = f(x)$ is given by $y_p = u_1 y_1 + u_2 y_2$.

6. A general solution of $ay'' + by' + cy = f(x)$ is given by $y = y_h + y_p$.

☰ *Higher-Order Nonhomogeneous Equations*

Nonhomogeneous higher-order linear equations with constant coefficients can be solved through Variation of Parameters as well. In general, if we are given the nonhomogeneous equation

$$y^{(n)}(x) + a_{n-1}(x)\, y^{(n-1)}(x) + \cdots + a_1(x)\, y'(x) + a_0(x)\, y(x) = \sum_{k=0}^{n} a_k(x)\, y^{(k)}(x) = g(x)$$

and a fundamental set of solutions $y_1(x), y_2(x), \ldots, y_n(x)$ of the associated homogeneous equations

$$y^{(n)}(x) + a_{n-1}(x)\, y^{(n-1)}(x) + \cdots + a_1(x)\, y'(x) + a_0(x)\, y(x) = \sum_{k=0}^{n} a_k(x)\, y^{(k)}(x) = 0,$$

then we can extend the method for second-order equations to find $u_1(x), u_2(x), \ldots, u_n(x)$ such that

$$y_p(x) = u_1(x)\, y_1(x) + u_2(x)\, y_2(x) + \cdots + u_n(x)\, y_n(x) = \sum_{i=1}^{n} u_i(x)\, y_i(x)$$

is a particular solution of the nonhomogeneous equation. If

$$u_1'(x)\, y_1(x) + u_2'(x)\, y_2(x) + \cdots + u_n'(x)\, y_n(x) = \sum_{i=1}^{n} u_i'(x)\, y_i(x) = 0,$$

then

$$y_p^{(m)}(x) = u_1(x)\, y_1^{(m)}(x) + u_2(x)\, y_2^{(m)}(x) + \cdots + u_n(x)\, y_n^{(m)}(x) = \sum_{i=1}^{n} u_i(x)\, y_i^{(m)}(x)$$

for $m = 0, 1, 2, \ldots, n - 1$, and if

$$u_1'(x)\, y_1^{(m-1)}(x) + u_2'(x)\, y_2^{(m-1)}(x) + \cdots + u_n'(x)\, y_n^{(m-1)}(x) = \sum_{i=1}^{n} u_i'(x)\, y_i^{(m-1)}(x) = 0$$

for $m = 1, 2, \ldots, n - 1$, then

$$y_p^{(n)}(x) = \sum_{i=1}^{n} u_i(x)\, y_i^{(n)}(x) + \sum_{i=1}^{n} u_i'(x)\, y_i^{(n-1)}(x).$$

Therefore, we obtain the system of n equations

$$\begin{cases} \displaystyle\sum_{i=1}^{n} y_i(x)\, u_i'(x) = 0 \\[2mm] \displaystyle\sum_{i=1}^{n} y_i'(x)\, u_i'(x) = 0 \\[2mm] \qquad\vdots \\[2mm] \displaystyle\sum_{i=1}^{n} y_i^{(n-1)}(x)\, u_i'(x) = g(x) \end{cases}$$

which, using Cramer's Rule, can be solved for $u_1'(x), u_2'(x), \ldots, u_n'(x)$.

Let $W_m(y_1(x), y_2(x), \ldots, y_n(x))$ denote the determinant of the matrix obtained by replacing the mth column of

$$\begin{pmatrix} y_1(x) & y_2(x) & \cdots & y_n(x) \\ y_1'(x) & y_2'(x) & \cdots & y_n'(x) \\ \vdots & \vdots & \vdots & \vdots \\ y_1^{(n-1)}(x) & y_2^{(n-1)}(x) & \cdots & y_n^{(n-1)}(x) \end{pmatrix}$$

by the column

$$\begin{pmatrix} 0 \\ 0 \\ \vdots \\ 0 \\ g(x) \end{pmatrix}.$$

Then, by Cramer's Rule,

$$u_i'(x) = \frac{g(x)\, W_i(y_1(x), y_2(x), \ldots, y_n(x))}{W(y_1(x), y_2(x), \ldots, y_n(x))}$$

for $i = 1, 2, \ldots, n$, and

$$u_i(x) = \int \frac{g(x)\, W_i(y_1(x), y_2(x), \ldots, y_n(x))}{W(y_1(x), y_2(x), \ldots, y_n(x))}\, dx.$$

Thus,

$$y_p(x) = \sum_{i=1}^{n} u_i(x)\, y_i(x) = \sum_{i=1}^{n} y_i(x) \int \frac{g(x)\, W_i(y_1(x), y_2(x), \ldots, y_n(x))}{W(y_1(x), y_2(x), \ldots, y_n(x))}\, dx$$

is a particular solution of the nonhomogeneous equation.

A general solution of the nonhomogeneous equation is given by $y(x) = y_h(x) + y_p(x)$, where $y_h(x)$ is a general solution of the associated homogeneous equation.

EXAMPLE: Solve $y''' + 3y'' + 2y' = \cos x$.

SOLUTION: We begin by finding a general solution of the corresponding homogeneous equation with characteristic equation $m^3 + 3m^2 + 2m = m(m + 1)(m + 2) = 0$. Hence, $y_h(x) = c_1 + c_2 e^{-x} + c_3 e^{-2x}$, and a fundamental set of solutions is $S = \{1, e^{-x}, e^{-2x}\}$. Therefore, we must solve the system

$$\begin{pmatrix} 1 & e^{-x} & e^{-2x} \\ 0 & -e^{-x} & -2e^{-2x} \\ 0 & e^{-x} & 4e^{-2x} \end{pmatrix} \begin{pmatrix} u_1'(x) \\ u_2'(x) \\ u_3'(x) \end{pmatrix} = \begin{pmatrix} 0 \\ 0 \\ \cos x \end{pmatrix},$$

where

$$W(S) = \begin{vmatrix} 1 & e^{-x} & e^{-2x} \\ 0 & -e^{-x} & -2e^{-2x} \\ 0 & e^{-x} & 4e^{-2x} \end{vmatrix} = -2e^{-3x}.$$

Using this with Cramer's Rule, we have

$$u_1'(x) = \frac{\begin{vmatrix} 0 & e^{-x} & e^{-2x} \\ 0 & -e^{-x} & -2e^{-2x} \\ \cos x & e^{-x} & 4e^{-2x} \end{vmatrix}}{-2e^{-3x}} = \frac{-e^{-3x}\cos x}{-2e^{-3x}} = \frac{1}{2}\cos x,$$

$$u_2'(x) = \frac{\begin{vmatrix} 1 & 0 & e^{-2x} \\ 0 & 0 & -2e^{-2x} \\ 0 & \cos x & 4e^{-2x} \end{vmatrix}}{-2e^{-3x}} = \frac{2e^{-2x}\cos x}{-2e^{-3x}} = -e^{x}\cos x,$$

and

$$u_3'(x) = \frac{\begin{vmatrix} 1 & e^{-x} & 0 \\ 0 & -e^{-x} & 0 \\ 0 & e^{-x} & \cos x \end{vmatrix}}{-2e^{-3x}} = \frac{-e^{-x}\cos x}{-2e^{-3x}} = \frac{1}{2}e^{2x}\cos x.$$

We can also solve the system

$$\begin{pmatrix} 1 & e^{-x} & e^{-2x} \\ 0 & -e^{-x} & -2e^{-2x} \\ 0 & e^{-x} & 4e^{-2x} \end{pmatrix} \begin{pmatrix} u_1'(x) \\ u_2'(x) \\ u_3'(x) \end{pmatrix} = \begin{pmatrix} 0 \\ 0 \\ \cos x \end{pmatrix}$$

by taking advantage of the `linsolve` command, which is contained in the `linalg` package. Generally, the command `linsolve(A,b)` returns the vector(s) `v` so that `Av = b`. If no solution exists, `NULL` is returned. After loading the `linalg` package by entering `with(linalg)`, we define `S` to be the list of functions corresponding to $S = \{1, e^{-x}, e^{-2x}\}$ and then use `Wronskian`, which is also contained in the `linalg` package, to define A to be the matrix

$$\begin{pmatrix} 1 & e^{-x} & e^{-2x} \\ 0 & -e^{-x} & -2e^{-2x} \\ 0 & e^{-x} & 4e^{-2x} \end{pmatrix}.$$

```
> with(linalg):
  S:=[1,exp(-x),exp(-2*x)]:
  A:=Wronskian(S,x);
```

$$A := \begin{bmatrix} 1 & e^{-x} & e^{-2x} \\ 0 & -e^{-x} & -2e^{-2x} \\ 0 & e^{-x} & 4e^{-2x} \end{bmatrix}$$

Next, we define **b** to be the vector

$$\begin{pmatrix} 0 \\ 0 \\ \cos x \end{pmatrix}$$

and use `linsolve` to find the vector

$$\begin{pmatrix} u_1'(x) \\ u_2'(x) \\ u_3'(x) \end{pmatrix},$$

naming the result **du**.

```
> b:=array([0,0,cos(x)]);
```

$$b := [0 \quad 0 \quad \cos(x)]$$

```
> du:=linsolve(A,b);
```

$$du := \left[\frac{1}{2}\cos(x) - \frac{\cos(x)}{3^{-x}} \frac{1}{2} \frac{\cos(x)}{e^{-2x}} \right]$$

Integration then gives us $u_1(x) = \frac{1}{2}\sin x$, $u_2(x) = \frac{1}{2}e^x(\cos x + \sin x)$, and $u_3(x) = \frac{1}{10}e^{2x}(2\cos x + \sin x)$, which we compute now using `map` and `int`.

```
> u:=map(int,du,x);
```

$$u := \left[\frac{1}{2}\sin(x) - \frac{1}{2}e^x\cos(x) - \frac{1}{2}e^x\sin(x)\frac{1}{5}e^{2x}\cos(x) + \frac{1}{10}e^{2x}\sin(x) \right]$$

Since

$$y_p(x) = u_1(x)\,y_1(x) + u_2(x)\,y_2(x) + u_3(x)\,y_3(x),$$

we find through substitution into this equation and simplification that $y_p(x) = -\frac{3}{10}\cos x + \frac{1}{10}\sin x$. We compute the same result using **dotprod**, which is contained in the **linalg** package, and **simplify**.

```
> step_1:=dotprod(S,u);
```

$$step\_1 := \frac{1}{2}\sin(x) + e^{-x}\left(-\frac{1}{2}e^x\cos(x) - \frac{1}{2}e^x\sin(x)\right) + e^{-2x}\left(\frac{1}{5}e^{2x}\cos(x) + \frac{1}{10}e^{2x}\sin(x)\right)$$

```
> yp:=x->simplify(step_1):
  yp(x);
```

$$\frac{1}{10}\sin(x) - \frac{3}{10}\cos(x)$$

Hence, a general solution is

$$y(x) = y_h(x) + y_p(x) = c_1 + c_2 e^{-x} + c_3 e^{-2x} - \frac{3}{10}\cos x + \frac{1}{10}\sin x.$$

As we have seen with previous examples, we can graph the solution for various values of the constants. For example, entering

```
> y:=x->dotprod([c1,c2,c3],S)+yp(x):
  y(x);
```

$$c1 + c2\,e^{-x} + c3\,e^{-2x} + \frac{1}{10}\sin(x) - \frac{3}{10}\cos(x)$$

defines y, and then entering

```
> to_graph:={seq(seq(seq(subs({c1=i,c2=j,c3=k},y(x)),
    i=-1..1),j=0..1),k=-1..0)}:
  plot(to_graph,x=0..2*Pi);
```

first defines **to_plot** to be the set of 12 functions obtained by replacing **c1** in $y(x)$ by i for $i = -1, 0$, and 1, **c2** in $y(x)$ by j for $j = 0$ and 1, and **c3** by k for $k = -1$ and 0, and then graphs the set of functions **to_plot** on the interval $[0, 2\pi]$.

We are often able to use **dsolve** to solve equations directly, which is especially useful when constructing a solution by traditional methods is tedious or time consuming.

EXAMPLE: Find a general solution of (a) $y'' - 4y = x^{-3}e^{-4x}$ and (b) $y^{(4)} + 8y'' + 16y = \dfrac{\sin 2x}{x}$. In each case, graph the solution for various values of the arbitrary constants.

SOLUTION: Entering

```
> y:='y':Sol:='Sol':to_graph:='to_graph':
  Sol:=dsolve(diff(y(x),x$2)-4*y(x)=exp(-4*x)/x^3,y(x));
```

$$Sol := y(x) = -\frac{1}{2}\left(-e^{-4x} + 9e^{2x}\,\text{Ei}(1,6x)\,x - e^{-2x}\,\text{Ei}(1,2x)\,x - 2\_C1\,e^{2x}x - 2\_C2\,e^{-2x}x\right)/x$$

finds a general solution of $y'' - 4y = x^{-3}e^{-4x}$ and names the result **Sol**. Note that the **Ei** function appearing in the result represents the **Exponential integral** function:

$$\text{Ei}(n,x) = \int_{t=1}^{\infty} \frac{e^{-xt}}{t^n}\,dt.$$

To graph the solution for various values of the arbitrary constants, we enter

```
> assign(Sol):
  j_vals:={-4,-2,0,2,4}:i_vals:={-2,0,2}:
  to_graph:={seq(seq(subs({_C1=i,_C2=j},y(x)),
   i=i_vals),j=j_vals)}:
```

which names $y(x)$ the solution obtained in Sol; defines j_vals to be the set of numbers $-4, -2,$ $0, 2,$ and 4; defines k_vals to be the set of numbers $-2, 0,$ and 2; and then defines to_graph to be the set of 15 functions obtained by replacing _C1 in $y(x)$ by each number in i_vals and replacing _C2 in $y(x)$ by each number in j_vals. The functions to_graph are then graphed on the interval $[0.01, 1]$ with plot. The option view=[0..1,-15..15] instructs Maple to show the interval corresponding to $[0, 1]$ on the x-axis (the horizontal axis) and the interval corresponding to $[-15, 15]$ on the y-axis (the vertical axis).

```
> plot(to_graph,x=0.01..1,view=[0..1,-15..15]);
```

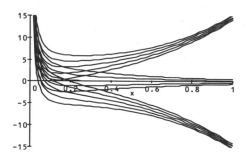

After clearing all prior definitions of y, entering

```
> y:='y':Sol:='Sol':
  Sol:=dsolve(diff(y(x),x$4)+8*diff(y(x),x$2)+16*y(x)=
  sin(2*x)/x,y(x));
```

$$Sol := y(x) = \frac{1}{32} \sin(2x)\, \mathrm{Si}(4x) - \frac{1}{8} \sin(2x) \sin(x) \cos(x)^3 + \frac{1}{16} \sin(x) \cos(x) \sin(2x)$$

$$+ \frac{1}{16} \sin(2x)\, x - \frac{1}{8} \cos(2x) \cos(x)^4 + \frac{1}{8} \cos(2x) \cos(x)^2 + \frac{1}{32} \cos(2x)\, \mathrm{Ci}(4x)$$

$$- \frac{1}{32} \cos(2x) \ln(x) + \frac{1}{16} \sin(2x)\, x\, \mathrm{Ci}(4x) - \frac{1}{16} \sin(2x)\, x \ln(x)$$

$$- \frac{1}{16} \mathrm{Si}(4x) \cos(2x) + \_C1 \sin(2x) + \_C2 \cos(2x) + \_C3 \sin(2x)\, x$$

$$+ \_C4 \cos(2x)\, x$$

finds a general solution of $y^{(4)} + 8y'' + 16y = \dfrac{\sin 2x}{x}$. Note that the functions Si and Ci appearing in the result represent the **Sine integral** and **Cosine integral** functions, respectively:

$$\mathsf{Si(x)} = \int_0^x \frac{\sin t}{t}\, dt$$

and, for real values of x,

$$\mathsf{Ci(x)} = \gamma + \ln x + \int_0^x \frac{\cos t - 1}{t}\, dt.$$

We graph the solutions for various values of the constants in the same manner as in (a).

```
> assign(Sol):
  i_vals:={0,4}:j_vals:={-3,0}:k_vals:={0,2}:l_vals:={-2,0}:
  to_graph:={seq(seq(seq(seq(subs({_C1=i,_C2=j,_C3=k,
   _C4=l},y(x)),i=i_vals),j=j_vals),k=k_vals),l=l_vals)}:
```

```
> plot(to_graph,x=0.01..2);
```

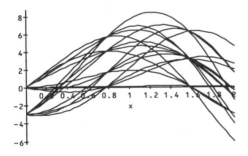

5

Applications of Higher-Order Differential Equations

In Chapter 4, we discussed several techniques for solving higher-order differential equations. In this chapter, we illustrate how some of these methods can be used to solve initial-value problems that model physical situations.

5.1 SIMPLE HARMONIC MOTION

Suppose that a mass is attached to an elastic spring that is suspended from a rigid support such as a ceiling. The mass causes the spring to stretch to a distance s from its natural position. The position in which it comes to rest is called the equilibrium position. According to Hooke's law, the spring exerts a restoring force in the upward direction that is proportional to the displacement of the spring. Mathematically, Hooke's law is stated as

DEFINITION	*Hooke's Law*
	$$F = ks,$$ where $k > 0$ is the constant of proportionality, or **spring constant**, and s is the displacement of the spring.

A spring has natural length b. When the mass is attached to the spring, it is stretched s units past its natural length to the equilibrium position $x = 0$. When the system is put into motion, the distance from $x = 0$ at time t is given by $x(t)$.

By Newton's Second Law of Motion, $F = ma = m\dfrac{d^2x}{dt^2}$, where m represents mass and a represents acceleration. If we assume that there are no other forces acting on the mass, then we determine the differential equation that models this situation in the following way:

$$m\frac{d^2x}{dt^2} = \sum (\text{forces acting on the system})$$
$$= -k(s + x) + mg$$
$$= -ks - kx + mg.$$

At equilibrium $ks = mg$, so after simplification, we obtain the differential equation

$$m\frac{d^2x}{dt^2} = -kx \quad \text{or} \quad m\frac{d^2x}{dt^2} + kx = 0.$$

The two initial conditions used with this problem are the initial position $x(0) = \alpha$ and initial velocity $\dfrac{dx}{dt}(0) = \beta$. Hence, the function $x(t)$ that describes the displacement of the mass with respect to the equilibrium position is found by solving the initial-value problem

$$\begin{cases} m\dfrac{d^2x}{dt^2} + kx = 0 \\[2mm] x(0) = \alpha, \dfrac{dx}{dt}(0) = \beta \end{cases}$$

The solution $x(t)$ to this problem represents the position of the mass at time t. Based on the assumptions made in deriving the differential equation (the positive direction is down), positive values of $x(t)$ indicate that the mass is beneath the equilibrium position, while negative values of $x(t)$ indicate that the mass is above the equilibrium position. We investigate solutions to this initial-value problem under varying conditions. The units we will encounter in these problems are as follows.

System	Force	Mass	Length	k (Spring Constant)	Time
English	pounds (lb)	slugs	feet (ft)	lb/ft	seconds (sec)
Metric	newtons (N)	kilograms (kg)	meters (m)	N/m	seconds

EXAMPLE: A mass weighing 60 lb stretches a spring 6 in. Determine the function $x(t)$ that describes the motion of the mass if the mass is released with zero initial velocity 12 in below the equilibrium position.

SOLUTION: First, the spring constant k must be determined from the given information. By Hooke's law, $F = ks$, so we have $60 = k(0.5)$. Therefore, $k = 120$ lb/ft. Next, the mass m must be determined using $F = mg$. In this case, $60 = m(32)$, so $m = 15/8$ slugs. Since $k/m = 64$ and 12 in is equivalent to 1 ft, the initial-value problem that needs to be solved is

$$\begin{cases} \dfrac{d^2x}{dt^2} + 64x = 0 \\[2mm] x(0) = 1, \dfrac{dx}{dt}(0) = 0. \end{cases}$$

This problem is now solved with **dsolve**, and the resulting output is named **DE1**. $x(t)$ is named the result obtained in **DE1** with **assign**.

```
> x:='x':
  DE1:=dsolve({diff(x(t),t$2)+64*x(t)=0,
   x(0)=1,D(x)(0)=0},x(t));
```

$DE1 := x(t) = \cos(8t)$

```
> assign(DE1):
  x(t);
```

$\cos(8t)$

We then graph the solution on the interval $[0, \pi/2]$ with **plot**.

```
> plot(x(t),t=0..Pi/2);
```

In order to better understand the relationship between the formula obtained in this example and the motion of the mass on the spring, an alternate approach is taken here. We begin by assigning n and eps the values 15 and 0.1.

```
> n:=15:
  eps:=0.1:
```

Next, we define the procedure spring. Given t0, the procedure spring first declares the variables xt0 and pts local to the procedure spring, then defines xt0 to be the value of $x(t)$ if $t = $ t0, pts to be the list of points corresponding to $(0, xt0)$, $\left(-eps, xt0 + \dfrac{1-xt0}{n}\right)$, $\left(eps, xt0 + 2 - \dfrac{1-xt0}{n}\right)$, ..., $\left(eps(-1)^{n-1}, xt0 + (n-1)\dfrac{1-xt0}{n}\right)$, $(0,1)$, and then uses plot to display the list of points pts. Note that in the resulting plot, the list of points are connected with line segments so the result **looks** like a spring.

```
> spring:=proc(t0)
    local xt0,pts;
    xt0:=evalf(subs(t=t0,x(t)));
    pts:=[[0,xt0],
        seq([eps*(-1)^m,xt0+m*(1-xt0)/n],m=1..n-1),
          [0,1]];
    plot(pts,view=[-1..1,-1.2..1.2],
      xtickmarks=2,ytickmarks=2);
    end:
```

Next, we load the **plots** package, define k_vals to be the list of numbers $k\dfrac{4}{49}$ for $k = 0, 1, 2, \dots, 49$, and then use seq to generate the list of graphs spring(k) for the values of k in k_vals, naming the resulting list of graphics objects to_animate.

```
> with(plots):
  k_vals:=seq(k*4/49,k=0..49):
  to_animate:=[seq(spring(k),k=k_vals)]:
```

The 50 graphs in to_animate are then animated using the **display** function, which is contained in the **plots** package, together with the option insequence=true.

```
> display(to_animate,insequence=true);
```

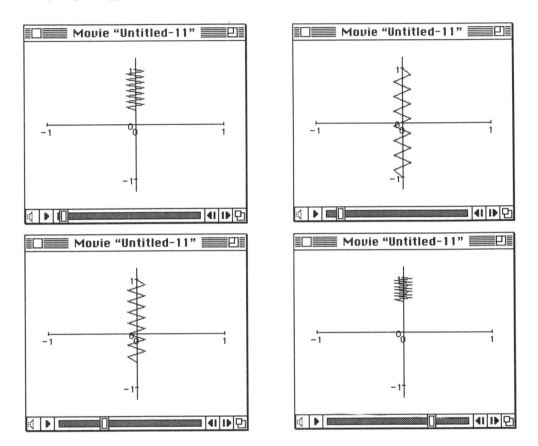

Notice that the position function $x(t) = \cos 8t$ indicates that the spring-mass system never comes to rest once it is set into motion. The solution is periodic, so the mass moves vertically, retracing its motion, as shown in the figures. Hence, motion of this type is called **simple harmonic motion**.

EXAMPLE: An object with mass $m = 1$ slug is attached to a spring with spring constant $k = 4$. (a) Determine the position function of the object if $x(0) = \alpha$ and $x'(0) = 0$. Plot the solution for $\alpha = 1, 4, -2$. (b) Determine the position function of the object if $x(0) = 0$ and $x'(0) = \beta$. Plot the solution for $\beta = 1, 4, -2$.

SOLUTION: For (a), the initial-value problem we need to solve is

$$\begin{cases} \dfrac{d^2x}{dt^2} + 4x = 0 \\ x(0) = \alpha, x'(0) = 0 \end{cases},$$

for $\alpha = 1, 4, -2$.

We now determine the solution to each of the three problems with **dsolve**. For example, entering

```
> x:='x':
  DE2:=dsolve({diff(x(t),t$2)+4*x(t)=0,
   x(0)=1,D(x)(0)=0},x(t));
```

$$DE2 := x(t) = \cos(2t)$$

solves

$$\begin{cases} \dfrac{d^2x}{dt^2} + 4x = 0 \\ x(0) = 1, x'(0) = 0 \end{cases}$$

and names the result **DE2**. Note that the formula for the solution can be extracted from **DE2** with **rhs** as follows:

```
> rhs(DE2);
```

$$\cos(2t)$$

Similarly, entering

```
> DE3:=dsolve({diff(x(t),t$2)+4*x(t)=0,
   x(0)=4,D(x)(0)=0},x(t));
  DE4:= dsolve({diff(x(t),t$2)+4*x(t)=0,
   x(0)=-2,D(x)(0)=0},x(t));
```

$$DE3 := x(t) = 4\cos(2t)$$
$$DE4 := x(t) = -2\cos(2t)$$

solves

$$\begin{cases} \dfrac{d^2x}{dt^2} + 4x = 0 \\ x(0) = 4, x'(0) = 0 \end{cases}$$

and

$$
\begin{cases}
\dfrac{d^2x}{dt^2} + 4x = 0 \\[2mm]
x(0) = -2, x'(0) = 0
\end{cases},
$$

naming the results **DE3** and **DE4**. We graph the solutions on the interval $[0, \pi]$ with **plot**. Note that **map(rhs,{DE2,DE3,DE4})** applies **rhs** to the set **{DE2,DE3,DE4}**; the result is the set of solutions.

```
> plot(map(rhs,{DE2,DE3,DE4}),t=0..Pi);
```

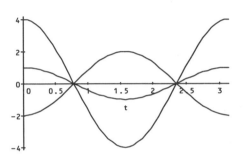

We see that the initial position affects only the amplitude of the function (and direction in the case of the negative initial position). The mass passes through the equilibrium ($x = 0$) at the same time in all three cases. For (b), we need to solve the initial-value problem

$$
\begin{cases}
\dfrac{d^2x}{dt^2} + 4x = 0 \\[2mm]
x(0) = 0, x'(0) = \beta
\end{cases}
$$

for $\beta = 1, 4, -2$.

In this case, we define the procedure **d** that, given β, returns the solution of

$$
\begin{cases}
\dfrac{d^2x}{dt^2} + 4x = 0 \\[2mm]
x(0) = 0, x'(0) = \beta
\end{cases}.
$$

```
> d:=proc(beta)
    dsolve({diff(x(t),t$2)+4*x(t)=0,x(0)=0,D(x)(0)=beta},x(t))
    end:
```

Thus, entering

> `d(1);`

$$x(t) = \frac{1}{2}\sin(2t)$$

solves

$$\begin{cases} \dfrac{d^2x}{dt^2} + 4x = 0 \\ x(0) = 0, x'(0) = 1 \end{cases},$$

entering

> `d(4);`

$$x(t) = 2\sin(2t)$$

solves

$$\begin{cases} \dfrac{d^2x}{dt^2} + 4x = 0 \\ x(0) = 0, x'(0) = 4 \end{cases},$$

and entering

> `d(-2);`

$$x(t) = -\sin(2t)$$

solves

$$\begin{cases} \dfrac{d^2x}{dt^2} + 4x = 0 \\ x(0) = 0, x'(0) = -2 \end{cases}.$$

All three solutions are graphed together on the interval $[0, \pi]$ with **plot** in the same manner as in (a).

```
> plot(map(rhs,{d(1),d(4),d(-2)}),t=0..Pi);
```

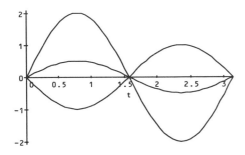

Notice that varying the initial velocity affects the amplitude (and direction in the case of the negative initial velocity) of each function. The mass passes through the equilibrium ($x = 0$) at the same time in all three cases.

5.2 DAMPED MOTION

Since the differential equation derived in Section 5.1 disregarded all retarding forces acting on the motion of the mass, a more realistic model that takes these forces into account is needed. Studies in mechanics reveal that resistive forces due to damping are proportional to a power of the velocity of the motion. Hence,

$$F_R = c\frac{dx}{dt} \qquad \text{or} \qquad F_R = c\left(\frac{dx}{dt}\right)^2,$$

where $c > 0$, is typically used to represent the damping force. We follow similar procedures that lead to the differential equation in Section 5.1 that modeled simple harmonic motion to determine the differential equation. Assuming that $F_R = c\dfrac{dx}{dt}$, we have, after summing the forces acting on the spring-mass system,

$$m\frac{d^2x}{dt^2} = -c\frac{dx}{dt} - kx \qquad \text{or} \qquad m\frac{d^2x}{dt^2} + c\frac{dx}{dt} + kx = 0.$$

Thus, the position function is found by solving the initial-value problem

$$\begin{cases} m\dfrac{d^2x}{dt^2} + c\dfrac{dx}{dt} + kx = 0 \\[2ex] x(0) = \alpha, \dfrac{dx}{dt}(0) = \beta. \end{cases}$$

From our experience with second-order ordinary differential equations with constant coefficients in Chapter 4, the solutions to initial-value problems of this type greatly depend on the values of m, k, and c.

Suppose that we assume that solutions of the differential equation have the form $x(t) = e^{rt}$. Note that m is not used in the exponent as it was in Chapter 4 to avoid confusion with the mass m. Otherwise, this calculation is identical to those followed in Chapter 4. Since $\dfrac{dx}{dt} = re^{rt}$ and $\dfrac{d^2x}{dt^2} = r^2 e^{rt}$, we have by substitution into the differential equation $mr^2 e^{rt} + cre^{rt} + ke^{rt} = 0$, so $e^{rt}(mr^2 + cr + k) = 0$.

The solutions to the characteristic (or auxiliary) equation are

$$r = \frac{-c \pm \sqrt{c^2 - 4mk}}{2a}.$$

Hence, the solution depends on the value of the quantity $(c^2 - 4mk)$. In fact, problems of this type are characterized by the value of $(c^2 - 4mk)$ as follows:

$\bullet$

CASE 1 $c^2 - 4mk > 0$

This situation is said to be **overdamped** since the damping coefficient c is large in comparison with the spring constant k.

$\bullet$

CASE 2 $c^2 - 4mk = 0$

This situation is described as **critically damped** since the resulting motion is oscillatory with a slight decrease in the damping coefficient c.

$\bullet$

CASE 3 $c^2 - 4mk < 0$

This situation is called **underdamped**, because the damping coefficient c is small in comparison with the spring constant k.

$\bullet$

EXAMPLE: A 8-lb weight is attached to a spring of length 4 ft. At equilibrium, the spring has length 6 ft. Determine the position function if $F_R = 2\dfrac{dx}{dt}$ and (a) the mass is released from its equilibrium position with an upward initial velocity of 1 ft/sec and (b) the mass is released 6 in above the equilibrium with an initial velocity of 5 ft/sec in the upward direction.

SOLUTION: Notice that $s = 6 - 4 = 2$ and that $F = 4$. Hence, we find the spring constant with $8 = k(2)$, so $k = 4$. Also, the mass of the object is $m = \dfrac{8}{32} = \dfrac{1}{4}$ slug. Therefore, the differential equation that models this spring-mass system is $\dfrac{1}{4}\dfrac{d^2x}{dt^2} + 2\dfrac{dx}{dt} + 4x = 0$ or $\dfrac{d^2x}{dt^2} + 8\dfrac{dx}{dt} + 16x = 0$.

(a) The initial conditions in this case are $x(0) = 0$ and $\dfrac{dx}{dt}(0) = 1$. We solve the initial-value problem and see that the solution is $x(t) = te^{-4t}$.

```
> x:='x':
  Eq:=diff(x(t),t$2)+8*diff(x(t),t)+16*x(t)=0;
```

$$Eq := \left(\frac{\partial^2}{\partial t^2}x(t)\right) + 8\left(\frac{\partial}{\partial t}x(t)\right) + 16x(t) = 0$$

```
> DE:=dsolve({Eq,x(0)=0,D(x)(0)=1},x(t));
```

$$DE := x(t) = e^{-4t}t$$

The graph of the solution is generated here with **plot**. Notice that it is always positive and approaches zero as t approaches infinity due to the damping.

```
> assign(DE):
  plot(x(t),t=0..3);
```

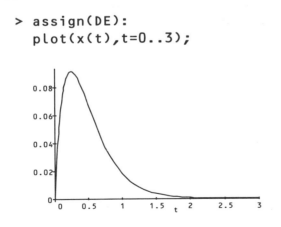

To understand further the relationship between the formula obtained and the motion of the spring, we redefine the command **spring**, which was defined in the previous section.

```
> spring:='spring':
  eps:=0.1:n:=15:
  spring:=proc(t0)
    local xt0,pts;
    xt0:=evalf(subs(t=t0,x(t)));
    pts:=[[0,xt0],
        seq([eps*(-1)^m,xt0+m*(0.1-xt0)/n],m=1..n-1),
          [0,0.1]];
    plot(pts,view=[-1..1,0..0.1],xtickmarks=2,ytickmarks=2);
  end:
```

We then use a **for** loop to generate nine graphs illustrating the motion of the spring. Alternatively, the graphs could be animated in the same manner as illustrated in Section 5.1.

```
> for k from 0 to 1.75 by 1.75/8 do spring(k) od;
```

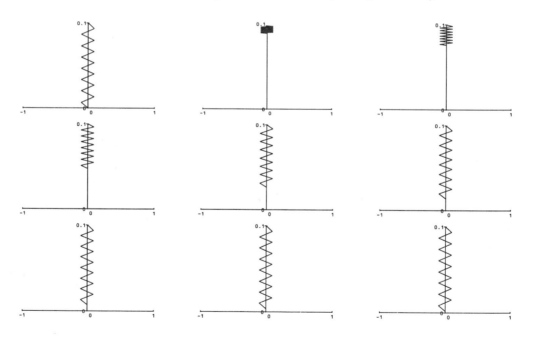

(b) In this case, $x(0) = -0.5$ and $\dfrac{dx}{dt}(0) = 5$. First, we use **dsolve** to solve the equation

```
> x:='x':
  DE:=dsolve({Eq,x(0)=-0.5,D(x)(0)=5},x(t));
```

$$DE := x(t) = -.5e^{-4t} + 3.0e^{-4t}t$$

and then graph the solution on the interval $[0, 2]$.

```
> assign(DE):
  plot(x(t),t=0..2);
```

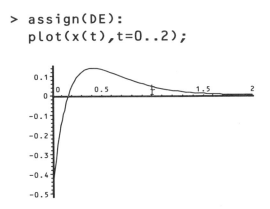

This graph indicates the importance of the initial conditions on the resulting motion. In this case, the position is negative initially, but the positive initial velocity causes the function to become positive before approaching zero.

EXAMPLE: A 32-lb weight stretches a spring 8 ft. If the resistive force due to damping is $F_R = 5\dfrac{dx}{dt}$, then determine the position function if the mass is released from 1 ft below the equilibrium position with (a) an upward velocity of 1 ft/sec and (b) an upward velocity of 6 ft/sec.

SOLUTION: Since $F = 32$, the spring constant is found with $32 = k(8)$, so $k = 4$ lb/ft. Also, $m = \dfrac{32}{32} = 1$ slug. Therefore, the differential equation that models this situation is $\dfrac{d^2x}{dt^2} + 5\dfrac{dx}{dt} + 4x = 0$.

```
> x:='x':
  DE:=diff(x(t),t$2)+5*diff(x(t),t)+4*x(t)=0;
```

$$DE := \left(\frac{\partial^2}{\partial t^2}x(t)\right) + 5\left(\frac{\partial}{\partial t}x(t)\right) + 4x(t) = 0$$

Since $c^2 - 4mk = 5^2 - 4(1)(4) = 9 > 0$, the system is overdamped.

The initial position is $x(0) = 1$, while the initial velocity in (a) is $\dfrac{dx}{dt}(0) = -1$. Using **dsolve**,

```
> DEq:=dsolve({DE,x(0)=1,D(x)(0)=-1},x(t));
```

$$DEq := x(t) = e^{-t}$$

we see that the solution of the initial-value problem is $x(t) = e^{-t}$, which is graphed as follows:

```
> assign(DEq):
  plot(x(t),t=0..5);
```

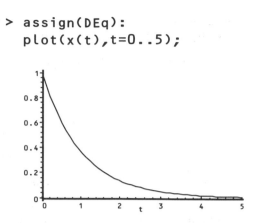

Notice that the solution is always positive and, due to the damping, approaches zero as t approaches infinity. For (b) we use the initial velocity $\frac{dx}{dt}(0) = -6$ and use **dsolve** to see that the solution to the initial-value problem is $x(t) = \frac{5}{3}e^{-4t} - \frac{2}{3}e^{-t}$.

```
> x:='x':
  DEq:=dsolve({Eq,x(0)=1,D(x)(0)=-6},x(t));
```

$$DEq := x(t) = -\frac{2}{3}e^{-t} + \frac{5}{3}e^{-4t}$$

```
> assign(DEq):
  plot(x(t),t=0..5);
```

These results indicate the importance of the initial conditions on the resulting motion. In this case, the position is positive initially, but the larger negative initial velocity causes the function to become negative before it approaches zero.

EXAMPLE: A 16-lb weight stretches a spring 2 ft. Determine the position function if the resistive force due to damping is $F_R = \dfrac{1}{2}\dfrac{dx}{dt}$ and the mass is released from the equilibrium position with a downward velocity of 1 ft/sec.

SOLUTION: Since $F = 16$, the spring constant is determined with $16 = k(2)$. Hence, $k = 8$ lb/ft. Also, $m = \dfrac{16}{32} = \dfrac{1}{2}$ slug. Therefore, the differential equation is $\dfrac{1}{2}\dfrac{d^2x}{dt^2} + \dfrac{1}{2}\dfrac{dx}{dt} + 8x = 0$ or $\dfrac{d^2x}{dt^2} + \dfrac{dx}{dt} + 16x = 0$. Notice that because $c^2 - 4mk = \left(\dfrac{1}{2}\right)^2 - 4\left(\dfrac{1}{2}\right)(8) = -\dfrac{63}{4} < 0$, the spring-mass system is underdamped. The initial position is $x(0) = 0$, ,and the initial velocity is $\dfrac{dx}{dt}(0) = 1$. Thus, we need to solve the initial-value problem

$$\begin{cases} \dfrac{d^2x}{dt^2} + \dfrac{dx}{dt} + 16x = 0 \\ x(0) = 0, \dfrac{dx}{dt}(0) = 1 \end{cases},$$

which is solved now with **dsolve**.

```
> Sol:=dsolve({diff(x(t),t$2)+diff(x(t),t)+16*x(t)=0,
  x(0)=0,D(x)(0)=1},x(t));
```

$$Sol := x(t) = \frac{2}{21}\sqrt{7}\,e^{-\frac{1}{2}t}\sin\left(\frac{3}{2}\sqrt{7}\,t\right)$$

Solutions of this type have several interesting properties. First, the trigonometric component of the solution causes the motion to oscillate. Also, the exponential portion forces the solution to approach zero as t approaches infinity. These qualities are illustrated in the following graph.

```
> assign(Sol):
  Plot_1:=plot(x(t),t=0..2*Pi):
  with(plots):
  display({Plot_1});
```

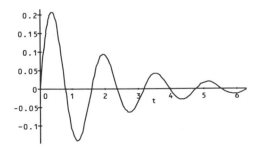

Physically, the position of the mass in this case oscillates about the equilibrium position and eventually comes to rest in the equilibrium position. Of course, with our model the position function $x(t) \to 0$ as $t \to \infty$, but there is no value of $t = T$ such that $x(t) = 0$ for $t > T$ as we might expect from the physical situation. Hence, our model only approximates the behavior of the mass. Notice also that the solution is bounded above and below by the exponential term of the solution and its reflection through the horizontal axis. This is illustrated with the simultaneous display of these functions.

```
> Plot_2:=plot({2/21*7^(1/2)*exp(-1/2*t),
   -2/21*7^(1/2)*exp(-1/2*t)},t=0..2*Pi):
  display({Plot_1,Plot_2});
```

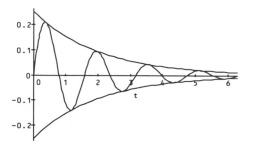

Other questions of interest include: (1) When does the mass first pass through its equilibrium point? (2) What is the maximum displacement of the spring?

The time at which the mass passes through $x = 0$ can be determined in several ways. The solution equals zero at the time that $\sin x$ first equals zero after $t = 0$ because the solution involves the sine function. We use **solve** to find the exact time at which the mass first passes through its equilibrium point. Then we use **fsolve** to approximate the time.

```
> solve(3/2*sqrt(7)*t=Pi,t);
```

$$\frac{2}{21}\pi\sqrt{7}$$

```
> fsolve(x(t)=0,t=0.6..0.9);
```

.7916069411

Similarly, the maximum displacement of the spring is found by finding the first value of t for which $x'(t) = 0$ as done here with **solve**.

```
> CP:=solve(diff(x(t),t)=0);
```

$$CP := \frac{2}{21}\arctan\left(3\sqrt{7}\right)\sqrt{7}$$

An approximation of the result given is obtained with `evalf`.

```
> CP1:=evalf(CP);
```

$$CP1 := .3642238255$$

The maximum displacement is then given by evaluating $x(t)$ for the value obtained with `evalf`.

```
> subs(t=CP1,x(t));
```

$$\frac{2}{21}\sqrt{7}\,e^{-.1821119128}\sin\left(.5463357383\sqrt{7}\right)$$

```
> evalf(subs(t=CP1,x(t)));
```

$$.2083770137$$

Another interesting characteristic of solutions to underdamped problems is the time between successive maxima of the solution, called the **quasiperiod**. This quantity is found by by first determining the time at which the second maximum occurs with `fsolve`. Then the difference between these values of t is taken to obtain the value of 1.58321.

```
> CP2:=fsolve(diff(x(t),t)=0,t=1.5..2.0);
```

$$CP2 := 1.947437708$$

```
> CP2-CP1;
```

$$1.583213883$$

To investigate the solution further, an animation can be used with the `spring` command defined earlier. Here, we redefine `spring`

```
> spring:='spring':
  n:=15:eps:=0.1:
  spring:=proc(t0)
     local xt0,pts;
     xt0:=evalf(subs(t=t0,x(t)));
     pts:=[[0,xt0],
        seq([eps*(-1)^m,xt0+m*(0.25-xt0)/n],m=1..n-1),
          [0,0.25]];
     plot(pts,view=[-1..1,-0.25..0.25],
       xtickmarks=2,ytickmarks=2);
     end:
```

and then use a `for` loop to generate 16 graphs of the spring.

```
> for k from 0 to 4 by 4/15 do spring(k) od;
```

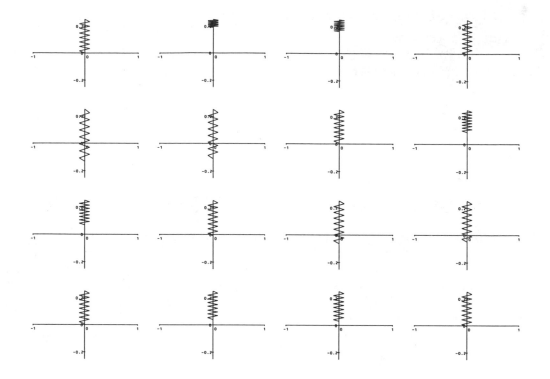

Alternatively, we can generate a list of graphs and animate the result. For example, entering

```
> with(plots):
  k_vals:=seq(k*6/59,k=0..59):
  to_animate:=[seq(spring(k),k=k_vals)]:
```

loads the **plots** package, defines **k_vals** to be the list of numbers $k\frac{6}{59}$, and then defines **to_animate** to be the list of graphs **spring(k)** for the values of k in **k_vals**. These 60 graphs are then animated using **display**, which is contained in the **plots** package, together with the option **insequence=true**.

```
> display(to_animate,insequence=true);
```

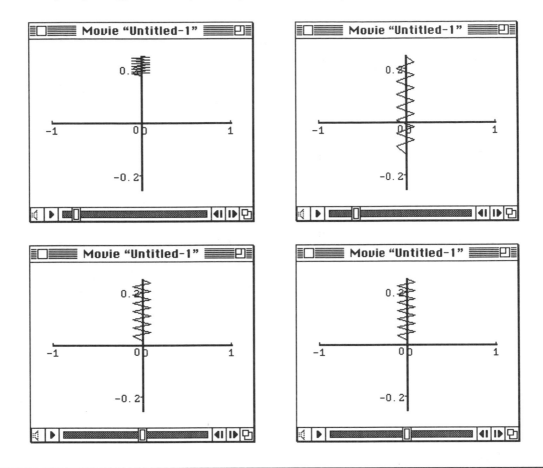

EXAMPLE: Suppose that we have the initial-value problem

$$\begin{cases} \dfrac{d^2x}{dt^2} + c\dfrac{dx}{dt} + 6x = 0 \\ x(0) = 0, x'(0) = 1 \end{cases},$$

where $c = 2\sqrt{6}, 4\sqrt{6}$, and $\sqrt{6}$. Determine how the value of c affects the solution of the initial-value problem.

SOLUTION: We begin by using `proc` to define the procedure `d`. Given `c`, `d(c)` solves the initial value problem.

$$\begin{cases} \dfrac{d^2x}{dt^2} + c\dfrac{dx}{dt} + 6x = 0 \\ x(0) = 0, x'(0) = 1 \end{cases}.$$

```
> x:='x':
  d:=proc(c)
     dsolve({diff(x(t),t$2)+c*diff(x(t),t)+6*x(t)=0,
     x(0)=0,D(x)(0)=1},x(t))
     end:
```

We then use `d` to find the solution of the initial-value problem for each value of `c`.

```
> DE1:=d(2*sqrt(6));
```
$$DE1 := x(t) = e^{-\sqrt{2}\sqrt{3}t}t$$

```
> DE2:=d(4*sqrt(6));
```
$$DE2 := x(t) = \frac{1}{12}\sqrt{2}e^{-\sqrt{2}(2\sqrt{3}-3)t} - \frac{1}{12}\sqrt{2}e^{-\sqrt{2}(2\sqrt{3}+3)t}$$

```
> DE3:=d(sqrt(6));
```
$$DE3 := x(t) = \frac{1}{3}\sqrt{2}e^{-\frac{1}{2}\sqrt{2}\sqrt{3}t}\sin\left(\frac{3}{2}\sqrt{2}t\right)$$

Note that each case results in a different classification: $c = 2\sqrt{6}$, critically damped; $c = 4\sqrt{6}$, overdamped; and $c = \sqrt{6}$, underdamped.

All three solutions are graphed together on the interval $[0,4]$ using `plot`, `map`, and `rhs`. Note that since the values of c vary more widely than those considered in the previous example, the behavior of the solutions differs more obviously as well.

```
> plot(map(rhs,{DE1,DE2,DE3}),t=0..4);
```

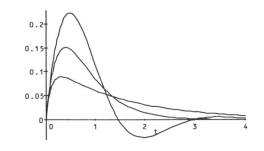

5.3 FORCED MOTION

In some cases, the motion of the spring is influenced by an external driving force, $f(t)$. Mathematically, this force is included in the differential equation that models the situation as follows:

$$m\frac{d^2x}{dt^2} = -kx - c\frac{dx}{dt} + f(t).$$

The resulting initial-value problem is

$$\begin{cases} m\dfrac{d^2x}{dt^2} + c\dfrac{dx}{dt} + kx = f(t) \\ x(0) = \alpha, \dfrac{dx}{dt}(0) = \beta. \end{cases}$$

Therefore, differential equations modeling forced motion are nonhomogeneous and require the Method of Undetermined Coefficients or Variation of Parameters for solution. We first consider forced motion that is undamped.

EXAMPLE: An object of mass $m = 1$ slug is attached to a spring with spring constant $k = 4$. Assuming there is no damping and that the object is released from rest in the equilibrium position, determine the position function of the object if it is subjected to an external force of (a) $f(t) = 0$, (b) $f(t) = 1$, (c) $f(t) = \cos t$, and (d) $f(t) = \sin t$.

SOLUTION: First, we note that we must solve the initial-value problem

$$\begin{cases} \dfrac{d^2x}{dt^2} + 4x = f(t) \\[2mm] x(0) = 0, \dfrac{dx}{dt}(0) = 0 \end{cases}$$

for each of the forcing functions in (a), (b), and (c). Since we will be solving this initial-value problem for various forcing functions, we begin by defining the procedure **fm**. Given **f(t)**, **fm(f(t))** solves the initial-value problem

$$\begin{cases} \dfrac{d^2x}{dt^2} + 4x = f(t) \\[2mm] x(0) = 0, \dfrac{dx}{dt}(0) = 0 \end{cases}$$

```
> x:='x':
  fm:=proc(f)
     dsolve({diff(x(t),t$2)+4*x(t)=f,x(0)=0,D(x)(0)=0},x(t))
     end:
```

(a) Using **fm**, we have that

```
> DE1:=fm(0);
```

$$DE1 := x(t) = 0$$

The solution is $x(t) = 0$. Physically, this solution indicates that because there is no forcing function, no initial displacement from the equilibrium position, and no initial velocity, the object does not move from the equilibrium position.

(b) Using **fm**, we have that

```
> DE2:=fm(1);
```

$$DE2 := x(t) = \frac{1}{4} - \frac{1}{4}\cos(2t)$$

$x(t) = -\dfrac{1}{4}\cos 2t + \dfrac{1}{4}$. Notice from the graph of this function, generated as follows with **plot**, that the object never moves above the equilibrium position. (Negative values of x indicate that the mass is below the equilibrium position.)

```
> Plot_2:=plot(rhs(DE2),t=0..2*Pi):
  with(plots):
  display({Plot_2});
```

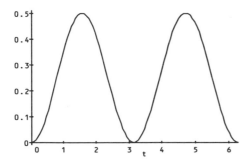

(c) Entering

```
> DE3:=fm(cos(t));
```

$$DE3 := x(t) = \frac{1}{3}\sin(2t)\sin(t)\cos(t)^2 + \frac{1}{6}\sin(2t)\sin(t) + \frac{1}{3}\cos(t)^3\cos(2t) - \frac{1}{3}\cos(2t)$$

followed by

```
> step_2:=simplify(DE3);
```

$$step\_2 := x(t) = -\frac{2}{3}\cos(t)^2\frac{1}{3} + \frac{1}{3}\cos(t)$$

shows that $x(t) = -\frac{1}{3}\cos 2t + \frac{1}{3}\cos t$. The graph of one period of this solution is generated now with plot. In this case, the mass passes through the equilibrium position twice (near $t = 2$ and $t = 4$) over the period and returns to the equilibrium without passing through it at $t = 0$ and at $t = 2\pi$.

```
> Plot_3:=plot(rhs(step_2),t=0..2*Pi):
  display({Plot_3});
```

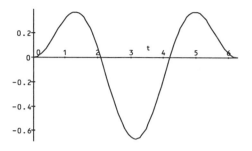

(d) We proceed as in (c). First, entering

```
> DE4:=fm(sin(t));
```

$$DE4 := x(t) = -\frac{1}{3}\sin(2t)\cos(t)^3 + \frac{1}{2}\sin(2t)\cos(t) + \frac{1}{3}\cos(2t)\sin(t)\cos(t)^2$$

$$-\frac{1}{3}\cos(2t)\sin(t) - \frac{1}{6}\sin(2t)$$

and then entering

```
> step_2:=simplify(DE4);
```

$$step\_2 := x(t) = \frac{1}{3}\sin(t) - \frac{1}{3}\sin(t)\cos(t)$$

solves the initial-value problem. Again, the forcing function is a periodic function of time. The motion shown in **Plot_4** points out that the value of the forcing function is different from the external force in (c) at different values of time. Thus, the resulting motion is not the same as that of part (c).

```
> Plot_4:=plot(rhs(step_2),t=0..2*Pi):
  display({Plot_4});
  display({Plot_2,Plot_3,Plot_4});
```

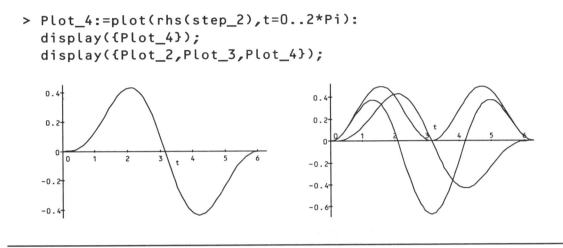

When we studied nonhomogeneous equations, we considered equations in which the nonhomogeneous function was a solution of the corresponding homogeneous equation. This situation is modeled by the initial-value problem

$$\begin{cases} \dfrac{d^2x}{dt^2} + \omega^2 x = F_1 \cos \omega t + F_2 \sin \omega t + G(t) \\ x(0) = \alpha, \dfrac{dx}{dt}(0) = \beta \end{cases}$$

where F_1 and F_2 are constants and G is any function of t. (Note that one of the constants F_1 and F_2 can equal zero and G can be identically the zero function.) In this case, we say that ω is the **natural frequency of the system**, because the homogeneous solution is $x_h(t) = c_1 \cos \omega t + c_2 \sin \omega t$. In the case of this initial-value problem, the forced frequency equals the natural frequency.

EXAMPLE: Investigate the effect that the forcing functions $f(t) = \cos 2t$ and $f(t) = \sin 2t$ have on the solution of the initial-value problem

$$\begin{cases} \dfrac{d^2x}{dt^2} + 4x = f(t) \\ x(0) = 0, \dfrac{dx}{dt}(0) = 0 \end{cases}$$

SOLUTION: We take advantage of the function **fm** defined in the previous example. (a) This situation is interesting because the forcing function is a component of the homogeneous solution, $x_h(t) = c_1 \cos 2t + c_2 \sin 2t$. Therefore, the nonperiodic function $t \sin t$ appears in the solution as follows:

```
> DE5:=fm(cos(2*t));
```

$$DE5 := x(t) = \frac{1}{2}\sin(2t)\sin(t)\cos(t)^3 - \frac{1}{4}\sin(2t)\sin(t)\cos(t) + \frac{1}{2}\sin(2t)t$$

$$+ \frac{1}{2}\cos(2t)\cos(t)^4 - \frac{1}{2}\cos(2t)\cos(t)^2$$

```
> step_2:=simplify(DE5);
```

$$step\_2 := x(t) = \frac{1}{2}\sin(t)\cos(t)t$$

Notice that the amplitude increases as t increases as illustrated in `Plot_5`. This indicates that the spring-mass system will encounter a serious problem in that the mass will eventually hit its support (i.e., ceiling or beam) or its lower boundary (i.e., the ground or floor).

```
> Plot_5:=plot(rhs(step_2),t=0..2*Pi):
  display({Plot_5});
```

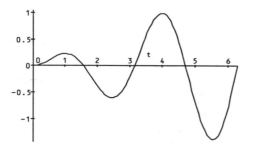

(b) In this case, the forcing function is also a solution to the corresponding homogeneous equation.

```
> DE6:=fm(sin(2*t));
```

$$DE6 := x(t) = -\frac{1}{2}\sin(2t)\cos(t)^4 + \frac{1}{2}\sin(2t)\cos(t)^2 + \frac{1}{2}\cos(2t)\sin(t)\cos(t)^3$$

$$- \frac{1}{4}\cos(2t)\sin(t)\cos(t) - \frac{1}{4}\cos(2t)t$$

```
> step_2:=simplify(DE6);
```

$$step\_2 := x(t) = \frac{1}{4}\sin(t)\cos(t) - \frac{1}{2}t\cos(t)^2 + \frac{1}{4}t$$

Hence, the nonperiodic functions appearing in the solution cause the values of x in `Plot_6` to increase without bound as t increases in the same way as in the preceding case.

```
> Plot_6:=plot(rhs(step_2),t=0..2*Pi):
  display({Plot_6});
  display({Plot_5,Plot_6});
```

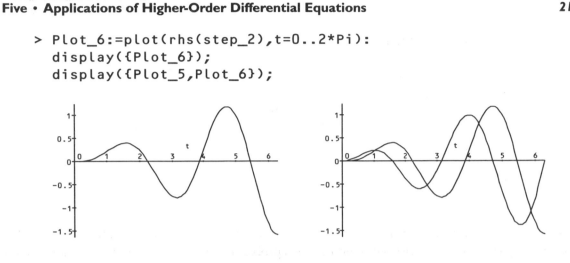

The phenomenon illustrated in the previous example is called **resonance** and can be extended to other situations such as vibrations in an aircraft wing, skyscraper, glass, or bridge. Some of the sources of excitation that lead to the vibration of these structures include unbalanced rotating devices, vortex shedding, strong winds, rough surfaces, and moving vehicles. Therefore, the engineer has to overcome many problems when structures and machines are subjected to forced vibrations.

EXAMPLE: How does slightly changing the value of the argument of the forcing function $f(t) = \cos 2t$ change the solution of the initial-value problem given in the previous example? Use the functions (a) $f(t) = \cos(1.9t)$ and (b) $f(t) = \cos(2.1t)$ with the initial-value problem.

SOLUTION: As in the previous example, we take advantage of the function **fm** used there. Thus, if $f(t) = \cos(1.9t)$, we have

```
DE7:=fm(cos(1.9*t));
```

$$
\begin{aligned}
DE7 := x(t) = {} & 2.500000000 \sin(2.t)\sin(0.1000000000t) \\
& + 0.06410256410 \sin(2.t)\sin(3.900000000t) \\
& + 0.06410256410 \cos(2.t)\cos(3.900000000t) \\
& + 2.500000000 \cos(2.t)\cos(0.1000000000t) - 2.564102564 \cos(2.t)
\end{aligned}
$$

and if $f(t) = \cos(2.1t)$, we have

```
DE8:=fm(cos(2.1*t));
```

$$
\begin{aligned}
DE8 := x(t) = {} & 2.500000000 \sin(2.t)\sin(0.1000000000t) \\
& + 0.06097560976 \sin(2.t)\sin(4.100000000t) \\
& + 0.06097560976 \cos(2.t)\cos(4.100000000t) \\
& - 2.500000000 \cos(2.t)\cos(0.1000000000t) + 2.439024390 \cos(2.t)
\end{aligned}
$$

Note that each solution is periodic and bounded because the solutions involve sines and cosines only. These solutions are then graphed to reveal the behavior of the curves. If the solutions are plotted only over a small interval, then resonance seems to be present.

```
> plot(rhs(DE7),t=0..2*Pi);
  plot(rhs(DE8),t=0..2*Pi);
```

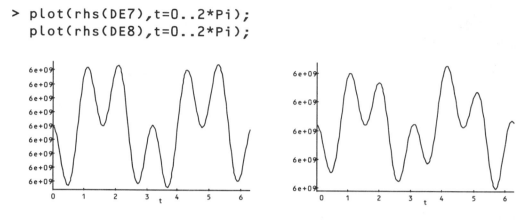

However, the functions obtained with **fm** clearly indicate that there is no resonance. This is further indicated with the second plot.

```
> plot(rhs(DE7),t=0..20*Pi);
  plot(rhs(DE8),t=0..20*Pi);
```

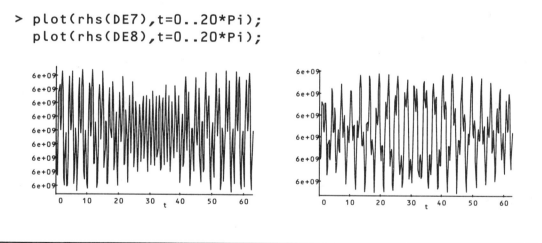

Let us investigate in detail initial-value problems of the form

$$\begin{cases} \dfrac{d^2x}{dt^2} + \omega^2 x = F\cos\beta t, \ \omega \neq \beta \\ x(0) = 0, \ \dfrac{dx}{dt}(0) = 0 \end{cases}$$

The homogeneous solution is given by $x_h(t) = c_1 \cos \omega t + c_2 \sin \omega t$. Using the Method of Undetermined Coefficients, the particular solution is given by $x_p(t) = A \cos \beta t + B \sin \beta t$. The corresponding derivatives of this solution are

$$\frac{dx_p}{dt}(t) = -A\beta \sin \beta t + B\beta \cos \beta t \text{ and } \frac{d^2 x_p}{dt^2}(t) = -A\beta^2 \cos \beta t - B\beta^2 \sin \beta t.$$

Substitution into the nonhomogeneous equation $\frac{d^2 x}{dt^2} + \omega^2 x = F \cos \beta t$ and equating the corresponding coefficients yields

$$A = \frac{F}{\omega^2 - \beta^2} \text{ and } B = 0.$$

Therefore, a general solution is

$$x(t) = c_1 \cos \omega t + c_2 \sin \omega t + \frac{F}{\omega^2 - \beta^2} \cos \beta t.$$

Application of the initial conditions yields the solution

$$x(t) = \frac{F}{\omega^2 - \beta^2}(\cos \beta t - \cos \omega t).$$

Using the trigonometric identity $\frac{1}{2}[\cos(A - B) - \cos(A + B)] = \sin A \sin B$, we have

$$x(t) = \frac{-2F}{\omega^2 - \beta^2} \sin \frac{(\omega + \beta)t}{2} \sin \frac{(\omega - \beta)t}{2}.$$

These solutions are of interest because of their motion. Notice that the solution can be represented as

$$x(t) = A(t) \sin \frac{(\omega + \beta)t}{2}, \text{ where } A(t) = \frac{-2F}{\omega^2 - \beta^2} \sin \frac{(\omega - \beta)t}{2}.$$

Therefore, when the quantity $(\omega - \beta)$ is small, $(\omega + \beta)$ is relatively large in comparison. Hence, the function $\sin \frac{(\omega + \beta)t}{2}$ oscillates quite frequently because it has period $\frac{\pi}{\omega + \beta}$. Meanwhile, the function $\sin \frac{(\omega - \beta)t}{2}$ oscillates relatively slowly because it has period $\frac{\pi}{|\omega - \beta|}$. Therefore, the function $\frac{-2F}{\omega^2 - \beta^2} \sin \frac{(\omega - \beta)t}{2}$ forms an **envelope** for the solution.

EXAMPLE: Solve the initial-value problem

$$\begin{cases} \dfrac{d^2x}{dt^2} + 4x = f(t) \\ x(0) = 0, \dfrac{dx}{dt}(0) = 0 \end{cases}$$

with (a) $f(t) = \cos 3t$ and (b) $f(t) = \cos 5t$.

SOLUTION: Again, we use `fm` to solve the initial-value problem in each case. For (a), entering

```
> DE9:=fm(cos(3*t));
```

$$DE9 := x(t) = \frac{4}{5}\sin(2t)\sin(t)\cos(t)^4 - \frac{3}{5}\sin(2t)\sin(t)\cos(t)^2 + \frac{3}{10}\sin(2t)\sin(t)$$

$$+ \frac{4}{5}\cos(2t)\cos(t)^5 - \cos(t)^3\cos(2t) + \frac{1}{5}\cos(2t)$$

```
> step_2:=simplify(DE9);
```

$$step\_2 := x(t) = -\frac{4}{5}\cos(t)^3 + \frac{2}{5}\cos(t)^2 - \frac{1}{5} + \frac{3}{5}\cos(t)$$

solves the initial-value problem, and the solution is then graphed in `Plot_9`.

```
> Plot_9:=plot(rhs(step_2),t=0..4*Pi):
  display({Plot_9});
```

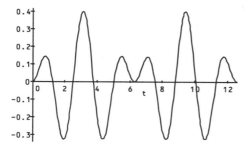

Using the formula obtained earlier for the functions that "envelope" the solution, we have $x(t) = \dfrac{2}{5}\sin\dfrac{t}{2}$ and $x(t) = -\dfrac{2}{5}\sin\dfrac{t}{2}$. The graph of the solution together with the envelope is shown as follows:

```
> Plot_App_9:=plot({2/5*sin(t/2),-2/5*sin(t/2)},t=0..4*Pi):
  display({Plot_9,Plot_App_9});
```

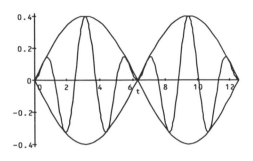

(b) In the same way, the solution of the initial-value problem is found with fm.

```
> DE10:=fm(cos(5*t));
```

$$DE10 := x(t) = \frac{25}{21}\sin(2t)\sin(t)\cos(t)^2 - \frac{5}{42}\sin(2t)\sin(t) + \frac{16}{7}\sin(2t)\cos(t)^6\sin(t)$$

$$- \frac{20}{7}\sin(2t)\sin(t)\cos(t)^4 + \frac{5}{3}\cos(t)^3\cos(2t) + \frac{16}{7}\cos(2t)\cos(t)^7$$

$$- 4\cos(2t)\cos(t)^5 + \frac{1}{21}\cos(2t)$$

```
> step_2:=simplify(DE10);
```

$$step\_2 := x(t) = -\frac{16}{21}\cos(t)^5 + \frac{20}{21}\cos(t)^3 + \frac{2}{21}\cos(t)^2 - \frac{1}{21} - \frac{5}{21}\cos(t)$$

The graph of the solution with the envelope functions $x(t) = \frac{2}{21}\sin\frac{3t}{2}$ and $x(t) = -\frac{2}{21}\sin\frac{3t}{2}$ is as follows:

```
> Plot_10:=plot(rhs(step_2),t=0..2*Pi):
  Plot_App_10:=plot({-2/21*sin(3*t/2),2/21*sin(3*t/2)},
  t=0..2*Pi):
  display({Plot_10,Plot_App_10});
```

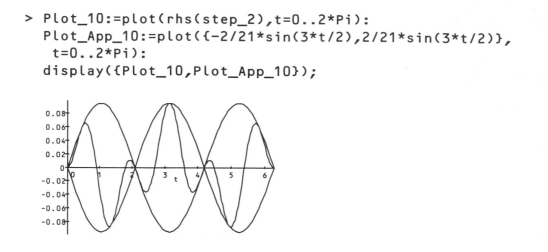

Oscillations like those illustrated in the previous example are called **beats** because of the periodic variation of amplitude. This phenomenon is commonly encountered when two musicians (especially bad ones) try to tune their instruments simultaneously or when two tuning forks with almost equivalent frequencies are played at the same time.

EXAMPLE: Investigate the effect that the forcing function $f(t) = e^{-t} \cos 2t$ has on the initial-value problem

$$\begin{cases} \dfrac{d^2x}{dt^2} + 4x = f(t) \\ x(0) = 0, \dfrac{dx}{dt}(0) = 0 \end{cases}$$

SOLUTION: The initial-value problem is solved with **fm**.

```
> DE11:=fm(exp(-t)*cos(2*t));
```

$$DE11 := x(t) = -\frac{1}{68} \sin(2t)e^{-t} \cos(4t) + \frac{1}{17} \sin(2t)e^{-t} \sin(4t) - \frac{1}{4}e^{-t} \sin(2t)$$

$$+ \frac{1}{17} \cos(2t)e^{-t} \cos(4t) + \frac{1}{68} \cos(2t)e^{-t} \sin(4t) + \frac{9}{34} \sin(2t) - \frac{1}{17} \cos(2t)$$

Notice that the effect of terms involving the exponential function diminishes as t increases. In this case, the forcing function $f(t) = e^{-t} \cos 2t$ approaches zero as t increases. Hence, over time, the solution of the nonhomogeneous equation approaches that of the homogeneous equation as illustrated in the plot.

```
> plot(rhs(DE10),t=0..4*Pi);
```

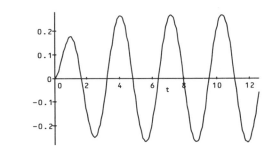

We now consider spring problems that involve forces due to damping as well as external forces. In particular, consider the following initial-value problem:

$$\begin{cases} m\dfrac{d^2x}{dt^2} + c\dfrac{dx}{dt} + kx = \rho\cos\lambda t \\ x(0) = \alpha,\ \dfrac{dx}{dt}(0) = \beta \end{cases}$$

Problems of this nature have solutions of the form $x(t) = h(t) + s(t)$, where $\lim_{t\to\infty} h(t) = 0$ and $s(t) = c_1 \cos\lambda t + c_2 \sin\lambda t$. The function $h(t)$ is called the **transient** solution, while $s(t)$ is known as the **steady-state** solution. Hence, as t approaches infinity, the solution $x(t)$ approaches the steady-state solution. Note that the steady-state solution simply corresponds to the particular solution obtained through the Method of Undetermined Coefficients or Variation of Parameters.

EXAMPLE: Solve the initial-value problem

$$\begin{cases} \dfrac{d^2x}{dt^2} + 4\dfrac{dx}{dt} + 13x = \cos t \\ x(0) = 0,\ \dfrac{dx}{dt}(0) = 1 \end{cases}$$

that models the motion of an object of mass $m = 1$ attached to a spring with spring constant $k = 13$ that is subjected to a resistive force of $F_R = 4\dfrac{dx}{dt}$ and an external force of $f(t) = \cos t$. Identify the transient and steady-state solutions.

SOLUTION: First, dsolve is used to obtain the solution of this nonhomogeneous problem. (The Method of Undetermined Coefficients could be used to find this solution as well.)

```
> DEq1:=dsolve({diff(x(t),t$2)+4*diff(x(t),t)+13*x(t)=cos(t),
  x(0)=0,D(x)(0)=1},x(t));
```

$$DEq1 := x(t) = \frac{1}{24}\sin(3t)\cos(2t) + \frac{1}{24}\sin(3t)\sin(2t) + \frac{1}{60}\sin(3t)\cos(4t)$$

$$+ \frac{1}{30}\sin(3t)\sin(4t) + \frac{1}{30}\cos(3t)\cos(4t) - \frac{1}{60}\cos(3t)\sin(4t)$$

$$+ \frac{1}{24}\cos(3t)\cos(2t) - \frac{1}{24}\cos(3t)\sin(2t) + \frac{11}{40}e^{-2t}\sin(3t) - \frac{3}{40}e^{-2t}\cos(3t)$$

```
> step_2:=expand(DEq1);
```

$$step\_2 := x(t) = -\frac{11}{40}\frac{\sin(t)}{(e^t)^2} - \frac{3}{10}\frac{\cos(t)^3}{(e^t)^2} + \frac{9}{40}\frac{\cos(t)}{(e^t)^2} + \frac{11}{10}\frac{\sin(t)\cos(t)^2}{(e^t)^2}$$

$$- \frac{7}{15}\sin(t)^2\cos(t)^3 + \frac{1}{20}\sin(t)^2\cos(t) + \frac{1}{40}\sin(t) + \frac{1}{40}\cos(t)$$

$$+ \frac{16}{15}\sin(t)^2\cos(t)^5 - \frac{23}{15}\cos(t)^5 + \frac{31}{60}\cos(t)^3 + \frac{16}{15}\cos(t)^7$$

The solution is then graphed over the interval $[0, 5\pi]$ in Plot_1 to illustrate the behavior of this solution.

```
> Plot_1:=plot(rhs(step_2),t=0..5*Pi):
  display({Plot_1});
```

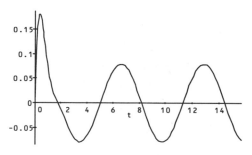

The transient solution is $e^{-2t}\left(-\dfrac{3}{40}\cos 3t + \dfrac{11}{40}\sin 3t\right)$, while the steady-state solution is $\dfrac{3}{40}\cos t + \dfrac{1}{40}\sin t$. We graph the steady-state solution over the same interval so that it can be displayed with Plot_1.

```
> SS_Plot:=plot(1/40*(3*cos(t)+sin(t)),t=0..5*Pi):
  display({Plot_1,SS_Plot});
```

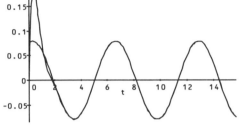

Notice that the two curves appear to be identical for $t > 2.5$. The reason for this is shown in the subsequent plot of the transient solution that becomes quite small near $t = 2.5$.

```
> plot(1/40*exp(-2*t)*(-3*cos(3*t)+11*sin(3*t)),t=0..Pi);
```

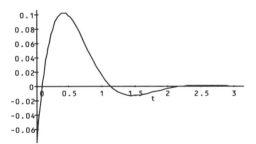

Notice also that the steady-state solution corresponds to the particular solution to the nonhomogeneous differential equation as verified here by defining $ss(t)$

```
> ss:=t->1/40*(3*cos(t)+sin(t)):
```

and then computing and simplifying $ss''(t) + 4ss'(t) + 13ss(t)$.

```
> diff(ss(t),t$2)+4*diff(ss(t),t)+13*ss(t);
```

$\cos(t)$

5.4 OTHER APPLICATIONS

☰ L-R-C Circuits

Second-order nonhomogeneous ordinary linear differential equations arise in the study of electrical circuits after the application of Kirchhoff's law. Suppose that $I(t)$ is the current in the L-R-C series electrical circuit shown here, where L, R, and C represent the inductance, resistance, and capacitance of the circuit, respectively.

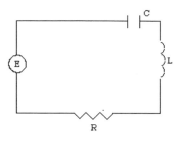

The voltage drops across the circuit elements shown in the following table have been obtained from experimental data, where Q is the charge of the capacitor and $\dfrac{dQ}{dt} = I$.

Circuit Element	Voltage Drop
Inductor	$L\dfrac{dI}{dt}$
Resistor	RI
Capacitor	$\dfrac{1}{C}Q$

Our goal is to model this physical situation with an initial-value problem so that we can determine the current and charge in the circuit. For convenience, the terminology used in this section is summarized in the following table.

Electrical Quantities	Units
Inductance (I)	Henrys (H)
Resistance (R)	Ohms (Ω)
Capacitance (C)	Farads (F)
Charge (Q)	Coulombs (C)
Current (I)	Amperes (A)

The physical principle needed to derive the differential equation that models the L-R-C series circuit, Kirchoff's law, is stated as follows:

DEFINITION | *Kirchoff's law*
The sum of the voltage drops across the circuit elements is equivalent to the voltage $E(t)$ impressed on the circuit.

Applying Kirchoff's law yields the differential equation $L\frac{dI}{dt} + RI + \frac{1}{C}Q = E(t)$. Using the fact that $\frac{dQ}{dt} = I$, we also have $\frac{d^2Q}{dt^2} = \frac{dI}{dt}$. Therefore, the equation becomes $L\frac{d^2Q}{dt^2} + R\frac{dQ}{dt} + \frac{1}{C}Q = E(t)$, which can be solved by the method of undetermined coefficients or the method of variation of parameters. Hence, if the initial charge and current are $Q(0) = Q_0$ and $I(0) = \frac{dQ}{dt}(0) = I_0$, then we must solve the initial-value problem

$$\begin{cases} L\frac{d^2Q}{dt^2} + R\frac{dQ}{dt} + \frac{1}{C}Q = E(t) \\ Q(0) = Q_0, I(0) = \frac{dQ}{dt}(0) = I_0 \end{cases}$$

for the charge $Q(t)$. This solution can then be differentiated to find the current $I(t)$.

EXAMPLE: Consider the L-R-C circuit with $L = 1$ henry, $R = 40$ ohms, $C = 4000$ farads, and $E(t) = 24$ volts. Determine the current in this circuit if there is zero initial current and zero initial charge.

SOLUTION: Using the indicated values, the initial-value problem that we must solve is

$$\begin{cases} \frac{d^2Q}{dt^2} + 40\frac{dQ}{dt} + 4000Q = 24 \\ Q(0) = 0, I(0) = \frac{dQ}{dt}(0) = 0 \end{cases}.$$

`dsolve` is used to obtain the solution to this nonhomogeneous problem in `Cir1`.

```
> Cir1:=dsolve({
    diff(q(t),t$2)+40*diff(q(t),t)+4000*q(t)=24,
    q(0)=0,D(q)(0)=0},q(t));
```

$$Cir1 := q(t) = \frac{3}{500} - \frac{1}{500}e^{-20t}\sin(60t) - \frac{3}{500}e^{-20t}\cos(60t)$$

These results indicate that in time the charge approaches the constant value of $\frac{3}{500}$, which is known as the **steady-state charge**. Also, due to the exponential term, the current approaches zero as t increases. This limit is indicated in the graph of $q(t)$ as well.

```
> assign(Cir1):
  plot(q(t)t=0..0.35);
```

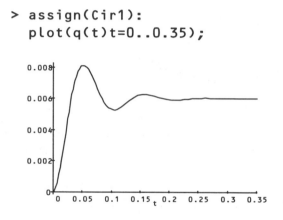

The current $i(t)$ for this circuit is obtained by differentiating the solution $q(t)$. This function is graphed as follows:

```
> dq:=diff(q(t),t);
```

$$dq := \frac{2}{5}e^{-20t}\sin(60t)$$

```
> plot(dq,t=0..0.35);
```

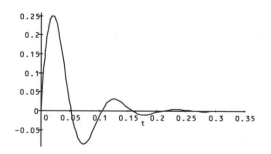

☰ *Deflection of a Beam*

An important mechanical model involves the deflection of a long beam that is supported at one or both ends as shown in the following figure.

Assuming that in its undeflected form the beam is horizontal, then the deflection of the beam can be expressed as a function of x.

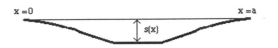

In particular, let $s(x)$ represent this deflection with x being the distance from one end of the beam and s the measurement of the vertical deflection from the equilibrium position. The initial-value problem that models this situation is derived as follows.

Let $m(x)$ equal the turning moment of the force relative to the point x and $w(x)$ represent the weight distribution of the beam. These two functions are related by the equation

$$\frac{d^2m}{dx^2} = w.$$

Also, the turning moment is proportional to the curvature of the beam. Hence,

$$m = \frac{EI}{\left(\sqrt{1 + \left(\frac{ds}{dt}\right)^2}\right)^3} \frac{d^2s}{dt^2},$$

where E and I are constants related to the composition of the beam and the shape and size of a cross section of the beam, respectively. Notice that this equation is, unfortunately, nonlinear. However, this difficulty is overcome with an approximation. For small values of s, the denominator of the right-hand side of the equation can be approximated by the constant 1. Therefore, the equation is simplified to

$$m = EI\frac{d^2s}{dt^2}.$$

This equation is linear and can be differentiated twice to obtain

$$\frac{d^2m}{dt^2} = EI\frac{d^4s}{dt^4}.$$

This equation can then be used with the preceding equation relating $m(x)$ and $w(x)$ to obtain the single fourth-order linear nonhomogeneous differential equation

$$EI\frac{d^4s}{dx^4} = w(x).$$

Boundary conditions for this problem may vary. In most cases, two conditions are given for each end of the beam. Some of these conditions that are specified in pairs at $x = a$ include: $s(a) = 0$, $\frac{ds}{dt}(a) = 0$ (fixed end); $\frac{d^2s}{dt^2}(a) = 0$, $\frac{d^3s}{dt^3}(a) = 0$ (free end); $s(a) = 0$, $\frac{d^2s}{dt^2}(a) = 0$ (simple support); and $\frac{ds}{dt}(a) = 0$, $\frac{d^3s}{dt^3}(a) = 0$ (sliding clamped end).

The following example investigates the effects that a nonconstant weight distribution function $w(x)$ has on the solution to these boundary-value problems.

EXAMPLE: Solve the beam equation over the interval $0 \le x \le 1$, assuming that the weight distribution $w(x)$ is constant and the following boundary conditions are used: $s(0) = 0$, $\frac{ds}{dt}(0) = 0$ (fixed end at $x = 0$); and (a) $s(1) = 0$, $\frac{d^2s}{dt^2}(1) = 0$ (simple support at $x = 1$); (b) $\frac{d^2s}{dt^2}(1) = 0$; $\frac{d^3s}{dt^3}(1) = 0$ (free end at $x = 1$); (c) $\frac{ds}{dt}(1) = 0$, $\frac{d^3s}{dt^3}(1) = 0$ (sliding clamped end at $x = 1$); and (d) $s(1) = 0$, $\frac{ds}{dt}(1) = 0$ (fixed end at $x = 1$).

SOLUTION: dsolve is used to obtain the solution to this nonhomogeneous problem. In DE1, the solution that depends on the parameters w, e, and i is given.

```
> DE1:=dsolve({e*i*diff(s(x),x$4)=w,
   s(0)=0,D(s)(0)=0,s(1)=0,(D@@2)(s)(1)=0},s(x));
```

$$DE1 := s(x) = \frac{1}{24}\frac{x^4 w}{ei} + \frac{1}{16}\frac{x^2 w}{ei} - \frac{5}{48}\frac{x^3 w}{ei}$$

The formula is then assigned the name $s(x)$ with assign. Later, we will graph the solution for the particular values of $e = 1$, $i = 1$, and $w = 48$, so we define to_plot1 to be the value of $s(x)$ for these values.

```
> assign(DE1);
  to_plot1:=subs({e=1,i=1,w=48},s(x));
```

$$to\_plot1 := 2x^4 + 3x^2 - 5x^3$$

Similar steps are followed to determine the solution to each of the other three boundary-value problems. The corresponding functions to be graphed are named `to_plot2`, `to_plot3`, and `to_plot4`.

```
> s:='s':
  DE2:=dsolve({e*i*diff(s(x),x$4)=w,
   s(0)=0,D(s)(0)=0,(D@@3)(s)(1)=0,
   (D@@2)(s)(1)=0},s(x));
```

$$DE2 := s(x) = \frac{1}{24}\frac{x^4 w}{ei} + \frac{1}{4}\frac{x^2 w}{ei} - \frac{1}{6}\frac{x^3 w}{ei}$$

```
> assign(DE2);
  to_plot2:=subs({e=1,i=1,w=48},s(x));
```

$$to\_plot2 := 2x^4 + 12x^2 - 8x^3$$

```
> s:='s':
  DE3:=dsolve({e*i*diff(s(x),x$4)=w,s(0)=0,D(s)(0)=0,
   (D@@3)(s)(1)=0,(D)(s)(1)=0},s(x));
```

$$DE3 := s(x) = \frac{1}{24}\frac{x^4 w}{ei} + \frac{1}{6}\frac{x^2 w}{ei} - \frac{1}{6}\frac{x^3 w}{ei}$$

```
> assign(DE3):
  to_plot3:=subs({e=1,i=1,w=48},s(x));
```

$$to\_plot3 := 2x^4 + 8x^2 - 8x^3$$

```
> s:='s':
  DE4:=dsolve({e*i*diff(s(x),x$4)=w,s(0)=0,D(s)(0)=0,
   s(1)=0,(D)(s)(1)=0},s(x));
```

$$DE4 := s(x) = \frac{1}{24}\frac{x^4 w}{ei} + \frac{1}{24}\frac{x^2 w}{ei} - \frac{1}{12}\frac{x^3 w}{ei}$$

```
> assign(DE4):
  to_plot4:=subs({e=1,i=1,w=48},s(x));
```

$$to\_plot4 := 2x^4 + 8x^2 - 8x^3$$

In order to compare the effects that the varying boundary conditions have on the resulting solutions, all four functions are graphed together with **plot** on the interval $[0, 1]$.

```
> plot({to_plot1,to_plot2,to_plot3,to_plot4},x=0..1);
```

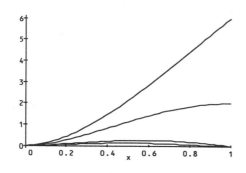

5.5 THE PENDULUM PROBLEM

Suppose that a mass m is attached to the end of a rod of length L, the weight of which is negligible.

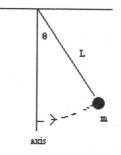

We want to determine the equation that describes the motion of the mass in terms of the displacement $\theta(t)$, which is measured counterclockwise in radians from the axis shown above. This is possible if we are given an initial position and an initial velocity of the mass. A force diagram for this situation is shown as follows:

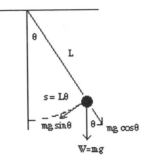

Notice that the forces are determined with trigonometry using this diagram. Here, $\cos \theta = \dfrac{mg}{x}$

and $\sin \theta = \dfrac{mg}{y}$, so we obtain the forces $x = mg \cos \theta$ and $y = mg \sin \theta$, indicated as follows:

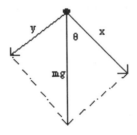

Since the momentum of the mass is given by $m\dfrac{ds}{dt}$, the rate of change of the momentum is

$$\frac{d}{dt}\left(m\frac{ds}{dt}\right) = m\frac{d^2s}{dt^2}$$

(where s represents the length of the arc formed by the motion of the mass). The force $mg \sin \theta$ acts in the opposite direction of the motion of the mass, so we have the equation

$$m\frac{d^2s}{dt^2} = -mg \sin \theta.$$

(Notice that the force $mg \cos \theta$ is offset by the force of constraint in the rod, so mg and $mg \cos \theta$ cancel each other in the sum of the forces.) Using the relationship from geometry between the length of the arc, the length of the rod, and the angle θ, $s = L\theta$, we have the relationship

$$\frac{d^2s}{dt^2} = \frac{d^2}{dt^2}(L\theta) = L\frac{d^2\theta}{dt^2}.$$

Hence, the displacement $\theta(t)$ satisfies $mL\dfrac{d^2\theta}{dt^2} = -mg\sin\theta$, or

$$mL\frac{d^2\theta}{dt^2} + mg\sin\theta = 0,$$

which is a nonlinear equation. However, because we are concerned only with small displacements, we note from the Maclaurin series for $\sin\theta$, $\sin\theta = \theta - \dfrac{\theta^3}{3!} + \dfrac{\theta^5}{5!} - \cdots$, that for small values of θ, $\sin\theta \approx \theta$. Therefore, we obtain the linear equation $mL\dfrac{d^2\theta}{dt^2} + mg\theta = 0$, or

$$\frac{d^2\theta}{dt^2} + \frac{g}{L}\theta = 0,$$

which approximates the original problem. If the initial displacement (position of the mass) is given by $\theta(0) = \theta_0$ and the initial velocity (the velocity with which the mass is set into motion) is given by $\dfrac{d\theta}{dt}(0) = v_0$, then we have the initial-value problem

$$\begin{cases} \dfrac{d^2\theta}{dt^2} + \dfrac{g}{L}\theta = 0 \\[2mm] \theta(0) = \theta_0, \quad \dfrac{d\theta}{dt}(0) = v_0 \end{cases}$$

to find the displacement function $\theta(t)$.

Suppose that $\omega^2 = \dfrac{g}{L}$ so that the differential equation becomes $\dfrac{d^2\theta}{dt^2} + \omega^2\theta = 0$. Hence, functions of the form $\theta(t) = C_1\cos\omega t + C_2\sin\omega t$, where $\omega = \sqrt{\dfrac{g}{L}}$, satisfy the equation $\dfrac{d^2\theta}{dt^2} + \dfrac{g}{L}\theta = 0$. Hence, if we use the conditions $\theta(0) = \theta_0$ and $\dfrac{d\theta}{dt}(0) = v_0$, we find that the function $\theta(t) = \theta_0\cos\omega t + \dfrac{v_0}{\omega}\sin\omega t$ satisfies the equation as well as the initial displacement and velocity conditions. As we did with the position of the spring-mass systems, we can write this function solely in terms of a cosine function with a phase shift. Hence,

$$\theta(t) = \sqrt{\theta_0^2 + \frac{v_0^2}{\omega^2}}\cos(\omega t - \phi),$$

where

$$\phi = \cos^{-1}\left(\frac{\theta_0}{\sqrt{\theta_0^2 + \frac{v_0^2}{\omega^2}}}\right)$$

and

$$\omega = \sqrt{\frac{g}{L}}.$$

Note that the approximate period of $\theta(t)$ is

$$T = \frac{2\pi}{\omega} = 2\pi\sqrt{\frac{L}{g}}.$$

EXAMPLE: Determine the displacement of a pendulum of length $L = 32$ ft if $\theta(0) = 0$ and $\frac{d\theta}{dt}(0) = 2$ using both the linear and nonlinear models. What is the period? If the pendulum is part of a clock that ticks once for each time the pendulum makes a complete swing, how many ticks does the clock make in 1 min?

SOLUTION: The linear initial-value problem that models this situation is

$$\begin{cases} \dfrac{d^2\theta}{dt^2} + \theta = 0 \\[2mm] \theta(0) = 0, \dfrac{d\theta}{dt}(0) = 2 \end{cases}$$

because $\dfrac{g}{L} = \dfrac{32}{32} = 1$.

A general solution of this differential equation is

```
> HM:=dsolve(diff(x(t),t$2)+x(t)=0,x(t));
```

$$HM := x(t) = \_C1\sin(t) + \_C2\cos(t)$$

and application of the initial conditions $\theta(0) = a$ and $\theta'(0) = b$ yields the solution

```
> x:='x':
  Eq:=dsolve({diff(x(t),t$2)+x(t)=0,
  x(0)=a,D(x)(0)=b},x(t));latex(");
```

$$Eq := x(t) = b\sin(t) + a\cos(t)$$

For the values of $a = 0$ and $b = 2$, we obtain

```
> assign(Eq);
  pen:=subs({a=0,b=2},x(t));
```

$$pen := 2\sin(t)$$

The period of this function is

$$T = 2\pi\sqrt{\frac{L}{g}} = 2\pi\sqrt{\frac{32 \text{ ft}}{32 \text{ ft/s}^2}} = 2\pi \sec.$$

Therefore, the number of ticks made by the clock per minute is calculated with the conversion $\frac{1 \text{ rev}}{2\pi \sec} \times \frac{1 \text{ tick}}{1 \text{ rev}} \times \frac{60 \sec}{1 \min} = 9.55$ ticks/min. Hence, the clock makes approximately 9.55 ticks in 1 min. To solve the nonlinear equation, we begin by defining **Eqn**

```
> Eqn:=diff(x(t),t$2)+sin(x(t))=0;
```

$$Eqn := \left(\frac{d^2}{dt^2}x(t)\right) + \sin(x(t)) = 0$$

and then using **dsolve** together with the **numeric** option to generate a numerical solution to the initial-value problem, naming the result **Soln_1**.

```
> Soln_1:=dsolve({Eqn,x(0)=0,D(x)(0)=2},x(t),numeric);
  Soln_1 := proc(t) 'dsolve/numeric/result2'
      (t,4703672,[2]) end
```

To graph the solution given in **Soln_1**, we use the **odeplot** command, which is contained in the **plots** package. We first load the **plots** package and then use **odeplot** to graph the solution given in **Soln_1** on the interval $[0, 20]$.

```
> with(plots);
  Plot_1:=odeplot(Soln_1,[t,x(t)],0..20):
  display({Plot_1});
```

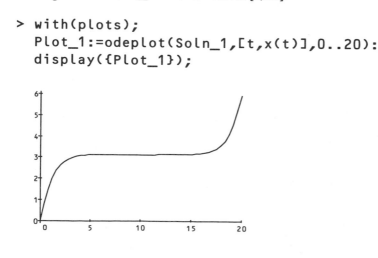

The solution **pen** is also graphed on the interval $[0, 20]$, naming the result **Plot_2**, and then both **Plot_1** and **Plot_2** are displayed together with **display**. We see that these functions vary greatly as t increases.

```
> Plot_2:=plot(pen,t=0..20):
  display({Plot_1,Plot_2});
```

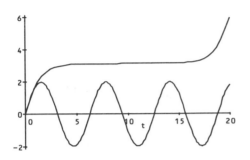

However, for small displacements, the two solutions yield similar results as illustrated here in **Plot_12** and **Plot_22**.

```
> Plot_12:=odeplot(Soln_1,[t,x(t)],0..1.5):
  Plot_22:=plot(pen(t,0,2),t=0..1.5):
  display({Plot_12,Plot_22});
```

Suppose that the pendulum undergoes a damping force that is proportional to the instantaneous velocity. Hence, the force due to damping is given as

$$F_R = b\frac{d\theta}{dt}.$$

Incorporating this force into the sum of the forces acting on the pendulum, we have the nonlinear equation $L\dfrac{d^2\theta}{dt^2} + b\dfrac{d\theta}{dt} + g\sin\theta = 0$. Again, using the approximation $\sin\theta \approx \theta$ for small values of θ, we obtain the linear equation $L\dfrac{d^2\theta}{dt^2} + b\dfrac{d\theta}{dt} + g\theta = 0$ that approximates the situation. Thus, we solve the initial-value problem

$$\begin{cases} L\dfrac{d^2\theta}{dt^2} + b\dfrac{d\theta}{dt} + g\theta = 0 \\[2mm] \theta(0) = \theta_0,\ \dfrac{d\theta}{dt}(0) = v_0 \end{cases}$$

to find the displacement function $\theta(t)$.

We investigate properties of solutions of this problem in the following example.

EXAMPLE: A pendulum of length $L = \dfrac{8}{5}$ ft is subjected to the resistive force $F_R = \dfrac{32}{5}\dfrac{d\theta}{dt}$ due to damping. Determine the displacement function if $\theta(0) = 1$ and $\dfrac{d\theta}{dt}(0) = 2$.

SOLUTION: The initial-value problem that models this situation is

$$\begin{cases} \dfrac{8}{5}\dfrac{d^2\theta}{dt^2} + \dfrac{32}{5}\dfrac{d\theta}{dt} + 32\theta = 0 \\[2mm] \theta(0) = 1,\ \dfrac{d\theta}{dt}(0) = 2 \end{cases}.$$

Simplifying the differential equation, we obtain $\dfrac{d^2\theta}{dt^2} + 4\dfrac{d\theta}{dt} + 20\theta = 0$, and then using **dsolve**, we find the solution

```
> Sol:=dsolve({diff(theta(t),t$2)+4*diff(theta(t),t)+
  20*theta(t)=0,
  theta(0)=1,D(theta)(0)=2},theta(t));latex(");
```

$Sol := \theta(t) = e^{-2t}\sin(4t) + e^{-2t}\cos(4t)$

that is then graphed with **plot**.

```
> assign(Sol):
  plot(theta(t),t=0..2);
```

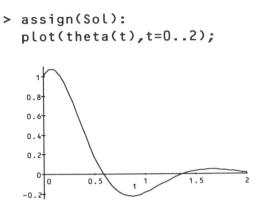

Notice that the damping causes the displacement of the pendulum to decrease over time.

To see the pendulum move, we define the procedure `pendulum`. Given `t0`, `pendulum` declares the variables `pt1` and `xt0` local to the procedure `pendulum`, defines `xt0` to be the value of $\theta(t)$ for $t = $ `t0`, defines `pt1` to be the point $\left(\frac{8}{5}\cos\left(\frac{3\pi}{2} + xt0\right), \frac{8}{5}\sin\left(\frac{3\pi}{2} + xt0\right)\right)$, and then displays the line segment connecting $(0,0)$ and `pt1` on the rectangle $[-2,2] \times [-2,0]$. The result looks like the pendulum at time $t = $ `t0`.

```
> pendulum:=proc(t0)
    local pt1,xt0;
    xt0:=evalf(subs(t=t0,theta(t)));
    pt1:=[8/5*cos(3*Pi/2+xt0),8/5*sin(3*Pi/2+xt0)];
    plot([[0,0],pt1],xtickmarks=2,ytickmarks=2,
         view=[-2..2,-2..0]);
    end:
```

To watch the pendulum move for $0 \le t \le 2$, we define `k_vals` to be 50 equally spaced numbers between 0 and 2 and then define `to_animate` to be the list of graphs `pendulum(k)` for k in `k_vals`. The resulting list of graphs is animated using the `display` function, which is contained in the `plots` package, together with the option `insequence=true`. Several frames from the resulting animation are shown as follows:

```
> k_vals:=seq(k*2/49,k=0..49):
  to_animate:=[seq(pendulum(k),k=k_vals)]:
  with(plots):
  display(to_animate,insequence=true);
```

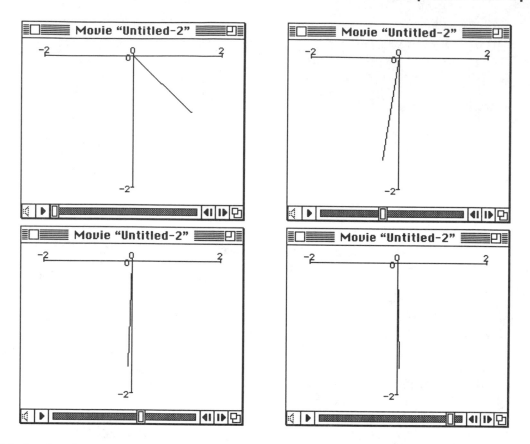

Notice that from our approximate solution, the displacement of the pendulum becomes very close to zero near $t = 2$, which was our observation from the graph of $\theta(t) = e^{-2t}(\cos 4t + 2\sin 4t)$.

Our last example investigates the properties of the nonlinear differential equation.

EXAMPLE: Graph the solution to the initial-value problem

$$\begin{cases} \dfrac{d^2\theta}{dt^2} + 0.5\dfrac{d\theta}{dt} + \sin\theta = 0 \\[2mm] \theta(0) = \theta_0, \ \dfrac{d\theta}{dt}(0) = v_0 \end{cases}$$

subject to the following initial conditions.

θ_0	v_0	θ_0	v_0	θ_0	v_0	θ_0	v_0
-1	0	$-.5$	0	$.5$	0	1	0
0	-2	0	-1	0	1	0	2
1	1	1	-1	-1	1	-1	-1
1	2	1	3	-1	4	-1	5
-1	2	-1	3	1	-4	1	-5

SOLUTION: After clearing all prior definitions of theta, if any, we define Eq to be

$$\frac{d^2\theta}{dt^2} + 0.5\frac{d\theta}{dt} + \sin\theta = 0.$$

```
>  theta:='theta':
   Eq:=diff(theta(t),t$2)+.5*diff(theta(t),t)+sin(theta(t))=0;
```

$$Eq := \left(\frac{\partial^2}{\partial t^2}\theta(t)\right) + 0.5\left(\frac{\partial}{\partial t}\theta(t)\right) + \sin(\theta(t)) = 0$$

To avoid retyping the same commands, we define the procedure s. Given an ordered pair pair, s returns a numerical solution of Eq that satisfies the initial conditions $\theta(0)$ equals the first coordinate of pair and $\theta'(0)$ equals the second coordinate of pair.

```
>  s:=proc(pair)
   dsolve({Eq,theta(0)=pair[1],D(theta)(0)=pair[2]},
    theta(t),numeric)
   end:
```

For example, we define t1 to be the list of ordered pairs corresponding to the initial conditions

θ_0	v_0	θ_0	v_0	θ_0	v_0	θ_0	v_0
-1	0	$-.5$	0	$.5$	0	1	0

and then use map to apply s to each ordered pair in t1, naming the resulting list of processes to_graph.

```
> t1:=[[-1,0],[-.5,0],[.5,0],[1,0]]:
  to_graph:=map(s,t1);

  to_graph:=[proc(t) 'dsolve/numeric/result2'
                    (t,4936284,[2]) end,
             proc(t) 'dsolve/numeric/result2'
                    (t,4532124,[2]) end,
             proc(t) 'dsolve/numeric/result2'
                    (t,4798856,[2]) end,
             proc(t) 'dsolve/numeric/result2'
                    (t,4612784,[2]) end]
```

Each of the processes in **to_graph** is then graphed using the **odeplot** command, which is contained in the **plots** package, on the interval $[0, 15]$, and all four graphs contained in the list **to_show** are displayed together using **display**, which is also contained in the **plots** package.

```
> with(plots):
  to_show:=map(odeplot,to_graph,[t,theta(t)],0..15):
  display(to_show);
```

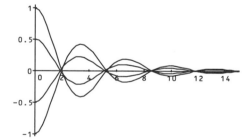

Similarly, entering

```
> t2:=[[0,-2],[0,-1],[0,1],[0,2]]:
  to_graph:=map(s,t2):
  to_show:=map(odeplot,to_graph,[t,theta(t)],0..15):
  display(to_show);
```

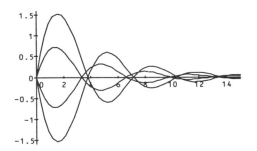

defines **t2** to be the list of ordered pairs corresponding to the initial conditions

θ_0	v_0	θ_0	v_0	θ_0	v_0	θ_0	v_0
0	−2	0	−1	0	1	0	2

to_graph to be the resulting list of processes obtained by applying **s** to **t2**, and **to_show** to be the list of graphs of each of the processes in **to_graph**; and displays the list of graphs **to_show**.

The solutions that satisfy the remaining initial conditions are graphed in the same manner as before. Thus, entering

```
> t3:=[[1,1],[1,-1],[-1,1],[-1,-1]]:
  to_graph:=map(s,t3):
  to_show:=map(odeplot,to_graph,[t,theta(t)],0..15):
  display(to_show);
```

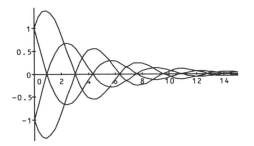

displays the graphs of the solutions that satisfy the initial conditions

θ_0	v_0	θ_0	v_0	θ_0	v_0	θ_0	v_0
1	1	1	−1	−1	1	−1	−1

Entering

```
> t4:=[[1,2],[1,3],[-1,4],[-1,5]]:
  to_graph:=map(s,t4):
  to_show:=map(odeplot,to_graph,[t,theta(t)],0..15):
  display(to_show);
```

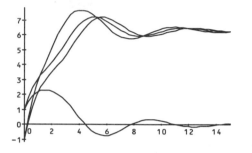

and

```
> t5:=[[-1,2],[-1,3],[1,-4],[1,-5]]:
  to_graph:=map(s,t5):
  to_show:=map(odeplot,to_graph,[t,theta(t)],0..15):
  display(to_show);
```

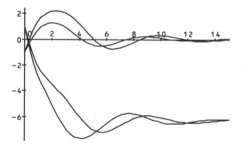

displays the graphs of the solutions that satisfy the initial conditions

θ_0	v_0	θ_0	v_0	θ_0	v_0	θ_0	v_0
1	2	1	3	−1	4	−1	5
−1	2	−1	3	1	−4	1	−5

6

Ordinary Differential Equations with Nonconstant Coefficients

In previous chapters, we have studied techniques used to solve ordinary differential equations with constant coefficients and typical applications of such differential equations. In some cases, similar techniques can be used to solve differential equations with nonconstant coefficients. In other cases, different techniques must be used.

6.1 CAUCHY-EULER EQUATIONS

Solving an arbitrary differential equation is a formidable, if not impossible, task, particularly when the coefficients are not constants. However, we are able to solve certain equations with variable coefficients using techniques similar to those discussed previously.

> **DEFINITION** | *Cauchy-Euler Equation*
>
> A **Cauchy-Euler** differential equation is an equation of the form
>
> $$a_n x^n y^{(n)} + a_{n-1} x^{n-1} y^{(n-1)} + \cdots + a_1 x y' + a_0 y = g(x),$$
>
> where $a_0, a_1, a_2, \ldots, a_n$ are constants.

☰ Second-Order Cauchy-Euler Equations

We begin our study by considering the second-order homogeneous Cauchy-Euler equation

$$ax^2y'' + bxy' + cy = 0.$$

Notice that because the coefficient of y'' is zero when $x = 0$, we must restrict our domain to either $x > 0$ or $x < 0$ to ensure that the theory of second-order equations stated in Section 4.1 holds.

Let $y = x^m$ for some constant m. Substitution of $y = x^m$ with derivatives $y' = mx^{m-1}$ and $y'' = m(m-1)y^{m-2}$ yields

$$ax^2y'' + bxy' + cy = ax^2m(m-1)x^{m-2} + bxmx^{m-1} + cx^m = x^m(am(m-1) + bm + c) = 0.$$

Then $y = x^m$ is a solution of $ax^2y'' + bxy' + cy = 0$ when m satisfies $am(m-1) + bm + c = 0$, which is called the **auxiliary** (or **characteristic**) **equation** associated with the Cauchy-Euler equation of order 2. The solutions of the auxiliary equation completely determine the general solution of the homogeneous Cauchy-Euler equation of order 2.

Let m_1 and m_2 denote two solutions of the auxiliary (or characteristic) equation. Notice that the roots of the equation

$$am(m-1) + bm + c = am^2 + (b-a)m + c = 0$$

are

$$m = \frac{-(b-a) \pm \sqrt{(b-a)^2 - 4ac}}{2a}.$$

Hence, we can obtain two real roots, one repeated real root, or a complex conjugate pair depending on the values of a, b, and c. We state here a general solution that corresponds to the different types of roots.

1. If $m_1 \neq m_2$ are real, then a general solution is $y = c_1x^{m_1} + c_2x^{m_2}$.
2. If m_1, m_2 are real and $m_1 = m_2$, then a general solution is $y = c_1x^{m_1} + c_2x^{m_1}\ln x$.
3. If $m_1 = \overline{m_2} = \alpha + i\beta$, $\beta \neq 0$, then a general solution is $y = x^\alpha[c_1\cos(\beta\ln x) + c_2\sin(\beta\ln x)]$.

EXAMPLE: Solve $3x^2y'' - 2xy' + 2y = 0$.

SOLUTION: If $y = x^m$, $y' = mx^{m-1}$, and $y'' = m(m-1)y^{m-2}$, then substitution into the differential equation yields

$$3x^2y'' - 2xy' + 2y = 3x^2m(m-1)x^{m-2} - 2xmx^{m-1} + 2x^m = x^m(3m(m-1) - 2m + 2) = 0.$$

Hence, the auxiliary equation is

$$3m(m-1) - 2m + 2 = 3m^2 - 5m + 2 = (3m - 2)(m - 1) = 0$$

with roots $m_1 = \dfrac{2}{3}$ and $m_2 = 1$. Therefore, a general solution is $y = c_1 x^{2/3} + c_2 x$. We obtain the same results with **dsolve**. For example, entering

```
> y:='y':
  Gen_Sol:=dsolve(3*x^2*diff(y(x),x$2)-
   2*x*diff(y(x),x)+2*y(x)=0,y(x));
```

$Gen\_Sol := y(x) = x^{2/3}(\_C1 + \_C2x^{1/3})$

finds a general solution of the equation, naming the result **Gen_Sol**, and then entering

```
> assign(Gen_Sol):
  vals:={-2,0,2}:
  to_graph:={seq(seq(subs({_C1=i,_C2=j},y(x)),
   i=vals),j=vals)}:
  plot(to_graph,x=0..8);
```

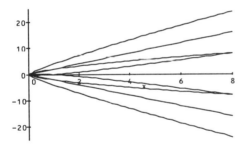

names $y(x)$ the result obtained in **Gen_Sol**; defines **vals** to be the set of numbers $-2, 0$, and 2 and **to_graph** to be the set of functions obtained by replacing _C1 in $y(x)$ by each number in **vals**; and _C2 in $y(x)$ by each number in **vals**; and graphs the set of functions **to_graph** on the interval $[0, 8]$.

EXAMPLE: Solve $x^2 y'' - xy' + y = 0$.

SOLUTION: In this case, the auxiliary equation is

$$m(m-1) - m + 1 = m^2 - 2m + 1 = (m-1)^2 = 0$$

with root $m_1 = m_2 = 1$ of multiplicity 2. Hence, a general solution is $y = c_1 x + c_2 x \ln x$.

As in the previous example, we see that we obtain the same results with `dsolve`.

```
> y:='y':
  Gen_Sol:=dsolve(x^2*diff(y(x),x$2)-
   x*diff(y(x),x)+y(x)=0,y(x));
```

$Gen\_Sol := y(x) = \_C1\,x + \_C2\,x\ln(x)$

```
> assign(Gen_Sol):
  to_graph:={seq(seq(subs({_C1=i,_C2=j},
   y(x)),i=-1..1),j=-1..1)}:
  plot(to_graph,x=0..8);
```

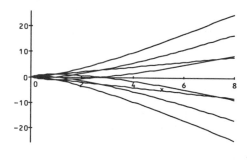

EXAMPLE: Solve $x^2 y'' - 5xy' + 10y = 0$.

SOLUTION: The auxiliary equation is given by

$$m(m-1) - 5m + 10 = m^2 - 6m + 10 = (m-1)^2 = 0$$

with complex conjugate roots $m = \frac{-4\pm\sqrt{36-40}}{2} = 3 \pm i$. Thus, a general solution is

$$y = x^3[c_1 \cos(\ln x) + c_2 \sin(\ln x)].$$

As in the previous examples, we see that equivalent results to those obtained here are returned by `dsolve`. For example, entering

```
> y:='y':
  Sol:=dsolve({
  x^2*diff(y(x),x$2)-5*x*diff(y(x),x)+10*y(x)=0,y(1)=A,
   D(y)(1)=B},y(x));
```

$Sol := y(x) = (-3A + B)x^3 \sin\left(\ln(x)\right) + Ax^3 \cos\left(\ln(x)\right)$

solves the initial-value problem

$$\begin{cases} x^2y'' - 5xy' + 10y = 0 \\ y(1) = A, y'(1) = B \end{cases}$$

and names the resulting output Sol. These solutions are then graphed for various initial conditions by entering

```
> assign(Sol):
  vals:={-2,0,2}:
  to_graph:={seq(seq(subs({A=i,B=j},y(x)),i=vals),j=vals)}:
  plot(to_graph,x=0..30);
```

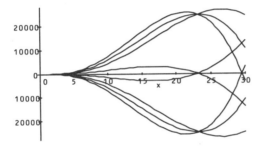

which names $y(x)$ the result given in Sol; defines vals to be the set of numbers -2, 0, and 2 and to_graph to be the set of nine functions obtained by replacing A in $y(x)$ by each number in vals and B in $y(x)$ by each number in vals; and then graphs to_graph on the interval $[0, 30]$.

≡ Higher-Order Cauchy-Euler Equations

The auxiliary equation of higher-order Cauchy-Euler equations is defined in the same way, and solutions of higher-order homogeneous Cauchy-Euler equations are determined in the same manner as solutions of higher-order homogeneous differential equations with constant coefficients.

EXAMPLE: Solve $2x^3y''' - 4x^2y'' - 20xy' = 0, x > 0$.

SOLUTION: In this case, if we assume that $y = x^m$ for $x > 0$, we have the derivatives $y' = mx^{m-1}$, $y'' = m(m - 1)x^{m-2}$, and $y''' = m(m - 1)(m - 2)x^{m-3}$. Substitution into the differential equation and simplification then yields $(2m^3 - 10m^2 - 12m)x^m = 0$.

```
> i:='i':y:='y':
  Eq:=2*x^3*diff(y(x),x$3)-4*x^2*diff(y(x),x$2)-
  20*x*diff(y(x),x)=0;
```

$$Eq := 2x^3 \left(\frac{\partial^3}{\partial x^3} y(x) \right) - 4x^2 \left(\frac{\partial^2}{\partial x^2} y(x) \right) - 20x \left(\frac{\partial}{\partial x} y(x) \right) = 0$$

```
> step_1:=eval(subs(y(x)=x^m,Eq));
```

$$step\_1 := 2x^3 \left(\frac{x^m m^3}{x^3} - 3\frac{x^m m^2}{x^3} + 2\frac{x^m m}{x^3} \right) - 4x^2 \left(\frac{x^m m^2}{x^2} - \frac{x^m m}{x^2} \right) - 20x^m m = 0$$

```
> step_2:=factor(step_1);
```

$$step\_2 := 2x^m m(m + 1)(m - 6) = 0$$

We must solve

$$(2m^3 - 10m^2 - 12m) = m(2m + 2)(m - 6) = 0$$

for m because $x^m \neq 0$.

```
> Aux_Sols:=solve(step_2,m);
```

$$Aux\_Sols := 0, -1, 6$$

We see that the solutions are $m_1 = 0$, $m_2 = -1$, and $m_3 = 6$, so a general solution is $y(x) = c_1 + c_2 x^{-1} + c_3 x^6$.

```
> i:='i':
  Gen_Sol:=sum(c[i]*x^Aux_Sols[i],i=1..3);
```

$$Gen\_Sol := c_{[1]} + \frac{c_{[2]}}{x} + c_{[3]} x^6$$

We can graph this solution for various values of the arbitrary constants in the same way as we graph solutions of other equations.

```
> to_graph:={seq(seq(subs({c[1]=0,c[2]=i,c[3]=j},Gen_Sol),
    i=-1..1),j=-1..1)}:
  plot(to_graph,x=0..2,-15..15);
```

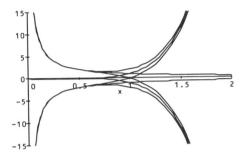

As you probably expect, **dsolve** can be used to find a general solution directly.

```
> dsolve(Eq,y(x));
```

$$y(x) = \frac{\_C1\,x + \_C2\,x^7 + \_C3}{x}$$

EXAMPLE: Solve the initial-value problem

$$\begin{cases} x^4 y^{(4)} + 4x^3 y''' + 11x^2 y'' - 9xy' + 9y = 0, x > 0 \\ y(1) = 1, y'(1) = -9, y''(1) = 27, y'''(1) = 1 \end{cases}.$$

SOLUTION: Substitution of $y(x) = x^m, x > 0$ into the differential equation results in

$$x^4 y^{(4)} + 4x^3 y''' + 11x^2 y'' - 9xy' + 9y = 0,$$

and simplification leads to the equation

$$(m^4 - m^3 + 8m^2 - 9m - 9)x^m = 0.$$

```
> y:='y':
  Eq:=x^4*diff(y(x),x$4)+4*x^3*diff(y(x),x$3)+
   11*x^2*diff(y(x),x$2)-9*x*diff(y(x),x)+
   9*y(x)=0;
```

$$Eq := x^4\left(\frac{\partial^4}{\partial x^4}y(x)\right) + 4x^3\left(\frac{\partial^3}{\partial x^3}y(x)\right) + 11x^2\left(\frac{\partial^2}{\partial x^2}y(x)\right) - 9x\left(\frac{\partial}{\partial x}y(x)\right) + 9y(x) = 0$$

```
> step_1:=factor(eval(subs(y(x)=x^m,Eq)));
```

$$step\_1 := x^m(m^2 + 9)(m - 1)^2 = 0$$

We solve

$$(m^4 - m^3 + 8m^2 - 9m - 9) = (m^2 + 9)(m - 1)^2 = 0.$$

```
> Aux_Sols:=solve(step_1,m);
```

$$Aux\_Sols := 3I, -3I, 1, 1$$

Hence, $m = \pm 3i$, and $m = 1$ is a root of multiplicity 2, so a general solution is $y(x) = c_1 \cos(3\ln x) + c_2 \sin(3\ln x) + c_3 x + c_4 x \ln x$

```
> y:=x->c[1]*cos(3*ln(x))+c[2]*sin(3*ln(x))+
   c[3]*x+c[4]*x*ln(x);
```

$$y := x \to c_{[1]} \cos(3\ln(x)) + c_{[2]} \sin(3\ln(x)) + c_{[3]}x + c_{[4]}x\ln(x)$$

with derivatives

```
> Diff(y(x),x)=diff(y(x),x);
  Diff(y(x),x$2)=diff(y(x),x$2);
```

$$\frac{\partial}{\partial x}c_{[1]} \cos(3\ln(x)) + c_{[2]} \sin(3\ln(x)) + c_{[3]}x + c_{[4]}x\ln(x) =$$

$$-3\frac{c_{[1]} \sin(3\ln(x))}{x} + 3\frac{c_{[2]} \cos(3\ln(x))}{x} + c_{[3]} + c_{[4]}\ln(x) + c_{[4]}$$

$$\frac{\partial^2}{\partial x^2}c_{[1]} \cos(3\ln(x)) + c_{[2]} \sin(3\ln(x)) + c_{[3]}x + c_{[4]}x\ln(x) =$$

$$-9\frac{c_{[1]} \cos(3\ln(x))}{x^2} + 3\frac{c_{[1]} \sin(3\ln(x))}{x^2} - 9\frac{c_{[2]} \sin(3\ln(x))}{x^2}$$

$$- 3\frac{c_{[2]} \cos(3\ln(x))}{x^2} + \frac{c_{[4]}}{x}$$

and

```
> Diff(y(x),x$3)=diff(y(x),x$3);
```

$$\frac{\partial^3}{\partial x^3} c_{[1]} \cos\left(3\ln(x)\right) + c_{[2]} \sin\left(3\ln(x)\right) + c_{[3]}x + c_{[4]}x\ln(x) =$$

$$21\frac{c_{[1]}\sin\left(3\ln(x)\right)}{x^3} + 27\frac{c_{[1]}\cos\left(3\ln(x)\right)}{x^3}$$

$$-21\frac{c_{[2]}\cos\left(3\ln(x)\right)}{x^3} + 27\frac{c_{[2]}\sin\left(3\ln(x)\right)}{x^3} - \frac{c_{[4]}}{x^2}$$

Substitution of the initial conditions then yields the system of equations

$$\{c_1 + c_3 = 1, c_3 + c_4 + 3c_2 = -9, c_4 - 9c_1 - 3c_2 = 27, -c_4 + 27c_1 - 21c_2 = 1\}$$

```
> Sys_of_Eqs:={y(1)=1,D(y)(1)=-9,(D@@2)(y)(1)=27,
  (D@@3)(y)(1)=1};
```

$$Sys\_of\_Eqs := \{3c_{[2]} + c_{[3]} + c_{[4]} = -9, 27c_{[1]} - 21c_{[2]} - c_{[4]} = 1,$$
$$c_{[1]} + c_{[3]} = 1, -9c_{[1]} - 3c_{[2]} + c_{[4]} = 27\}$$

that has the solution $\left\{c_1 = -\dfrac{12}{5}, c_2 = -\dfrac{89}{30}, c_3 = \dfrac{17}{5}, c_4 = -\dfrac{7}{2}\right\}$.

```
> Vals:=solve(Sys_of_Eqs);
```

$$Vals := \left\{c_{[3]} = \frac{17}{5}, c_{[1]} = -\frac{12}{5}, c_{[4]} = -\frac{7}{2}, c_{[2]} = -\frac{89}{30}\right\}$$

Therefore, the solution of the initial-value problem is

$$y(x) = -\frac{12}{5}\cos(3\ln x) - \frac{89}{30}\sin(3\ln x) + \frac{17}{5}x - \frac{7}{2}x\ln x.$$

```
> assign(Vals):
  y(x);
```

$$-\frac{12}{5}\cos\left(3\ln(x)\right) - \frac{89}{30}\sin\left(3\ln(x)\right) + \frac{17}{5}x - \frac{7}{2}x\ln(x)$$

We graph this solution with **plot**.

```
> plot(y(x),x=0..1);
```

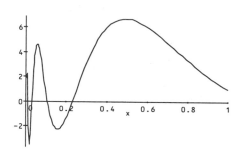

As expected, we see that **dsolve** can be used to solve the initial-value problem directly.

```
> y:='y':
  dsolve({Eq,y(1)=1,D(y)(1)=-9,(D@@2)(y)(1)=27,
  (D@@3)(y)(1)=1},y(x));
```

$$y(x) = -\frac{12}{5}\cos(3\ln(x)) - \frac{89}{30}\sin(3\ln(x)) + \frac{17}{5}x - \frac{7}{2}x\ln(x)$$

☰ Variation of Parameters

Of course, Cauchy-Euler equations can be nonhomogeneous. In some cases the method of variation of parameters can be used to solve nonhomogeneous Cauchy-Euler equations. We illustrate this procedure in the following example.

EXAMPLE: Solve $x^2 y'' - xy' + 5y = x$.

SOLUTION: We first note that **dsolve** can be used to find a general solution of the equation directly.

```
> x:='x':y:='y':
  Sol:=dsolve(x^2*diff(y(x),x$2)-x*diff(y(x),x)+5*y(x)=1/x,
  y(x));
```

$$Sol := y(x) = \frac{1}{8}\frac{1 - 8\_C1\,x^2\sin(2\ln(x)) + 8\_C2\,x^2\cos(2\ln(x))}{x}$$

Alternatively, we can use Maple to help us implement the method of variation of parameters. We begin by finding a general solution to the corresponding homogeneous equation $x^2 y'' - xy' + 5y = 0$ with dsolve, naming the result SolHom.

```
> SolHom:=dsolve(x^2*diff(y(x),x$2)-x*diff(y(x),x)+
    5*y(x)=0,y(x));
```

$$SolHom := y(x) = -\_C1\, x \sin(2 \ln(x)) + \_C2\, x \cos(2 \ln(x))$$

We see that the homogeneous solution is $y_h(x) = x(c_1 \cos(2 \ln x) + c_2 \sin(2 \ln x))$. A fundamental set of solutions for the corresponding homogeneous equation is $S = \{x \cos(2 \ln x)\, x \sin(2 \ln x)\}$, and the Wronskian is

$$W(S) = \begin{vmatrix} x \cos(2 \ln x) & x \sin(2 \ln x) \\ -x \sin(2 \ln x) \cdot \dfrac{2}{x} + (2 \ln x) & x \cos(2 \ln x) \cdot \dfrac{2}{x} + \sin(2 \ln x) \end{vmatrix}$$

$$= 2x[\cos^2(2 \ln x) + \sin^2(2 \ln x)] = 2x.$$

```
> y1:=x->x*cos(2*ln(x)):
  y2:=x->x*sin(2*ln(x)):
```

```
> with(linalg):
  wr:=det(Wronskian([y1(x),y2(x)],x));
```

$$wr := 2x \cos(2 \ln(x))^2 + 2x \sin(2 \ln(x))^2$$

```
wr:=simplify(wr);
```

$$wr := 2x$$

Using the method of variation of parameters, we have (with integration by the u-substitution of $u = 2 \ln x$ in both cases)

$$u_1(x) = -\int \frac{x \sin(2 \ln x) \cdot \frac{1}{x}}{2x}\, dx = \frac{1}{4} \cos(2 \ln x) \text{ and } u_2(x) = \int \frac{x \cos(2 \ln x) \cdot \frac{1}{x}}{2x}\, dx = \frac{1}{4} \sin(2 \ln x).$$

```
> f:=x->1/x;
```

$$f := x \rightarrow \frac{1}{x}$$

```
> u1p:=-y2(x)*f(x)/wr:
  u2p:=y1(x)*f(x)/wr:
```

```
> u1:=x->int(u1p,x):
  u1(x);
```

$$\frac{1}{4}\cos(2\ln(x))$$

```
> u2:=x->int(u2p,x):
  u2(x);
```

$$\frac{1}{4}\sin(2\ln(x))$$

Therefore, a general solution is

$$y(x) = y_h(x) + y_p(x) = x(c_1\cos(2\ln x) + c_2\sin(2\ln x)) + \frac{1}{4}x\cos^2(2\ln x) + \frac{1}{4}x\sin^2(2\ln x)$$

$$= x(c_1\cos(2\ln x) + c_2\sin(2\ln x)) + \frac{1}{4}x.$$

```
> yp:=x->y1(x)*u1(x)+y2(x)*u2(x):
  yp(x);
  yc:=x->c[1]*y1(x)+c[2]*y2(x):
  yc(x);
  y:=x->yc(x)+yp(x):
  y(x);
```

$$\frac{1}{4}x\cos(2\ln(x))^2 + \frac{1}{4}x\sin(2\ln(x))^2$$

$$c_{[1]}x\cos(2\ln(x)) + c_{[2]}x\sin(2\ln(x))$$

$$c_{[1]}x\cos(2\ln(x)) + c_{[2]}x\sin(2\ln(x)) + \frac{1}{4}x\cos(2\ln(x))^2 + \frac{1}{4}x\sin(2\ln(x))^2$$

As in previous examples, we graph this general solution for various values of the arbitrary constants.

```
> vals:={-8,0,8}:
  to_graph:={seq(seq(subs({c[1]=i,c[2]=j},y(x)),i=vals),
   j=vals)}:
  plot(to_graph,x=0..2);
```

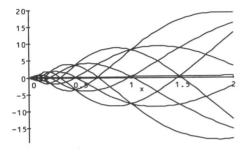

6.2 POWER SERIES REVIEW

In the previous section, we saw that techniques similar to those learned in earlier chapters can be used to solve some differential equations with nonconstant coefficients. Often, however, other methods must be used when the equation is not in the form of a Cauchy-Euler equation or when some other transformation cannot be performed to obtain an equation that we can solve with the techniques we have learned. In many cases, power series can be used to construct solutions of differential equations with nonconstant coefficients.

Before discussing how we can use power series to construct solutions of some differential equations, we will briefly review the basic properties of power series that will be used in later sections. These properties and proofs of the major theorems are found in most calculus books.

≡ Basic Definitions and Theorems

DEFINITION

> *Power Series*
>
> Let x_0 be a number. A **power series** in $(x - x_0)$ is a series of the form $\displaystyle\sum_{n=0}^{\infty} a_n(x - x_0)^n$, where a_n is a number for all values of n.

DEFINITION

> **Radius of Convergence**
>
> The power series $\sum\limits_{n=0}^{\infty} a_n(x - x_0)^n$ always converges for $x = x_0$. If there is a positive number h so that the power series $\sum\limits_{n=0}^{\infty} a_n(x - x_0)^n$ converges absolutely for all values of x in the interval $(x_0 - h, x_0 + h)$ and diverges for all values of x in the interval $(-\infty, x_0 - h) \cup (x_0 + h, +\infty)$, then the power series has **radius of convergence** h. In this case, the power series may or may not converge for $x = x_0 - h$ and may not converge for $x = x_0 + h$. If the power series converges absolutely for all values of x, the power series has infinite radius of convergence. If a power series has radius of convergence $h > 0$, then the power series is said to be **analytic**.

In calculus, we learn that many functions can be approximated by Taylor and Maclaurin polynomials, provided that the necessary derivatives exist. In fact, Taylor's theorem tells us under what conditions we can approximate a given function by Taylor or Maclaurin polynomials.

DEFINITION

> **nth-Degree Taylor and Maclaurin Polynomials**
>
> The **nth-degree Taylor polynomial for** f about $x = x_0$ is
>
> $$\sum_{k=0}^{n} \frac{f^{(k)}(x_0)}{k!}(x - x_0)^k;$$
>
> the **nth-degree Maclaurin polynomial for** f is the nth-degree Taylor polynomial for f about $x = 0$:
>
> $$\sum_{k=0}^{n} \frac{f^{(k)}(0)}{k!}x^k.$$

· ·

THEOREM   *Taylor's Theorem*

Let f be a function with derivatives of all orders on an interval (a, b) and let $a < x_0 < b$.

If $x \in (a, b)$ and $x \neq x_0$, there is a number z between x and x_0 so that

$$f(x) = \sum_{k=0}^{n} \frac{f^{(k)}(x_0)}{k!}(x - x_0)^k + R_n(x),$$

where $R_n(x) = \dfrac{f^{(n+1)}(z)}{(n + 1)!}(x - x_0)^{n+1}$. Moreover, if $\lim_{n \to \infty} R_n(x) = 0$ for every x in the interval (a, b), then

$$f(x) = \sum_{k=0}^{\infty} \frac{f^{(k)}(x_0)}{k!}(x - x_0)^k.$$

· ·

EXAMPLE: Let $f(x) = e^x$. (a) Find the Maclaurin series for $f(x)$; (b) find the nth Maclaurin polynomial for $f(x)$; and (c) show that $e^x = \sum_{k=0}^{\infty} \dfrac{1}{k!}x^k$ for every value of x.

SOLUTION: Note that $f^{(k)}(0) = 1$ for every k because $f^{(k)}(x) = e^x$, so the Maclaurin series for $f(x)$ is $\sum_{k=0}^{\infty} \dfrac{f^{(k)}(0)}{k!}x^k = \sum_{k=0}^{\infty} \dfrac{1}{k!}x^k$. The nth Maclaurin polynomial is $\sum_{k=0}^{n} \dfrac{1}{k!}x^k$, and by Taylor's theorem,

$$R_n(x) = \frac{f^{(n+1)}(z)}{(n + 1)!}x^{n+1} = \frac{e^z}{(n + 1)!}x^{n+1}.$$

If $x < 0$, then since z is between x and 0, $e^z \leq 1$, and $|R_n(x)| = \left| \dfrac{e^z}{(n + 1)!}x^{n+1} \right| \leq \dfrac{|x|^{n+1}}{(n + 1)!}$ so that $\lim_{n \to \infty} |R_n(x)| \leq \lim_{n \to \infty} \dfrac{|x|^{n+1}}{(n + 1)!} = 0$. If $x > 0$, then $e^z \leq e^x$ and $\lim_{n \to \infty} |R_n(x)| \leq e^x \lim_{n \to \infty} \dfrac{x^{n+1}}{(n + 1)!} = 0$. In any case, $\lim_{n \to \infty} R_n(x) = 0$, so $e^x = \sum_{k=0}^{\infty} \dfrac{1}{k!}x^k$ for every value of x. These results are now verified with Maple. Entering

```
> series(exp(x),x=0,11);
```

$$1 + x + \frac{1}{2}x^2 + \frac{1}{6}x^3 + \frac{1}{24}x^4 + \frac{1}{120}x^5 + \frac{1}{720}x^6 + \frac{1}{5040}x^7 + \frac{1}{40320}x^8$$
$$+ \frac{1}{362880}x^9 + \frac{1}{3628800}x^{10} + O(x^{11})$$

computes the first 11 terms of the Maclaurin series for e^x. Note that the 11th term represents the omitted higher-order terms of the series. Similarly, we can compute the exact value of $\sum_{k=0}^{\infty} \frac{1}{k!}x^k$ with sum. As expected, the result is e^x.

```
> Sum(x^k/k!,k=0..infinity)=sum(x^k/k!,k=0..infinity);
```

$$\sum_{k=0}^{\infty} \frac{x^k}{k!} = e^x$$

Constructing a Taylor series for a given function is a tedious task. However, in some cases, we can do this rather easily.

DEFINITION

Geometric Series

A **geometric series** is a series of the form $\sum_{n=0}^{\infty} ar^n$. If $|r| < 1$, the geometric series converges and $\sum_{n=0}^{\infty} ar^n = \frac{a}{1-r}$. If $|r| \geq 1$, the geometric series diverges.

EXAMPLE: Find the Maclaurin series for $f(x) = \frac{x}{2 - 3x^2}$, and find the Taylor series for $g(x) = \frac{1}{3x - 5}$ about $x = 1$.

SOLUTION: In this case, we rewrite the function to find the Maclaurin series. Hence,

$$f(x) = \frac{x}{2 - 3x^2} = x\frac{1}{2 - 3x^2} = \frac{x}{2}\frac{1}{1 - \frac{3}{2}x^2} = \frac{x}{2}\sum_{n=0}^{\infty}\left(\frac{3}{2}x^2\right)^n = \sum_{n=0}^{\infty}\frac{3^n}{2^{n+1}}x^{2n+1}.$$

The first few terms of the Maclaurin series are obtained with series.

```
> series(x/(2-3*x^2),x=0);
```

$$\frac{1}{2}x + \frac{3}{4}x^3 + \frac{9}{8}x^5 + O(x^7)$$

Similarly,

$$g(x) = \frac{1}{3x - 5} = \frac{1}{3(x - 1) + 3 - 5} = \frac{1}{3(x - 1) - 2} = -\frac{1}{2}\frac{1}{1 - \frac{3}{2}(x - 1)}$$

$$= -\frac{1}{2}\sum_{n=0}^{\infty}\left(\frac{3}{2}(x - 1)\right)^n = \sum_{n=0}^{\infty} -\frac{3^n}{2^{n+1}}(x - 1)^n.$$

Frequently, the same results returned by **series** are obtained with **taylor**. For example, entering

```
> taylor(1/(3*x-5),x=1,8);
```

$$-\frac{1}{2} - \frac{3}{4}(x - 1) - \frac{9}{8}(x - 1)^2 - \frac{27}{16}(x - 1)^3 - \frac{81}{32}(x - 1)^4 - \frac{243}{64}(x - 1)^5$$

$$-\frac{729}{128}(x - 1)^6 - \frac{2187}{256}(x - 1)^7 + O((x - 1)^8)$$

computes the first eight terms of the Taylor series for $g(x) = \dfrac{1}{3x - 5}$ about $x = 1$. As with results returned by **series**, the last term of the output represents the omitted higher-order terms of the series.

Frequently used Maclaurin series are listed in the following table.

$\dfrac{1}{1 - x} = \displaystyle\sum_{n=0}^{\infty} x^n, x \in (-1, 1)$	$e^x = \displaystyle\sum_{n=0}^{\infty} \dfrac{1}{n!}x^n, x \in (-\infty, \infty)$
$\sin x = \displaystyle\sum_{n=0}^{\infty} \dfrac{(-1)^n x^{2n+1}}{(2n + 1)!}, x \in (-\infty, \infty)$	$\cos x = \displaystyle\sum_{n=0}^{\infty} \dfrac{(-1)^n x^{2n}}{(2n)!}, x \in (-\infty, \infty)$
$\ln(1 + x) = \displaystyle\sum_{n=1}^{\infty} \dfrac{(-1)^{n-1} x^n}{n}, x \in (-1, 1]$	$\tan^{-1} x = \displaystyle\sum_{n=0}^{\infty} \dfrac{(-1)^n x^{2n+1}}{2n + 1}, x \in [-1, 1]$
$\sinh x = \displaystyle\sum_{n=0}^{\infty} \dfrac{x^{2n+1}}{(2n + 1)!}, x \in (-\infty, \infty)$	$\cosh x = \displaystyle\sum_{n=0}^{\infty} \dfrac{x^{2n}}{(2n)!}, x \in (-\infty, \infty)$

A power series may be differentiated and integrated term-by-term on its interval of convergence. We state this more precisely in the following theorems.

· ·

THEOREM   *Term-by-Term Differentiation*

If the power series $\sum\limits_{n=0}^{\infty} a_n(x - x_0)^n$ has a radius of convergence $h > 0$ (h may be $+\infty$),

then the function $f(x) = \sum\limits_{n=0}^{\infty} a_n(x - x_0)^n$ has derivatives of all orders on its interval

of convergence, and $f'(x) = \sum\limits_{n=0}^{\infty} na_n(x - x_0)^{n-1}$, $f''(x) = \sum\limits_{n=0}^{\infty} n(n - 1)a_n(x - x_0)^{n-2}$,

and so on.

· ·

THEOREM   *Term-by-Term Integration*

If the power series $\sum\limits_{n=0}^{\infty} a_n(x - x_0)^n$ has radius of convergence $h > 0$ (h may be

$+\infty$), then the series $\sum\limits_{n=0}^{\infty} \dfrac{a_n}{n + 1}(x - x_0)^{n+1}$ has radius of convergence h. In fact, if

$f(x) = \sum\limits_{n=0}^{\infty} a_n(x - x_0)^n$, then $\int f(x)\,dx = \sum\limits_{n=0}^{\infty} \dfrac{a_n}{n + 1}(x - x_0)^{n+1} + c$.

· ·

EXAMPLE: (a) Use the geometric series $\dfrac{1}{1 - x} = \sum\limits_{n=0}^{\infty} x^n, x \in (-1, 1)$ to find the function that has

as its Maclaurin series $1 + 2x + 3x^2 + 4x^3 + \cdots = \sum\limits_{n=0}^{\infty} (n + 1)x^n$. (b) Use the Maclaurin series for

$f(x) = \dfrac{1}{1 + x^2}$ to obtain the Maclaurin series for $f(x) = \tan^{-1} x$.

SOLUTION: (a) First, notice that $1 + 2x + 3x^2 + 4x^3 + \cdots = \dfrac{d}{dx}(1 + x + x^2 + x^3 + x^4 + \cdots)$.

Therefore, $1 + 2x + 3x^2 + 4x^3 + \cdots = \dfrac{d}{dx}\left(\dfrac{1}{1 - x}\right) = \dfrac{1}{(1 - x)^2}$, so the Maclaurin series for $\dfrac{1}{(1 - x)^2}$

is $\sum\limits_{n=0}^{\infty} (n + 1)x^n$. By the ratio test, this series converges for $x \in (-1, 1)$. We graph $\dfrac{1}{(1 - x)^2}$ with

the polynomial of degree 8 obtained from $\sum\limits_{n=0}^{\infty} (n + 1)x^n$ on the interval $[-0.75, 0.75]$ by first

defining **ser_approx** to be the first nine terms of the Maclaurin series for $\dfrac{1}{(1 - x)^2}$. Note that

ser_approx cannot be graphed with **plot** because of the O-term, indicating the omitted
higher-order terms of the series.

> ```
> ser_approx:=series(1/(1-x)^2,x=0,9);
> ```

$$ser\_approx := 1 + 2x + 3x^2 + 4x^3 + 5x^4 + 6x^5 + 7x^6 + 8x^7 + 9x^8 + O(x^9)$$

We remove the O-term from `ser_approx` using `convert` together with the `polynom` option. We name the resulting polynomial of degree 8 `poly_approx`

> ```
> poly_approx:=convert(ser_approx,polynom);
> ```

$$poly\_approx := 1 + 2x + 3x^2 + 4x^3 + 5x^4 + 6x^5 + 7x^6 + 8x^7 + 9x^8$$

and then graph $\dfrac{1}{(1-x)^2}$ and `poly_approx` with `plot`.

> ```
> plot({1/(1-x)^2,poly_approx},x=-0.75..0.75);
> ```

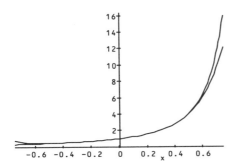

Notice that the approximation is not as accurate near the boundary of this interval.

(b) The Maclaurin series for $f(x) = \dfrac{1}{1 + x^2}$ is determined as follows:

$$\frac{1}{1 + x^2} = \frac{1}{1 - (-x^2)} = \sum_{n=0}^{\infty}(-x^2)^n = \sum_{n=0}^{\infty}(-1)^n x^{2n}.$$

Notice that $\displaystyle\int \frac{1}{1 + x^2}dx = \tan^{-1}x + c$. Therefore,

$$\tan^{-1}x + c = \int \sum_{n=0}^{\infty}(-1)^n x^{2n}\,dx = \sum_{n=0}^{\infty}\frac{(-1)^n x^{2n+1}}{2n + 1}.$$

To determine the value of the constant of integration, we substitute $x = 0$ into the equation to obtain

$$\tan^{-1}0 + c = \sum_{n=0}^{\infty}\frac{(-1)^n 0^{2n+1}}{2n + 1} = 0.$$

Thus, $c = 0$, and the Maclaurin series for $f(x) = \tan^{-1} x$ is $\displaystyle\sum_{n=0}^{\infty} \frac{(-1)^n x^{2n+1}}{2n + 1}$. By using the ratio

test for series, we can determine that this series converges for $x \in [-1, 1]$. As in (a), we graph

$f(x) = \tan^{-1} x$ along with the polynomial of degree 9 obtained from the series $\displaystyle\sum_{n=0}^{\infty} \frac{(-1)^n x^{2n+1}}{2n + 1}$.

Notice that the accuracy of the approximation decreases near the endpoints of the interval $[-1, 1]$.

```
> poly_approx:=convert(series(arctan(x),x=0,10),polynom);
```

$$poly\_approx := x - \frac{1}{3}x^3 + \frac{1}{5}x^5 - \frac{1}{7}x^7 + \frac{1}{9}x^9$$

```
> plot({arctan(x),poly_approx},x=-1.01..1.01);
```

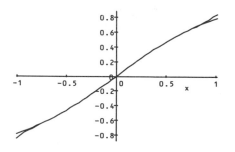

≡ *Reindexing a Power Series*

In the next sections, the ability to manipulate the index of a series will be important. In many cases, the power series must be modified before they can be combined. We demonstrate this process in the next example.

EXAMPLE: Simplify the sum

$$\sum_{n=2}^{\infty} n(n - 1)a_n x^{n-2} - \sum_{n=1}^{\infty} 5n a_n x^n + \sum_{n=0}^{\infty} 10 a_n x^n.$$

SOLUTION: We begin by determining the first term in each series. The series $\displaystyle\sum_{n=2}^{\infty} n(n - 1)a_n x^{n-2}$

begins with x^0, $\displaystyle\sum_{n=1}^{\infty} 5n a_n x^n$ begins with x, and $\displaystyle\sum_{n=0}^{\infty} 10 a_n x^n$ begins with x^0. Hence, in order to write the sum as a single series, we must pull off enough terms in the first and third series so that they all begin with x. In this case, we accomplish our goal by pulling off the first term in both of these series. Thus,

$$2a_2x^0 + \sum_{n=3}^{\infty} n(n-1)a_nx^{n-2} - \sum_{n=1}^{\infty} 5na_nx^n + 10a_0x^0 + \sum_{n=1}^{\infty} 10a_nx^n.$$

Now, all three series in sigma notation begin with x, but we must change the index in the first sum so that it starts with $n = 1$ like the other two. Substituting $(n + 2)$ for each occurrence of n in $\sum_{n=3}^{\infty} n(n-1)a_nx^{n-2}$, we obtain

$$\sum_{n+2=3}^{\infty} (n+2)(n+2-1)a_{n+2}x^{n+2-2} = \sum_{n=1}^{\infty} (n+2)(n+1)a_{n+2}x^n.$$

Therefore

$$2a_2x^0 + \sum_{n=3}^{\infty} n(n-1)a_nx^{n-2} - \sum_{n=1}^{\infty} 5na_nx^n + 10a_0x^0 + \sum_{n=1}^{\infty} 10a_nx^n$$

$$= 2a_2 + 10a_0 + \sum_{n=1}^{\infty} [(n+2)(n+1)a_{n+2} + (10-5n)a_n]x^n.$$

6.3 POWER SERIES SOLUTIONS ABOUT ORDINARY POINTS

We saw in Section 6.2 that Maclaurin and Taylor polynomials can be used to approximate functions. This idea can be extended to approximating the solution of a differential equation. First, we introduce some necessary terminology.

DEFINITION

Standard Form and Ordinary and Singular Points

Let f be a function with derivatives of all orders on an interval (a, b), and let $a < x_0 < b$.

Consider the equation $a_2(x)y''(x) + a_1(x)y'(x) + a_0(x)y(x) = 0$, and let $p(x) = \dfrac{a_1(x)}{a_2(x)}$ and $q(x) = \dfrac{a_0(x)}{a_2(x)}$. Then $a_2(x)y''(x) + a_1(x)y'(x) + a_0(x)y(x) = 0$ is equivalent to $y''(x) + p(x)y'(x) + q(x)y(x) = 0$, which is called the **standard form** of the equation. A number x_0 is an **ordinary point** means that both $p(x)$ and $q(x)$ are analytic at x_0. If x_0 is not an ordinary point, then x_0 is called a **singular point**.

When x_0 is an ordinary point of the differential equation $y''(x) + p(x)y'(x) + q(x)y(x) = 0$, we can write $p(x) = \sum_{n=0}^{\infty} b_n (x - x_0)^n$, where $b_n = \dfrac{p^{(n)}(x_0)}{n!}$, and $q(x) = \sum_{n=0}^{\infty} c_n(x - x_0)^n$, where $c_n = \dfrac{q^{(n)}(x_0)}{n!}$. Substitution into the equation $y''(x) + p(x)y'(x) + q(x)y(x) = 0$ results in

$$y''(x) + y'(x) \sum_{n=0}^{\infty} b_n(x - x_0)^n + y(x) \sum_{n=0}^{\infty} c_n(x - x_0)^n = 0.$$

If we assume that $y(x)$ is analytic at x_0, then we can write $y(x) = \sum_{n=0}^{\infty} a_n(x - x_0)^n$. Since a power series can be differentiated term-by-term, we can compute the first and second derivatives of y and substitute back into the equation to calculate the coefficients a_n. Thus, we can obtain a power series solution of the equation. We now provide a summary of this method.

☰ Power Series Solution Method about an Ordinary Point

1. Assume that $y = \sum_{n=0}^{\infty} a_n(x - x_0)^n$.

2. After taking the appropriate derivatives, substitute $y = \sum_{n=0}^{\infty} a_n(x - x_0)^n$ into the differential equation.

3. Find the unknown series coefficients a_n.

4. Apply any given initial conditions.

Since the differentiation of power series is necessary in this method for solving differential equations, we should make a few observations about this procedure. Consider the Maclaurin series $y = \sum_{n=0}^{\infty} a_n x^n$. Term-by-term differentiation of this series yields $y' = \sum_{n=0}^{\infty} na_n x^{n-1}$. Notice, however, that with the initial value of $n = 0$, the first term of the series is 0. Hence, we rewrite the series in its equivalent form

$$y' = \sum_{n=1}^{\infty} na_n x^{n-1}.$$

Similarly,

$$y'' = \sum_{n=1}^{\infty} n(n - 1)a_n x^{n-2} = \sum_{n=2}^{\infty} n(n - 1)a_n x^{n-2}.$$

We make use of these derivatives throughout the section.

EXAMPLE: (a) Find a general solution of $(4 - x^2)\dfrac{dy}{dx} + y(x) = 0$. (b) Solve the initial-value problem

$$\begin{cases} (4 - x^2)\dfrac{dy}{dx} + y(x) = 0 \\ y(0) = 1 \end{cases}.$$

SOLUTION: Let $y(x) = \displaystyle\sum_{n=0}^{\infty} a_n x^n$. Then term-by-term differentiation yields $y'(x) = \dfrac{dy}{dx} = \displaystyle\sum_{n=1}^{\infty} n a_n x^{n-1}$, and substitution into the differential equation gives us

$$(4 - x^2)\frac{dy}{dx} + y(x) = (4 - x^2) \sum_{n=1}^{\infty} n a_n x^{n-1} + \sum_{n=0}^{\infty} a_n x^n$$

$$= \sum_{n=1}^{\infty} 4n a_n x^{n-1} - \sum_{n=1}^{\infty} n a_n x^{n+1} + \sum_{n=0}^{\infty} a_n x^n = 0.$$

Note that the first term in these three series involves x^0, x^2, and x^0, respectively. Hence, if we pull off the first two terms in the first and third series, all three series will begin with x^2. Doing this, we have

$$(4a_1 + a_0) + (8a_2 + a_1)x + \sum_{n=3}^{\infty} 4n a_n x^{n-1} - \sum_{n=1}^{\infty} n a_n x^{n+1} + \sum_{n=2}^{\infty} a_n x^n = 0.$$

Unfortunately, the indices of these three series do not match, so we must change two of the three to match the third. Substitution of $(n + 1)$ for n in $\displaystyle\sum_{n=3}^{\infty} 4n a_n x^{n-1}$ yields

$$\sum_{n+1=3}^{\infty} 4(n + 1)a_{n+1} x^{n+1-1} = \sum_{n=2}^{\infty} 4(n + 1)a_{n+1} x^n.$$

Similarly, substitution of $(n - 1)$ for n in $\displaystyle\sum_{n=1}^{\infty} n a_n x^{n+1}$ yields

$$\sum_{n-1=1}^{\infty} (n - 1)a_{n-1} x^{n-1+1} = \sum_{n=2}^{\infty} (n - 1)a_{n-1} x^n.$$

Therefore, after combining the three series, we have the equation

$$(4a_1 + a_0) + (8a_2 + a_1)x + \sum_{n=2}^{\infty} (a_n + 4(n + 1)a_{n+1} - (n - 1)a_{n-1})x^n = 0.$$

Since the sum of the term on the left-hand side of the equation is zero, each coefficient must be zero. Equating the coefficients of x^0 and x to zero yields $a_1 = \dfrac{-a_0}{4}$ and $a_2 = \dfrac{-a_1}{8} = \dfrac{a_0}{32}$. When the series coefficient $a_n + 4(n+1)a_{n+1} - (n-1)a_{n-1}$ is set to zero, we obtain the recurrence relation $a_{n+1} = \dfrac{(n-1)a_{n-1} - a_n}{4(n+1)}$ for the indices in the series, $n \geq 2$. We use this formula to determine the values of a_n for $n = 2, 3, \ldots, 10$ and give these values in the following table.

n	a_n	n	a_n	n	a_n
0	a_0	4	$\dfrac{11}{2048}a_0$	8	$\dfrac{1843}{8388608}a_0$
1	$-\dfrac{1}{4}a_0$	5	$-\dfrac{31}{8192}a_0$	9	$-\dfrac{4859}{33554432}a_0$
2	$\dfrac{1}{32}a_0$	6	$\dfrac{69}{65536}a_0$	10	$\dfrac{12767}{268435456}a_0$
3	$-\dfrac{3}{28}a_0$	7	$-\dfrac{187}{262144}a_0$		

Therefore,

$$y = a_0 - \frac{1}{4}a_0 x + \frac{1}{32}a_0 x^2 - \frac{3}{28}a_0 x^3 + \frac{11}{2048}a_0 x^4 - \frac{31}{8192}a_0 x^5 + \frac{69}{65536}a_0 x^6$$
$$- \frac{187}{262144}a_0 x^7 + \frac{1843}{8388608}a_0 x^8 - \frac{4859}{33554432}a_0 x^9 + \frac{12767}{268435456}a_0 x^{10} + \cdots.$$

(b) When we apply the initial condition $y(0) = 1$, we substitute the $x = 0$ into the solution obtained in (a). Hence, $a_0 = 1$, so the series solution of the initial-value problem is

$$y = 1 - \frac{1}{4}x + \frac{1}{32}x^2 - \frac{3}{28}x^3 + \frac{11}{2048}x^4 - \frac{31}{8192}x^5 + \frac{69}{65536}x^6$$
$$- \frac{187}{262144}x^7 + \frac{1843}{8388608}x^8 - \frac{4859}{33554432}x^9 + \frac{12767}{268435456}x^{10} + \cdots.$$

Notice that the equation $(4 - x^2)\dfrac{dy}{dx} + y = 0$ is separable, so we can compute the solution directly with separation of variables by rewriting the equation as $\dfrac{-dy}{y} = \dfrac{dx}{4 - x^2}$. Integrating yields

$\ln(y) = \dfrac{\ln|x - 2| - \ln|x + 2|}{4} = \ln\left|\dfrac{x - 2}{x + 2}\right|^{1/4} + c$. Applying the initial condition $y(0) = 1$, we obtain $y = \left|\dfrac{x - 2}{x + 2}\right|^{1/4}$. We can approximate the solution of the problem by taking a finite number of terms of the series solution. We now use dsolve to find the exact solution to the problem, naming the resulting output Ex_Sol.

```
> Ex_Sol:=dsolve({(4-x^2)*diff(y(x),x)+y(x)=0,
  y(0)=1},y(x));
```

$$Ex\_Sol := y(x) = -\frac{1}{2}\frac{(x-2)^{1/4}(-2)^{3/4}2^{1/4}}{(x+2)^{1/4}}$$

Often, the command dsolve together with the series option can be used to generate a power series solution of a differential equation. Note that the variable Order has default value 6. Thus, to increase the number of terms displayed in the series solution, we increase the value of Order as the following input. We name the result of using dsolve together with the series option Ser_Sol.

```
> Order:=11:
  Ser_Sol:=dsolve({(4-x^2)*diff(y(x),x)+y(x)=0,
  y(0)=1},y(x),series);
```

$$Ser\_Sol := y(x) = 1 - \frac{1}{4}x + \frac{1}{32}x^2 - \frac{3}{128}x^3 + \frac{11}{2048}x^4 - \frac{31}{8192}x^5 + \frac{69}{65536}x^6 - \frac{187}{262144}x^7$$
$$+ \frac{1843}{8388608}x^8 - \frac{4859}{33554432}x^9 + \frac{12767}{268435456}x^{10} + O(x^{11})$$

In order to graph the approximating polynomial of degree 10, we must remove the O-term from Ser_Sol which we do with convert together with the polynom option, naming the resulting output Ser_Approx.

```
> Ser_Approx:=convert(rhs(Ser_Sol),polynom);
```

$$Ser\_Approx := 1 - \frac{1}{4}x + \frac{1}{32}x^2 - \frac{3}{128}x^3 + \frac{11}{2048}x^4 - \frac{31}{8192}x^5 + \frac{69}{65536}x^6 - \frac{187}{262144}x^7$$
$$+ \frac{1843}{8388608}x^8 - \frac{4859}{33554432}x^9 + \frac{12767}{268435456}x^{10}$$

We graph the exact solution and the approximate solution on the interval $[0, 2]$ with plot.

```
> plot({rhs(Ex_Sol),Ser_Approx},x=0..2);
```

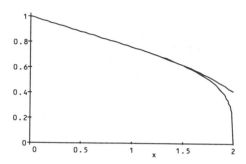

Notice the accuracy of the approximation decreases near $x = 2$, which is a singular point of the differential equation.

The following theorem explains where the approximation of the solution of the differential equation by the series is valid.

• •

THEOREM   *Convergence of the Power Series Solution*

Let $x = x_0$ be an ordinary point of the differential equation

$$a_2(x)y''(x) + a_1(x)y'(x) + a_0(x)y(x) = 0,$$

and suppose that D is the distance from $x = x_0$ to the closest singular point of the equation. Then the power series solution $y = \sum_{n=0}^{\infty} a_n(x - x_0)^n$ converges at least on the interval $(x_0 - D, x_0 + D)$.

• •

This theorem indicates that the approximation may not be as accurate near singular points of the equation. Hence, we understand why the approximation in the previous example breaks down near $x = 2$, the closest singular point to the ordinary point $x = 0$.

Of course, $x = 0$ is not an ordinary point for every differential equation. However, because the series $y = \sum_{n=0}^{\infty} a_n(x - x_0)^n$ is easier to work with when $x_0 = 0$, we can always make a transformation so that we can use $y = \sum_{n=0}^{\infty} a_n x^n$ to solve any linear equation. For example, suppose that $x = x_0$ is an ordinary point of a linear equation. If we make the change of variable $t = x - x_0$, then $t = 0$ corresponds to $x = x_0$. Hence, $t = 0$ is an ordinary point of the transformed equation. We illustrate this procedure in the following example.

EXAMPLE: Solve $xy'' + y = 0$.

SOLUTION: Notice that in standard form the equation becomes $y'' + \dfrac{1}{x}y = 0$. Because $\dfrac{1}{x}$ is not analytic at $x = 0$, this equation has a singular point at $x = 0$. All other values of x are classified as ordinary points, so we can select one to use in our power series solution. Choosing $x = 1$, we consider the power series $y = \displaystyle\sum_{n=0}^{\infty} a_n(x - 1)^n$. However, with the change of variable $t = x - 1$, we have that $t = 0$ corresponds to $x = 1$. Therefore, by changing variables, we can use the series $y = \displaystyle\sum_{n=0}^{\infty} a_n t^n$. Notice that $\dfrac{dy}{dx} = \dfrac{dy}{dt}\dfrac{dt}{dx} = \dfrac{dy}{dt}$ and

$$\frac{d^2y}{dx^2} = \frac{d}{dx}\left(\frac{dy}{dx}\right) = \frac{d}{dt}\left(\frac{dy}{dx}\right)\frac{dt}{dx} = \frac{d}{dt}\left(\frac{dy}{dt}\right) = \frac{d^2y}{dt^2},$$

so with these substitutions into $xy'' + y = 0$, we obtain $(t + 1)\dfrac{d^2y}{dt^2} + y = 0$. Hence, we assume that $y = \displaystyle\sum_{n=0}^{\infty} a_n t^n$. Substitution of this function into the transformed equation yields

$$(t + 1)\sum_{n=2}^{\infty} n(n - 1)a_n t^{n-2} + \sum_{n=0}^{\infty} a_n t^n = 0.$$

Simplification then gives us

$$\sum_{n=2}^{\infty} n(n - 1)a_n t^{n-1} + 2a_2 t^0 + \sum_{n=3}^{\infty} n(n - 1)a_n t^{n-2} + a_0 t^0 + \sum_{n=1}^{\infty} a_n t^n = 0.$$

In this case, we must change the index in two of the three series. If we substitute $(n + 1)$ for n in $\displaystyle\sum_{n=3}^{\infty} n(n - 1)a_n t^{n-2}$ and $(n - 1)$ for n in $\displaystyle\sum_{n=1}^{\infty} a_n t^n$, we obtain

$$(2a_2 + a_0) + \sum_{n=2}^{\infty} [n(n - 1)a_n + n(n + 1)a_{n+1} + a_{n-1}]t^{n-1} = 0.$$

Equating the coefficients to zero, we determine that $a_2 = -\dfrac{a_0}{2}$ and $a_{n+1} = \dfrac{-a_{n-1} - n(n - 1)a_n}{n(n + 1)}$ for the indices in the series, $n \geq 2$. If we replace each n in $a_{n+1} = \dfrac{-a_{n-1} - n(n - 1)a_n}{n(n + 1)}$ by $n - 1$, we obtain $a_n = \dfrac{-a_{n-2} - (n-2)(n-1)a_{n-1}}{n(n-1)}$ for $n \geq 2$. We first use `proc` to define this recursively defined function

```
> a:='a':
  a:=proc(n) option remember;
     (-a(n-2)-(n-1)*(n-2)*a(n-1))/((n-1)*n)
     end:
  a(0):=a_0:
  a(1):=a_1:
```

and then use `seq` and `array` to compute a_n for $n = 1, 2, 3, \ldots, 10$.

```
> array([seq([n,a(n)],n=1..10)]);
```

$$
\begin{bmatrix}
1 & a\_1 \\[2ex]
2 & -\dfrac{1}{2}a\_0 \\[2ex]
3 & -\dfrac{1}{6}a\_1 + \dfrac{1}{6}a\_0 \\[2ex]
4 & -\dfrac{1}{24}a\_0 + \dfrac{1}{12}a\_1 \\[2ex]
5 & -\dfrac{1}{24}a\_1 + \dfrac{1}{60}a\_0 \\[2ex]
6 & -\dfrac{7}{720}a\_0 + \dfrac{1}{40}a\_1 \\[2ex]
7 & -\dfrac{17}{1008}a\_1 + \dfrac{11}{1680}a\_0 \\[2ex]
8 & -\dfrac{191}{40320}a\_0 + \dfrac{41}{3360}a\_1 \\[2ex]
9 & -\dfrac{3359}{362880}a\_1 + \dfrac{163}{45360}a\_0 \\[2ex]
10 & -\dfrac{1463}{518400}a\_0 + \dfrac{1319}{181440}a\_1
\end{bmatrix}
$$

Hence,

$$
y = a_0 + a_1 t - \frac{a_0}{2}t^2 + \frac{a_0 - a_1}{6}t^3 + \frac{a_1 - a_0}{12}t^4 + \frac{a_0 - a_1}{24}t^5 + \frac{a_1 - a_0}{40}t^6 + \cdots
$$

$$
= a_0\left(1 - \frac{t^2}{2} + \frac{t^3}{6} - \frac{t^4}{12} + \frac{t^5}{24} - \frac{t^6}{40} + \cdots\right) + a_1\left(t - \frac{t^3}{6} + \frac{t^4}{12} - \frac{t^5}{24} + \frac{t^6}{40} + \cdots\right).
$$

Returning to the original variable, we have

$$y = a_0 \left(1 - \frac{(x-1)^2}{2} + \frac{(x-1)^3}{6} - \frac{(x-1)^4}{12} + \frac{(x-1)^5}{24} - \frac{(x-1)^6}{40} + \cdots \right)$$
$$+ a_1 \left((x-1) - \frac{(x-1)^3}{6} + \frac{(x-1)^4}{12} - \frac{(x-1)^5}{24} + \frac{(x-1)^6}{40} + \cdots \right).$$

As in the previous example, we can obtain the same results using **dsolve** together with the **series** option. Alternatively, we can take advantage of the command **powsolve**, which is contained in the **powseries** package. We load the **powseries** package by entering **with(powseries)**. Note that because a semicolon is included at the end of the command, a list of the functions contained in the **powseries** package is returned.

```
> with(powseries);
```

$$[add, compose, evalpow, inverse, multconst, multiply, negative, powcreate, powdiff, powexp,$$
$$powint, powlog, powpoly, powsolve, quotient, reversion, subtract, tpsform]$$

We then use **powsolve** to find a power series solution of $(t + 1)\dfrac{d^2y}{dt^2} + y = 0$, naming the resulting procedure **ser_y**.

```
> ser_y:=powsolve((t+1)*diff(y(t),t$2)+y(t)=0):
```

The first few terms of the power series are viewed using **tpsform**, which is also contained in the **powseries** package.

```
> ser_in_t:=tpsform(ser_y,t,11);
```

$$ser\_in\_t := C0 + C1\,t - \frac{1}{2}C0\,t^2 + \left(\frac{1}{6}C0 - \frac{1}{6}C1\right)t^3 + \left(-\frac{1}{24}C0 + \frac{1}{12}C1\right)t^4$$
$$+ \left(\frac{1}{60}C0 - \frac{1}{24}C1\right)t^5 + \left(-\frac{7}{720}C0 + \frac{1}{40}C1\right)t^6 + \left(\frac{11}{1680}C0 - \frac{17}{1008}C1\right)t^7$$
$$+ \left(-\frac{191}{40320}C0 + \frac{41}{3360}C1\right)t^8 + \left(\frac{163}{45360}C0 - \frac{3359}{362880}C1\right)t^9$$
$$+ \left(-\frac{1463}{518400}C0 + \frac{1319}{181440}C1\right)t^{10} + O(t^{11})$$

To obtain this solution, we replace the t in **ser_in_t** by $x - 1$ with **subs**.

```
> ser_in_x:=subs(t=x-1,ser_in_t);
```

$$ser\_in\_x := C0 + C1(x - 1) - \frac{1}{2}C0(x - 1)^2 + \left(\frac{1}{6}C0 - \frac{1}{6}C1\right)(x - 1)^3$$

$$+ \left(-\frac{1}{24}C0 + \frac{1}{12}C1\right)(x - 1)^4 + \left(\frac{1}{60}C0 - \frac{1}{24}C1\right)(x - 1)^5$$

$$+ \left(-\frac{7}{720}C0 + \frac{1}{40}C1\right)(x - 1)^6 + \left(\frac{11}{1680}C0 - \frac{17}{1008}C1\right)(x - 1)^7$$

$$+ \left(-\frac{191}{40320}C0 + \frac{41}{3360}C1\right)(x - 1)^8 + \left(\frac{163}{45360}C0 - \frac{3359}{362880}C1\right)(x - 1)^9$$

$$+ \left(-\frac{1463}{518400}C0 + \frac{1319}{181440}C1\right)(x - 1)^{10} + O((x - 1)^{11})$$

Power series solutions can be used to solve equations that involve functions that must be expressed as a power series. In the following example we see that **dsolve** cannot find an explicit solution, but it can generate a power series solution.

EXAMPLE: Consider the initial-value problem $y'' + f(x)y' + y(x) = 0$, $y(0) = 1$, $y'(0) = -1$, where

$$f(x) = \begin{cases} \dfrac{\sin x}{x}, & \text{if } x \neq 0 \\ 1, & \text{if } x = 0 \end{cases}.$$

Find a power series solution of the equation and graph an approximation of the solution on an interval.

SOLUTION: Since $\sin x = \sum_{n=0}^{\infty} \dfrac{(-1)^n x^{2n+1}}{(2n + 1)!}$ for all values of x, $f(x) = \sum_{n=0}^{\infty} \dfrac{(-1)^n x^{2n}}{(2n + 1)!}$. After defining **DEQ** to be the differential equation $y'' + f(x)y' + y(x) = 0$,

```
> DEQ:=diff(y(x),x$2)+sin(x)/x*diff(y(x),x)+y(x)=cos(x);
```

$$DEQ := \left(\frac{\partial^2}{\partial x^2}y(x)\right) + \frac{\sin(x)\left(\frac{\partial}{\partial x}y(x)\right)}{x} + y(x) = \cos(x)$$

we attempt to use **dsolve** to find a solution of the equation but are unsuccessful.

```
> dsolve({DEQ,y(0)=1,D(y)(0)=-1},y(x));
```

Conversely, we see that using **dsolve** together with the **series** option generates a series solution of the initial-value problem.

```
> Ser_1:=dsolve({DEQ,y(0)=1,D(y)(0)=-1},y(x),series);
```

$$Ser\_1 := y(x) = 1 - x + \frac{1}{2}x^2 - \frac{7}{72}x^4 + \frac{1}{36}x^5 + O(x^6)$$

We obtain more terms of the series solution by increasing the `Order` variable, with default value 6, to 16.

```
> Order:=16:
  Ser_2:=dsolve({DEQ,y(0)=1,D(y)(0)=-1},y(x),series);
```

$$Ser\_2 := y(x) = 1 - x + \frac{1}{2}x^2 - \frac{7}{72}x^4 + \frac{1}{36}x^5 + \frac{1}{3600}x^6 - \frac{277}{113400}x^7 + \frac{1741}{2540160}x^8$$

$$+ \frac{179}{19051200}x^9 - \frac{25853}{489888000}x^{10} + \frac{1147403}{94303440000}x^{11} + \frac{34187}{43908480000}x^{12}$$

$$- \frac{19136521}{19418964364800}x^{13} + \frac{239615237}{1460434510080000}x^{14}$$

$$+ \frac{84186614203}{3092470075094400000}x^{15} + O(x^{16})$$

We remove the O-term from `Ser_2` using `convert` together with the `polynom` option, naming the resulting polynomial `Poly_Approx`, which we graph on the interval $[0, 3]$ with `plot`.

```
> Poly_Approx:=convert(rhs(Ser_2),polynom);
```

$$Poly\_Approx := 1 - x + \frac{1}{2}x^2 - \frac{7}{72}x^4 + \frac{1}{36}x^5 + \frac{1}{3600} - \frac{277}{113400}x^7 + \frac{1741}{2540160}x^8$$

$$+ \frac{179}{19051200}x^9 - \frac{25853}{489888000}x^{10} + \frac{1147403}{94303440000}x^{11}$$

$$+ \frac{34187}{43908480000}x^{12} - \frac{19136521}{19418964364800}x^{13}$$

$$+ \frac{239615237}{1460434510080000}x^{14} + \frac{84186614203}{3092470075094400000}x^{15}$$

```
> plot(Poly_Approx,x=0..3);
```

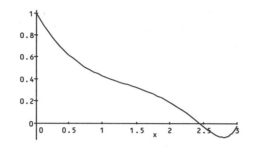

6.4 POWER SERIES SOLUTIONS ABOUT REGULAR SINGULAR POINTS

In Section 6.3, we used a power series expansion about an ordinary point to find (or approximate) the solution of a differential equation. We noted that these series solutions may not converge near the singular points of the equation.

☰ Regular and Irregular Singular Points

In this section, we investigate the problem of obtaining a series expansion about a singular point. We begin with the following classification of singular points.

DEFINITION	*Singular Points*
	Let x_0 be singular point of $y''(x) + p(x)y'(x) + q(x)y(x) = 0$. x_0 is a **regular singular point** of the equation if both $(x - x_0)p(x)$ and $(x - x_0)^2q(x)$ are analytic at $x = x_0$. If x_0 is not a regular singular point, then x_0 is called an **irregular singular point** of the equation.

Sometimes this definition is difficult to apply. Therefore, we supply the following equivalent definition for polynomial coefficients p and q of the equation $y''(x) + p(x)y'(x) + q(x)y(x) = 0$.

DEFINITION | *Singular Points of Equations with Polynomial Coefficients*

Suppose that $p(x)$ and $q(x)$ are polynomials with no common factors. If, after reducing $p(x)$ and $q(x)$ to lowest terms, the highest power of $(x - x_0)$ in the denominator of $p(x)$ is 1 and the highest power of $(x - x_0)$ in the denominator of $q(x)$ is 2, then x_0 is a **regular singular point** of the equation. Otherwise, it is an **irregular singular point**.

EXAMPLE: Classify the singular points of each of the following equations: (a) $y'' + \dfrac{y'}{x} + \left(1 - \dfrac{\mu^2}{x^2}\right)y = 0$, (b) $(x^2 - 16)^2 y'' + (x - 4)y' + y = 0$, and (c) $x^2 y'' + xy' + \left(x^2 - \mu^2\right)y = 0$ (the Bessel equation).

SOLUTION: (a) $x = 0$ is a singular point of this equation because $p(x) = \dfrac{1}{x}$ is not analytic at $x = 0$. Since $xp(x) = 1$ and $x^2 q(x) = x^2\left(1 - \dfrac{\mu^2}{x^2}\right) = x^2 - \mu^2$, $x = 0$ is a regular singular point.

(b) In standard form, the equation is

$$y'' + \frac{x - 4}{(x^2 - 16)^2}y' + \frac{1}{(x^2 - 16)^2}y = 0$$

or

$$y'' + \frac{1}{(x - 4)(x + 4)^2}y' + \frac{1}{(x - 4)^2(x + 4)^2}y = 0.$$

Thus, the singular points are $x = 4$ and $x = -4$. For $x = 4$, we have

$$(x - 4)p(x) = (x - 4)\frac{1}{(x - 4)(x + 4)^2} = \frac{1}{(x + 4)^2}$$

and

$$(x - 4)^2 q(x) = (x - 4)^2 \frac{1}{(x - 4)^2(x + 4)^2} = \frac{1}{(x + 4)^2}.$$

Both of these functions are analytic at $x = 4$, so $x = 4$ is a regular singular point. For $x = -4$,

$$(x + 4)p(x) = (x + 4)\frac{1}{(x - 4)(x + 4)^2} = \frac{1}{(x - 4)(x + 4)},$$

which is not analytic at $x = -4$. Thus, $x = -4$ is an irregular singular point.

(c) Written in standard form, the Bessel equation is

$$y'' + \frac{1}{x}y' + \frac{(x^2 - \mu^2)}{x^2}y = 0.$$

Hence, $x = 0$ is a singular point. Because the functions $xp(x) = x\frac{1}{x} = 1$ and

$$x^2 q(x) = x^2 \frac{(x^2 - \mu^2)}{x^2} = x^2 - \mu^2$$

are analytic at $x = 0$, we classify $x = 0$ as a regular singular point.

☰ Method of Frobenius

In Section 6.2, we discussed how a power series expansion about an ordinary point can be used to solve a differential equation. Here, we show that a power series expansion about a regular singular point can be used to solve an equation.

· ·

THEOREM   *Method of Frobenius*

Let x_0 be a regular singular point of $y''(x) + p(x)y'(x) + q(x)y(x) = 0$. Then this differential equation has at least one solution of the form

$$y = \sum_{n=0}^{\infty} a_n(x - x_0)^{n+r},$$

where r is a constant that must be determined. This solution is convergent at least on some interval $|x - x_0| < R$.

· ·

EXAMPLE: Find a general solution of $xy'' + (1 + x)y' - \dfrac{1}{16x}y = 0$.

SOLUTION: First, we note that in standard form this equation is $y'' + \dfrac{1 + x}{x}y' - \dfrac{1}{16x^2}y = 0$.

Thus, $x = 0$ is a singular point. Moreover, because $xp(x) = x\dfrac{1 + x}{x} = 1 + x$ and $x^2q(x) =$

$x^2\left(-\dfrac{1}{16x^2}\right) = -16$ are both analytic at $x = 0$, we classify $x = 0$ as a regular singular point. By

the earlier theorem, there is at least one solution of the form $y = \displaystyle\sum_{n=0}^{\infty} a_n x^{n+r}$. Differentiating this

function, we obtain $y' = \displaystyle\sum_{n=0}^{\infty} a_n(n + r)x^{n+r-1}$ and $y'' = \displaystyle\sum_{n=0}^{\infty} a_n(n + r)(n + r - 1)x^{n+r-2}$. Substituting

this power series into the differential equation yields

$$x\sum_{n=0}^{\infty} a_n(n + r)(n + r - 1)x^{n+r-2} + (1 + x)\sum_{n=0}^{\infty} a_n(n + r)x^{n+r-1} - \frac{1}{16x}\sum_{n=0}^{\infty} a_n x^{n+r} = 0$$

and

$$\sum_{n=0}^{\infty} a_n(n + r)(n + r - 1)x^{n+r-1} + \sum_{n=0}^{\infty} a_n(n + r)x^{n+r-1} + \sum_{n=0}^{\infty} a_n(n + r)x^{n+r} - \sum_{n=0}^{\infty} \frac{1}{16}a_n x^{n+r-1} = 0.$$

Notice that the first term in three of the four series is x^{r-1}. However, the first term in

$$\sum_{n=0}^{\infty} a_n(n + r)x^{n+r}$$

is x^r, so we must pull off the first terms in the other three series so that they match. Hence,

$$\left[r(r - 1) + r - \frac{1}{16}\right]a_0 x^{r-1} + \sum_{n=1}^{\infty} a_n(n + r)(n + r - 1)x^{n+r-1} + \sum_{n=1}^{\infty} a_n(n + r)x^{n+r-1}$$

$$+ \sum_{n=0}^{\infty} a_n(n + r)x^{n+r} - \sum_{n=1}^{\infty} \frac{1}{16}a_n x^{n+r-1} = 0.$$

Changing the index in the third series by substituting $(n - 1)$ for each occurrence of n, we have

$$\sum_{n-1=0}^{\infty} a_{n-1}(n - 1 + r)x^{n-1+r} = \sum_{n=1}^{\infty} a_{n-1}(n + r - 1)x^{n+r-1}.$$

After simplification, we have

$$\left[r(r - 1) + r - \frac{1}{16} \right] a_0 x^{r-1} + \sum_{n=1}^{\infty} \left\{ \left[(n + r)(n + r - 1) + (n + r) - \frac{1}{16} \right] a_n \right.$$

$$+ (n + r - 1)a_{n-1} \} x^{n+r-1} = 0.$$

We equate the coefficients to zero to find the coefficients and the value of r. Assuming that $a_0 \neq 0$ so that the first term of our series solution is not zero, we have from the first term the equation

$$r(r - 1) + r - \frac{1}{16} = 0,$$

called the **indicial equation**, because it yields the value of r. In this case,

$$r^2 - r + r - \frac{1}{16} = r^2 - \frac{1}{16} = \left(r + \frac{1}{4} \right)\left(r - \frac{1}{4} \right) = 0,$$

so the roots are $r_1 = \frac{1}{4}$ and $r_2 = -\frac{1}{4}$. Thus, the differential equation has solutions

$$y = \sum_{n=0}^{\infty} a_n x^{n+1/4} = x^{1/4} \sum_{n=0}^{\infty} a_n x^n \quad \text{and} \quad y = \sum_{n=0}^{\infty} b_n x^{n-1/4} = x^{-1/4} \sum_{n=0}^{\infty} b_n x^n.$$

Starting with the larger of the two roots, if $r_1 = \frac{1}{4}$ and we assume that $y = \sum_{n=0}^{\infty} a_n x^{n+1/4} = x^{1/4} \sum_{n=0}^{\infty} a_n x^n$, then we have $\left[\left(n + \frac{1}{4} \right)\left(n + \frac{1}{4} - 1 \right) + \left(n + \frac{1}{4} \right) - \frac{1}{16} \right] a_n + \left(n + \frac{1}{4} - 1 \right) a_{n-1} = 0$, so

$$a_n = \frac{-\left(n - \frac{3}{4} \right) a_{n-1}}{\left(n + \frac{1}{4} \right)\left(n - \frac{3}{4} \right) + \left(n + \frac{1}{4} \right) - \frac{1}{16}} = \frac{(3 - 4n)a_{n-1}}{2(2n^2 + n)}, \quad n \geq 1.$$

After defining the recursively defined function a, several of these coefficients are calculated using `seq` and `array`.

```
> a:='a':
  a:=proc(n) option remember;
    (3-4*n)*a(n-1)/(2*(2*n^2+n))
    end:
  a(0):=a0:
  array([seq([n,a(n)],n=0..10)]);
```

$$
\begin{bmatrix}
0 & a0 \\
1 & -\dfrac{1}{6}a0 \\
2 & \dfrac{1}{24}a0 \\
3 & -\dfrac{1}{112}a0 \\
4 & \dfrac{13}{8064}a0 \\
5 & -\dfrac{221}{887040}a0 \\
6 & \dfrac{17}{506880}a0 \\
7 & -\dfrac{17}{4257792}a0 \\
8 & \dfrac{29}{68124672}a0 \\
9 & -\dfrac{29}{706019328}a0 \\
10 & \dfrac{1073}{296528117760}a0
\end{bmatrix}
$$

In this case, the solution is $y_1 = a_0 x^{1/4}\left(1 - \dfrac{1}{6}x + \dfrac{1}{24}x^2 - \dfrac{7}{112}x^3 + \dfrac{11}{8064}x^4 - \cdots\right)$. If $r_2 = -\dfrac{1}{4}$ and

we assume that $y = \displaystyle\sum_{n=0}^{\infty} b_n x^{n_1/4} = x^{-1/4}\sum_{n=0}^{\infty} b_n x^n$, we have

$$
\left[\left(n - \frac{1}{4}\right)\left(n - \frac{1}{4} - 1\right) + \left(n - \frac{1}{4}\right) - \frac{1}{16}\right]b_n + \left(n - \frac{1}{4} - 1\right)b_{n-1} = 0,
$$

so

$$
b_n = \frac{-\left(n - \frac{5}{4}\right)b_{n-1}}{\left(n - \frac{1}{4}\right)\left(n - \frac{5}{4}\right) + \left(n - \frac{1}{4}\right) - \frac{1}{16}} = \frac{(5 - 4n)b_{n-1}}{2(2n^2 - n)}, n \geq 1.
$$

The values of several coefficients are determined with this formula in the same manner as for a.

```
> b:=proc(n) option remember;
  (5-4*n)*b(n-1)/(2*(2*n^2-n))
  end:
b(0):=b0:
array([seq([n,b(n)],n=0..10)]);
```

$$
\begin{bmatrix}
0 & b0 \\
1 & \dfrac{1}{2}b0 \\
2 & -\dfrac{1}{8}b0 \\
3 & \dfrac{7}{240}b0 \\
4 & -\dfrac{11}{1920}b0 \\
5 & \dfrac{11}{11520}b0 \\
6 & -\dfrac{19}{138240}b0 \\
7 & \dfrac{437}{25159680}b0 \\
8 & -\dfrac{437}{223641600}b0 \\
9 & \dfrac{13547}{68434329600}b0 \\
10 & -\dfrac{713}{39105331200}b0
\end{bmatrix}
$$

Therefore, the solution obtained with $r_2 = -\dfrac{1}{4}$ is

$$
y_2(x) = b_0 x^{-1/4} \left(1 + \frac{1}{2}x - \frac{1}{8}x^2 + \frac{7}{240}x^3 - \frac{11}{1920}x^4 - \cdots \right).
$$

Therefore, a general solution of the differential equation is

$$
y(x) = c_1 x^{1/4} \left(1 - \frac{1}{6}x + \frac{1}{24}x^2 - \frac{7}{112}x^3 + \frac{11}{8064}x^4 - \cdots \right)
$$
$$
+ c_2 x^{-1/4} \left(1 + \frac{1}{2}x - \frac{1}{8}x^2 + \frac{7}{240}x^3 - \frac{11}{1920}x^4 - \cdots \right),
$$

where c_1 and c_2 are arbitrary constants. Notice that these two solutions are linearly independent because they are not scalar multiples of one another. We obtain equivalent results using dsolve together with the series option as follows:

```
> dsolve(x*diff(y(x),x$2)+(1+x)*diff(y(x),x)-
      1/(16*x)*y(x)=0,y(x),series);
```

$$y(x) = \_C1\left(1 + \frac{1}{2}x - \frac{1}{8}x^2 + \frac{7}{240}x^3 - \frac{11}{1920}x^4 + \frac{11}{11520}x^5 + O(x^6)\right)\Big/x^{1/4}$$

$$+ \_C2x^{1/4}\left(1 - \frac{1}{6}x + \frac{1}{24}x^2 - \frac{1}{112}x^3 + \frac{13}{8064}x^4 - \frac{221}{887040}x^5 + O(x^6)\right)$$

In the previous example, we found the **indicial equation** by direct substitution of the power series solution into the differential equation. In order to derive a general formula for the indicial equation, suppose that $x = 0$ is a regular singular point of the differential equation $y'' + p(x)y' + q(x)y = 0$. Then the functions $xp(x)$ and $x^2q(x)$ are analytic, which means that both of these functions have a power series in x with a positive radius of convergence. Hence,

$$xp(x) = p_0 + p_1x + p_2x^2 + \cdots \qquad \text{and} \qquad x^2q(x) = q_0 + q_1x + q_2x^2 + \cdots.$$

Therefore,

$$p(x) = \frac{p_0}{x} + p_1 + p_2x + p_3x^2 + p_4x^3 + \cdots \quad \text{and} \quad q(x) = \frac{q_0}{x^2} + \frac{q_1}{x} + q_2 + q_3x + q_4x^2 + q_5x^3 + \cdots.$$

Substitution of these series into the differential equation $y'' + p(x)y' + q(x)y = 0$ and multiplying through by the first term in the power series for $p(x)$ and $q(x)$, we see that the lowest term in the series is x^{n+r-2}.

$$\left(\sum_{n=0}^{\infty} a_n(n+r)(n+r-1)x^{n+r-2}\right) + \left(\sum_{n=0}^{\infty} a_np_0(n+r)x^{n+r-2}\right)$$

$$+ (p_1 + p_2x + p_3x^2 + p_4x^3 + \cdots)\left(\sum_{n=0}^{\infty} a_n(n+r)x^{n+r-1}\right) + \left(\sum_{n=0}^{\infty} a_nq_0x^{n+r-2}\right)$$

$$+ \left(\frac{q_1}{x} + q_2 + q_3x + q_4x^2 + q_5x^3 + \cdots\right)\left(\sum_{n=0}^{\infty} a_nx^{n+r}\right) = 0.$$

Then, with $n = 0$, we find that the coefficient of x^{r-2} is

$$-ra_0 + r^2a_0 + ra_0p_0 + a_0q_0 = a_0(r^2 + (p_0 - 1)r + q_0) = a_0(r(r-1) + p_0r + q_0).$$

Thus, for any equation of the form $y'' + p(x)y' + q(x)y = 0$ with regular singular point $x = 0$, we have the indicial equation $r(r - 1) + p_0 r + q_0 = 0$.

The values of r are called the **exponents**, or **indicial roots**. Generally, several situations can arise when finding the roots of the indicial equation: (1) the roots are distinct and differ by a fractional value, (2) the roots are distinct and differ by an integer value, and (3) the roots are equal. We discuss the following situations.

If $x = 0$ is a regular singular point of $y'' + p(x)y' + q(x)y = 0$, the roots of the indicial equation $r^2 + (p_0 - 1)r + q_0 = 0$ are $r_1 = \dfrac{1 - p_0 + \sqrt{1 - 2p_0 + p_0^2 - 4q_0}}{2}$ and $r_2 = \dfrac{1 - p_0 - \sqrt{1 - 2p_0 + p_0^2 - 4q_0}}{2}$,

where $r_1 \geq r_2$ and $r_1 - r_2 = \sqrt{1 - 2p_0 + p_0^2 - 4q_0}$.

· ·

CASE I   If $r_1 \neq r_2$ and $r_1 - r_2 = \sqrt{1 - 2p_0 + p_0^2 - 4q_0}$ is not an integer, then there are two linearly independent solutions of the equation of the form $y_1(x) = x^{r_1} \sum\limits_{n=0}^{\infty} a_n x^n$ and

$y_2(x) = x^{r_2} \sum\limits_{n=0}^{\infty} b_n x^n$.

· ·

CASE 2   If $r_1 \neq r_2$ and $r_1 - r_2 = \sqrt{1 - 2p_0 + p_0^2 - 4q_0}$ is an integer, then there are two linearly independent solutions of the equation of the form $y_1(x) = x^{r_1} \sum\limits_{n=0}^{\infty} a_n x^n$ and $y_2(x) =$

$cy_1(x) \ln(x) + x^{r_2} \sum\limits_{n=0}^{\infty} b_n x^n$.

· ·

CASE 3   If $r_1 - r_2 = \sqrt{1 - 2p_0 + p_0^2 - 4q_0} = 0$, then there are two linearly independent solutions of the problem of the form $y_1(x) = x^{r_1} \sum\limits_{n=0}^{\infty} a_n x^n$ and $y_2(x) = y_1(x) \ln x +$

$x^{r_1} \sum\limits_{n=0}^{\infty} b_n x^n$. In any case, if $y_1(x)$ is a solution of the equation, a second linearly independent solution is given by $y_2(x) = y_1(x) \int \dfrac{e^{-\int p(x)\,dx}}{[y_1(x)]^2}\,dx$, which is obtained through reduction of order.

· ·

Hint: When solving a differential equation in Case II, first attempt to find a general solution using $y = x^{r_2} \sum\limits_{n=0}^{\infty} a_n x^n = \sum\limits_{n=0}^{\infty} a_n x^{n+r_2}$, where r_2 is the smaller of the two roots. A general solution can sometimes be found with this procedure. However, if the contradiction $a_0 = 0$ is reached, then find solutions of the form $y_1(x) = x^{r_1} \sum_{n=0}^{\infty} a_n x^n$ and $y_2(x) = c y_1(x) \ln(x) + x^{r_2} \sum_{n=0}^{\infty} b_n x^n$.

The following examples do not illustrate the possibility of complex-valued roots of the indicial equation. When this occurs, the equation is solved using the procedures of Case I. The solutions that are obtained are complex, so they can be transformed into real solutions by taking the appropriate linear combinations such as those discussed for complex-valued roots of the characteristic equation of Cauchy-Euler differential equations.

Also, we have not mentioned if a solution can be found with an expansion about an irregular singular point. If $x = x_0$ is an irregular singular point of $y'' + p(x)y' + q(x)y = 0$, then there may or may not be a solution of the form $y = \sum_{n=0}^{\infty} a_n (x - x_0)^{n-r}$.

☰ Indicial Roots that Differ by an Integer

EXAMPLE: Find a general solution of $xy'' + 3y' - y = 0$ using a series expansion about the regular singular point $x = 0$.

SOLUTION: In standard form, this equation is $y'' + \dfrac{3}{x} y' - \dfrac{1}{x} y = 0$. Hence, $xp(x) = x\left(\dfrac{3}{x}\right) = 3$ and $x^2 q(x) = x^2\left(-\dfrac{1}{x}\right) = -x$, so $p_0 = 3$ and $q_0 = 0$. Thus, the indicial equation is $r(r - 1) + 3r = r^2 + 2r = r(r + 2) = 0$ with roots $r_1 = 0$ and $r_2 = -2$. (Notice that we always use r_1 to denote the larger root.) We can verify that there is no series solution corresponding to $r_2 = -2$, so we assume there is a solution of the form $y = \sum_{n=0}^{\infty} a_n x^n$ corresponding to $r_1 = 0$ with derivatives $y' = \sum_{n=1}^{\infty} n a_n x^{n-1}$ and $y'' = \sum_{n=2}^{\infty} n(n - 1) a_n x^{n-2}$. Substitution into the differential equation and simplifying the result yields

$$(3a_1 - a_0)x^0 + \sum_{n=2}^{\infty} \{[n(n - 1) + 3n] a_n - a_{n-1}\}x^{n-1} = 0.$$

Equating the coefficients to zero, we have $a_1 = \dfrac{1}{3} a_0$ and $a_n = \dfrac{a_{n-1}}{n(n-1) + 3n} = \dfrac{a_{n-1}}{n^2 + 2n}$, $n \geq 2$. After defining a, we use this formula together with seq and array to calculate several coefficients

```
> a:='a':
  a:=proc(n) option remember;
    a(n-1)/(n^2+2*n)
    end:
  a(0):=a0:
  array([seq([n,a(n)],n=0..10)]);
```

$$
\begin{bmatrix}
0 & a0 \\
1 & \dfrac{1}{3}a0 \\
2 & \dfrac{1}{24}a0 \\
3 & \dfrac{1}{360}a0 \\
4 & \dfrac{1}{8640}a0 \\
5 & \dfrac{1}{302400}a0 \\
6 & \dfrac{1}{14515200}a0 \\
7 & \dfrac{1}{914457600}a0 \\
8 & \dfrac{1}{73156608000}a0 \\
9 & \dfrac{1}{7242504192000}a0 \\
10 & \dfrac{1}{869100503040000}a0
\end{bmatrix}
$$

and use them to form $y_1(x) = a_0 \left(1 + \dfrac{x}{3} + \dfrac{x^2}{24} + \dfrac{x^3}{360} + \cdots \right)$.

To determine the second linearly independent solution, we assume that $y_2(x) = y_1(x) \ln x + \sum_{n=0}^{\infty} b_n x^{n-2}$ and substitute this formula into the differential equation to find the coefficients b_n. Because the derivatives of $y_2(x)$ are

$$
y_2'(x) = \frac{y_1(x)}{x} + y_1'(x) \ln x + \sum_{n=0}^{\infty} (n-2) b_n x^{n-3}
$$

and

$$y_2''(x) = \frac{-y_1(x)}{x^2} + \frac{2y_1'(x)}{x} + y_1''(x)\ln x + \sum_{n=0}^{\infty}(n-2)(n-3)b_n x^{n-4},$$

substitution into the differential equation yields

$$\frac{2y_1(x)}{x} + 2y_1'(x) + \sum_{n=0}^{\infty}(n-2)(n-3)b_n x^{n-3} + \sum_{n=0}^{\infty}3(n-2)b_n x^{n-3} - \sum_{n=0}^{\infty}b_n x^{n-2}$$

$$+ \underbrace{\left[xy_1'(x) + 3y_1'(x) - y_1(x)\right]}_{\substack{=0 \text{ because } y_1 \text{ is a solution} \\ \text{of the differential equation}}}\ln x = 0.$$

Simplifying this expression gives us

$$\frac{2y_1(x)}{x} + 2y_1'(x) + 6b_0 x^{-3} - 6b_0 x^{-3}$$

$$+ \sum_{n=1}^{\infty}(n-2)(n-3)b_n x^{n-3} + \sum_{n=1}^{\infty}3(n-2)b_n x^{n-3} - \sum_{n=1}^{\infty}b_n x^{n-2} = 0$$

$$\frac{2y_1(x)}{x} + 2y_1'(x) + \sum_{n=1}^{\infty}[(n-2)nb_n - b_{n-1}]x^{n-3} = 0.$$

Now let $a_0 = 1$, so $y_1(x) = 1 + \dfrac{x}{3} + \dfrac{x^2}{24} + \dfrac{x^3}{360} + \cdots$ and $y_1'(x) = \dfrac{1}{3} + \dfrac{x}{12} + \dfrac{x^2}{120} + \cdots$. Substitution into the previous equation then yields

$$\frac{2}{x} + \frac{4}{3} + \frac{x}{4} + \frac{7x^2}{40} + \cdots + \sum_{n=1}^{\infty}[(n-2)nb_n - b_{n-1}]x^{n-3} = 0$$

$$\frac{2}{x} + \frac{4}{3} + \frac{x}{4} + \frac{7x^2}{40} + \cdots + [-b_1 - b_0]x^{-2} - b_1 x^{-1}$$

$$+ (3b_3 - b_2)x^0 + (8b_4 - b_3)x + (15b_5 - b_4)x^2 = 0,$$

so we have the sequence of equations $-b_1 - b_0 = 0$, $-b_1 + 2 = 0$, $3b_3 - b_2 + \dfrac{4}{3} = 0$, $8b_4 - b_3 + \dfrac{1}{4} = 0$, $15b_5 - b_4 + \dfrac{7}{40} = 0$, ... Solving these equations, we see that $b_1 = 2$ and $b_0 = -2$. However, the other coefficients depend on the value of b_2. We compute several of these coefficients as follows.

n	b_n
0	-2
1	2
2	b_2
3	$\dfrac{3b_2 - 4}{9}$
4	$\dfrac{12b_2 - 25}{288}$
5	$\dfrac{30b_2 - 201}{720}$

Hence,

$$y_2(x) = y_1(x)\ln x + x^{-2}\left(-2 + 2x + b_2 x^2 + \frac{3b_2 - 4}{9}x^3 + \frac{12b_2 - 25}{288}x^4 + \frac{30b_2 - 201}{720}x^5 + \cdots\right),$$

where b_2 is arbitrary. A general solution, therefore, is given by $y(x) = c_1 y_1(x) + c_2 y_2(x)$, where c_1 and c_2 are arbitrary.

As in the previous example, we see that we obtain the same result using **dsolve** together with the **series** option.

```
> dsolve(x*diff(y(x),x$2)+3*diff(y(x),x)-y(x)=0,y(x),series);
```

$$y(x) = \_C1\left(1 + \frac{1}{3}x + \frac{1}{24}x^2 + \frac{1}{360}x^3 + \frac{1}{8640}x^4 + \frac{1}{302400}x^5 + O(x^6)\right)$$

$$+ \_C2\left(\frac{\ln(x)\left(x^2 + \frac{1}{3}x^3 + \frac{1}{24}x^4 + \frac{1}{360}x^5 + O(x^6)\right)}{x^2}\right.$$

$$\left. + \frac{-2 + 2x - \frac{4}{9}x^3 - \frac{25}{288}x^4 - \frac{157}{21600}x^5 + O(x^6)}{x^2}\right)$$

☰ Equal Indicial Roots

We now demonstrate a method for solving a differential equation with equal indicial roots.

EXAMPLE: Find a general solution of $xy'' + (2 - x)y' + \dfrac{1}{4x}y = 0$ by using a series about the regular singular point $x = 0$.

SOLUTION: In standard form, the equation is $y'' + \dfrac{2 - x}{x}y' + \dfrac{1}{4x^2}y = 0$. Since $xp(x) = x\left(\dfrac{2 - x}{x}\right) = 2 - x$ and $x^2q(x) = x^2\left(\dfrac{1}{4x^2}\right) = \dfrac{1}{4}$, $p_0 = 2$ and $q_0 = \dfrac{1}{4}$. Thus, the indicial equation is $r^2 + (2 - 1)r + \dfrac{1}{4} = 0$, which has equal roots $r_1 = -\dfrac{1}{2}$ and $r_2 = -\dfrac{1}{2}$. Then there is a solution of the form $y_1(x) = x^{-1/2}\displaystyle\sum_{n=0}^{\infty} a_nx^n = \sum_{n=0}^{\infty} a_nx^{n-1/2}$. Replacing y in the equation by y_1 and simplifying yields

$$\sum_{n=0}^{\infty}\left(n - \frac{1}{2}\right)\left(n - \frac{3}{2}\right)a_nx^{n-3/2} + \sum_{n=0}^{\infty}2\left(n - \frac{1}{2}\right)a_nx^{n-3/2} - \sum_{n=0}^{\infty}\left(n - \frac{1}{2}\right)a_nx^{n-12} + \sum_{n=0}^{\infty}\frac{a_n}{4}x^{n-3/2}$$

$$= \sum_{n=1}^{\infty}\left[\left(n - \frac{1}{2}\right)\left(n - \frac{3}{2}\right)a_n + 2\left(n - \frac{1}{2}\right)a_n - \left(n - \frac{3}{2}\right)a_{n-1} + \frac{a_n}{4}\right]x^{n-3/2}$$

$$= \sum_{n=1}^{\infty}\left[a_nn^2 - \frac{2n - 3}{2}a_{n-1}\right]x^{n-3/2}$$

$$= 0.$$

Then, equating coefficients, we find that $a_n = a_{n-1}\dfrac{2n - 3}{2n^2}$. Therefore, for $a_0 \neq 0$, we obtain the coefficients as follows:

```
> a:='a':
  a:=proc(n) option remember;
    a(n-1)*(2*n-3)/(2*n^2);
    end:
  a(0):=a0:
  array([seq([n,a(n)],n=0..10)]);
```

$$
\begin{bmatrix}
0 & a0 \\
1 & -\dfrac{1}{2}a0 \\
2 & -\dfrac{1}{16}a0 \\
3 & -\dfrac{1}{96}a0 \\
4 & -\dfrac{5}{3072}a0 \\
5 & -\dfrac{7}{30720}a0 \\
6 & -\dfrac{7}{245760}a0 \\
7 & -\dfrac{11}{3440640}a0 \\
8 & -\dfrac{143}{440401920}a0 \\
9 & -\dfrac{143}{4756340736}a0 \\
10 & -\dfrac{2431}{951268147200}a0
\end{bmatrix}
$$

Using these coefficients,

$$
y_1(x) = a_0 \left(x^{-1/2} - \frac{1}{2}x^{1/2} - \frac{1}{16}x^{3/2} - \frac{1}{96}x^{5/2} - \frac{5}{3072}x^{7/2} - \frac{7}{30720}x^{9/2} - \cdots \right).
$$

Because the roots of the indicial equation are equal, there is a second linearly independent solution of the form $y_2(x) = y_1(x)\ln x + \sum_{n=1}^{\infty} b_n x^{n-1/2}$. Substituting y_2 into the differential equation and simplifying the result yields

$$\frac{y_1(x)}{x} + 2y_1'(x) - y_1(x) + \sum_{n=1}^{\infty} n^2 b_n x^{n-3/2} - \sum_{n=1}^{\infty} \left(n - \frac{1}{2}\right) b_n x^{n-1/2}$$

$$= \frac{y_1(x)}{x} + 2y_1'(x) - y_1(x) + b_1 x^{-1/2} + \sum_{n=2}^{\infty} \left[n^2 b_n - \left(n - \frac{3}{2}\right) b_{n-1}\right] x^{n-3/2}$$

$$= 0.$$

Hence,

$$b_1 x^{-1/2} + \sum_{n=2}^{\infty} \left[n^2 b_n - \left(n - \frac{3}{2}\right) b_{n-1}\right] x^{n-3/2} = y_1(x) - \frac{y_1(x)}{x} - 2y_1'(x).$$

Because

$$y_1(x) = a_0 \left(x^{-1/2} - \frac{1}{2}x^{1/2} - \frac{1}{16}x^{3/2} - \frac{1}{96}x^{5/2} - \frac{5}{3072}x^{7/2} - \frac{7}{30720}x^{9/2} + \cdots\right),$$

we obtain

$$b_1 x^{-1/2} + \sum_{n=2}^{\infty} \left[n^2 b_n - \left(n - \frac{3}{2}\right) b_{n-1}\right] x^{n-3/2} = a_0 \left(2x^{-1/2} - \frac{1}{4}x^{1/2} + \frac{1}{384}x^{5/2} + \frac{1}{1536}x^{7/2} + \cdots\right).$$

Equating the coefficients yields $b_1 = 2a_0, -\dfrac{b_1}{2} + 4b_2 = -\dfrac{a_0}{4}, -\dfrac{3b_2}{2} + 9b_3 = 0, -\dfrac{5b_3}{2} + 16b_4 = \dfrac{a_0}{384},$

$-\dfrac{7b_4}{2} + 25b_5 = \dfrac{a_0}{1536}, \ldots$ Therefore,

$$y_2(x) = y_1(x)\ln x + \sum_{n=1}^{\infty} b_n x^{n-1/2}$$

$$= y_1(x)\ln x + a_0 \left(2x^{1/2} + \frac{3}{16}x^{3/2} + \frac{1}{32}x^{5/2} + \frac{31}{6144}x^{7/2} + \frac{3}{4096}x^{9/2} + \cdots\right),$$

so a general solution is $y(x) = c_1 y_1(x) + c_2 y_2(x)$, where c_1 and c_2 are arbitrary constants.

As in the previous examples, we see that **dsolve** together with the **series** option produces an equivalent result.

```
> dsolve(x*diff(y(x),x$2)+(2-x)*diff(y(x),x)+
  1/(4*x)*y(x)=0,y(x),series);
```

$$y(x) = \_C1\left(1 - \frac{1}{2}x - \frac{1}{16}x^2 - \frac{1}{96}x^3 - \frac{5}{3072}x^4 - \frac{7}{30720}x^5 + O(x^6)\right)/\sqrt{x}$$

$$+ \_C2\left(\ln(x)\left(1 - \frac{1}{2}x - \frac{1}{16}x^2 - \frac{1}{96}x^3 - \frac{5}{3072}x^4 - \frac{7}{30720}x^5 + O(x^6)\right)/\sqrt{x}\right.$$

$$\left. + \frac{2x + \frac{3}{16}x^2 + \frac{1}{32}x^3 + \frac{31}{6144}x^4 + \frac{3}{4096}x^5 + O(x^6)}{\sqrt{x}}\right)$$

6.5 SOME SPECIAL EQUATIONS

The techniques of solving differential equations with power series expansions about ordinary and regular singular points can also be used to solve several special ordinary differential equations. In addition to their historical significance, these equations are important because we find them in many problems in applied mathematics and physics such as those to determine the motion of a circular drumhead and to find the steady-state temperature on the surface of a sphere.

☰ Legendre's Equation

We begin our discussion with **Legendre's equation,**

$$(1 - x^2)y'' - 2xy' + k(k + 1)y = 0,$$

where k is a constant, which is named after the French mathematician Adrien Marie Legendre.

EXAMPLE: Find a general solution of the Legendre's equation $(1 - x^2)y'' - 2xy' + k(k + 1)y = 0$.

SOLUTION: In standard form, the equation is $y'' - \frac{2x}{1 - x^2}y' + \frac{k(k + 1)}{1 - x^2} = 0$. We could search for power series solutions about the regular singular points $x = \pm 1$. However, because $x = 0$ is an ordinary point, there is a solution of the form $y = \sum_{n=0}^{\infty} a_n x^n$. This solution will converge at least on the interval $(-1, 1)$ because the closest singular points to $x = 0$ are $x = \pm 1$. Substitution of this function and its derivatives $y' = \sum_{n=1}^{\infty} na_n x^{n-1}$ and $y'' = \sum_{n=2}^{\infty} n(n - 1)a_n x^{n-2}$ into the differential equation yields

$$[2a_2 + k(k + 1)a_0]x^0 + [-2a_1 + k(k + 1)a_1 + 6a_3]x$$

$$+ \sum_{n=4}^{\infty} n(n - 1)a_n x^{n-2} - \sum_{n=2}^{\infty} n(n - 1)a_n x^n - \sum_{n=2}^{\infty} 2na_n x^n + \sum_{n=2}^{\infty} k(k + 1)a_n x^n = 0.$$

After substituting $(n + 2)$ for each occurrence of n in the first series, simplifying, and equating the coefficients to zero, we find that

$$a_2 = -\frac{k(k + 1)}{2}a_0, \quad a_3 = -\frac{k(k + 1) - 2}{6}a_1 = -\frac{(k - 1)(k + 2)}{6}a_1$$

and

$$a_{n+2} = \frac{n(n - 1) + 2n - k(k + 1)}{(n + 2)(n + 1)}a_n = \frac{(n - k)(n + k + 1)}{(n + 2)(n + 1)}a_n, \qquad n \geq 2.$$

Hence, we have the two linearly independent solutions

$$y_1(x) = a_0 \left(1 - \frac{k(k + 1)}{2!}x^2 + \frac{(2 - k)(3 + k)k(k + 1)}{4!}x^4 \right.$$

$$\left. - \frac{(4 - k)(5 + k)(2 - k)(3 + k)k(k + 1)}{6!}x^6 + \cdots \right)$$

and

$$y_1(x) = a_0 \left(1 - \frac{k(k + 1)x^2}{2!} + \frac{(2 - k)(3 + k)k(k + 1)x^4}{4!} \right.$$

$$\left. - \frac{(4 - k)(5 + k)(2 - k)(3 + k)k(k + 1)x^6}{6!} + \cdots \right)$$

Thus, a general solution is given by

$$y = a_0 \left(1 - \frac{k(k + 1)}{2!}x^2 + \frac{(2 - k)(3 + k)k(k + 1)}{4!}x^4 - \frac{(4 - k)(5 + k)(2 - k)(3 + k)k(k + 1)}{6!}x^6 + \cdots \right)$$

$$+ a_1 \left(x - \frac{(k - 1)(k + 2)}{6}x^3 + \frac{(3 - k)(4 + k)(k - 1)(k + 2)}{5!}x^5 \right.$$

$$\left. - \frac{(5 - k)(6 + k)(3 - k)(4 + k)(k - 1)(k + 2)}{7!}x^7 + \cdots \right).$$

We obtain equivalent results using $\mathtt{dsolve}$ together with the $\mathtt{series}$ option as follows:

```
> y:='y':k:='k':
  dsolve((1-x^2)*diff(y(x),x$2)-2*x*diff(y(x),x)+
  k*(k+1)*y(x)=0,y(x),series);
```

$$y(x) = y(0) + \mathrm{D}(y)(0)x - \frac{1}{2}k(k+1)y(0)x^2$$

$$+ \left(-\frac{1}{6}k^2\mathrm{D}(y)(0) - \frac{1}{6}k\mathrm{D}(y)(0) + \frac{1}{3}\mathrm{D}(y)(0)\right)x^3$$

$$+ \frac{1}{24}(-6 + k^2 + k)k(k+1)y(0)x^4$$

$$+ \frac{1}{120}(-12 + k^2 + k)\mathrm{D}(y)(0)(k^2 + k - 2)x^5 + O(x^6)$$

An interesting observation from the general solution to Legendre's equation is that the series solutions terminate for integer values of k. If k is an even integer, then the first series terminates, while if k is an odd integer, the second series terminates. Because these polynomials are useful and are encountered in numerous applications, we have a special notation for them: $P_n(x)$ is called the **Legendre polynomial of degree** n and represents the nth-degree polynomial solution to Legendre's equation. Maple contains definitions of the Legendre polynomials in the $\mathtt{orthopoly}$ package. We load the $\mathtt{orthopoly}$ package by entering $\mathtt{with(orthopoly)}$.

```
> with(orthopoly);
```

$$[G, H, L, P, T, U]$$

After the $\mathtt{orthopoly}$ package has been loaded, the command $\mathtt{P(n,x)}$ returns the Legendre polynomial of degree n. We now use $\mathtt{array}$ and $\mathtt{seq}$ to generate the first 10 Legendre polynomials.

```
> Legendre_Polys:=array([seq([n,P(n,x)],n=1..10)]);
```

$Legendre\_Polys :=$

$[1, x]$

$\left[2, \dfrac{3}{2}x^2 - \dfrac{1}{2}\right]$

$\left[3, \dfrac{5}{2}x^3 - \dfrac{3}{2}x\right]$

$\left[4, \dfrac{35}{8}x^4 - \dfrac{15}{4}x^2 + \dfrac{3}{8}\right]$

$\left[5, \dfrac{63}{8}x^5 - \dfrac{35}{4}x^3 + \dfrac{15}{8}x\right]$

$\left[6, \dfrac{231}{16}x^6 - \dfrac{315}{16}x^4 + \dfrac{105}{16}x^2 - \dfrac{5}{16}\right]$

$\left[7, \dfrac{429}{16}x^7 - \dfrac{693}{16}x^5 + \dfrac{315}{16}x^3 - \dfrac{35}{16}x\right]$

$\left[8, \dfrac{6435}{128}x^8 - \dfrac{3003}{32}x^6 + \dfrac{3465}{64}x^4 - \dfrac{315}{32}x^2 + \dfrac{35}{128}\right]$

$\left[9, \dfrac{12155}{128}x^9 - \dfrac{6435}{32}x^7 + \dfrac{9009}{64}x^5 - \dfrac{1155}{32}x^3 + \dfrac{315}{128}x\right]$

$\left[10, \dfrac{46189}{256}x^{10} - \dfrac{109395}{256}x^8 + \dfrac{45045}{128}x^6 - \dfrac{15015}{128}x^4 + \dfrac{3465}{256}x^2 - \dfrac{63}{256}\right]$

≡ *The Gamma Function*

One of the more useful functions is the **gamma function**.

DEFINITION	*Gamma Function* The gamma function, denoted $\Gamma(x)$, is given by $$\Gamma(x) = \int_0^\infty e^{-u}u^{x-1}du, \qquad x > 0.$$ Notice that because integration is with respect to u, the result is a function of x.

EXAMPLE: Evaluate $\Gamma(1)$.

SOLUTION: $\Gamma(1) = \int_0^\infty e^{-u}u^{1-1}du = \int_0^\infty e^{-u}du = \lim_{b\to\infty}[-e^{-u}]_0^b = \lim_{b\to\infty}(-e^{-b} + 1) = 1.$

A useful property associated with the gamma function is $\Gamma(x + 1) = x\Gamma(x)$. Hence, if x is an integer, then using this property, we have $\Gamma(2) = \Gamma(1 + 1) = \Gamma(1)$, $\Gamma(3) = \Gamma(2 + 1) = 2\Gamma(2) = 2$, $\Gamma(4) = \Gamma(3 + 1) = 3\Gamma(3) = 3 \cdot 2$, $\Gamma(5) = \Gamma(4 + 1) = 4\Gamma(4) = 4 \cdot 3 \cdot 2, \ldots$ Therefore, for the integer n, $\Gamma(n + 1) = n!$. This property will be used in solving the following equation.

☰ Bessel's Equation

Another important equation is **Bessel's equation,**

$$x^2y'' + xy' + (x^2 - \mu^2)y = 0,$$

where $\mu \geq 0$ is a constant, which is named after the German astronomer Friedrich Wilhelm Bessel. To use a power series method to solve Bessel's equation, we first write the equation in standard form as

$$y'' + \frac{1}{x}y' + \frac{x^2 - \mu^2}{x^2}y = 0,$$

so $x = 0$ is a regular singular point. Using the method of Frobenius, we assume that there is a solution of the form $y = \sum_{n=0}^{\infty} a_n x^{n+r}$. We determine the value(s) of r with the indicial equation. Because $xp(x) = x\left(\frac{1}{x}\right) = 1$ and $x^2q(x) = x^2\left(\frac{x^2 - \mu^2}{x^2}\right) = x^2 - \mu^2$, $p_0 = 1$ and $q_0 = -\mu^2$. Hence, the indicial equation is

$$r(r - 1) + p_0 r + q_0 = r(r - 1) + r - \mu^2 = r^2 - \mu^2 = 0$$

with roots $r_1 = \mu$ and $r_2 = -\mu$. Therefore, we assume that $y = \sum_{n=0}^{\infty} a_n x^{n+\mu}$ with derivatives $y' = \sum_{n=0}^{\infty}(n + \mu)a_n x^{n+\mu-1}$ and $y'' = \sum_{n=0}^{\infty}(n + \mu)(n + \mu - 1)a_n x^{n+\mu-2}$. Substitution into Bessel's equation and simplifying the result yields

$$[\mu(\mu - 1) + \mu - \mu^2]a_0 x^\mu + [(1 + \mu)\mu + (1 + \mu) - \mu^2]a_1 x^{\mu+1}$$

$$+ \sum_{n=2}^{\infty}\{[(n + \mu)(n + \mu - 1) + (n + \mu) - \mu^2]a_n + a_{n-2}\}x^{n+\mu} = 0$$

Notice that the coefficient of $a_0 x^\mu$ is zero because $r_1 = \mu$ is a root of the indicial equation. After simplifying the other coefficients and equating them to zero, we have $(1 + 2\mu)a_1 = 0$ and

$$a_n = -\frac{a_{n-2}}{(n + \mu)(n + \mu - 1) + (n + \mu) - \mu^2} = -\frac{a_{n-2}}{n(n + 2\mu)}, \qquad n \geq 2.$$

From the first equation, $a_1 = 0$. Therefore, from $a_n = -\dfrac{a_{n-2}}{n(n + 2\mu)}$, $n \geq 2$, $a_n = 0$ for all odd n. The

coefficients that correspond to even indices are given by $a_{2n} = \dfrac{(-1)^n a_0}{2^{2n}(1 + \mu)(2 + \mu)(3 + \mu) \cdots (n + \mu)}$,

$n \geq 2$. Our solution can then be written as

$$y = \sum_{n=0}^{\infty} a_{2n} x^{2n+\mu} = \sum_{n=0}^{\infty} \frac{(-1)^n x^{2n+\mu}}{2^{2n}(1 + \mu)(2 + \mu)(3 + \mu) \cdots (n + \mu)}$$

$$= \sum_{n=0}^{\infty} \frac{(-1)^n 2^\mu}{(1 + \mu)(2 + \mu)(3 + \mu) \cdots (n + \mu)} \left(\frac{x}{2}\right)^{2n+\mu}.$$

If μ is an integer, then by using the gamma function, $\Gamma(x)$, we can write this solution as

$$y = \sum_{n=0}^{\infty} \frac{(-1)^n}{n! \, \Gamma(1 + \mu + n)} \left(\frac{x}{2}\right)^{2n+\mu}.$$

This function, denoted $J_\mu(x)$, is called the **Bessel function of the first kind of order μ**. For the other root $r_2 = -\mu$, a similar derivation yields a second solution,

$$y = \sum_{n=0}^{\infty} \frac{(-1)^n}{n! \, \Gamma(1 - \mu + n)} \left(\frac{x}{2}\right)^{2n-\mu},$$

which is the **Bessel function of the first kind of order $-\mu$** and is denoted $J_{-\mu}(x)$. Now we must determine if the functions $J_\mu(x)$ and $J_{-\mu}(x)$ are linearly independent. Notice that if $\mu = 0$, then these two functions are the same. If $\mu > 0$, then $r_1 - r_2 = \mu - (-\mu) = 2\mu$. If 2μ is not an integer, then by the method of Frobenius, the two solutions $J_\mu(x)$ and $J_{-\mu}(x)$ are linearly independent. Also, we can show that if 2μ is an odd integer, then $J_\mu(x)$ and $J_{-\mu}(x)$ are linearly independent. In both of these cases, a general solution is given by $y = c_1 J_\mu(x) + c_2 J_{-\mu}(x)$. The built-in command **BesselJ(mu,x)** represents $J_\mu(x)$. The graphs of the functions $J_\mu(x)$, $\mu = 0, 1, 2, 3$ are generated using **BesselJ** and **plot**. Notice that these functions have numerous zeros.

```
> to_plot:={seq(BesselJ(n,x),n=0..3)}:
  plot(to_plot,x=0..20);
```

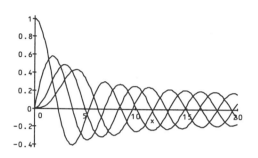

Suppose that μ is not an integer. We define the **Bessel function of the second kind of order** μ by the linear combination of the functions $J_\mu(x)$ and $J_{-\mu}(x)$. This function, denoted by $Y_\mu(x)$, is given by

$$Y_\mu(x) = \frac{\cos \mu\pi J_\mu(x) - J_{-\mu}(x)}{\sin \mu\pi}.$$

We can show that $J_\mu(x)$ and $Y_\mu(x)$ are linearly independent, so a general solution of Bessel's equation of order μ can be represented by $y = c_1 J_\mu(x) + c_2 Y_\mu(x)$. The built-in function **BesselY(mu,x)** represents $Y_\mu(x)$. We use **BesselY** and **plot** to generate the graphs of the functions $Y_\mu(x)$, $\mu = 0, 1, 2$.

```
> to_plot:={seq(BesselY(n,x),n=0..2)}:
  plot(to_plot,x=0..20,-3..1);
```

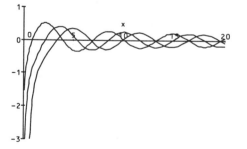

Notice that $\lim_{x \to 0^+} Y_\mu(x) = -\infty$. This property will be important in some applications in later chapters. We can use **dsolve** to obtain the same results as those obtained above. For example, entering

```
> dsolve(x^2*diff(y(x),x$2)+x*diff(y(x),x)+
  (x^2-mu^2)*y(x)=0,y(x),series);
```

$$y(x) = \_C1\, x^{-\mu}\left(1 + \frac{1}{(4\mu-4)}x^2 + \frac{1}{(8\mu-16)(4\mu-4)}x^4 + O(x^6)\right)$$

$$+ \_C2\, x^{\mu}\left(1 + \frac{1}{(-4\mu-4)}x^2 + \frac{1}{(-8\mu-16)(-4\mu-4)}x^4 + O(x^6)\right)$$

returns a **series** solution, while entering

```
> dsolve(x^2*diff(y(x),x$2)+x*diff(y(x),x)+
  (x^2-mu^2)*y(x)=0,y(x));
```

$$y(x) = \_C1\mathrm{BesselJ}(\mu, x) + \_C2\mathrm{BesselY}(\mu, x)$$

returns a solution in terms of **BesselJ** and **BesselY**.
   A more general form of Bessel's equation is expressed in the form

$$x^2 y'' + xy' + (\lambda^2 x^2 - \mu^2)y = 0.$$

We can show that a general solution of this parametric Bessel's equation that is defined on the interval $0 < x < \infty$ is

$$y = c_1 J_\mu(\lambda x) + c_2 Y_\mu(\lambda x).$$

This is verified with **dsolve**.

```
> x:='x':y:='y':
  dsolve(x^2*diff(y(x),x$2)+x*diff(y(x),x)+
  (lambda^2*x^2-mu^2)*y(x)=0,y(x));
```

$$y(x) = \_C1\,\mathrm{BesselJ}(\mu, \lambda x) + \_C2\,\mathrm{BesselY}(\mu, \lambda x)$$

EXAMPLE: Find a general solution of the equations (a) $x^2 y'' + xy' + (x^2 - 16)y = 0$ and (b) $x^2 y'' + xy' + (9x^2 - 4)y = 0$.

SOLUTION: (a) In this case, $\mu = 4$. Hence, $y = c_1 J_4(x) + c_2 Y_4(x)$.

```
> Sol_1:=dsolve(x^2*diff(y(x),x$2)+x*diff(y(x),x)+
  (x^2-16)*y(x)=0,y(x));
```

$$Sol\_1 := y(x) = \_C1\,\mathrm{BesselJ}(4, x) + \_C2\,\mathrm{BesselY}(4, x)$$

We graph this solution on $[0, 20]$ for various choices of the arbitrary constants with `plot`.

```
> assign(Sol_1):
  to_plot:={seq(seq(subs({_C1=i,_C2=j},y(x)),
   i=0..1),j=-1..0)}:
  plot(to_plot,x=0..20,-1/2..3/4);
```

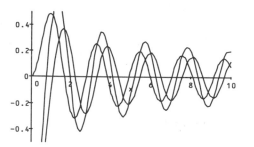

(b) Using the parametric Bessel's equation with $\lambda = 3$ and $\mu = 2$, we have $y = c_1 J_2(3x) + c_2 Y_2(3x)$.

```
> y:='y':
  Sol_2:=dsolve(x^2*diff(y(x),x$2)+x*diff(y(x),x)+
   (9*x^2-4)*y(x)=0,y(x));
```

$$Sol\_2 := y(x) = \_C1\,\text{BesselJ}(2, 3x) + \_C2\,\text{BesselY}(2, 3x)$$

As in (a), we graph this solution for several choices of the arbitrary constants with `plot`.

```
> assign(Sol_2):
  to_plot:={seq(seq(subs({_C1=i,_C2=j},y(x)),
   i=0..1),j=0..1)};
  plot(to_plot,x=0..10,-1/2..1/2);
```

7

Introduction to the Laplace Transform

In previous chapters we have investigated solving the equation

$$a_n y^{(n)} + a_{n-1} y^{(n-1)} + \cdots + a_1 y' + a_0 y = f(x)$$

for y. We have seen that when the coefficients $a_n, a_{n-1}, \ldots, a_0$ are numbers, we can find a general solution of the equation by first finding a general solution of the corresponding homogeneous equation

$$a_n y^{(n)} + a_{n-1} y^{(n-1)} + \cdots + a_1 y' + a_0 y = 0$$

and then finding a particular solution of

$$a_n y^{(n)} + a_{n-1} y^{(n-1)} + \cdots + a_1 y' + a_0 y = f(x).$$

In the case when the coefficients $a_n, a_{n-1}, \ldots, a_0$ are not constants, the situation is more difficult. In particular cases, like a Cauchy-Euler equation, similar techniques can be used. In other cases, we might be able to use power series to find a solution. In all of these cases, however, the function $f(x)$ has typically been a smooth function. In cases when $f(x)$ is not a smooth function, like when $f(x)$ is a piecewise-defined function, we cannot easily use methods like variation of parameters to solve the equation

$$a_n y^{(n)} + a_{n-1} y^{(n-1)} + \cdots + a_1 y' + a_0 y = f(x).$$

In this chapter, we discuss a technique that transforms the equation

$$a_n y^{(n)} + a_{n-1} y^{(n-1)} + \cdots + a_1 y' + a_0 y = f(x)$$

into an algebraic equation that can often be solved so that a solution of the differential equation can be obtained.

7.1 THE LAPLACE TRANSFORM: PRELIMINARY DEFINITIONS AND NOTATION

We are already familiar with several operations on functions. In previous courses, we have learned to add, subtract, multiply, divide, compose, differentiate, and integrate functions. In this section, we introduce another operation on functions, the **Laplace transform**, and discuss several of its properties.

DEFINITION

> *Laplace Transform*
>
> Let $f(t)$ be a function defined on the interval $[0, +\infty)$. The **Laplace transform** of $f(t)$ is the function (of s)
>
> $$L\{f(t)\} = \int_0^\infty e^{-st} f(t) \, dt.$$

Because the Laplace transform yields a function of s, we often use the notation $L\{f(t)\} = F(s)$ to denote the Laplace transform of $f(t)$. Hence, we use the capital letter to denote the Laplace transform of the functions named with the corresponding small letter.

We begin our discussion of the Laplace transform by finding the transform of several functions that can be used to find the Laplace transform of many other functions without using the definition.

EXAMPLE: Compute $L\{f\}(s)$ if $f(t) = 1$.

SOLUTION:

$$L\{f\}(s) = \int_0^\infty e^{-st} dt = \lim_{M\to\infty} \int_0^M e^{-st} dt = \lim_{M\to\infty} \left[-\frac{e^{-st}}{s} \right]_{t=0}^{t=M} = -\frac{1}{s} \lim_{M\to\infty} [e^{-sM} - 1]$$

$$= -\frac{1}{s}(0 - 1) = \frac{1}{s}, \ s > 0.$$

Notice that in order for $\lim_{M\to\infty}[e^{-sM}] = 0$, we must require that $s > 0$ (otherwise, the limit does not exist). After using **assume** to instruct Maple to assume that s is positive, Maple is able to compute this improper integral with **int**. Note that the $\sim$ symbol appearing in the result indicates that we have made assumptions about s.

```
> assume(s>0):
  int(exp(-s*t),t=0..infinity);
```

$$\frac{1}{s \sim}$$

EXAMPLE: Compute $L\{f(t)\}$ if $f(t) = e^{at}$.

SOLUTION:

$$L\{f(t)\} = \int_0^\infty e^{-st} f(t)\, dt = \int_0^\infty e^{-st} e^{at} dt = \int_0^\infty e^{-(s-a)t} dt = \lim_{M \to \infty} \left[-\frac{e^{-(s-a)t}}{s-a} \right]_{t=0}^{t=M}$$

$$= -\lim_{M \to \infty} \left(\frac{e^{-(s-a)M}}{s-a} - \frac{1}{s-a} \right) = -\left(0 - \frac{1}{s-a} \right) = \frac{1}{s-a}, \ s > a.$$

Notice that we must require that $s > a$ so that $\lim\limits_{M \to \infty} e^{-(s-a)M} = 0$.

We are often able to use the built-in command **laplace** to compute the Laplace transform of a function. For example, entering

```
> a:='a':s:='s':
  laplace(exp(a*t),t,s);
```

$$\frac{1}{(s-a)}$$

computes the Laplace transform of $f(t) = e^{at}$.

This formula can now be used to avoid using the definition.

EXAMPLE: Compute (a) $L\{e^{-3t}\}$ and (b) $L\{e^{5t}\}$.

SOLUTION: Because we have derived the general formula $L\{e^{at}\} = \dfrac{1}{s-a}$, we find that

(a) $L\{e^{-3t}\} = \dfrac{1}{s-(-3)} = \dfrac{1}{s+3}, s > -3$, and (b) $L\{e^{5t}\} = \dfrac{1}{s-5}, s > 5$.

In most cases, using the definition of the Laplace transform to calculate the Laplace transform of a function is a difficult and time-consuming task.

EXAMPLE: Compute (a) $L\{t^3\}$, (b) $L\{\sin(at)\}$, and (c) $L\{\cos(at)\}$.

SOLUTION: To compute $L\{t^3\}$ by hand requires the application of integration by parts three times. Instead, we proceed with **int**. After defining **step_1** to be $\int_0^M t^3 e^{-st}\, dt$,

```
> s:='s':
  step_1:=int(t^3*exp(-s*t),t=0..M);
```

$$step\_1 := -\frac{e^{-sM}(6sM + 6 + 3s^2 M^2 + s^3 M^3)}{s^4} + 6\frac{1}{s^4}$$

we attempt to use **limit** to calculate $\lim_{M\to\infty} \int_0^M t^3 e^{-st}\, dt$ but are unsuccessful because Maple does not know that we are assuming that s is positive.

```
> limit(step_1,M=infinity);
```

$$\lim_{M\to\infty} -\frac{e^{-sM}(6sM + 6 + 3s^2 M^2 + s^3 M^3)}{s^4} + 6\frac{1}{s^4}$$

However, after we use **assume** to instruct Maple to assume that s is positive, we are able to compute $\lim_{M\to\infty} \int_0^M t^3 e^{-st}\, dt$. Note that the $\sim$ symbol appearing in the result indicates that we have made assumptions about s.

```
> assume(s>0);
  limit(step_1,M=infinity);
```

$$6\frac{1}{s\sim^4}$$

Each integral that results when computing $L\{\sin(at)\}$ and $L\{\cos(at)\}$ using the definition of the Laplace transform requires the use of integration by parts twice. Instead, we use **laplace** to compute each Laplace transform.

```
> s:='s':
  laplace(sin(a*t),t,s);
```

$$\frac{a}{s^2 + a^2}$$

```
> laplace(cos(a*t),t,s);
```

$$\frac{s}{s^2 + a^2}$$

As we can see, the definition of the Laplace transform is difficult to apply in most cases. Hence, we now discuss the **linearity property** that enables us to use the transforms that we have found thus far to find the Laplace transform of other functions.

• •

THEOREM   *Linearity Property*

Let a and b be constants, and suppose that the Laplace transform of the functions f and g exists. Then

$$L\{af(t) + bg(t)\} = aL\{f(t)\} + bL\{g(t)\}.$$

• •

EXAMPLE: Calculate (a) $L\{6\}$ and (b) $L\{5 - 2e^{-t}\}$.

SOLUTION: Using the results obtained in previous examples, we have for (a)

$$L\{6\} = 6L\{1\} = 6\left(\frac{1}{s}\right) = \frac{6}{s}$$

and for (b)

$$L\{5 - 2e^{-t}\} = 5L\{1\} - 2L\{e^{-t}\} = 5\left(\frac{1}{s}\right) - 2\left(\frac{1}{s - (-1)}\right) = \frac{5}{s} - \frac{2}{s + 1}.$$

☰ Exponential Order, Jump Discontinuities, and Piecewise Continuous Functions

In calculus, we saw that in some cases, improper integrals diverge, meaning that the limit of the definite integral does not exist. Hence, we may believe that the Laplace transform may not exist for some functions. Therefore, we present the following definitions and theorems so that we can better understand the types of functions for which the Laplace transform exists.

DEFINITION

> *Exponential Order*
>
> A function f is of **exponential order** b if there are numbers $b, M > 0$, and $T > 0$ such that
>
> $$|f(t)| \le Me^{bt}$$
>
> for $t > T$.

Note: This definition is the same as showing that

$$\lim_{t \to \infty} \frac{f(t)}{e^{bt}} = 0.$$

In the following sections, we will see that the Laplace transform is particularly useful in solving equations involving piecewise or recursively defined functions.

DEFINITION

> *Jump Discontinuity*
>
> A function f has a **jump discontinuity** at $t = c$ on the closed interval $[a, b]$ if the one-sided limits $\lim_{t \to c^+} f(t)$ and $\lim_{t \to c^-} f(t)$ are finite, but unequal, values. f has a jump discontinuity at $t = a$ if $\lim_{t \to a^+} f(t)$ is a finite value. f has a jump discontinuity at $t = b$ if $\lim_{t \to b^-} f(t)$ is a finite value.

DEFINITION

> *Piecewise Continuous*
>
> A function f is **piecewise continuous on the finite interval** $[a, b]$ if f is continuous at every point in $[a, b]$ except at finitely many points at which f has a jump discontinuity.
>
> A function f is **piecewise continuous on** $[0, \infty)$ if f is piecewise continuous on $[0, N]$ for all $N > 0$.

THEOREM

Sufficient Condition for Existence of L{f(t)}

Suppose that f is a piecewise continuous function on the interval $[0, \infty)$ and that it is of exponential order b for $t > T$. Then $L\{f(t)\}$ exists for $s > b$.

EXAMPLE: Find the Laplace transform of

$$f(t) = \begin{cases} -1, 0 \le t \le 4 \\ 1, t > 4 \end{cases}.$$

SOLUTION: Because f is a piecewise continuous function on $[0, \infty)$ and of exponential order, $L\{f(t)\}$ exists. We simply use the definition and evaluate the integral using the sum of two integrals.

$$L\{f(t)\} = \int_0^\infty f(t)e^{-st}dt = \int_0^4 (-1)e^{-st}dt + \int_4^\infty e^{-st}dt = \left[\frac{e^{-st}}{s}\right]_{t=0}^{t=4} + \lim_{M\to\infty}\left[-\frac{e^{-st}}{s}\right]_{t=4}^{t=M}$$

$$= \frac{1}{s}(e^{-4s} - 1) - \frac{1}{s}\lim_{M\to\infty}(e^{-Ms} - e^{-4s}) = \frac{1}{s}(2e^{-4s} - 1).$$

We can define and graph this piecewise-defined function as follows. Note that because f is defined using **proc**, we must use operator notation when graphing f.

```
> f:=proc(t) if t>=0 and t<=4 then -1 else 1 fi end:
  plot(f,0..8);
```

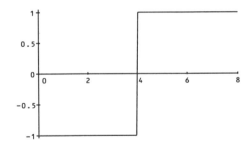

However, the **laplace** command is unable to compute the Laplace transform of f when f is defined in this manner. To compute the Laplace transform using Maple, we take advantage of the **Heaviside** function. The **Heaviside** function is defined by

$$\text{Heaviside(t)} = \begin{cases} 0, & \text{if } t < 0 \\ 1, & \text{if } t \ge 0 \end{cases},$$

so f is given by $f(t) = $ **Heaviside(t-4)-Heaviside(4-t)**.

```
> plot(Heaviside(t-4)-Heaviside(4-t),t=0..8);
```

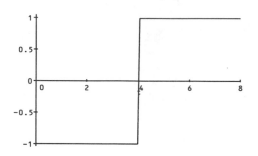

Laplace is able to compute the Laplace transform of functions defined in terms of **Heaviside**. Thus, to use **laplace** to compute the Laplace transform of $f(t) =$ **Heaviside(t-4)-Heaviside(4-t)** we enter

```
> laplace(Heaviside(t-4)-Heaviside(4-t),t,s);
```

$$2\frac{e^{-4s}}{s} - \frac{1}{s}$$

which gives us the same result as that given before.

≡ *Properties of the Laplace Transform*

The definition of the Laplace transform is not easy to apply. Therefore, we would like to discuss several properties of the Laplace transform so that we can make numerous transformations without having to use the definition. Most of the properties follow directly from our knowledge of integrals. We begin with the shifting, or translation, property.

THEOREM   *Shifting Property*

If $L\{f(t)\} = F(s)$ exists for $s > b$, then

$$L\{e^{at}f(t)\} = F(s - a).$$

EXAMPLE: Find the Laplace transform of (a) $e^{-2t} \cos t$ and (b) $4te^{3t}$.

SOLUTION: (a) In this case, $f(t) = \cos t$ and $a = -2$. Then, $F(s) = L\{\cos t\} = \dfrac{s}{s^2 + 1}$, so we replace each s with $(s - a) = (s + 2)$. Therefore,

$$L\{e^{-2t} \cos t\} = \frac{s + 2}{(s + 2)^2 + 1} = \frac{s + 2}{s^2 + 4s + 5}.$$

In this case, we see that `laplace` is successful in computing the Laplace transform.

> `laplace(exp(-2*t)*cos(t),t,s);`

$$\frac{s + 2}{(s + 2)^2 + 1}$$

(b) Using the linearity property, we have $L\{4te^{3t}\} = 4L\{te^{3t}\}$. Hence, $f(t) = t$ and $a = 3$, so we replace s in $F(s) = L\{t\} = \dfrac{1}{s^2}$ by $(s - a) = (s - 3)$. Therefore, $L\{te^{3t}\} = \dfrac{1}{(s - 3)^2}$, so

$$L\{4te^{3t}\} = \frac{4}{(s - 3)^2}.$$

Identical results are obtained with `laplace`.

> `laplace(4*t*exp(3*t),t,s);`

$$4\frac{1}{(s - 3)^2}$$

In order to use the Laplace transform to solve differential equations, we will need to be able to compute the Laplace transform of the derivatives of an arbitrary function, provided the Laplace transform of such a function exists.

• •

THEOREM   *Laplace Transform of the First Derivative*

Suppose that f is a piecewise continuous function on the interval $[0, \infty)$ and that it is of exponential order b for $t > T$. Then for $s > b$,

$$L\{f'(t)\} = sL\{f(t)\} - f(0).$$

Using induction, we can easily verify the following corollary.

• •

COROLLARY   *Laplace Transform of Higher Derivatives*

More generally, if $f^{(i)}(t)$ is a continuous function on $[0, \infty)$ for $i = 0, 1, \dots, n - 1$ and $f^{(n)}(t)$ is piecewise continuous on $[0, \infty)$ and of exponential order b, then for $s > b$,

$$L\{f^{(n)}(t)\} = s^n L\{f(t)\} - s^{n-1} f(0) - \dots - s f^{(n-2)}(0) - f^{(n-1)}(0).$$

• •

We will use this theorem and corollary in solving initial-value problems. However, we can also use them to find the Laplace transform of a function when we know the Laplace transform of the derivative of the function. We illustrate this in the following example.

EXAMPLE: Find $L\{\sin^2 kt\}$.

SOLUTION: We can use this corollary to find the Laplace transform of $f(t) = \sin^2 kt$. Notice that $f'(t) = 2k \sin kt \cos kt = k \sin 2kt$. Then, because $L\{f'(t)\} = sL\{f(t)\} - f(0)$ and $L\{f'(t)\} = L\{k \sin 2kt\} = k \dfrac{2k}{s^2 + (2k)^2} = \dfrac{2k^2}{s^2 + 4k^2}$, we have $\dfrac{2k^2}{s^2 + 4k^2} = sL\{f(t)\} - 0$. Therefore, $L\{f(t)\} = \dfrac{2k^2}{s(s^2 + 4k^2)}$. As in previous examples, we see that the same results are obtained with `laplace`.

```
> laplace(sin(k*t)^2,t,s);
```

$$2 \frac{k^2}{s(s^2 + 4k^2)}$$

• •

THEOREM   *Derivatives of the Laplace Transform*

Suppose that $F(s) = L\{f(t)\}$ where f is a piecewise continuous function on $[0, \infty)$ and of exponential order b. Then for $s > b$,

$$L\{t^n f(t)\} = (-1)^n \frac{d^n F}{ds^n}(s).$$

• •

EXAMPLE: Find the Laplace transform of (a) $f(t) = t\cos 2t$ and (b) $f(t) = t^2 e^{-3t}$.

SOLUTION: (a) In this case, $n = 1$ and $F(s) = L\{\cos 2t\} = \dfrac{s}{s^2 + 4}$. Then

$$L\{t\cos 2t\} = (-1)\frac{d}{ds}\left[\frac{s}{s^2 + 4}\right] = -\frac{(s^2 + 4) - s(2s)}{(s^2 + 4)^2} = \frac{s^2 - 4}{(s^2 + 4)^2}.$$

```
> simplify(laplace(t*cos(2*t),t,s));
```

$$\frac{s^2 - 4}{(s^2 + 4)^2}$$

(b) Because $n = 2$ and $F(s) = L\{e^{-3t}\} = \dfrac{1}{s + 3}$, we have

$$L\{t^2 e^{-3t}\} = (-1)^2 \frac{d^2}{ds^2}\left[\frac{1}{s + 3}\right] = \frac{2}{(s + 3)^3}.$$

```
> laplace(t^2*exp(-3*t),t,s);
```

$$2\frac{1}{(s + 3)^3}$$

EXAMPLE: Find $L\{t^n\}$.

SOLUTION: Using the previous theorem with $L\{t^n\} = L\{t^n \cdot 1\}$, we have $f(t) = 1$. Then, $F(s) = L\{1\} = \dfrac{1}{s}$. Calculating the derivatives of F, we obtain

$$\frac{dF}{ds}(s) = -\frac{1}{s^2}$$

$$\frac{d^2F}{ds^2}(s) = \frac{2}{s^3}$$

$$\frac{d^3F}{ds^3}(s) = -\frac{3 \cdot 2}{s^4}$$

$$\vdots$$

$$\frac{d^nF}{ds^n}(s) = (-1)^n \frac{n!}{s^{n+1}}.$$

Therefore,

$$L\{t^n\} = L\{t^n \cdot 1\} = (-1)^n(-1)^n \frac{n!}{s^{n+1}} = (-1)^{2n} \frac{n!}{s^{n+1}} = \frac{n!}{s^{n+1}}.$$

We see that we obtain equivalent results with Maple.

```
> laplace(t^n,t,s);
```

$$\frac{\Gamma(n + 1)}{s^{n+1}}$$

EXAMPLE: Let $f(t) = (3t - 1)^3$. Compute the Laplace transform of $f(t)$, $f'(t)$, and $f''(t)$.

SOLUTION: Since $f(t) = (3t - 1)^3 = 27t^3 - 27t^2 + 9t - 1$ and $L\{t^n\} = \frac{n!}{s^{n+1}}$,

$$L\{f(t)\} = 27\frac{3!}{s^4} - 27\frac{2!}{s^3} + 9\frac{1}{s^2} - \frac{1}{s} = \frac{162 - 54s + 9s^2 - s^3}{s^4}.$$

By the previous theorem, $L\{f'(t)\} = sL\{f(t)\} - f(0)$. Hence,

$$L\{f'(t)\} = s\frac{162 - 54s + 9s^2 - s^3}{s^4} - f(0) = \frac{162 - 54s + 9s^2 - s^3}{s^3} + 1 = \frac{9(18 - 6s + s^2)}{s^3}.$$

With the corollary, $L\{f''(t)\} = s^2L\{f(t)\} - sf(0) - f'(0)$. Then, since $f'(t) = 9(3t - 1)^2$,

$$L\{f''(t)\} = s^2\frac{162 - 54s + 9s^2 - s^3}{s^4} - sf(0) - f'(0) = \frac{54(3 - s)}{s^2}.$$

Using the properties of the Laplace transform, we can compute the Laplace transform of a large number of frequently encountered functions. The following table lists the Laplace transform of several frequently encountered functions.

$f(t)$	$F(s) = L\{f(t)\}$	$f(t)$	$F(s) = L\{f(t)\}$
1	$\dfrac{1}{s}, s > 0$	$t^n, n = 1, 2, \ldots$	$\dfrac{n!}{s^{n+1}}, s > 0$
e^{at}	$1s - a, s > a$	$t^n e^{at}, \; n = 1, 2, \ldots$	$\dfrac{n!}{(s-a)^{n+1}}$
$\sin kt$	$\dfrac{k}{s^2 + k^2}$	$e^{at} \sin kt$	$\dfrac{k}{(s-a)^2 + k^2}$
$\cos kt$	$\dfrac{s}{s^2 + k^2}$	$e^{at} \cos kt$	$\dfrac{s-a}{(s-a)^2 + k^2}$
$\sinh kt$	$\dfrac{k}{s^2 - k^2}$	$e^{at} \sinh kt$	$\dfrac{k}{(s-a)^2 - k^2}$
$\cosh kt$	$\dfrac{s}{s^2 - k^2}$	$e^{at} \cosh kt$	$\dfrac{s-a}{(s-a)^2 - k^2}$

7.2 THE INVERSE LAPLACE TRANSFORM

In the previous section, we were concerned with finding the Laplace transform of a given function using either the definition of the Laplace transform or one of its numerous properties. At that time, we discussed the sufficient conditions for the existence of the Laplace transform. In this section, we reverse this process in that we will be given a function $F(s)$ and will want to find a function $f(t)$ such that $Lf(t) = F(s)$.

DEFINITION

Inverse Laplace Transform

The **inverse Laplace transform** of the function $F(s)$ is the unique continuous function $f(t)$ on $[0, \infty)$ that satisfies $L\{f(t)\} = F(s)$. We denote the inverse Laplace transform of $F(s)$ as

$$f(t) = L^{-1}\{F(s)\}.$$

If the only functions that satisfy this relationship are discontinuous on $[0, \infty)$, then we choose a piecewise continuous function on $[0, \infty)$ to be $L^{-1}\{F(s)\}$.

The table of Laplace transforms listed in the previous section is useful in finding the inverse Laplace transform of a given function.

EXAMPLE: Find the inverse Laplace transform of (a) $F(s) = \dfrac{1}{s - 6}$, (b) $F(s) = \dfrac{2}{s^2 + 4}$,

(c) $F(s) = \dfrac{6}{s^4}$, and (d) $F(s) = \dfrac{6}{(s + 2)^4}$.

SOLUTION: (a) Because $L\{e^{6t}\} = \dfrac{1}{s - 6}$, $L^{-1}\left\{\dfrac{1}{s - 6}\right\} = e^{6t}$. (b) Since $L\{\sin 2t\} = \dfrac{2}{s^2 + 2^2} = \dfrac{2}{s^2 + 4}$,

$L^{-1}\left\{\dfrac{2}{s^2 + 4}\right\} = \sin 2t$. (c) Note that $L\{t^3\} = \dfrac{3!}{s^4} = \dfrac{6}{s^4}$. Hence, $L^{-1}\left\{\dfrac{6}{s^4}\right\} = t^3$. (d) Notice that

$F(s) = \dfrac{6}{(s + 2)^4}$ is obtained from $F(s) = \dfrac{6}{s^4}$ by substituting $(s + 2)$ for s. Therefore, by the shifting

property, $L\{e^{-2t}t^3\} = \dfrac{6}{(s + 2)^4}$, so $L^{-1}\left\{\dfrac{6}{(s + 2)^4}\right\} = e^{-2t}L^{-1}\left\{\dfrac{6}{s^4}\right\} = e^{-2t}t^3$.

In each of these cases, we can take advantage of the miscellaneous library command `invlaplace` to compute the inverse Laplace transform. Be sure to enter `readlib(laplace)` before using the `invlaplace` command. For example, entering

```
> readlib(laplace):
  invlaplace(1/(s-6),s,t);
```
e^{6t}

computes the inverse Laplace transform of $F(s) = \dfrac{1}{s - 6}$. Similarly, entering

```
> map(invlaplace,[2/(s^2+4),6/s^4,6/(s+2)^4],s,t);
```
$[\sin(2t), t^3, t^3e^{-2t}]$

computes the inverse Laplace transform of $F(s) = \dfrac{2}{s^2 + 4}$, $F(s) = \dfrac{6}{s^4}$, and $F(s) = \dfrac{6}{(s + 2)^4}$.

$$\cdots\cdots\cdots\cdots\cdots\cdots\cdots$$

THEOREM *Linearity Property of the Inverse Laplace Transform*

Suppose that $L^{-1}\{F(s)\}$ and $L^{-1}\{G(s)\}$ exist and are continuous on $[0, \infty)$. Also, suppose that a and b are constants. Then

$$L^{-1}\{aF(s) + bG(s)\} = aL^{-1}\{F(s)\} + bL^{-1}\{G(s)\}.$$

$$\cdots\cdots\cdots\cdots\cdots\cdots\cdots$$

EXAMPLE: Find the inverse Laplace transform of (a) $F(s) = \dfrac{1}{s^3}$, (b) $F(s) = -\dfrac{7}{s^2 + 16}$, and (c) $F(s) = \dfrac{5}{s} - \dfrac{2}{s - 10}$.

SOLUTION: (a) $L^{-1}\left\{\dfrac{1}{s^3}\right\} = L^{-1}\left\{\dfrac{1}{2}\dfrac{2}{s^3}\right\} = \dfrac{1}{2}L^{-1}\left\{\dfrac{2}{s^3}\right\} = \dfrac{1}{2}t^2.$

(b) $L^{-1}\left\{-\dfrac{7}{s^2 + 16}\right\} = -7L^{-1}\left\{\dfrac{1}{s^2 + 16}\right\} = -7L^{-1}\left\{\dfrac{1}{4}\dfrac{4}{s^2 + 4^2}\right\} = -\dfrac{7}{4}L^{-1}\left\{\dfrac{4}{s^2 + 4^2}\right\} = -\dfrac{7}{4}\sin 4t.$

(c) $L^{-1}\left\{\dfrac{5}{s} - \dfrac{2}{s - 10}\right\} = 5L^{-1}\left\{\dfrac{1}{s}\right\} - 2L^{-1}\left\{\dfrac{1}{s - 10}\right\} = 5 - 2e^{10t}.$

As in the first example, we see that once we have loaded the `invlaplace` command by entering `readlib(laplace)`, we can use `invlaplace` to compute the inverse Laplace transform of each function.

```
> readlib(laplace):
  invlaplace(1/s^3,s,t);
  invlaplace(-7/(s^2+16),s,t);
  invlaplace(5/s-2/(s-10),s,t);
```

$$\dfrac{1}{2}t^2$$

$$-\dfrac{7}{4}\sin(4t)$$

$$5 - 2e^{10t}$$

Unfortunately, the functions $F(s)$ that are encountered do not have to be of the forms previously discussed. In these cases, we must make use of the linearity property to determine the inverse Laplace transform. For example, sometimes we must complete the square in the denominator of $F(s)$ before finding $L^{-1}\{F(s)\}$.

EXAMPLE: Determine $L^{-1}\left\{\dfrac{s}{s^2 + 2s + 5}\right\}$.

SOLUTION: Notice that all of the forms of $F(s)$ in the table of Laplace transforms involve a term of the form $s^2 + k^2$ in the denominator. However, through shifting, this term is replaced by $(s - a)^2 + k^2$. We obtain a term of this form in the denominator by completing the square. This yields

$$\frac{s}{s^2 + 2s + 5} = \frac{s}{(s^2 + 2s + 1) + 4} = \frac{s}{(s + 1)^2 + 4}.$$

Note that we can take advantage of the **completesquare** command that is contained in the **student** package to perform the operation of completing the square just encountered. We load the **student** package by entering **with(student)**. The commands contained in the **student** package are returned because a semicolon is included at the end of the command instead of a colon.

```
> with(student);
```

> [D, Doubleint, Int, Limit, Sum, Tripleint, changevar, combine, completesquare, distance, equate, extrema, integrand, intercept, intparts, isolate, leftbox, leftsum, makeproc, maximize, middlebox, middlesum, midpoint, minimize, powsubs, rightbox, rightsum, showtangent, simpson, slope, trapezoid, value]

Thus, entering

```
> completesquare(s^2+2*s+5,s);
```

> $(s + 1)^2 + 4$

shows us that $s^2 + 2s + 5 = (s + 1)^2 + 4$.

Because the variable appears in the numerator, we must write it in the form $(s + 1)$ to find the inverse Laplace transform. Doing this, we find that

$$\frac{s}{s^2 + 2s + 5} = \frac{s}{(s + 1)^2 + 4} = \frac{(s + 1) - 1}{(s + 1)^2 + 4}.$$

Hence,

$$L^{-1}\left\{\frac{s}{s^2 + 2s + 5}\right\} = L^{-1}\left\{\frac{(s + 1) - 1}{(s + 1)^2 + 4}\right\}$$

$$= L^{-1}\left\{\frac{(s + 1)}{(s + 1)^2 + 2^2}\right\} - \frac{1}{2}L^{-1}\left\{\frac{2}{(s + 1)^2 + 2^2}\right\}$$

$$= e^{-t}\cos 2t - \frac{1}{2}e^{-t}\sin 2t.$$

As in previous examples, we see that `invlaplace` quickly finds $L^{-1}\left\{\dfrac{s}{s^2 + 2s + 5}\right\}$.

```
> readlib(laplace):
  invlaplace(s/(s^2+2*s+5),s,t);
```

$$-\frac{1}{2}e^{-t}\sin(2t) + e^{-t}\cos(2t)$$

In other cases, **partial fractions** must be used to obtain terms for which the inverse Laplace transform can be found. Suppose that

$$F(s) = \frac{P(s)}{Q(s)},$$

where $P(s)$ and $Q(s)$ are polynomials of degree m and n, respectively. If $n > m$, then the method of partial fractions can be used to expand $F(s)$. Recall from calculus that three possible situations can be solved through partial fractions: (1) linear factors (nonrepeated), (2) repeated linear factors, and (3) irreducible quadratic factors. We illustrate these three cases in the following examples.

☰ Linear Factors (Nonrepeated)

In this case, $Q(s)$ can be written as a product of linear factors, so

$$Q(s) = (s - q_1)(s - q_2)\cdots(s - q_n),$$

where $q_1, q_2, \ldots, q_n$ are distinct numbers. Therefore, $F(s)$ can be written as

$$F(s) = \frac{A_1}{s - q_1} + \frac{A_2}{s - q_2} + \cdots + \frac{A_n}{s - q_n},$$

where $A_1, A_2, \ldots, A_n$ are constants that must be determined.

EXAMPLE: Find $L^{-1}\left\{\dfrac{3s - 4}{s(s - 4)}\right\}$.

SOLUTION: In this case, we have linear factors in the denominator. Hence, we write $F(s)$ as

$$\frac{3s - 4}{s(s - 4)} = \frac{A}{s} + \frac{B}{s - 4}.$$

Multiplying both sides of the equation by the denominator $s(s - 4)$, we have

$$3s - 4 = A(s - 4) + Bs = (A + B)s - 4A.$$

Equating the coefficients of s as well as the constant terms, we see that the following system of equations must be satisfied.

$$\begin{cases} A + B = 3 \\ -4A = -4 \end{cases}.$$

Maple can solve the system of equations $\begin{cases} A + B = 3 \\ -4A = -4 \end{cases}$ with **solve**, or we can solve the equation $3s - 4 = A(s - 4) + Bs = (A + B)s - 4A$ for A and B with **match** as follows:

```
> match(3*s-4=(A+B)*s-4*A,s,'Val'):
  Val;
```

$$\{A = 1, B = 2\}$$

Hence, $A = 1$ and $B = 2$. Therefore,

$$\frac{3s - 4}{s(s - 4)} = \frac{A}{s} + \frac{B}{s - 4} = \frac{1}{s} + \frac{2}{s - 4},$$

so

$$L^{-1}\left\{\frac{3s - 4}{s(s - 4)}\right\} = L^{-1}\left\{\frac{1}{s} + \frac{2}{s - 4}\right\} = 1 + 2e^{4t}.$$

In this case, we can use **convert** together with the **parfrac** option to compute the partial fraction decomposition directly,

```
> convert((3*s-4)/(s*(s-4)),parfrac,s);
```

$$s^{-1} + \frac{2}{s - 4}$$

or we can simply use **invlaplace** as shown in the previous examples.

```
> readlib(laplace):
  invlaplace((3*s-4)/(s*(s-4)),s,t);
```

$$1 + 2e^{4t}$$

☰ Repeated Linear Factors

Suppose that $(s - q)$ is a factor of $Q(s)$ of multiplicity k. Then the partial fraction expansion of $F(s)$ that corresponds to this factor is

$$\frac{A_1}{s - q} + \frac{A_2}{(s - q)^2} + \cdots + \frac{A_k}{(s - q)^k},$$

where $A_1, A_2, \ldots, A_k$ are constants that must be found.

EXAMPLE: Calculate $L^{-1}\left\{\dfrac{5s^2 + 20s + 6}{s^3 + 2s^2 + s}\right\}$.

SOLUTION: After using `convert` together with the `parfrac` option,

```
> convert((5*s^2+20*s+6)/(s^3+2*s^2+s),parfrac,s);
```

$$6\frac{1}{s} + 9\frac{1}{(s + 1)^2} - \frac{1}{(s + 1)}$$

we see that

$$\frac{5s^2 + 20s + 6}{s^3 + 2s^2 + s} = \frac{6}{s} - \frac{1}{s + 1} + \frac{9}{(s + 1)^2}.$$

Therefore,

$$L^{-1}\left\{\frac{5s^2 + 20s + 6}{s^3 + 2s^2 + s}\right\} = L^{-1}\left\{\frac{6}{s} - \frac{1}{s + 1} + \frac{9}{(s + 1)^2}\right\}$$

$$= 6L^{-1}\left\{\frac{1}{s}\right\} - L^{-1}\left\{\frac{1}{s + 1}\right\} + 9L^{-1}\left\{\frac{1}{(s + 1)^2}\right\}$$

$$= 6 - e^{-t} + 9te^{-t}.$$

As expected, we obtain the same results using `invlaplace`.

```
> readlib(laplace):
  invlaplace((5*s^2+20*s+6)/(s^3+2*s^2+s),s,t);
```

$$6 + 9te^{-t} - e^{-t}$$

☰ Irreducible Quadratic Factors

Suppose that $(s - a)^2 + b^2$ is a factor of $Q(s)$ that cannot be reduced to linear factors. If this irreducible quadratic is a factor of $Q(s)$ of multiplicity k, then the partial fraction expansion of $F(s)$ corresponding to $(s - a)^2 + b^2$ is

$$\frac{A_1 + B_1}{(s - a)^2 + b^2} + \frac{A_2 + B_2}{[(s - a)^2 + b^2]^2} + \cdots + \frac{A_k + B_k}{[(s - a)^2 + b^2]^k}.$$

EXAMPLE: Find $L^{-1}\left\{\dfrac{2s^3 - 4s - 8}{(s^2 - s)(s^2 + 4)}\right\}$.

SOLUTION: As in the previous example, we use **convert** together with the **parfrac** option

```
> convert((2*s^3-4*s-8)/((s^2-s)*(s^2+4)),parfrac,s);
```

$$2\frac{1}{s} - 2\frac{1}{s - 1} + 2\frac{2 + s}{s^2 + 4}$$

to obtain

$$\frac{2s^3 - 4s - 8}{s(s - 1)(s^2 + 4)} = \frac{2}{s} - \frac{2}{s - 1} + \frac{2s + 4}{s^2 + 4}.$$

Thus,

$$L^{-1}\left\{\frac{2s^3 - 4s - 8}{s(s - 1)(s^2 + 4)}\right\} = 2L^{-1}\left\{\frac{1}{s}\right\} - 2L^{-1}\left\{\frac{1}{s - 1}\right\} + 2L^{-1}\left\{\frac{s}{s^2 + 4}\right\} + 2L^{-1}\left\{\frac{2}{s^2 + 4}\right\}$$

$$= 2 - 2e^t + 2\cos 2t + 2\sin 2t.$$

The same result is obtained with **invlaplace**.

```
> readlib(laplace):
  invlaplace((2*s^3-4*s-8)/((s^2-s)*(s^2+4)),s,t);
```

$$2 - 2e^t + 2\sin(2t) + 2\cos(2t)$$

☰ Laplace Transform of an Integral

We have seen that the Laplace transform of the derivatives of a given function can be found from the Laplace transform of the function. Similarly, the Laplace transform of the integral of a given function can also be obtained from the Laplace transform of the function, as stated in the following theorem.

• •

THEOREM   *Laplace Transform of an Integral*

Suppose that $F(s) = L\{f(t)\}$, where f is a piecewise continuous function on $[0, \infty)$ and of exponential order b. Then for $s > b$,

$$L\left\{\int_0^t f(\alpha)\,d\alpha\right\} = \frac{L\{f(t)\}}{s}.$$

In other words,

$$L^{-1}\left\{\frac{L\{f(t)\}}{s}\right\} = \int_0^t f(\alpha)\,d\alpha.$$

• •

EXAMPLE: Compute $L^{-1}\left\{\dfrac{1}{s(s+2)}\right\}$.

SOLUTION: In this case, $\dfrac{1}{s(s+2)} = \dfrac{1/(s+2)}{s}$, so $L\{f(t)\} = \dfrac{1}{s+2}$. Therefore, $f(t) = L^{-1}\left\{\dfrac{1}{s+2}\right\} = e^{-2t}$. With the previous theorem, we then have

$$L^{-1}\left\{\frac{1}{s(s+2)}\right\} = \int_0^t e^{-2\alpha}\,d\alpha = \frac{1 - e^{-2t}}{2}.$$

Note that the same result is obtained through a partial fraction expansion of $\dfrac{1}{s(s+2)}$: because

$$\frac{1}{s(s+2)} = \frac{1}{2s} - \frac{1}{2(2+s)}, L^{-1}\left\{\frac{1}{s(s+2)}\right\} = L^{-1}\left\{\frac{1}{2s} - \frac{1}{2(2+s)}\right\} = \frac{1}{2} - \frac{1}{2}e^{-2t}.$$

● ●

THEOREM   Suppose that f is a piecewise continuous function on $[0, \infty)$ and of exponential order b. Then

$$\lim_{s \to \infty} F(s) = \lim_{s \to \infty} L\{f(t)\} = 0.$$

● ●

This theorem is useful in determining if the inverse Laplace transform of a function $F(s)$ exists.

EXAMPLE: Determine if the inverse Laplace transform exists for (a) $F(s) = \dfrac{2s}{s - 6}$ and

(b) $F(s) = \dfrac{s^3}{s^2 + 16}$.

SOLUTION: In both cases, we find $\lim\limits_{s \to \infty} F(s)$. If this value is not zero, then $L^{-1}\{F(s)\}$ cannot be

found. (a) $\lim\limits_{s \to \infty} F(s) = \lim\limits_{s \to \infty} \dfrac{2s}{s - 6} = 2 \neq 0$, so $L^{-1}\left\{\dfrac{2s}{s - 6}\right\}$ does not exist. (b) $\lim\limits_{s \to \infty} F(s) = \lim\limits_{s \to \infty} \dfrac{s^3}{s^2 + 16} =$

$\infty \neq 0$. Thus, $L^{-1}\left\{\dfrac{s^3}{s^2 + 16}\right\}$ does not exist.

7.3 SOLVING INITIAL-VALUE PROBLEMS WITH THE LAPLACE TRANSFORM

Laplace transforms can be used to solve certain initial-value problems. Typically, when we use Laplace transforms to solve an initial-value problem for a function y, we use the following steps:

1. Compute the Laplace transform of each term in the differential equation.
2. Solve the resulting equation for $L\{y(t)\}$.
3. Determine y by computing the inverse Laplace transform of $L\{y(t)\}$. The advantage of this method is that through the use of the property

$$L\{f^{(n)}(t)\} = s^n L\{f(t)\} - s^{n-1} f(0) - \cdots s f^{(n-2)}(0) - f^{(n-1)}(0),$$

we change the differential equation to an algebraic equation.

EXAMPLE: Solve the initial-value problem $y' - 4y = e^{4t}$, $y(0) = 0$.

SOLUTION: We begin by taking the Laplace transform of both sides of the differential equation. Because $L\{y'\} = sY(s) - y(0) = sY(s)$, we have

$$L\{y' - 4y\} = L\{e^{4t}\}$$

$$L\{y'\} - 4L\{y\} = \frac{1}{s-4}$$

$$sY(s) - 4Y(s) = \frac{1}{s-4}$$

$$(s-4)Y(s) = \frac{1}{s-4}.$$

Equivalent results are obtained using `laplace` in `step_1`.

> `step_1:=laplace(diff(y(t),t)-4*y(t)=exp(4*t),t,s);`

$step\_1 := \text{laplace}(y(t), t, s)s - y(0) - 4\,\text{laplace}(y(t), t, s) = \dfrac{1}{(s-4)}$

We then apply the initial condition to `step_1` with `subs` and name the result `step_2`.

> `step_2:=subs(y(0)=0,step_1);`

$step\_2 := \text{laplace}(y(t), t, s)s - 4\,\text{laplace}(y(t), t, s) = \dfrac{1}{(s-4)}$

Solving for $Y(s)$ yields

$$Y(s) = \frac{1}{(s-4)^2}.$$

The same result is obtained with `solve`.

> `step_3:=solve(step_2,laplace(y(t),t,s));`

$step\_3 := (s-4)^{-2}$

Hence, by using the shifting property with $Lt = \dfrac{1}{s^2}$, we have

$$y(t) = L^{-1}\left\{\frac{1}{(s-4)^2}\right\} = te^{4t}.$$

Identical results are obtained using `invlaplace`.

```
> readlib(laplace):
  Sol:=invlaplace(step_3,s,t);
```

$$Sol := te^{4t}$$

We then graph this solution with `plot`.

```
> plot(Sol,t=0..1);
```

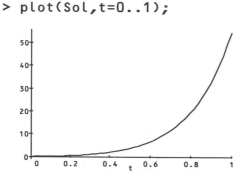

We can also use `dsolve` to solve the initial-value problem directly.

```
> dsolve({diff(y(t),t)-4*y(t)=exp(4*t),y(0)=0},y(t));
```

$$y(t) = te^{4t}$$

As we can see from the previous example, Laplace transforms are useful in solving nonhomogeneous equations. Hence, problems in Chapter 4 for which the methods of undetermined coefficients or variation of parameters were difficult to apply may be more easily solved through the method of Laplace transforms.

EXAMPLE: Use Laplace transforms to solve $y'' + 4y = e^{-t}\cos 2t$ subject to $y(0) = 0$ and $y'(0) = -1$.

SOLUTION: As in the previous example, we proceed by computing the Laplace transform of each side of the equation with `laplace`

```
> step_1:=laplace(diff(y(t),t$2)+4*y(t)=
  exp(-t)*cos(2*t),t,s)
```

$$step\_1 := (\text{laplace}\,(y(t), t, s)\,s - y(0))s - D(y)(0) + 4, \text{laplace}\,(y(t), t, s) = \frac{s+1}{(s+1)^2 + 4}$$

and then applying the initial conditions $y(0) = 0$ and $y'(0) = -1$ with **subs**, naming the result **step_2**.

```
> step_2:=subs({y(0)=0,D(y)(0)=-1},step_1);
```

$$step\_2 := \text{laplace}\,(y(t), t, s)\,s^2 + 1 + 4\,\text{laplace}\,(y(t), t, s) = \frac{s+1}{(s+1)^2 + 4}$$

Next, we solve **step_2** for the Laplace transform of $y(t)$ and simplify the result, naming the resulting output **step_3**.

```
> step_3:=simplify(solve(step_2,laplace(y(t),t,s)));
```

$$step\_3 := -\frac{s^2 + s + 4}{(s^2 + 2s + 5)(s^2 + 4)}$$

Then we use **invlaplace** to compute the inverse Laplace transform of **step_3** and name the result **Sol**.

```
> readlib(laplace):
  Sol:=invlaplace(step_3,s,t);latex(");
```

$$Sol := -\frac{4}{17}e^{-t}\sin(2t) + \frac{1}{17}e^{-t}\cos(2t) - \frac{4}{17}\sin(2t) - \frac{1}{17}\cos(2t)$$

The result means that

$$y(t) = -\frac{4}{17}\sin(2t) - \frac{1}{17}\cos(2t) + \frac{1}{17}e^{-t}\cos(2t) - \frac{4}{17}e^{-t}\sin(2t).$$

Last, we use **plot** to graph the solution obtained in **Sol** on the interval $[0, 2\pi]$.

```
> plot(Sol,t=0..2*Pi);
```

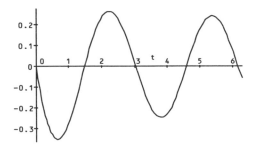

As we have seen in many previous examples, **dsolve** is able to find an equivalent solution to the initial-value problem.

```
> dsolve({diff(y(t),t$2)+4*y(t)=exp(-t)*cos(2*t),
  y(0)=0,D(y)(0)=-1},y(t));
```

$$y(t) = -\frac{2}{17}\sin(2t)\cos(t)^4 + \frac{8}{17}e^{-t}\sin(2t)\cos(t)^3\sin(t)$$

$$+ \frac{2}{17}e^{-t}\sin(2t)\cos(t)^2 - \frac{4}{17}e^{-t}\sin(2t)\cos(t)\sin(t)$$

$$- \frac{9}{34}e^{-t}\sin(2t) + \frac{1}{17}e^{-t}\cos(2t)\cos(4t) + \frac{1}{68}e^{-t}\cos(2t)\sin(4t)$$

$$- \frac{4}{17}\sin(2t) - \frac{1}{17}\cos(2t)$$

Higher-order initial-value problems can be solved with the method of Laplace transforms as well.

EXAMPLE: Solve $y''' + y'' - 6y' = \sin 4t$, $y(0) = 2$, $y'(0) = 0$, $y''(0) = -1$.

SOLUTION: We begin by defining **EQ** to be the equation $y''' + y'' - 6y' = \sin 4t$.

```
> EQ:=diff(y(t),t$3)+diff(y(t),t$2)-6*diff(y(t),t)=sin(4*t);
```

$$EQ := \left(\frac{\partial^3}{\partial t^3}y(t)\right) + \left(\frac{\partial^2}{\partial t^2}y(t)\right) - 6\left(\frac{\partial}{\partial t}y(t)\right) = \sin(4t)$$

We first note that **dsolve** is able to find an explicit solution of the initial-value problem quickly.

```
> dsolve({EQ,y(0)=2,D(y)(0)=0,(D@@2)(y)(0)=-1},y(t));
```

$$y(t) = \frac{1}{3}\cos(t)^4 + \frac{13}{6} - \frac{1}{3}\cos(t)^2 - \frac{23}{750}\cos(4t) - \frac{1}{500}\sin(4t) - \frac{2}{25}e^{2t} - \frac{7}{125}e^{-3t}$$

Alternatively, we can use Maple to implement the steps encountered when solving the equation using the method of Laplace transforms.

Taking the Laplace transform of both sides of the equation, we find

```
> step_1:=laplace(EQ,t,s);
```

$step\_1 := ((\text{laplace}(y(t), t, s)s - y(0))s - D(y)(0))s - (D^{(2)})(y)(0)$
$\qquad + (\text{laplace}(y(t), t, s)s - y(0))s - D(y)(0) - 6\,\text{laplace}(y(t), t, s)s$
$\qquad + 6y(0) = \dfrac{4}{s^2 + 16}$

and then we apply the initial conditions, naming the result **step_2**.

```
> step_2:=subs({y(0)=2,D(y)(0)=0,
  (D@@2)(y)(0)=-1},step_1);
```

$step\_2 := (\text{laplace}(y(t), t, s)s - 2)s^2 + 13 + (\text{laplace}(y(t), t, s)s - 2)s$
$\qquad - 6\,\text{laplace}(y(t), t, s)s = 4\dfrac{1}{s^2 + 16}$

Solving for $Y(s) = $ **laplace(y(t),t,s)**, we obtain

```
> step_3:=simplify(solve(step_2,laplace(y(t),t,s)));
```

$step\_3 := \dfrac{2s^4 + 19s^2 - 204 + 2s^3 + 32s}{(s^2 + 16)s(s^2 + s - 6)}$

and computing the inverse Laplace transform of **step_3** with **invlaplace** yields the solution of the initial-value problem.

```
> readlib(laplace):
  Sol:=invlaplace(step_3,s,t);
```

$Sol := \dfrac{17}{8} - \dfrac{7}{125}e^{-3t} - \dfrac{2}{25}e^{2t} - \dfrac{1}{500}\sin(4t) + \dfrac{11}{1000}\cos(4t)$

Last, a graph of the solution is generated with **plot**.

```
> plot(Sol,t=-3/2..3/2);
```

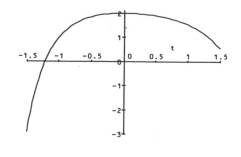

Some initial-value problems that involve differential equations with nonconstant coefficients can also be solved with the method of Laplace transforms as illustrated in the following example. However, Laplace transforms do not provide a general method for solving equations with nonconstant coefficients.

EXAMPLE: Solve $ty'' + (11t + 1)y' + (30t + 2)y = 0$, $y(0) = 1$, $y'(0) = -2$.

SOLUTION: First, we take the Laplace transform of both sides of the equation. Recall the property

$$L\{t^n f(t)\} = (-1)^n \frac{d^n F}{ds^n}(s) = (-1)^n \frac{d^n}{ds^n} L\{f(t)\}.$$

Thus, entering

```
> laplace(t*y(t),t,s);
  laplace(t*diff(y(t),t),t,s);
  laplace(t*diff(y(t),t$2),t,s);
```

$$-\frac{\partial}{\partial s}\, \text{laplace}\,(y(t), t, s)$$

$$-\left(\frac{\partial}{\partial s}\, \text{laplace}\,(y(t), t, s)\right)s - \text{laplace}\,(y(t), t, s)$$

$$-\left(\left(\frac{\partial}{\partial s}\, \text{laplace}\,(y(t), t, s)\right)s + \text{laplace}\,(y(t), t, s)\right)s - \text{laplace}\,(y(t), t, s)s$$

$$+ y(0)$$

shows us that

$$L\{ty(t)\} = -\frac{d}{ds}L\{y(t)\} = -\frac{d}{ds}Y(s) = -Y'(s),$$

$$L\{ty'(t)\} = -s\frac{d}{ds}L\{y(t)\} - L\{y(t)\} = -s\frac{d}{ds}Y(s) - Y(s) = -sY'(s) - Y(s),$$

and

$$L\{ty''(t)\} = -sY'(s) - 2sY(s) + y(0).$$

To solve the initial-value problem, we begin by defining EQ to be the equation $ty'' + (11t + 1)y' + (30t + 2)y = 0$.

```
> EQ:=expand(t*diff(y(t),t$2)+(11*t+1)*diff(y(t),t)+
  (30*t+2)*y(t))=0;
```

$$EQ := t\left(\frac{\partial^2}{\partial t^2}y(t)\right) + 11\left(\frac{\partial}{\partial t}y(t)\right)t + \left(\frac{\partial}{\partial t}y(t)\right) + 30y(t)t + 2y(t) = 0$$

In this case, we see that **dsolve** is unable to solve the initial-value problem.

```
> dsolve({EQ,y(0)=1,D(y)(0)=-2},y(t));
```

In subsequent sections, we will see that the **laplace** option enables **dsolve** to use Laplace transforms to solve differential equations. However, even when we include the **laplace** option, **dsolve** is unable to solve this initial-value problem.

```
> dsolve({EQ,y(0)=1,D(y)(0)=-2},y(t),laplace);
```

To solve the problem we proceed as in the previous examples. First, we compute the Laplace transform of each side of the equation, naming the resulting output **step_1**.

```
> step_1:=laplace(EQ,t,s);
```

$$step\_1 := -\left(\left(\frac{\partial}{\partial s}\text{laplace}\,(y(t), t, s)\right)s + \text{laplace}\,(y(t), t, s)\right)s$$

$$- 11\left(\frac{\partial}{\partial s}\text{laplace}\,(y(t), t, s)\right)s - 9\,\text{laplace}\,(y(t), t, s)$$

$$- 30\frac{\partial}{\partial s}\text{laplace}\,(y(t), t, s) = 0$$

Then we apply the initial conditions and name the result **step_2**.

```
> step_2:=subs({y(0)=1,D(y)(0)=-2},step_1);
```

$$step\_2 := -\left(\left(\frac{\partial}{\partial s} \text{laplace}\,(y(t), t, s)\right)s + \text{laplace}\,(y(t), t, s)\right)s$$

$$-11\left(\frac{\partial}{\partial s} \text{laplace}\,(y(t), t, s)\right)s - 9\,\text{laplace}\,(y(t), t, s)$$

$$-30\frac{\partial}{\partial s} \text{laplace}\,(y(t), t, s) = 0$$

For convenience, we replace each occurrence of `laplace(y(t),t,s)`, which represents the Laplace transform of $y(t)$, in `step_2` by $Y(s)$.

```
> step_3:=subs(laplace(y(t),t,s)=Y(s),step_2);
```

$$step\_3 := -\left(\left(\frac{\partial}{\partial s} Y(s)\right)s + Y(s)\right)s - 11\left(\frac{\partial}{\partial s} Y(s)\right)s - 9Y(s) - 30\left(\frac{\partial}{\partial s} Y(s)\right) = 0$$

The equation in `step_3` is a first-order (separable) differential equation that we solve with `dsolve`, naming the result `step_4`.

```
> step_4:=dsolve(step_3,Y(s));
```

$$step\_4 := Y(s) = \frac{(s + 6)^3 \_C1}{s^4 + 20s^3 + 150s^2 + 500s + 625}$$

We then use `assign` to name $Y(s)$ the result we obtained and use `invlaplace` to compute the inverse Laplace transform of $Y(s)$.

```
> readlib(laplace):
  assign(step_4):
  step_5:=invlaplace(Y(s),s,t);
```

$$step\_5 := \frac{1}{6}t^3e^{-5t}\_C1 + \frac{3}{2}t^2e^{-5t}\_C1 + 3te^{-5t}\_C1 + \_C1\,e^{-5t}$$

To find the value of the arbitrary constant, $\_C1$, we evaluate `step_5` and the derivative of `step_5` for $t = 0$.

```
> eval(subs(t=0,step_5));
  eval(subs(t=0,diff(step_5,t)));
```

$$\_C1$$

$$-2\_C1$$

Because we must have that $y(0) = 1$ and $y'(0) = -2$, it follows that we must have $\_C1 = 1$. Therefore, the solution to the initial-value problem is obtained by replacing $\_C1$ in step_5 by 1.

> Sol:=subs(_C1=1,step_5);

$$Sol := \frac{1}{6}t^3 e^{-5t} + \frac{3}{2}t^2 e^{-5t} + 3t e^{-5t} + e^{-5t}$$

As in previous examples, we graph the solution using plot. In this case, the option -10..3 instructs Maple that the range displayed corresponds to the interval $[-10, 3]$.

> plot(Sol,t=-1..1,-10..3);

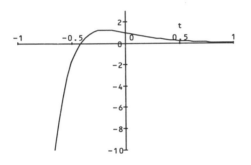

7.4 LAPLACE TRANSFORMS OF SEVERAL IMPORTANT FUNCTIONS

≡ Piecewise Defined Functions: The Unit Step Function

An important function in modeling many physical situations is the unit step function u.

DEFINITION | *Unit Step Function*

The unit step function $u(t - a)$, where a is a given number, is defined by

$$u(t - a) = \begin{cases} 0, 0 \le t < a \\ 1, t \ge a \end{cases}.$$

We can use the built-in function **Heaviside** to define the unit step function. The **Heaviside** function is defined by

$$\text{Heaviside(t)} = \begin{cases} 0, & \text{if } t < 0 \\ 1, & \text{if } t \geq 0 \end{cases}.$$

Thus, $u(t - a) =$ **Heaviside(t-a)**.

EXAMPLE: Sketch the graph of (a) $u(t - 5)$ and (b) $u(t)$.

SOLUTION: (a) In this case, $u(t - 5) = \begin{cases} 0, 0 \leq t < 5 \\ 1, t \geq 5 \end{cases}$, so the "jump" occurs at $t = 5$. (b) Here, $u(t) = u(t - 0)$, so $u(t) = 1$ for $t \geq 0$. Both of these functions are graphed as follows using **plot** and **Heaviside**.

```
> plot(Heaviside(t-5),t=0..10);
  plot(Heaviside(t),t=0..5);
```

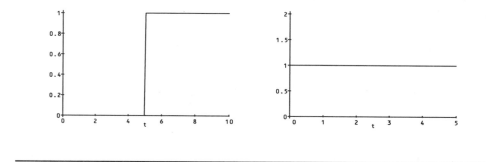

The unit step function is useful in defining functions that are piecewise continuous. For example, we can define the function

$$g(t) = \begin{cases} 0, 0 \leq t < a \\ h(t), a \leq t < b \\ 0, t \geq b \end{cases}$$

as $g(t) = h(t)[u(t - a) - u(t - b)]$.

Similarly, a function like

$$f(t) = \begin{cases} g(t), 0 \leq t < a \\ h(t), t \geq a \end{cases}$$

can be written as

$$f(t) = g(t)[u(t - 0) - u(t - a)] + h(t)u(t - a) = g(t)[1 - u(t - a)] + h(t)u(t - a).$$

The reason for writing piecewise continuous functions in terms of unit functions is that we encounter functions of this type in solving initial-value problems. Using our methods in Chapters 4 and 5, we had to solve the problem over each piece of the function. However, the method of Laplace transforms can be used to avoid these complicated calculations. Hence, we state the following theorem.

••••••••••••••••••••••••••

THEOREM   Suppose that $F(s) = L\{f(t)\}$ exists for $s > b \geq 0$. If a is a positive constant, then

$$L\{f(t - a) u(t - a)\} = e^{-st}F(s).$$

••••••••••••••••••••••••••

EXAMPLE: Find $L\{(t - 3)^5 u(t - 3)\}$.

SOLUTION: In this case, $a = 3$ and $f(t) = t^5$. Thus,

$$L\{(t - 3)^5 u(t - 3)\} = e^{-5s}L\{t^5\} = e^{-5s}\frac{5!}{s^6} = \frac{120}{s^6}e^{-5s}.$$

To use Maple, we note that `(t-3)^5*Heaviside(t-3)` represents $(t - 3)^5 u(t - 3)$, so entering

```
> laplace((t-3)^5*Heaviside(t-3),t,s);
```

$$120\frac{e^{-3s}}{s^6}$$

computes $L\{(t - 3)^5 u(t - 3)\}$.

In most cases, we must calculate $L\{g(t)u(t - a)\}$ instead of $L\{f(t - a)u(t - a)\}$. To solve this problem, we let $g(t) = f(t - a)$, so $f(t) = g(t + a)$. Therefore,

$$L\{g(t)u(t - a)\} = e^{-as}L\{g(t + a)\}.$$

EXAMPLE: Calculate $L\{\sin t\, u(t - \pi)\}$.

SOLUTION: In this case, $g(t) = \sin t$ and $a = \pi$. Thus,

$$L\{\sin t\, u(t - \pi)\} = e^{-\pi s}L\{\sin(t + \pi)\} = e^{-\pi s}L\{-\sin t\} = -e^{-\pi s}\frac{1}{s^2 + 1} = -\frac{e^{-\pi s}}{s^2 + 1}.$$

To use Maple, we note that `sint(t)*Heaviside(t-Pi)` represents $\sin t\, u(t - \pi)$. Thus, entering

> `laplace(sin(t)*Heaviside(t-Pi),t,s);`

$$-\frac{e^{-\pi s}}{s^2 + 1}$$

computes $L\{\sin t\, u(t - \pi)\}$.

· ·

THEOREM   Suppose that $F(s) = L\{f(t)\}$ exists for $s > b \geq 0$. If a is a positive constant and f is continuous on $[0, \infty)$, then

$$L^{-1}\{e^{-as}F(s)\} = f(t - a)\, u(t - a).$$

· ·

EXAMPLE: Find (a) $L^{-1}\left\{\dfrac{e^{-4s}}{s^3}\right\}$ and (b) $L^{-1}\left\{\dfrac{e^{-\pi s/2}}{s^2 + 16}\right\}$.

SOLUTION: (a) If we write the expression $\dfrac{e^{-4s}}{s}$ in the form $e^{-as}F(s)$, we see that $a = 4$ and $F(s) = \dfrac{1}{s^3}$. Hence,

$$f(t) = L^{-1}\left\{\frac{1}{s^3}\right\} = \frac{1}{2}t^2$$

and

$$L^{-1}\left\{\frac{e^{-4s}}{s^3}\right\} = f(t - 4)\, u(t - 4) = \frac{1}{2}(t - 4)^2 u\,(t - 4).$$

(b) In this case, $a = \dfrac{\pi}{2}$ and $F(s) = \dfrac{1}{s^2 + 16}$. Then

$$f(t) = L^{-1}\left\{\frac{1}{s^2 + 16}\right\} = \frac{1}{4}\sin 4t$$

and

$$L^{-1}\left\{\frac{e^{-\pi s/2}}{s^2 + 16}\right\} = f\left(t - \frac{\pi}{2}\right)u\left(t - \frac{\pi}{2}\right) = \frac{1}{4}\sin 4\left(t - \frac{\pi}{2}\right)u\left(t - \frac{\pi}{2}\right)$$

$$= \frac{1}{4}\sin(4t - 2\pi)u\left(t - \frac{\pi}{2}\right) = \frac{1}{4}\sin 4t\, u\left(t - \frac{\pi}{2}\right).$$

For each of (a) and (b), the same results are obtained using `invlaplace`.

```
> readlib(laplace):
  invlaplace(exp(-4*s)/s^3,s,t);
  invlaplace(exp(-Pi*s/2)/(s^2+16),s,t);
```

$$\frac{1}{2}\text{Heaviside}(t - 4)(t - 4)^2$$

$$\frac{1}{4}\text{Heaviside}\left(t - \frac{\pi}{2}\right)\sin(4t - 2\pi)$$

≡ Solving Initial-Value Problems

With the unit step function, we can solve initial-value problems that involve piecewise continuous functions.

EXAMPLE: Solve $y'' + 9y = \begin{cases} 1, 0 \le t < \pi \\ 0, t \ge \pi \end{cases}$ subject to $y(0) = y'(0) = 0$.

SOLUTION: In order to solve this initial-value problem, we must compute $L\{f(t)\}$, where $f(t) = \begin{cases} 1, 0 \le t < \pi \\ 0, t \ge \pi \end{cases}$. Since this is a piecewise continuous function, we write it in terms of the unit step function as

$$f(t) = 1[u(t - 0) - u(t - \pi)] + 0[u(t - \pi)] = u(t) - u(t - \pi).$$

Then

$$L\{f(t)\} = L\{1 - u(t - \pi)\} = \frac{1}{s} - \frac{e^{-\pi s}}{s}.$$

Hence,

$$L\{y''\} + 9L\{y\} = L\{f(t)\}$$

$$s^2 Y(s) - sy(0) - y'(0) + 9Y(s) = \frac{1}{s} - \frac{e^{-\pi s}}{s}$$

$$(s^2 + 9)Y(s) = \frac{1}{s} - \frac{e^{-\pi s}}{s}$$

$$Y(s) = \frac{1}{s(s^2 + 9)} - \frac{e^{-\pi s}}{s(s^2 + 9)}.$$

The same steps are performed with Maple. First, we define $f(t) = \begin{cases} 1, 0 \le t < \pi \\ 0, t \ge \pi \end{cases}$ with Heaviside and then define EQ to be the equation $y'' + 9y = \begin{cases} 1, 0 \le t < \pi \\ 0, t \ge \pi \end{cases}$.

```
> f:=t->Heaviside(t)-Heaviside(t-Pi):
  EQ:=diff(y(t),t$2)+9*y(t)=f(t);
```

$$EQ := \left(\frac{\partial^2}{\partial^2} y(t) \right) + 9y(t) = \text{Heaviside}(t) - \text{Heaviside}(t - \pi)$$

Next, we use laplace to compute the Laplace transform of each side of the equation, naming the resulting equation step_1,

```
> step_1:=laplace(EQ,t,s);
```

$$step\_1 := (\text{laplace}(y(t), t, s)s - y(0))s - D(y)(0) + 9, \text{laplace}(y(t), t, s)$$

$$= s^{-1} - \frac{e^{-\pi s}}{s}$$

apply the initial conditions, naming the result step_2,

```
> step_2:=subs({y(0)=0,D(y)(0)=0},step_1);
```

$$step\_2 := \text{laplace}(y(t), t, s)s^2 + 9 \text{laplace}(y(t), t, s) = \frac{1}{s} - 1 - \frac{e^{-\pi s}}{s}$$

and solve step_2 for laplace(y(t),t,s), naming the result step_3.

```
> step_3:=simplify(solve(step_2,laplace(y(t),t,s)));
```

$$step\_3 := -\frac{-1 + e^{-\pi s}}{s(s^2 + 9)}$$

Then

$$y(t) = L^{-1}\{Y(s)\} = L^{-1}\left\{\frac{1}{s(s^2 + 9)}\right\} - L^{-1}\left\{\frac{e^{-\pi s}}{s(s^2 + 9)}\right\}.$$

Consider $L^{-1}\left\{\dfrac{e^{-\pi s}}{s(s^2 + 9)}\right\}$. In the form of $L^{-1}\{e^{-as}F(s)\}$, $a = \pi$ and $F(s) = \dfrac{1}{s(s^2 + 9)}$. $f(t) = L^{-1}\{F(s)\}$ can be found with either a partial fraction expansion or with the formula

$$f(t) = L^{-1}\left\{\frac{1}{s(s^2 + 9)}\right\} = \int_0^t L^{-1}\left\{\frac{1}{s^2 + 9}\right\}d\alpha = \int_0^t \frac{1}{3}\sin 3\alpha\, d\alpha = -\frac{1}{3}\left[\frac{\cos 3\alpha}{3}\right]_0^t = \frac{1}{9} - \frac{1}{9}\cos 3t.$$

Then

$$L^{-1}\left\{\frac{e^{-\pi s}}{s(s^2 + 9)}\right\} = \left[\frac{1}{9} - \frac{1}{9}\cos 3(t - \pi)\right]u(t - \pi)$$

$$= \left[\frac{1}{9} - \frac{1}{9}\cos(3t - 3\pi)\right]u(t - \pi) = \left[\frac{1}{9} + \frac{1}{9}\cos 3t\right]u(t - \pi).$$

Combining these results yields the solution

$$y(t) = L^{-1}\{Y(s)\} = L^{-1}\left\{\frac{1}{s(s^2 + 9)}\right\} - L^{-1}\left\{\frac{e^{-\pi s}}{s(s^2 + 9)}\right\}$$

$$= \frac{1}{9} - \frac{1}{9}\cos 3t - \left[\frac{1}{9} + \frac{1}{9}\cos 3t\right]u(t - \pi).$$

Notice that we can rewrite this solution as the piecewise-defined function

$$y(t) = \begin{cases} \dfrac{1}{9} - \dfrac{1}{9}\cos 3t, & 0 \le t < \pi \\[2mm] -\dfrac{2}{9}\cos 3t & \ge \pi \end{cases}.$$

We now use Maple to compute the inverse Laplace transform with `invlaplace` to obtain the solution, named `Sol`.

```
> readlib(laplace):
  Sol:=invlaplace(step_3,s,t);
```

$$Sol := \frac{1}{9} - \frac{1}{9}\cos(3t) - \text{Heaviside}(t - \pi)\left(\frac{1}{9} + \frac{1}{9}\cos(3t)\right)$$

We then graph with the solution **plot**.

```
> plot(Sol,t=0..3*Pi);
```

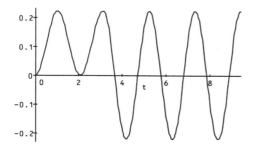

An equivalent result is obtained using **dsolve**.

```
> Alt_Sol:=dsolve({EQ,y(0)=0,D(y)(0)=0},y(t));
```

$$Alt\_Sol := y(t) = \int_0^t \frac{1}{3}\cos(3u)\,\text{Heaviside}(u) - \frac{1}{3}\cos(3u)\,\text{Heaviside}(u - \pi)\,du\,\sin(3t)$$

$$+ \int_0^t -\frac{1}{3}\sin(3u)\,\text{Heaviside}(u) + \frac{1}{3}\sin(3u)\,\text{Heaviside}(u - \pi)\,du\,\cos(3t)$$

$$- \int_0^0 \frac{1}{3}\cos(3u)\,\text{Heaviside}(u) - \frac{1}{3}\cos(3u)\,\text{Heaviside}(u - \pi)\,du\,\sin(3t)$$

$$- \int_0^0 -\frac{1}{3}\sin(3u)\,\text{Heaviside}(u) + \frac{1}{3}\sin(3u)\,\text{Heaviside}(u - \pi)\,du\,\cos(3t)$$

We can simplify this result somewhat with **eval**.

```
> eval(rhs(Alt_Sol));
```

$$\int_0^t \frac{1}{3}\cos(3u)\,\text{Heaviside}(u) - \frac{1}{3}\cos(3u)\,\text{Heaviside}(u - \pi)\,du\,\sin(3t)$$

$$+ \int_0^t -\frac{1}{3}\sin(3u)\,\text{Heaviside}(u) + \frac{1}{3}\sin(3u)\,\text{Heaviside}(u - \pi)\,du\,\cos(3t)$$

When the laplace option is included in the dsolve command, Maple uses Laplace transforms to solve the problem. However, in this case we see that when we use dsolve together with the laplace option, an incorrect result is returned.

```
> dsolve({EQ,y(0)=0,D(y)(0)=0},y(t),laplace);
```

$$y(t) = \frac{2}{9} - \frac{2}{9}\cos(3t)$$

☰ Periodic Functions

Another type of function that is encountered in many areas of applied mathematics is the **periodic** function.

DEFINITION

> *Periodic Function*
>
> A function f is **periodic** if there is a positive number T such that
>
> $$f(t + T) = f(t)$$
>
> for all $t \geq 0$. The minimum value of T that satisfies this equation is called the period of f.

Due to the nature of periodic functions, we can simplify the calculation of the Laplace transform of these functions as indicated in the following theorem.

THEOREM

Laplace Transform of Periodic Functions

Suppose that f is a periodic function of period T and that f is piecewise continuous on $[0, \infty)$. Then, the $L\{f(t)\}$ exists for $s > 0$ and is determined with the definite integral

$$L\{f(t)\} = \frac{1}{1 - e^{-sT}} \int_0^T e^{-st} f(t)\, dt.$$

EXAMPLE: Find the Laplace transform of the periodic function $f(t) = t$, $0 \le t < 1$, and $f(t + 1) = f(t)$.

SOLUTION: The period of f is $T = 1$. Hence, through integration by parts,

$$L\{f(t)\} = \frac{1}{1-e^{-s}} \int_0^1 e^{-st} t\, dt = \frac{1}{1-e^{-s}} \left\{ \left[-\frac{te^{-st}}{s} \right]_0^1 + \int_0^1 \frac{e^{-st}}{s} dt \right\}$$

$$= \frac{1}{1-e^{-s}} \left\{ -\frac{e^{-s}}{s} - \left[\frac{e^{-st}}{s^2} \right]_0^1 \right\} = \frac{1}{1-e^{-s}} \left[-\frac{e^{-s}}{s} + \frac{1-e^{-s}}{s^2} \right] = \frac{1-(s+1)e^{-s}}{s^2(1-e^{-s})}.$$

Alternatively, using Maple we enter

```
> simplify(1/(1-exp(-s))*int(t*exp(-s*t),t=0..1));
```

$$\frac{e^{-s}s + e^{-s} - 1}{(-1+e^{-s})s^2}$$

to compute the Laplace transform. Alternatively, note that

$$f(t) = t(u(t-1) - u(t)) + (t-1)(u(t-2) - u(t-1)) + (t-2)(u(t-3) - u(t-2)) + \cdots$$

$$= t - u(t-1) - u(t-2) - u(t-3) - u(t-4) - \cdots$$

$$= t - \sum_{n=1}^{\infty} u(t-n)$$

so

$$L\{f(t)\} = L\{t\} - L\left\{ \sum_{n=1}^{\infty} u(t-n) \right\} = L\{t\} - \sum_{n=1}^{\infty} L\{u(t-n)\}.$$

After using **assume** to instruct Maple to assume that $n \ge 0$ and $s > 0$ we use **Heaviside** and **laplace**

```
> assume(n>=0):
  assume(s>0):
  nth_term:=laplace(Heaviside(t-n),t,s)
```

$$nth\_term := \frac{e^{-n\sim s\sim}}{s\sim}$$

to see that $L\{u(t-n)\} = \dfrac{e^{-ns}}{s}$. Note that the $\sim$ operator in the result indicates that we have made assumptions about n and s. Next, we use **sum** and **simplify** to calculate

$$L\{f(t)\} = L\{t\} - \sum_{n=1}^{\infty} L\{u(t-n)\} = \frac{1}{s^2} - \sum_{n=1}^{\infty} \frac{e^{-ns}}{s}$$

$$= \frac{1}{s}\left(\frac{1}{s} - \sum_{n=1}^{\infty}(e^{-s})^n\right) = \frac{1}{s}\left(\frac{1}{s} - \frac{e^{-s}}{1-e^{-s}}\right).$$

```
> simplify(laplace(t,t,s)-sum(nth_term,n=1..infinity));
```

$$-\frac{1 - e^{s\sim} + s\sim}{s\sim^2(-1+e^{s\sim})}$$

To graph the periodic function f, we proceed in two steps. If $g(t)$ is defined on the interval $[0, p]$, then the periodic extension of g with period p, f, is defined with the following procedure.

```
> f:=proc(t)
    local a,b;
    a:=trunc(t) mod p;
    b:=frac(t);
    g(a+b);
    end:
```

Thus, to graph f, we first define $g(t) = t$ and then use the previous procedure with $p = 1$ to define f.

```
> g:=t->t:
    f:=proc(t)
    local a,b;
    a:=trunc(t) mod 1;
    b:=frac(t);
    g(a+b);
    end:
```

Because f has been defined as a procedure, we use operator notation to graph f.

```
> plot(f,0..5);
```

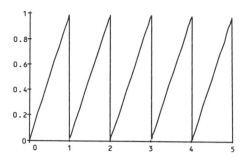

Laplace transforms can now be used to solve initial-value problems with periodic forcing functions more easily.

Periodic functions are encountered in many initial-value problems as we will see in Chapter 8. We illustrate how problems of this type are solved with Laplace transforms in the following example.

EXAMPLE: Solve $y'' + y = f(t)$ subject to $y(0) = y'(0) = 0$ if $f(t) = \begin{cases} \sin t, 0 \le t < \pi \\ 0, \pi \le t < 2\pi \end{cases}$ and $f(t + 2\pi) = f(t)$. (Note: f is known as the **half-wave rectification** of $\sin t$.)

SOLUTION: To graph f, we begin defining $g(t) = \begin{cases} \sin t, 0 \le t < \pi \\ 0, \pi \le t < 2\pi \end{cases}$, which is represented by `sin(t)*Heaviside(Pi-t)`.

```
> g:=t->sin(t)*Heaviside(Pi-t):
```

Then

$$f(t) = g(t)(u(2\pi - t) - u(-t)) + g(t - 2\pi)(u(4\pi - t) - u(2\pi - t)) + \cdots$$

$$= \sum_{n=0}^{\infty} g(t - 2n\pi)(u(2(n + 1)\pi - t) - u(2n\pi - t)).$$

Thus, the graph of f on the interval $[0, 2k\pi]$, where k represents a positive integer, is obtained by graphing

$$\sum_{n=0}^{k-1} g(t - 2n\pi)(u(2(n + 1)\pi - t) - u(2n\pi - t))$$

on the interval $[0, 2k\pi]$. For convenience, we define `nth_term` to be $g(t - 2n\pi)(u(t - 2(n + 1)\pi) - u(t - 2n\pi))$.

```
> nth_term:=n->g(t-2*n*Pi)*
    (Heaviside(2*(n+1)*Pi-t)-Heaviside(2*n*Pi-t)):
```

Then, to graph f on the interval $[0, 10\pi]$, we first enter

```
> to_graph:=sum(nth_term(n),n=0..4);
```

$$
\begin{aligned}
to\_graph := {} & \sin(t)\,\text{Heaviside}(\pi - t)(\,\text{Heaviside}(2\pi - t) - \text{Heaviside}(-t)) \\
& + \sin(t)\,\text{Heaviside}(3\pi - t)(\,\text{Heaviside}(4\pi - t) - \text{Heaviside}(2\pi - t)) \\
& + \sin(t)\,\text{Heaviside}(5\pi - t)(\,\text{Heaviside}(6\pi - t) - \text{Heaviside}(4\pi - t)) \\
& + \sin(t)\,\text{Heaviside}(7\pi - t)(\,\text{Heaviside}(8\pi - t) - \text{Heaviside}(6\pi - t)) \\
& + \sin(t)\,\text{Heaviside}(9\pi - t)(\,\text{Heaviside}(10\pi - t) - \text{Heaviside}(8\pi - t))
\end{aligned}
$$

and then

```
> plot(to_graph,t=0..10*Pi,0..2);
```

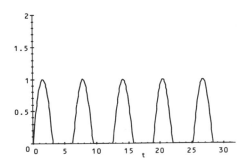

To solve the initial-value problem we must find $L\{f(t)\}$. Because the period is $T = 2\pi$, we have

$$
L\{f(t)\} = \frac{1}{1 - e^{-2\pi s}} \int_0^{2\pi} e^{-st} f(t)\, dt = \frac{1}{1 - e^{-2\pi s}} \left[\int_0^{\pi} e^{-st} \sin t\, dt + \int_0^{\pi} e^{-st}(0)\, dt \right]
$$

$$
= \frac{1}{1 - e^{-2\pi s}} \int_0^{\pi} e^{-st} \sin t\, dt.
$$

Using `int`

```
> s:='s':t:='t':
  step_1:=simplify(1/(1-exp(-2*Pi*s))*
   int(exp(-s*t)*sin(t),t=0..Pi));
```

$$step\_1 := -\frac{e^{-\pi s} + 1}{(-1 + e^{-2\pi s})(s^2 + 1)}$$

```
> lap_f:=normal(step_1,expanded);
```

$$lap\_f := \frac{e^{\pi s}}{e^{\pi s}s^2 + e^{\pi s} - s^2 - 1}$$

we see that $L\{f(t)\} = \dfrac{e^{\pi s}}{(e^{\pi s} - 1)(s^2 + 1)} = \dfrac{1}{(1 - e^{-\pi s})(s^2 + 1)}$.Alternatively, we can use

$$f(t) = \sum_{n=0}^{\infty} g(t - 2n\pi)(u(2(n + 1)\pi - t) - u(2n\pi - t))$$

to rewrite f as

$$f(t) = \sum_{n=0}^{\infty} (-1)^n \sin(t)\, u(t - n\pi).$$

Then

$$L\{f(t)\} = L\left\{ \sum_{n=0}^{\infty} (-1)^n \sin(t)\, u(t - n\pi) \right\} = \sum_{n=0}^{\infty} L\{(-1)^n \sin(t)\, u(t - n\pi)\}.$$

We use `laplace` and `Heaviside` to compute $L\{(-1)^n \sin(t)\, u\,(t - n\pi)\}$, naming the result `nth_lap`,

```
> assume(n>=0):
  assume(s>0):
  nth_lap:=laplace((-1)^n*sin(t)*Heaviside(t-n*Pi),t,s);
```

$$nth\_lap := \frac{e^{-n\sim \pi s\sim}(-1)^{n\sim}(s \sim \sin(n \sim \pi) + \cos(n \sim \pi))}{s \sim^2 + 1}$$

and then use `sum` to compute

$$\sum_{n=0}^{\infty} L\{(-1)^n \sin(t)\, u(t - n\pi)\}.$$

```
> sum(nth_lap,n=0..infinity);
```

$$\frac{e^{\pi s\sim}}{-s\sim^2 -1 + e^{\pi s\sim} s\sim^2 + e^{\pi s\sim}}$$

Taking the Laplace transform of both sides of the equation and solving for $Y(s)$ then gives us

$$L[y''] + L\{y\} = L\{f(t)\}$$

$$s^2 Y(s) - sy(0) - y'(0) + Y(s) = \frac{1}{(1 - e^{-\pi s})(s^2 + 1)}$$

$$Y(s) = \frac{1}{(1 - e^{-\pi s})(s^2 + 1)^2}.$$

Using lap_f, which was computed earlier, for $L\{f(t)\}$, we perform the same steps with Maple.

```
> first_step:=laplace(diff(y(t),t$2)+
  y(t),t,s)=lap_f;
```

$$\textit{first\_step} : = (\text{laplace}\,(y(t), t, s)s - y(0))s - D(y)(0) + \text{laplace}\,(y(t), t, s)$$

$$= \frac{e^{\pi s}}{e^{\pi s}s^2 + e^{\pi s} - s^2 - 1}$$

```
> second_step:=subs({y(0)=0,D(y)(0)=0},first_step);
```

$$\textit{second\_step} := \text{laplace}(y(t), t, s)s^2 + \text{laplace}(y(t), t, s) = \frac{e^{\pi s}}{e^{\pi s}s^2 + e^{\pi s} - s^2 - 1}$$

```
> find_inverse:=factor(
  solve(second_step,laplace(y(t),t,s)));
```

$$\textit{find\_inverse} := \frac{e^{\pi s}}{(s^2 + 1)^2(e^{\pi s} - 1)}$$

Recall from our work with the geometric series that if $|x| < 1$, then

$$\frac{1}{1 - x} = 1 + x + x^2 + x^3 + \cdots.$$

Because we do not know the inverse Laplace transform of $\dfrac{1}{(1 - e^{-\pi s})(s^2 + 1)}$, we must use a

geometric series expansion of $\dfrac{1}{1 - e^{-\pi s}}$ to obtain terms for which we can calculate the inverse

Laplace transform. This gives us

$$\frac{1}{1 - e^{-\pi s}} = 1 + e^{-\pi s} + e^{-2\pi s} + e^{-3\pi s} + \cdots,$$

so

$$Y(s) = (1 + e^{-\pi s} + e^{-2\pi s} + e^{-3\pi s} + \cdots)\frac{1}{(s^2 + 1)^2}$$

$$= \frac{1}{(s^2 + 1)^2} + \frac{e^{-\pi s}}{(s^2 + 1)^2} + \frac{e^{-2\pi s}}{(s^2 + 1)^2} + \frac{e^{-3\pi s}}{(s^2 + 1)^2} + \cdots.$$

Then

$$y(t) = L^{-1}\left\{\frac{1}{(s^2 + 1)^2} + \frac{e^{-\pi s}}{(s^2 + 1)^2} + \frac{e^{-2\pi s}}{(s^2 + 1)^2} + \frac{e^{-3\pi s}}{(s^2 + 1)^2} + \cdots\right\}$$

$$= L^{-1}\left\{\frac{1}{(s^2 + 1)^2}\right\} + L^{-1}\left\{\frac{e^{-\pi s}}{(s^2 + 1)^2}\right\} + L^{-1}\left\{\frac{e^{-2\pi s}}{(s^2 + 1)^2}\right\} + L^{-1}\left\{\frac{e^{-3\pi s}}{(s^2 + 1)^2}\right\} + \cdots.$$

Notice that $L^{-1}\left\{\frac{1}{(s^2 + 1)^2}\right\}$ is needed to find all of the other terms. Using `invlaplace`,

```
> readlib(laplace):
  invlaplace(1/(s^2+1)^2,s,t);
```

$$-\frac{1}{2}t\cos(t) + \frac{1}{2}\sin(t)$$

we have

$$L^{-1}\left\{\frac{1}{(s^2 + 1)^2}\right\} = \frac{1}{2}\sin t - \frac{1}{2}t\cos t.$$

Then

$$y(t) = \frac{1}{2}\{(\sin t - t\cos t) + [\sin(t - \pi) - (t - \pi)\cos(t - \pi)]u(t - \pi)$$

$$+ [\sin(t - 2\pi) - (t - 2\pi)\cos(t - 2\pi)]u(t - 2\pi)$$

$$+ [\sin(t - 3\pi) - (t - 3\pi)\cos(t - 3\pi)]u(t - 3\pi) + \cdots\}$$

$$= \frac{1}{2}\sum_{n=0}^{\infty}[\sin(t - n\pi) - (t - n\pi)\cos(t - n\pi)]u(t - n\pi).$$

To graph $y(t)$ on the interval $[0, k\pi]$, where k represents a positive integer, we note that

$$\frac{1}{2}[\sin(t - n\pi) - (t - n\pi)\cos(t - n\pi)]\,u(t - n\pi) = 0$$

for all values of t in $[0, k\pi]$ if $n \geq k$, so we need to graph

$$\frac{1}{2}\sum_{n=0}^{k-1}[\sin(t - n\pi) - (t - n\pi)\cos(t - n\pi)]\,u(t - n\pi).$$

For convenience, we define `nth_term` to represent

$$\frac{1}{2}[\sin(t - n\pi) - (t - n\pi)\cos(t - n\pi)]\,u(t - n\pi).$$

```
> nth_term:=n->1/2*(sin(t-n*Pi)-
  (t-n*Pi)*cos(t-n*Pi))*Heaviside(t-n*Pi):
```

Thus, to graph $y(t)$ on the interval $[0, 4\pi]$, we enter

```
> to_graph:=sum(nth_term(n),n=0..4);
```

$$to\_graph := \frac{1}{2}(\sin(t) - t\cos(t))\,\text{Heaviside}(t)$$

$$+ \frac{1}{2}(-\sin(t) + (t - \pi)\cos(t))\,\text{Heaviside}(t - \pi)$$

$$+ \frac{1}{2}(\sin(t) - (t - 2\pi)\cos(t))\,\text{Heaviside}(t - 2\pi)$$

$$+ \frac{1}{2}(-\sin(t) + (t - 3\pi)\cos(t))\,\text{Heaviside}(t - 3\pi)$$

$$+ \frac{1}{2}(\sin(t) - (t - 4\pi)\cos(t))\,\text{Heaviside}(t - 4\pi)$$

```
> plot(to_graph,t=0..5*Pi);
```

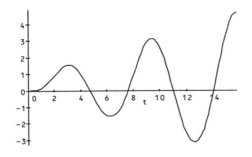

☰ *Impulse Functions: The Delta Function*

We now consider differential equations of the form $ax'' + bx' + cx = f(t)$, where f is "large" over the short interval centered at t_0, $t_0 - \alpha < t < t_0 + \alpha$, and zero otherwise. Hence, we define the impulse delivered by the function $f(t)$ as $I(t) = \int_{t_0 - \alpha}^{t_0 + \alpha} f(t)\,dt$, or since $f(t) = 0$ for t on $(-\infty, t_0 - \alpha) \cup (t_0 + \alpha, +\infty)$,

$$I(t) = \int_{-\infty}^{+\infty} f(t)\,dt.$$

In order to better understand the impulse function, we let f be defined in the following manner:

$$f(t) = \delta_\alpha(t - t_0) = \begin{cases} \dfrac{1}{2\alpha}, & t_0 - \alpha < t < t_0 + \alpha \\ 0, & \text{otherwise} \end{cases}$$

To graph $\delta_\alpha(t - t_0)$ for several values of α and $t_0 = 0$, we define **del** using **proc**.

```
> del:=proc(t,alpha)
    if t>-alpha and t<alpha then 1/(2*alpha) else 0 fi;
    end:
```

Usually, we use operator notation when graphing functions defined as procedures. In this case, we take advantage of **delayed evaluation**, represented by **'...'**, to graph **del** for various values of α. For example, entering

```
> plot('del(t,1/4)','t'=-1..1);
```

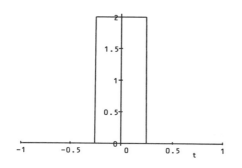

graphs $\delta_{1/4}(t)$ on the interval $[-1, 1]$. Similarly, to graph $\delta_i(t)$ for $i = 0.01, 0.02, 0.03, 0.04$, and 0.05, we first define **i_vals** using **seq** and then generate the set of graphs **to_display** that consists of plots of $\delta_i(t)$ on the interval $[-0.1, 0.1]$ for i values in **i_vals**.

```
> i_vals:=seq(0.01+0.01*i,i=0..4):
  to_display:={seq(plot('del(t,i)','t'=-0.1..0.1),i=i_vals)}:
```

The set of graphs `to_display` is shown together using the `display` command, which is contained in the `plots` package.

```
> with(plots):
  display(to_display);
```

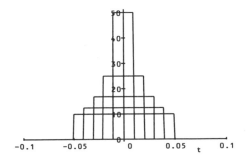

With this definition, the impulse is given by

$$I(t) = \int_{t_0-\alpha}^{t_0+\alpha} f(t)\, dt = \int_{t_0-\alpha}^{t_0+\alpha} \frac{1}{2\alpha}\, dt = \frac{1}{2\alpha}((t_0 + \alpha) - (t_0 - \alpha)) = \frac{1}{2\alpha}(2\alpha) = 1.$$

Notice that the value of this integral does not depend on α as long as α is not zero. We now try to create the idealized impulse function by requiring $\delta_\alpha(t - t_0)$ to act on smaller and smaller intervals. From the integral calculation, we have

$$\lim_{\alpha \to 0} I(t) = 1.$$

We also note that

$$\lim_{\alpha \to 0} \delta_\alpha(t - t_0) = 0, \ t \neq t_0.$$

We use these properties now to define the idealized unit impulse function as follows.

DEFINITION

> *Unit Impulse Function*
>
> The **idealized unit impulse function (Dirac delta function)** δ satisfies
>
> $$\delta(t - t_0) = 0, \ t \neq t_0$$
>
> $$\int_{-\infty}^{+\infty} \delta(t - t_0) \, dt = 1.$$
>
> The built-in function **Dirac(t)** represents the Dirac delta function.

We now state the following useful theorem involving the unit impulse function.

• •

THEOREM   Suppose that g is a bounded and continuous function. Then

$$\int_{-\infty}^{+\infty} \delta(t - t_0)g(t) \, dt = g(t_0).$$

The Laplace transform of $\delta(t - t_0)$ is found by using the function $\delta_\alpha(t - t_0)$ and L'Hôpital's rule.

• •

THEOREM   For $t_0 > 0$,

$$L\{\delta(t - t_0)\} = e^{-st_0}.$$

• •

EXAMPLE: Find (a) $L\{\delta(t - 1)\}$, (b) $L\{\delta(t - \pi)\}$, and (c) $L\{\delta(t)\}$.

SOLUTION: (a) In this case, $t_0 = 1$, so $L\{\delta(t - 1)\} = e^{-s}$. (b) With $t_0 = \pi$, $L\{\delta(t - \pi)\} = e^{-s\pi}$. (c) Because $t_0 = 0$, $L\{\delta(t)\} = L\{\delta(t - 0)\} = e^{-s(0)} = 1$. We obtain the same results using **Dirac** and **laplace** as follows:

```
> map(laplace,[Dirac(t-1),Dirac(t-Pi),Dirac(t)],t,s);
     [e^-s, e^-πs, 1]
```

EXAMPLE: Solve $y'' + y = \delta(t - \pi)$ subject to $y(0) = y'(0) = 0$.

SOLUTION: As in previous examples, we solve this initial-value problem by taking the Laplace transform of both sides of the differential equation. This yields

$$L\{y''\} + L\{y\} = L\{\delta(t - \pi)\}$$
$$s^2 Y(s) - sy(0) - y'(0) + Y(s) = e^{-\pi s}$$
$$(s^2 + 1)Y(s) = e^{-\pi s}$$
$$Y(s) = \frac{e^{-\pi s}}{s^2 + 1}.$$

Hence,

$$y(t) = L^{-1}\left\{\frac{e^{-\pi s}}{s^2 + 1}\right\}.$$

The same steps are carried out now with Maple.

> `step_1:=laplace(diff(y(t),t$2)+y(t)=Dirac(t-Pi),t,s);`

$step\_1 := (\text{laplace}(y(t), t, s)s - y(0))s - D(y)(0) + \text{laplace}(y(t), t, s) = e^{-\pi s}$

> `step_2:=subs({y(0)=0,D(y)(0)=0},step_1);`

$step\_2 := \text{laplace}(y(t), t, s)s^2 + \text{laplace}(y(t), t, s) = e^{-\pi s}$

> `step_3:=solve(step_2,laplace(y(t),t,s));`

$step\_3 := \dfrac{e^{-\pi s}}{s^2 + 1}$

Then, since $f(t) = L^{-1}\left\{\dfrac{1}{s^2 + 1}\right\} = \sin t$,

$$y(t) = L^{-1}\left\{\frac{e^{-\pi s}}{s^2 + 1}\right\} = \sin(t - \pi)\,u(t - \pi) = -\sin t\,u(t - \pi).$$

We obtain the same result, which we then graph, using `invlaplace`.

> `readlib(laplace):`
  `Sol:=invlaplace(step_3,s,t);`

$Sol := -\text{Heaviside}(t - \pi)\sin(t)$

```
> plot(Sol,t=0..3*Pi);
```

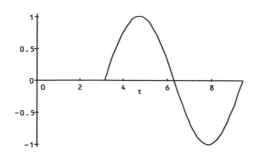

We can use **dsolve** to find the solution to the initial-value problem, although the form is different from that obtained previously.

```
> dsolve({diff(y(t),t$2)+y(t)=Dirac(t-Pi),
   y(0)=0,D(y)(0)=0},y(t));
```

$$y(t) = \int_0^t \text{Dirac}(u - \pi)\cos(u)\,du\,\sin(t) + \int_0^t -\sin(u)\,\text{Dirac}(u - \pi)\,du\,\cos(t)$$

In contrast, when we use **dsolve** together with the **laplace** option, we obtain the exact same result.

```
> dsolve({diff(y(t),t$2)+y(t)=Dirac(t-Pi),
   y(0)=0,D(y)(0)=0},y(t),laplace);
```

$$y(t) = -\text{Heaviside}(t - \pi)\sin(t)$$

The forcing function may involve a combination of functions, as illustrated in the following example.

EXAMPLE: Solve $y'' + 2y' + y = \delta(t) + \delta(t - 2\pi)$ subject to $y(0) = y'(0) = 0$.

SOLUTION: Using the method of Laplace transforms, we must calculate

$$L\{\delta(t) + \delta(t - 2\pi)\} = L\{\delta(t)\} + L\{\delta(t - 2\pi)\} = 1 + e^{-2\pi s}.$$

Then

$$L\{y''\} + 2L\{y'\} + L\{y\} = L\{\delta(t) + \delta(t - 2\pi)\}$$
$$s^2 Y(s) - sy(0) - y'(0) + 2sY(s) - y(0) + Y(s) = 1 + e^{-2\pi s}$$
$$Y(s) = \frac{1}{s^2 + 2s + 1} + \frac{e^{-2\pi s}}{s^2 + 2s + 1} = \frac{1}{(s + 1)^2} + \frac{e^{-2\pi s}}{(s + 1)^2}.$$

Because

$$L^{-1}\left\{\frac{1}{(s + 1)^2}\right\} = te^{-t},$$

we determine that

$$L^{-1}\left\{\frac{e^{-2\pi s}}{(s + 1)^2}\right\} = (t - 2\pi)e^{-(t-2\pi)}u(t - 2\pi).$$

Hence,

$$y(t) = L^{-1}\{Y(s)\} = te^{-t} + (t - 2\pi)e^{-(t-2\pi)}u(t - 2\pi).$$

After defining y, we graph y on the interval $[0, 3\pi]$. The graph illustrates the two shocks that occur at $t = 0$ and $t = 2\pi$.

```
> y:=t*exp(-t)+(t-2*Pi)*exp(-(t-2*Pi))*Heaviside(t-2*Pi):
  plot(y,t=0..4*Pi);
```

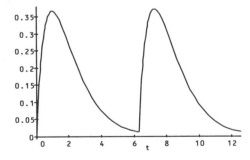

Equivalent results are obtained with Maple. After defining **EQ** to be the equation

$$y'' + 2y' + y = \delta(t) + \delta(t - 2\pi),$$

we define **step_1** to be the equation that results from taking the Laplace transform of both sides of **EQ**,

```
> EQ:=diff(y(t),t$2)+2*diff(y(t),t)+
  y(t)=Dirac(t)+Dirac(t-2*Pi):
  step_1:=Laplace(EQ,t,s);
```

$$step\_1 := (\text{laplace}(y(t), t, s)s - y(0))s - D(y)(0) + 2\,\text{laplace}\,(y(t), t, s)s - 2y(0)$$
$$+ \text{laplace}\,(y(t), t, s) = 1 + e^{-2\pi s}$$

step_2 to be the result of applying the initial conditions $y(0) = y'(0) = 0$ to **step_1**,

```
> step_2:=subs({y(0)=0,D(y)(0)=0},step_1);
```

$$step\_2 := \text{laplace}\,(y(t), t, s)s^2 + 2\,\text{laplace}\,(y(t), t, s)s + \text{laplace}\,(y(t), t, s) = 1 + e^{-2\pi s}$$

step_3 to be the result of solving **step_2** for **laplace(y(t),t,s)**, which represents the Laplace transform of $y(t)$,

```
> step_3:=solve(step_2,laplace(y(t),t,s));
```

$$step\_3 := -\frac{-1 - e^{-2\pi s}}{s^2 + 2s + 1}$$

and **Sol** to be the inverse Laplace transform of **step_3**.

```
> readlib(laplace):
  Sol:=invlaplace(step_3,s,t);
```

$$Sol := \left(\sum_{-r=\%1} \frac{e^{-rt}}{2 - r + 2} \right) + \text{Heaviside}(t - 2\pi) \left(\sum_{-r=\%1} \frac{e^{-4(t-2\pi)}}{2 - r + 2} \right)$$

$$\%1 := \text{RootOf}(-Z^2 + 2 - z + 1)$$

We also see that **dsolve** finds a solution of the initial-value problem

```
> dsolve({EQ,y(0)=0,D(y)(0)=0},y(t));
```

$$y(t) = \int_0^t (- \text{Dirac}(u) - \text{Dirac}(u - 2\pi))u\,e^u\,du\,e^{-t}$$

$$+ \int_0^t (\text{Dirac}(u) + \text{Dirac}(u - 2\pi))e^u\,du\,e^{-t}t$$

while **dsolve** together with the **laplace** option returns the same solution as that obtained in **Sol**.

> dsolve({EQ,y(0)=0,D(y)(0)=0},y(t),laplace);

$$y(t) = \left(\sum_{-r=\%1} \frac{e^{-rt}}{2 - r + 2} \right) + \text{Heaviside}(t - 2\pi) \left(\sum_{-r=\%1} \frac{e^{-r(t-2\pi)}}{2 - r + 2} \right)$$

$$\%1 := \text{RootOf}(-Z^2 + 2 - z + 1)$$

7.5 THE CONVOLUTION THEOREM

≡ The Convolution Theorem

In many cases, we are required to determine the inverse Laplace transform of a product of two functions. Just as in integral calculus when the integral of the product of two functions did not produce the product of the integrals, neither does the inverse Laplace transform of the product yield the product of the inverse Laplace transforms. Thus, we state the following theorem.

THEOREM   *Convolution Theorem*

Suppose that $f(t)$ and $g(t)$ are piecewise continuous on $[0, \infty)$ and both of exponential order b. Further suppose that $L\{f(t)\} = F(s)$ and $L\{g(t)\} = G(s)$. Then

$$L^{-1}\{F(s)G(s)\} = L^{-1}\{L\{(f * g)(t)\}\} = (f * g)(t) = \int_0^t f(t - v)g(v)\, dv$$

Note that $(f * g)(t) = \int_0^t f(t - v)g(v)\, dv$ is called the **convolution integral**.

EXAMPLE: Compute $(f * g)$ if $f(t) = e^{-t}$ and $g(t) = \sin t$. Verify the convolution theorem with these functions.

SOLUTION: We use the definition and a table of integrals (or integration by parts) to obtain

$$(f * g)(t) = \int_0^t f(t - v)g(v)\, dv = \int_0^t e^{-(t-v)} \sin v \, dv = e^{-t} \int_0^t e^v \sin v \, dv$$

$$= e^{-t} \left[\frac{e^v}{2} (\sin v - \cos v) \right]_0^t = e^{-t} \left[\frac{e^t}{2} (\sin t - \cos t) - \frac{1}{2} (\sin 0 - \cos 0) \right]$$

$$= \frac{1}{2} (\sin t - \cos t) + \frac{1}{2} e^{-t}.$$

The same results are obtained with Maple. After defining `convolution`, which computes $(f * g)$,

```
> convolution:=(f,g)->int(f(t-v)*g(v),v=0..t):
```

we define f and g

```
> f:='f':g:='g':
  f:=t->exp(-t):g:=t->sin(t):
```

and then use convolution to compute $(f * g)$.

```
> convolution(f,g);
```

$$-\frac{1}{2}\cos(t) + \frac{1}{2}\sin(t) + \frac{1}{2}e^{-t}$$

Note that $(g * f)$, computed here in `reverse`, is the same as $(f * g)$.

```
> reverse:=convolution(g,f);
```

$$reverse := -\frac{1}{2}\cos(t) + \frac{1}{2}\sin(t) + \frac{1}{2}e^{-t}$$

Now, according to the convolution theorem, $L\{f(t)\}L\{g(t)\} = (f * g)(t)$. In this example, we have

$$F(s) = L\{f(t)\} = L\{e^{-t}\} = \frac{1}{s+1} \text{ and } G(s) = L\{g(t)\} = L\sin(t) = \frac{1}{s^2+1}.$$

Hence, $L^{-1}\{F(s)G(s)\} = L^{-1}\left\{\frac{1}{s+1} \cdot \frac{1}{s^2+1}\right\}$ should equal $(f * g)$. We compute $L^{-1}\left\{\frac{1}{s+1} \cdot \frac{1}{s^2+1}\right\}$ with `invlaplace`.

```
> readlib(laplace):
  invlaplace(1/((s+1)*(s^2+1)),s,t);
```

$$\frac{1}{2}e^{-t} - \frac{1}{2}\cos(t) + \frac{1}{2}\sin(t)$$

Hence,

$$L^{-1}\left\{\frac{1}{s+1} \cdot \frac{1}{s^2+1}\right\} = \frac{1}{2}e^{-t} - \frac{1}{2}\cos t + \frac{1}{2}\sin t,$$

which is the same result as that obtained for $(f * g)$.

EXAMPLE: Use the convolution theorem to find the Laplace transform of

$$h(t) = \int_0^t \cos(t - v) \sin(v) \, dv.$$

SOLUTION: Notice that $h(t) = (f * g)(t)$, where $f(t) = \cos t$ and $g(t) = \sin t$. Therefore, by the convolution theorem, $L\{(f * g)(t)\} = F(s)G(s)$. Hence,

$$L\{h(t)\} = L\{f(t)\}L\{g(t)\} = L \cos t L \sin t = \left(\frac{s}{s^2 + 1}\right)\left(\frac{1}{s^2 + 1}\right) = \frac{s}{(s^2 + 1)^2}.$$

Remember that **Int** represents the inert form of the integration command **int**. Thus, to compute the Laplace transform of $h(t)$, we first define $h(t)$ by entering

```
> h:=t->Int(cos(t-v)*sin(v),v=0..t);
```

$$h := t \longrightarrow \int_0^t \cos(t - v) \sin(v) \, dv$$

and then compute the Laplace transform with **laplace**.

```
> laplace(h(t),t,s);
```

$$\frac{s}{(s^2 + 1)^2}$$

☰ Integral and Integrodifferential Equations

The convolution theorem is useful in solving numerous problems. In particular, this theorem can be employed to solve **integral equations**, equations that involve an integral of the unknown function. We illustrate this procedure in the following example.

EXAMPLE: Use the convolution theorem to solve the integral equation

$$h(t) = 4t + \int_0^t h(t - v) \sin(v) \, dv.$$

SOLUTION: We first note that the integral in this equation represents $(h * g)(t)$ for $g(t) = \sin t$. Therefore, if we apply the Laplace transform to both sides of the equation, we obtain

$$L\{h(t)\} = L\{4t\} + L\{h(t)\}L \sin(t),$$

or

$$H(s) = \frac{4}{s^2} + H(s)\frac{1}{s^2 + 1},$$

where $L\{h(t)\} = H(s)$.

```
> EQ:=h(t)=4*t+Int(h(t-v)*sin(v),v=0..t);
```

$$EQ := h(t) = 4t + \int_0^t h(t - v)\sin(v)\, dv$$

```
> step_1:=laplace(EQ,t,s);
```

$$step\_1 := \text{laplace}(h(t), t, s) = \frac{4}{s^2} + \frac{\text{laplace}(h(t), t, s)}{s^2 + 1}$$

Solving for $H(s)$, we have

$$H(s)\left(1 - \frac{1}{s^2 + 1}\right) = \frac{4}{s^2},$$

so

$$H(s) = \frac{4(s^2 + 1)}{s^4} = \frac{4}{s^2} + \frac{4}{4s^4}.$$

```
> step_2:=simplify(solve(step_1,laplace(h(t),t,s)));
```

$$step\_2 := \frac{4s^2 + 4}{s^4}$$

Then, by computing the inverse Laplace transform, we find that

$$h(t) = L^{-1}\left\{\frac{4}{s^2} + \frac{4}{s^4}\right\} = 4t + \frac{2}{3}t^3.$$

```
> readlib(laplace):
  Sol:=invlaplace(step_2,s,t);
```

$$Sol := 4t + \frac{2}{3}t^3$$

Laplace transforms are helpful in solving problems of other types as well. We now illustrate another important problem by solving an **integrodifferential equation**, an equation that involves a derivative as well as an integral of the unknown function.

EXAMPLE: Solve $\dfrac{dy}{dt} + y + \displaystyle\int_0^t y(u)\,du = 1$ subject to $y(0) = 0$.

SOLUTION: Because we must take the Laplace transform of both sides of this integrodifferential equation, we first compute

$$L\left\{\int_0^t y(u)\,du\right\} = L\{(1 * y)(t)\} = L\{1\}L\{y\} = \frac{Y(s)}{s}.$$

Hence,

$$L\left\{\frac{dy}{dt}\right\} + L\{y\} + L\left\{\int_0^t y(u)\,du\right\} = L\{1\}$$

$$sY(s) - y(0) + Y(s) + \frac{Y(s)}{s} = \frac{1}{s}$$

$$s^2 Y(s) + sY(s) + Y(s) = 1$$

$$Y(s) = \frac{1}{s^2 + s + 1}.$$

The same steps are carried out with Maple.

> `EQ:=diff(y(t),t)+y(t)+Int(y(u),u=0..t)=1;`

$$EQ := \left(\frac{\partial}{\partial t}y(t)\right) + y(t) + \int_0^t y(u)\,du = 1$$

> `step_1:=laplace(EQ,t,s);`

$$step\_1 := \text{laplace}\,(y(t), t, s)s - y(0) + \text{laplace}\,(y(t), t, s) + \frac{\text{laplace}\,(y(t), t, s)}{s} = \frac{1}{s}$$

> `step_2:=subs(y(0)=0,step_1);`

$$step\_2 := \text{laplace}\,(y(t), t, s)s + \text{laplace}\,(y(t), t, s) + \frac{\text{laplace}\,(y(t), t, s)}{s} = \frac{1}{s}$$

> `step_3:=simplify(solve(step_2,laplace(y(t),t,s)));`

$$step\_3 := (s^2 + s + 1)^{-1}$$

Because

$$Y(s) = \frac{1}{s^2 + s + 1} = \frac{1}{\left(s + \frac{1}{2}\right)^2 + \left(\frac{\sqrt{3}}{2}\right)^2}, \quad y(t) = \frac{2}{\sqrt{3}}e^{-t/2}\sin\frac{\sqrt{3}}{2}t.$$

The same solution, which is then graphed on the interval $[0, 3\pi]$ with plot, is found with invlaplace and named Sol.

```
> readlib(laplace):
  Sol:=invlaplace(step_3,s,t);
```

$$Sol := \frac{2}{3}e^{-\frac{t}{2}}\sin\left(\frac{1}{2}\sqrt{3}t\right)\sqrt{3}$$

```
> plot(Sol,t=0..3*Pi);
```

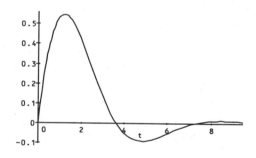

8

Applications of Laplace Transforms

8.1 SPRING-MASS SYSTEMS REVISITED

Laplace transforms are useful in solving the spring–mass systems that were discussed in earlier sections. Although the method of Laplace transforms can be used to solve all problems discussed in the section on applications of higher-order equations, this method is most useful in alleviating the difficulties associated with problems that involve piecewise-defined forcing functions. Hence, we investigate the use of Laplace transforms to solve the second-order initial-value problem that models the motion of a mass attached to the end of a spring. We found in Chapter 5 that this situation is modeled by the initial-value problem

$$\begin{cases} m\dfrac{d^2x}{dt^2} + c\dfrac{dx}{dt} + kx = 0 \\ x(0) = \alpha, \ \dfrac{dx}{dt}(0) = \beta \end{cases},$$

where m represents the mass, c the damping coefficient, and k the spring constant determined by Hooke's law. The following example demonstrates how the method of Laplace transforms is used to solve this type of initial-value problem when the forcing function is continuous.

EXAMPLE: An object with mass $m = 4$ is attached to the end of a spring with spring constant $k = 1$. Determine the motion of the mass if $c = 2$, $x(0) = -1$, $\dfrac{dx}{dt}(0) = 1$, and the forcing function $f(t)$ is given by (a) $f(t) = 0$ and (b) $f(t) = 1$.

SOLUTION: The initial-value problem that models this situation is

$$\begin{cases} 4\dfrac{d^2x}{dt^2} + 2\dfrac{dx}{dt} + x = 0 \\ x(0) = -1, \quad \dfrac{dx}{dt}(0) = 1 \end{cases}.$$

(a) If we take the Laplace transform of both sides of the differential equation, we obtain

$$4L\left\{\frac{d^2x}{dt^2}\right\} + 2L\left\{\frac{dx}{dt}\right\} + L\{x\} = L\{0\}$$

$$4[s^2X(s) - sx(0) - x'(0)] + 2[sX(s) - x(0)] + X(s) = 0$$

$$(4s^2 + 2s + 1)X(s) = -4s + 2.$$

Solving for $X(s)$, we have

$$X(s) = \frac{-4s + 2}{4s^2 + 2s + 1} = -\frac{s - \frac{1}{2}}{s^2 + \frac{s}{2} + \frac{1}{4}} = -\frac{s - 12}{\left(s + \frac{1}{4}\right)^2 + \frac{3}{16}} = -\frac{s + \frac{1}{4}}{\left(s + \frac{1}{4}\right)^2 + \frac{3}{16}} + \frac{\frac{3}{4}}{\left(s + \frac{1}{4}\right)^2 + \frac{3}{16}}.$$

Hence, by taking the inverse Laplace transform, we obtain

$$x(t) = -e^{-t/4}\left[\cos\frac{\sqrt{3}}{4}t - \sqrt{3}\sin\frac{\sqrt{3}}{4}t\right].$$

We see that the same results are obtained with **dsolve**. After defining **ODE** to be the equation $4\dfrac{d^2x}{dt^2} + 2\dfrac{dx}{dt} + x = 0$,

```
>  x:='x':
   ODE:=4*diff(x(t),t$2)+2*diff(x(t),t)+x(t)=0;
```

$$ODE := 4\left(\frac{\partial^2}{\partial t^2}x(t)\right) + 2\left(\frac{\partial}{\partial t}x(t)\right) + x(t) = 0$$

we use **dsolve** to solve the initial-value problem

$$\begin{cases} 4\dfrac{d^2x}{dt^2} + 2\dfrac{dx}{dt} + x = 0 \\ x(0) = -1, \quad \dfrac{dx}{dt}(0) = 1 \end{cases}$$

and name the result **Sol**.

```
> Sol:=dsolve({ODE,x(0)=-1,D(x)(0)=1},x(t));
```

$$Sol := x(t) = \sqrt{3}e^{-\frac{1}{4}t}\sin\left(\frac{1}{4}\sqrt{3}\,t\right) - e^{-\frac{1}{4}t}\cos\left(\frac{1}{4}\sqrt{3}\,t\right)$$

We then name $x(t)$ the result obtained in Sol with assign and graph $x(t)$ on the interval $[0, 6\pi]$ with plot.

```
> assign(Sol):
  plot(x(t),t=0..6*Pi);
```

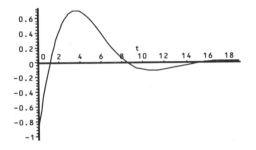

(b) In this case,

$$4[s^2X(s) - sx(0) - x'(0)] + 2[sX(s) - x(0)] + X(s) = \frac{1}{s}$$

$$(4s^2 + 2s + 1)X(s) = \frac{1}{s} - 4s + 2$$

$$X(s) = \frac{1}{s(4s^2 + 2s + 1)} + \frac{-4s + 2}{4s^2 + 2s + 1}.$$

Since $\dfrac{1}{s(4s^2 + 2s + 1)} = \dfrac{1}{s} + \dfrac{-4s + 2}{4s^2 + 2s + 1}$,

$$X(s) = \frac{1}{s} - \frac{4s + 2}{4s^2 + 2s + 1} + \left(\frac{-4s + 2}{4s^2 + 2s + 1}\right) = \frac{1}{s} - \frac{8s}{4s^2 + 2s + 1}.$$

Therefore, because

$$\frac{8s}{4s^2 + 2s + 1} = \frac{2s}{s^2 + \frac{s}{2} + \frac{1}{4}} = \frac{2s}{\left(s + \frac{1}{4}\right)^2 + \frac{3}{16}} = \frac{2\left(s + \frac{1}{4}\right)}{\left(s + \frac{1}{4}\right)^2 + \frac{3}{16}} - \frac{\frac{1}{2}}{\left(s + \frac{1}{4}\right)^2 + \frac{3}{16}},$$

$$x(t) = L^{-1}\{X(s)\} = 1 - 2e^{-t/4}\left[\cos\left(\frac{\sqrt{3}}{4}t\right) - \frac{1}{\sqrt{3}}\sin\left(\frac{\sqrt{3}}{4}t\right)\right].$$

As in (a), the same results are obtained with **dsolve**.

```
>  x:='x':
   ODE2:=4*diff(x(t),t$2)+2*diff(x(t),t)+x(t)=1;
   Sol2:=dsolve({ODE2,x(0)=-1,D(x)(0)=1},x(t));
```

$$ODE2 := 4\left(\frac{\partial^2}{\partial t^2}x(t)\right) + 2\left(\frac{\partial}{\partial t}x(t)\right) + x(t) = 1$$

$$Sol2 := x(t) = 1 + \frac{2}{3}\sqrt{3}e^{-\frac{1}{4}t}\sin\left(\frac{1}{4}\sqrt{3}t\right) - 2e^{-\frac{1}{4}t}\cos\left(\frac{1}{4}\sqrt{3}t\right)$$

We then graph the solution on the interval $[0, 6\pi]$.

```
>  assign(Sol2):
   plot(x(t),t=0..6*Pi);
```

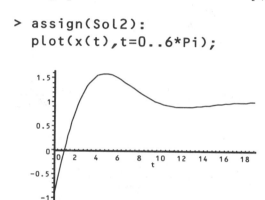

We now consider an initial-value problem that involves a discontinuous forcing function. Instead of having to solve two initial-value problems, we use the Laplace transform method.

EXAMPLE: Suppose that a mass with $m = 1$ is attached to a spring with spring constant $k = 1$. If there is no resistance due to damping, then determine the position of the mass if it is released from its equilibrium position and is subjected to the force

$$f(t) = \begin{cases} \sin t, & 0 \le t < \dfrac{\pi}{2} \\ 0, & t \ge \dfrac{\pi}{2} \end{cases}.$$

SOLUTION: In this case, the constants are $m = 1, c = 0$, and $k = 1$. The initial position is $x(0) = 0$ and the initial velocity is $\dfrac{dx}{dt}(0)$. Hence, the initial-value problem that models this situation is

$$\frac{d^2x}{dt^2} + x = \begin{cases} \sin t, \ 0 \le t < \dfrac{\pi}{2} \\ 0, \ t \ge \dfrac{\pi}{2} \end{cases}, x(0) = 0, \ \frac{dx}{dt}(0) = 0.$$

Because we will take the Laplace transform of both sides of the differential equation, we write $f(t)$ in terms of the unit step function. This gives us

$$f(t) = \sin t \left[u(t - 0) - u\left(t - \frac{\pi}{2}\right) \right] = \sin t \left[1 - u\left(t - \frac{\pi}{2}\right) \right].$$

Hence, we must compute

$$L\left\{ \sin t - \sin t\, u\left(t - \frac{\pi}{2}\right) \right\} = L\{\sin t\} - e^{-s\pi/2} L\left\{ \sin\left(t + \frac{\pi}{2}\right) \right\} = \frac{s}{s^2 + 1} - e^{-s\pi/2} \frac{1}{s^2 + 1}$$

and use this result in the following calculations:

$$L\left\{ \frac{d^2x}{dt^2} \right\} + L\{x\} = L\left\{ \sin t - \sin t\, u\left(t - \frac{\pi}{2}\right) \right\}$$

$$s^2 X(s) - sy(0) - y'(0) + X(s) = \frac{s}{s^2 + 1} - e^{-s\pi/2} \frac{1}{s^2 + 1}$$

$$X(s) = \frac{s}{(s^2 + 1)^2} + e^{-s\pi/2} \frac{1}{(s^2 + 1)^2}.$$

Now, in order to compute $x(t) = L^{-1}\{X(s)\}$, we must determine $L^{-1}\left\{ \dfrac{1}{(s^2 + 1)^2} \right\}$. Using **invlaplace**, we find that

$$L^{-1}\left\{ \frac{1}{(s^2 + 1)^2} \right\} = \frac{1}{2}(\sin t - t \cos t).$$

Hence, by using this formula and simplifying with trigonometric identities, we have

$$L^{-1}\left\{ e^{-\pi s/2} \frac{1}{(s^2 + 1)^2} \right\} = \frac{1}{2}\left[\sin\left(t - \frac{\pi}{2}\right) - \left(t - \frac{\pi}{2}\right) \cos\left(t - \frac{\pi}{2}\right) \right] u\left(t - \frac{\pi}{2}\right)$$

$$= \frac{1}{2}\left(-\cos t - \left(t - \frac{\pi}{2}\right) \sin t \right) u\left(t - \frac{\pi}{2}\right).$$

Again, with a table of Laplace transforms,

$$L^{-1}\left\{\frac{s}{(s^2+1)^2}\right\} = \frac{1}{2}\,t\sin t,$$

so

$$x(t) = L^{-1}\{X(s)\} = \frac{1}{2}\,t\sin t + \frac{1}{2}\left(-\cos t - \left(t - \frac{\pi}{2}\right)\sin t\right)u\left(t - \frac{\pi}{2}\right).$$

By eliminating the unit step function, we can write $x(t)$ in the form

$$x(t) = \begin{cases} \dfrac{1}{2}\,t\sin t,\ 0 \le t < \dfrac{\pi}{2} \\[2mm] \dfrac{\pi}{4}\sin t - \dfrac{1}{2}\cos t,\ t \ge \dfrac{\pi}{2} \end{cases}.$$

Notice that resonance occurs on the interval $0 \le t < \dfrac{\pi}{2}$. Then, for $t \ge \dfrac{\pi}{2}$, the motion is harmonic.

Hence, although the forcing function is zero for $t \ge \dfrac{\pi}{2}$, the mass continues to follow the path defined by $x(t)$ indefinitely. We graph this function in the following figure.

```
> x:='x':
  Sol:=dsolve({diff(x(t),t$2)+x(t)=
  sin(t)*Heaviside(Pi/2-t),x(0)=0,D(x)(0)=0},x(t));
```

$$Sol := x(t) = \frac{1}{2}\cos(t)\,\text{Heaviside}\left(\frac{1}{2}\pi - t\right)\pi - \cos(t)\,\text{Heaviside}\left(\frac{1}{2}\pi - t\right)t$$

$$+ \cos(t)\int_0^t \text{Heaviside}\left(\frac{1}{2}\pi - u\right)\cos(u)^2\,du$$

$$+ \int_0^t \cos(u)\sin(u)\,\text{Heaviside}\left(\frac{1}{2}\pi - u\right)du\,\sin(t) - \frac{1}{2}\pi\cos(t)$$

```
> assign(Sol):
  plot(x(t),t=0..2*Pi);
```

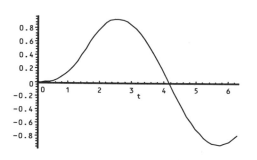

```
> x:='x':Sol:=dsolve({diff(x(t),t$2)+x(t)=
  sin(t)*Heaviside(Pi/2-t),x(0)=0,D(x)(0)=0},
  x(t),laplace);
```

$$Sol := x(t) = -\text{Heaviside}\left(t - \frac{1}{2}\pi\right)\left(\sum_{\_r=\%1} \frac{-r e^{\left(-r\left(t-\frac{1}{2}\pi\right)\right)}}{4\_r^3 + 4\_r}\right) + \left(\sum_{\_r=\%1} \frac{e^{(-rt)}}{4\_r^3 + 4\_r}\right)$$

$$\%1 := \text{RootOf}(\_Z^4 + 2\_Z^2 + 1)$$

EXAMPLE: Suppose that a mass with $m = 1$ is attached to a spring with spring constant $k = 13$. If the mass is subjected to the resistive force due to damping $F_R = 4\dfrac{dx}{dt}$, then determine the position of the mass if it is released from its equilibrium position and is subjected to the force

$$f(t) = \begin{cases} \sin t, & 0 \le t < \dfrac{\pi}{2} \\ 0, & t \ge \dfrac{\pi}{2} \end{cases}.$$

SOLUTION: In this case, the initial-value problem is

$$\frac{d^2x}{dt^2} + 4\frac{dx}{dr} + 13x = \begin{cases} \sin t, & 0 \le t < \dfrac{\pi}{2} \\ 0, & t \ge \dfrac{\pi}{2} \end{cases}, \quad x(0) = 0, \quad \frac{dx}{dt}(0) = 0.$$

As in the previous example, we write f in terms of the unit step function, so

$$f(t) = \sin t\left[u(t - 0) - u\left(t - \frac{\pi}{2}\right)\right] = \sin t\left[1 - u\left(t - \frac{\pi}{2}\right)\right]$$

with Laplace transform

$$L\left\{\sin t - \sin t\, u\left(t - \frac{\pi}{2}\right)\right\} = \frac{s}{s^2 + 1} - e^{-s\pi/2}\frac{1}{s^2 + 1}.$$

Taking the Laplace transform of both sides of the differential equation then yields

$$L\left\{\frac{d^2x}{dt^2}\right\} + 4L\left\{\frac{dx}{dt}\right\} + 13L\{x\} = L\left\{\sin t - \sin t\, u\left(t - \frac{\pi}{2}\right)\right\}$$

$$s^2X(s) - sX(0) - \frac{dx}{dt}(0) + 4X(s) - X(0) + X(s) = \frac{s}{s^2 + 1} - e^{-s\pi/2}\frac{1}{s^2 + 1}$$

$$X(s) = \frac{s}{(s^2 + 1)(s^2 + 4s + 13)} - e^{-s\pi/2}\frac{1}{(s^2 + 1)(s^2 + 4s + 13)}.$$

We can use a partial fraction expansion to find $L^{-1}\left\{\frac{s}{(s^2 + 1)(s^2 + 4s + 13)}\right\}$. If we assume that

$$\frac{s}{(s^2 + 1)(s^2 + 4s + 13)} = \frac{As + B}{s^2 + 1} + \frac{Cs + D}{s^2 + 4s + 13},$$ then we find that the system of equations $\{A + C = 0, 4A + B + D = 0, 13A + 4B + C = 1, 13B + D = 0\}$ must be satisfied. This system has solution $\left\{A = \frac{3}{40}, B = \frac{1}{40}, C = -\frac{3}{40}, D = -\frac{13}{40}\right\}$, so

$$L^{-1}\left\{\frac{s}{(s^2 + 1)(s^2 + 4s + 13)}\right\} = L^{-1}\left\{\frac{\frac{3}{40}s + \frac{1}{40}}{s^2 + 1} - \frac{\frac{3}{40}s + \frac{13}{40}}{s^2 + 4s + 13}\right\}$$

$$= \frac{3}{40}L^{-1}\left\{\frac{s}{s^2 + 1}\right\} + \frac{1}{40}L^{-1}\left\{\frac{1}{s^2 + 1}\right\}$$

$$- \frac{3}{40}L^{-1}\left\{\frac{s + 2}{(s + 2)^2 + 9}\right\} - \frac{7}{40}L^{-1}\left\{\frac{1}{(s + 2)^2 + 9}\right\}$$

$$= \frac{3}{40}\sin t + \frac{1}{40}\sin t - \frac{3}{40}e^{-2t}\cos 3t - \frac{7}{120}e^{-2t}\sin 3t.$$

In a similar manner, if we assume that

$$\frac{1}{(s^2 + 1)(s^2 + 4s + 13)} = \frac{As + B}{s^2 + 1} + \frac{Cs + D}{s^2 + 4s + 13},$$

then the system of equations $\{A + C = 0, 4A + B + D = 0, 13A + 4B + C = 0, 13B + D = 1\}$ must be satisfied. Because this system has solution $\left\{A = -\dfrac{1}{40}, B = \dfrac{3}{40}, C = \dfrac{1}{40}, D = \dfrac{1}{40}\right\}$,

$$
L^{-1}\left\{\frac{1}{(s^2 + 1)(s^2 + 4s + 13)}\right\} = L^{-1}\left\{\frac{-\frac{1}{40}s + \frac{3}{40}}{s^2 + 1} + \frac{\frac{1}{40}s + \frac{1}{40}}{s^2 + 4s + 13}\right\}
$$

$$
= -\frac{1}{40}L^{-1}\left\{\frac{s}{s^2 + 1}\right\} + \frac{3}{40}L^{-1}\left\{\frac{1}{s^2 + 1}\right\}
$$

$$
+ \frac{1}{40}L^{-1}\left\{\frac{s + 2}{(s + 2)^2 + 9}\right\} + \frac{1}{40}L^{-1}\left\{\frac{1}{(s + 2)^2 + 9}\right\}
$$

$$
= -\frac{1}{40}\cos t + \frac{3}{40}\sin t + \frac{1}{40}e^{-2t}\cos 3t - \frac{1}{120}e^{-2t}\sin 3t.
$$

Hence,

$$
x(t) = L^{-1}\{X(s)\} = \frac{3}{40}\sin t + \frac{1}{40}\cos t - \frac{3}{40}e^{-2t}\cos 3t - 7120e^{-2t}\sin 3t
$$

$$
- \left[-\frac{1}{40}\cos\left(t - \frac{\pi}{2}\right) + \frac{3}{40}\sin\left(t - \frac{\pi}{2}\right) + \frac{1}{40}e^{-2(t-\pi/2)}\cos 3\left(t - \frac{\pi}{2}\right)\right.
$$

$$
\left. - \frac{1}{120}e^{-2(t-\pi/2)}\sin 3\left(t - \frac{\pi}{2}\right)\right]u\left(t - \frac{\pi}{2}\right)
$$

$$
= \frac{3}{40}\sin t + \frac{1}{40}\cos t - \frac{3}{40}e^{-2t}\cos 3t - \frac{7}{120}e^{-2t}\sin 3t
$$

$$
- \left[-\frac{1}{40}\sin t - \frac{3}{40}\cos t - \frac{1}{40}e^{-2(t-\pi/2)}\sin 3t - \frac{1}{120}e^{-2(t-\pi/2)}\cos 3t\right]u\left(t - \frac{\pi}{2}\right).
$$

Notice that the effects of damping diminish rather quickly. We can see this from the formula for $x(t)$ because the sine and cosine terms do not cancel. Hence, the initial forcing function causes the motion of the object to continue. If this force were not present, then the damping would cause the motion of the object to stop. We obtain the same results using **dsolve** together with the **laplace** option.

```
> x:='x':
  Sol:=dsolve({diff(x(t),t$2)+4*diff(x(t),t)+13*x(t)=
  sin(t)*Heaviside(Pi/2-t),x(0)=0,D(x)(0)=0},
  x(t),laplace);
```

$$
Sol := x(t) = -\text{Heaviside}\left(t - \frac{1}{2}\pi\right)\left(\frac{1}{40}\sin\left(t - \frac{1}{2}\pi\right) + \frac{3}{40}\cos\left(t - \frac{1}{2}\pi\right)\right.
$$

$$
\left. - \frac{7}{120}e^{(-2t+\pi)}\sin\left(3t - \frac{3}{2}\pi\right) - \frac{3}{40}e^{(-2t+\pi)}\cos\left(3t - \frac{3}{2}\pi\right)\right) + \frac{3}{40}\sin(t)
$$

$$
- \frac{1}{40}\cos(t) - \frac{1}{120}e^{(-2t)}\sin(3t) + \frac{1}{40}e^{(-2t)}\cos(3t)
$$

```
> assign(Sol):
  plot(x(t),t=0..2*Pi);
```

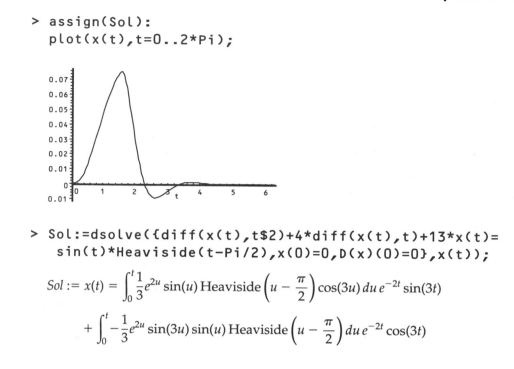

```
> Sol:=dsolve({diff(x(t),t$2)+4*diff(x(t),t)+13*x(t)=
  sin(t)*Heaviside(t-Pi/2),x(0)=0,D(x)(0)=0},x(t));
```

$$Sol := x(t) = \int_0^t \frac{1}{3} e^{2u} \sin(u) \operatorname{Heaviside}\left(u - \frac{\pi}{2}\right) \cos(3u) \, du \, e^{-2t} \sin(3t)$$

$$+ \int_0^t -\frac{1}{3} e^{2u} \sin(3u) \sin(u) \operatorname{Heaviside}\left(u - \frac{\pi}{2}\right) du \, e^{-2t} \cos(3t)$$

8.2 L-R-C CIRCUITS REVISITED

Laplace transforms can be used to solve the L-R-C circuits problems, as illustrated in the following figure, that were introduced earlier. Recall that the initial-value problem that is used to find the current is

$$\begin{cases} L\dfrac{d^2Q}{dt^2} + R\dfrac{dQ}{dt} + \dfrac{1}{C}Q = E(t) \\ Q(0) = Q_0, I(0) = \dfrac{dQ}{dt}(0) = I_0 \end{cases},$$

where $L, R,$ and C represent the inductance, resistance, and capacitance, respectively. Q is the charge of the capacitor and $\dfrac{dQ}{dt} = I$, where I is the current. $E(t)$ is the voltage supply.

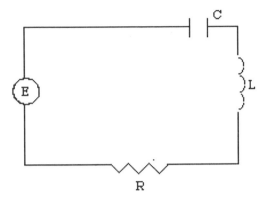

In particular, this method is most useful when the supplied voltage $v(t)$ is piecewise defined. This is demonstrated through the following example.

EXAMPLE: Suppose that we consider a circuit with a capacitor C, a resistor R, and a voltage supply $E(t) = \begin{cases} E_0, 0 \le t < 1 \\ 0, t \ge 1 \end{cases}$. If $L = 0$, then find $Q(t)$ and $I(t)$ if $Q(0) = 0$, $C = 10^{-2}$ farads, $R = 100\Omega$, and $E_0 = 100$ V.

SOLUTION: Because $L = 0$, we can state the first-order initial-value problem as

$$\begin{cases} 100\dfrac{dQ}{dt} + 100Q = \begin{cases} 100, 0 \le t < 1 \\ 0, t \ge 1 \end{cases} \\ Q(0) = 0 \end{cases}.$$

Since $E(t)$ is a piecewise continuous function, we defined it in terms of the unit step functions as

$$E(t) = 100[u(t - 0) - u(t - 1)] = 100[u(t) - u(t - 1)] = 100[1 - u(t - 1)].$$

Taking the Laplace transform of both sides of the differential equation yields

$$100sL\{Q(t)\} - Q(0) + 100L\{Q(t)\} = 100L\{1 - u(t - 1)\}$$

$$(s + 1)L\{Q(t)\} = \frac{1}{s} - \frac{e^{-s}}{s}$$

$$L\{Q(t)\} = \frac{1}{s(s + 1)} - \frac{e^{-s}}{s(s + 1)}.$$

The same steps are performed with Maple.

```
> f:=t->100*Heaviside(1-t):
  EQ:=100*diff(Q(t),t)+100*Q(t)=f(t):
  step_1:=laplace(EQ,t,s);
```

$$step\_1 := 100\,\text{laplace}\,(Q(t), t, s)s - 100Q(0) + 100\,\text{laplace}\,(Q(t), t, s)$$

$$= -100\frac{e^{-s}}{s} + 100\frac{1}{s}$$

```
> step_2:=subs(Q(0)=0,step_1);
```

$$step\_2 := 100\,\text{laplace}\,(Q(t), t, s)s + 100\,\text{laplace}\,(Q(t), t, s)$$

$$= -100\frac{e^{-s}}{s} + 100\frac{1}{s}$$

```
> step_3:=simplify(solve(step_2,laplace(Q(t),t,s)));
```

$$step\_3 := -\frac{e^{-s} - 1}{s(s + 1)}$$

Then

$$L^{-1}\left\{\frac{1}{s(s + 1)}\right\} = \int_0^t e^{-\alpha}d\alpha = -[e^{-\alpha}]_0^t = 1 - e^{-t},$$

so

$$L^{-1}\left\{\frac{e^{-s}}{s(s + 1)}\right\} = [1 - e^{-(t-1)}]u(t - 1).$$

As expected, `invlaplace` produces the same result, which we name `Sol`.

```
> readlib(laplace):
  Sol:=invlaplace(step_3,s,t);
```

$$Sol := -\,\text{Heaviside}(t - 1)(1 - e^{-t+1}) + 1 - e^{-t}$$

We then graph `Sol` and the derivative of `Sol` on the interval $[0, 2]$ with `plot`.

```
> plot(Sol,t=0..2);
  plot(diff(Sol,t),t=0..2);
```

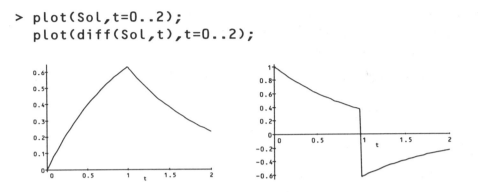

Therefore, $Q(t) = 1 - e^{-t} - [1 - e^{-(t-1)}]u(t - 1)$, which we can write as

$$Q(t) = \begin{cases} 1 - e^{-t}, 0 \le t < 1 \\ e^{-t}(e - 1), t > 1 \end{cases}.$$

From the graph, we see that after the voltage source is turned off at $t = 1$, then the charge approaches zero. Once we have determined the charge $Q(t)$, then we find $I(t)$ by calculating $\dfrac{dQ}{dt}$. Using the piecewise-defined form of Q, we have

$$I(t) = \frac{dQ}{dt} = \begin{cases} e^{-t}, 0 \le t < 1 \\ -e^{-t}(e - 1), t > 1 \end{cases}.$$

EXAMPLE: Consider the circuit with no capacitor, $R = 100\Omega$, and $L = 100$ H if $E(t) = \begin{cases} 100 \text{ V}, 0 \le t < 1 \\ 0, 1 \le t < 2 \end{cases}$ and $E(t + 2) = E(t)$. Find the current $I(t)$ if $I(0) = 0$.

SOLUTION: Because there is no capacitor, $C = 0$ and the differential equation that models the situation is $100\dfrac{d^2Q}{dt^2} + 100\dfrac{dQ}{dt} = E(t)$. Now, since $\dfrac{dQ}{dt} = I$, we can write this equation as $100\dfrac{dI}{dt} + 100I = E(t)$. Hence, the initial-value problem is

$$\begin{cases} 100\dfrac{dI}{dt} + 100I = E(t) \\ I(0) = 0 \end{cases}.$$

Notice that $E(t)$ is a periodic function, so we first compute $L\{E(t)\}$.

$$L\{E(t)\} = \frac{1}{1 - e^{-2s}} \int_0^2 e^{-st} E(t)\, dt = \frac{1}{1 - e^{-2s}} \int_0^1 100 e^{-st}\, dt = \frac{100}{1 - e^{-2s}} \left[-\frac{e^{-st}}{s} \right]_0^1 = \frac{100(1 - e^{-s})}{s(1 - e^{-2s})}$$

$$= \frac{100(1 - e^{-s})}{s(1 - e^{-s})(1 + e^{-s})} = \frac{100}{s(1 + e^{-s})}.$$

The same results are obtained using `int`.

```
> lap_lhs:=simplify(1/(1-exp(-2*s))*int(100*exp(-s*t),
  t=0..1));
```

$$lap\_lhs := 100 \frac{e^{-s} - 1}{(-1 + e^{-2s})s}$$

Then

$$100 L\left\{\frac{dI}{dt}\right\} + 100 L\{I\} = L\{E(t)\}$$

$$100 s L\{I\} - 100 I(0) + 100 L\{I\} = \frac{100}{s(1 + e^{-s})}$$

$$L\{I\} = \frac{1}{s(s + 1)(1 + e^{-s})}.$$

We obtain the same results with Maple. Note that we use i to represent $I(t)$ instead of I because I represents the imaginary number $\sqrt{-1}$.

```
> step_1:=laplace(diff(i(t),t)+100*i(t),t,s)=lap_lhs;
```

$$step\_1 := laplace\,(i(t), t, s)s - i(0) + 100\,laplace\,(i(t), t, s)$$

$$= 100 \frac{e^{-s} - 1}{(-1 + e^{-2s})s}$$

```
> step_2:=subs(i(0)=0,step_1);
  step_3:=simplify(solve(step_2,laplace(i(t),t,s)));
```

$$step\_2 := laplace\,(i(t), t, s)s + 100\,laplace\,(i(t), t, s) = 100 \frac{e^{-s} - 1}{(-1 + e^{-2s})s}$$

$$step\_3 := 100 \frac{e^{-s} - 1}{(-1 + e^{-2s})s(s + 100)}$$

As we did in Section 7.4, we write a power series expansion of $\dfrac{1}{1+e^{-s}}$. Since $\dfrac{1}{1-x} = 1 - x + x^2 - x^3 + \cdots$,

$$\frac{1}{1+e^{-s}} = 1 - e^{-s} + e^{-2s} - e^{-3s} + \cdots.$$

Thus,

$$L\{I\} = \frac{1}{s(s+1)}[1 - e^{-s} + e^{-2s} - e^{-3s} + \cdots] = \frac{1}{s(s+1)} - \frac{e^{-s}}{s(s+1)} + \frac{e^{-2s}}{s(s+1)} - \frac{e^{-3s}}{s(s+1)} + \cdots.$$

Because $L^{-1}\left\{\dfrac{1}{s(s+1)}\right\} = 1 - e^{-t}$,

```
> readlib(laplace):
  invlaplace(1/(s*(s+1)),s,t);
```

$$1 - e^{-t}$$

$$I(t) = (1 - e^{-t}) - (1 - e^{-(t-1)})\,u(t-1) + (1 - e^{-(t-2)})\,u(t-2) - (1 - e^{-(t-3)})\,u(t-3) + \cdots,$$

we can write this function as

$$I(t) = \begin{cases} 1 - e^{-t}, 0 \le t < 1 \\ -e^{-t} + e^{-(t-1)}, 1 \le t < 2 \\ 1 - e^{-t} + e^{-(t-1)} - e^{-(t-2)}, 2 \le t < 3 \\ -e^{-t} + e^{-(t-1)} - e^{-(t-2)} + e^{-(t-3)}, 3 \le t < 4 \\ \vdots \end{cases}.$$

To graph $I(t)$ on the interval $[0, n]$, we note that $u(t - n) = 0$ for $t \le n$, so the graph of $I(t)$ on the interval $[0, n]$ is the same as the graph of

$$(1 - e^{-t}) - (1 - e^{-(t-1)})\,u(t-1) + (1 - e^{-(t-2)})\,u(t-2) - (1 - e^{-(t-3)})\,u(t-3)$$
$$+ \cdots + (-1)^{n-1}(1 - e^{-[t-(n-1)]})\,u(t-(n-1)).$$

For convenience, we define **nth_term** to represent $(-1)^n(1 - e^{-(t-n)})\,u(t-n)$.

```
> nth_term:=n->(-1)^n*(1-exp(-(t-n)))*Heaviside(t-n):
```

Then, to graph $I(t)$ on the interval $[0, 4]$, we enter

```
> to_graph:=sum(nth_term(n),n=0..3);
```

$to\_graph := (1 - e^{-t})\,\text{Heaviside}(t) - (1 - e^{-t+1})\,\text{Heaviside}(t - 1)$
$\qquad + (1 - e^{-t+2})\,\text{Heaviside}(t - 2) - (1 - e^{-t+3})\,\text{Heaviside}(t - 3)$

and then use **plot**.

```
> plot(to_graph,t=0..4);
```

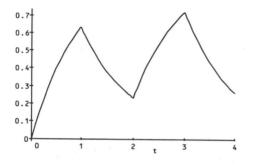

Notice that I increases over the intervals where $E(t) = 100$ and decreases on those where $E(t) = 0$.

We can consider the L-R-C circuit in terms of the **integrodifferential equation**

$$L\frac{dI}{dt} + RI + \frac{1}{C}\int_0^t I(\alpha)\,d\alpha = E(t),$$

which is useful when using the method of Laplace transforms to find the current.

EXAMPLE: Find the current $I(t)$ if $L = 1$ henry, $R = 6$ ohms, $C = \dfrac{1}{9}$ farad, $E(t) = 1$ volt, and $I(0) = 0$.

SOLUTION: In this case, we must solve the initial-value problem

$$\begin{cases} \dfrac{dI}{dt} + 6I + 9\displaystyle\int_0^t I(\alpha)\,d\alpha = 1 \\ I(0) = 0 \end{cases}.$$

Hence,

$$sL\{I\} - I(0) + 6L\{I\} + \frac{9I(s)}{s} = \frac{1}{s}$$

$$s^2L\{I\} + 6sL\{I\} + 9I(s) = 1$$

$$L\{I\} = \frac{1}{s^2 + 6s + 9} = \frac{1}{(s + 3)^2}.$$

We obtain the same results using Maple.

> `EQ:=diff(i(t),t)+6*i(t)+9*Int(i(alpha),alpha=0..t)=1;`

$$EQ := \left(\frac{\partial}{\partial t}i(t)\right) + 6i(t) + 9\int_0^t i(\alpha)\,d\alpha = 1$$

> `step_1:=laplace(EQ,t,s);`

$$step\_1 := \text{laplace}\,(i(t), t, s)s - i(0) + 6\,\text{laplace}\,(i(t), t, s) + 9\frac{\text{laplace}\,(i(t), t, s)}{s} = \frac{1}{2}$$

> `step_2:=subs(i(0)=0,step_1);`

$$step\_2 := \text{laplace}\,(i(t), t, s)s + 6\,\text{laplace}\,(i(t), t, s) + 9\frac{\text{laplace}\,(i(t), t, s)}{s} = \frac{1}{s}$$

> `step_3:=factor(solve(step_2,laplace(i(t),t,s)));`

$$step\_3 := (s + 3)^{-2}$$

Using `invlaplace` results in

> `readlib(laplace):`
> `Sol:=invlaplace(step_3,s,t);`

$$Sol := te^{-3t}$$

so

$$I(t) = L^{-1}\left\{\frac{1}{(s + 3)^2}\right\} = te^{-3t},$$

which we graph with `plot`.

```
> plot(Sol,t=0..3);
```

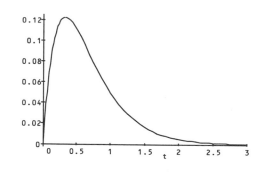

8.3 POPULATION PROBLEMS REVISITED

Laplace transforms can be used to solve the population problems that were discussed as applications of first-order equations. In this case, however, we focus our attention on those problems that include a nonhomogeneous forcing function. Laplace transforms are especially useful when dealing with piecewise-defined forcing functions, but they are useful in many other cases as well. We consider now a problem that involves a continuous forcing function used to describe the presence of immigration or emigration.

EXAMPLE: Let $x(t)$ represent the population of a certain country. The rate at which the population increases depends on the growth rate of the country as well as the rate at which people are being added to or subtracted from the population due to immigration or emigration. Hence, we consider the population problem

$$x' + kx = 1000(1 + a \sin t), x(0) = x_0.$$

Solve this problem using Laplace transforms with $k = 3$, $x_0 = 2000$, and $a = 0.25, 0.75$, and 1.25. Plot the solution in each case.

SOLUTION: Using the method of Laplace transforms, we begin by computing the Laplace transform of each side of the equation with `laplace`.

```
> EQ:=diff(x(t),t)+k*x(t)=1000*(1+a*sin(t)):
  step_1:=laplace(EQ,t,s);
```

$$step\_1 := \text{laplace}\,(x(t), t, s)s - x(0) + k\,\text{laplace}\,(x(t), t, s) = 1000\frac{s^2 + 1 + as}{s(s^2 + 1)}$$

We then use **solve** and **simplify** to solve **step_1** for laplace(y(t),t,s), which represents the Laplace transform of $y(t)$.

```
> step_2:=simplify(solve(step_1,laplace(x(t),t,s)));
```

$$step\_2 := \frac{x(0)s^3 + x(0)s + 1000s^2 + 1000 + 1000as}{s(s^2 + 1)(s + k)}$$

To find the solution, we use **invlaplace** and name the result **Gen_Sol**.

```
> readlib(laplace):
  Gen_Sol:=invlaplace(step_2,s,t);
```

$$Gen\_Sol := 1000\frac{1}{k} + 1000\frac{ak\sin(t)}{k^2 + 1} - 1000\frac{a\cos(t)}{k^2 + 1} + 1000\frac{kae^{-kt}}{k^3 + k} + \frac{kx(0)e^{-kt}}{k^3 + k} - 1000\frac{k^2e^{-kt}}{k^3 + k}$$
$$- 1000\frac{e^{-kt}}{k^3 + k} + \frac{k^3x(0)e^{-kt}}{k^3 + k}$$

Note that we are able to obtain the same result directly with **dsolve**.

```
> simplify(dsolve({EQ,x(0)=x0},x(t)));
```

$$x(t) = (x0k^3 + 1000k^2a\sin(t)e^{kt} - 1000k^2 + 1000k^2e^{kt} + x0k + 1000ak$$
$$- 1000a\cos(t)e^{kt}k - 1000 + 1000e^{kt})e^{-kt}k(k^2 + 1)$$

Suppose that $x_0 = 2000$ and $k = 3$. Then we can investigate the population for values of $a = 0.25$, 0.75, and 1.25 using **plot**. For example, entering

```
> to_plot_1:=subs({k=3,x(0)=2000,a=0.25},Gen_Sol);
```

$$to\_plot\_1 := \frac{1000}{3} + 75.0\sin(t) - 25.0\cos(t) + 1691.666667e^{-3t}$$

defines **to_plot_1** to be the expression obtained by substituting the values $k = 3, x(0) = 2000$, and $a = 0.25$ into **Gen_Sol**. We then graph **to_plot_1** on the interval $[0, 10]$.

```
> plot(to_plot_1,t=0..10);
```

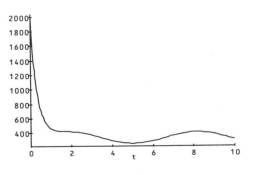

Similarly, we define `to_plot_2` and `to_plot_3` to be the expressions obtained by substituting $k = 3$, $x(0) = 2000$, and $a = 0.75$ and $a = 1.25$, respectively, into `Gen_Sol`. As with `to_plot_1`, both are then graphed with `plot`.

```
> to_plot_2:=subs({k=3,x(0)=2000,a=0.75},Gen_Sol):
  to_plot_3:=subs({k=3,x(0)=2000,a=1.25},Gen_Sol):
  plot(to_plot_2,t=0..10);
  plot(to_plot_3,t=0..10);
```

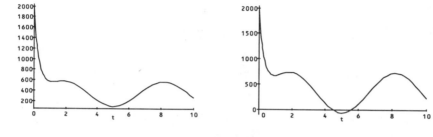

9
Systems of Ordinary Differential Equations

9.1 SYSTEMS OF EQUATIONS: THE OPERATOR METHOD

Up to this point, we have focused our attention on solving differential equations that involve one dependent variable. Unfortunately, many physical situations are modeled with more than one equation and involve more than one dependent variable. For example, if we want to determine the population of two interacting populations such as foxes and rabbits, we would have two dependent variables to represent the two populations where these populations depend on one independent variable that represents time.

DEFINITION | *System of Ordinary Differential Equations*

A **system** of ordinary differential equations is a simultaneous set of equations that involves two or more dependent variables. A **solution** of the system is a set of differentiable functions that satisfies each equation on some interval I.

☰ *Operator Notation*

One method for solving systems of linear equations involves the differential operator

$$D = \frac{d}{dt},$$

which was discussed with the annihilator method for solving nonhomogeneous higher-order linear equations with constant coefficients. Recall that the differential equation

$$a_n \frac{d^n y}{dt^n} + a_{n-1} \frac{d^{n-1} y}{dt^{n-1}} + \cdots + a_1 \frac{dy}{dt} + a_0 y = f(t)$$

is expressed in operator notation as

$$(a_n D^n + a_{n-1} D^{n-1} + \cdots + a_1 D + a_0)y = f(t).$$

Therefore, in a system of equations, each equation can be written in operator notation.

☰ Solution Method with Operator Notation

An advantage of operator notation is that systems of linear differential equations can be solved in much the same way as an algebraic system of equations. We represent the general form of a system in operator notation as

$$\begin{cases} L_1 x + L_2 y = f_1(t) \\ L_3 x + L_4 y = f_2(t) \end{cases},$$

where L_1, L_2, L_3, and L_4 are **linear differential operators with constant coefficients**. Applying the operator L_4 to the first equation and $-L_2$ yields the system

$$\begin{cases} L_1 L_4 x + L_2 L_4 y = L_4 f_1(t) \\ -L_2 L_3 x - L_2 L_4 y = -L_2 f_2(t) \end{cases}.$$

Now, because the operators have constant coefficients, $L_2 L_4 y = L_2 L_4 y$. Therefore, when these equations are added, we have $(L_1 L_4 - L_2 L_3)x = g_1(t)$, where $g_1(t) = L_4 f_1(t) - L_2 f_2(t)$. Similarly, we can eliminate y to obtain the equation $(L_1 L_4 - L_2 L_3)y = g_2(t)$, where $g_2(t) = L_1 f_2(t) - L_3 f_1(t)$.

Because the left-hand side of the equations for x and y are the same (except for the variable), the solutions x and y must depend on the same linearly independent functions, but not the same arbitrary constants.

The number of independent arbitrary constants in the solution will be the same as the order of the operator $L_1 L_4 - L_2 L_3$, which can be represented as the determinant

$$\begin{vmatrix} L_1 & L_2 \\ L_3 & L_4 \end{vmatrix} = L_1 L_4 - L_2 L_3.$$

EXAMPLE: Solve the system

$$\begin{cases} x' + y' = e^t + 2x + 4y \\ x' + y' - y = e^{4t} \end{cases}.$$

SOLUTION: We begin by writing the system in operator notation:

$$\begin{cases} (D-2)x + (D-4)y = e^t \\ Dx + (D-1)y = e^{4t} \end{cases}.$$

Applying $L_4 = (D - 1)$ to the first equation and $-L_2 = -(D - 4)$ to the second yields the system

$$\begin{cases} (D - 1)(D - 2)x + (D - 1)(D - 4)y = (D - 1)e^t = e^t - e^t = 0 \\ -(D - 4)Dx - (D - 4)(D - 1)y = -(D - 4)e^{4t} = -(4e^{4t} - 4e^{4t}) = 0 \end{cases}.$$

Adding then yields

$$(D - 1)(D - 2)x - (D - 4)Dx = [(D^2 - 3D + 2) - D^2 + 4D]x = (D + 2)x = 0.$$

Because the characteristic equation for the differential equation $(D + 2)x = 0$ is $m + 2 = 0$ with root $m = -2$, $x(t) = c_1 e^{-2t}$.

Next, we solve for $y(t)$ by repeating the elimination procedure. Applying $L_3 = D$ to the first equation and $-L_1 = -(D - 2)$ to the second yields the system

$$\begin{cases} D(D - 2)x + D(D - 4)y = De^t = e^t \\ -(D - 2)Dx - (D - 2)(D - 1)y = -(D - 2)e^{4t} = -4e^{4t} + 2e^{4t} = -2e^{4t} \end{cases}.$$

Adding the equations, we obtain the nonhomogeneous equation for y,

$$D(D - 4)y - (D - 2)(D - 1)y = e^t - 2e^{4t}.$$

The homogeneous solution is found by solving

$$D(D - 4)y - (D - 2)(D - 1)y = (D^2 - 4D - D^2 + 3D - 2)y = (-D - 2)y = 0$$

with characteristic equation $m + 2 = 0$. Therefore, $y_h(t) = k_1 e^{-2t}$. Using the method of undetermined coefficients, we assume a particular solution of the form $y_p(t) = Ae^t + Be^{4t}$. Substitution of $y_p(t)$ into the equation $(-D - 2)y = e^t - 2e^{4t}$ yields

$$(-D - 2)y_p(t) = -Dy_p(t) - 2y_p(t) = -Ae^t - 4Be^{4t} - 2Ae^t - 2Be^{4t} = -3Ae^t - 6Be^{4t} = e^t - 2e^{4t},$$

so $A = -\dfrac{1}{3}$ and $B = \dfrac{1}{3}$. Therefore,

$$y(t) = y_h(t) + y_p(t) = k_1 e^{-2t} - \frac{1}{3}e^t + \frac{1}{3}e^{4t}.$$

Notice that

$$\begin{vmatrix} L_1 & L_2 \\ L_3 & L_4 \end{vmatrix} = \begin{vmatrix} D - 2 & D - 4 \\ D & D - 1 \end{vmatrix} = (D^2 - 3D + 2) - (D^2 - 4D) = D + 2,$$

so there is only **one** arbitrary constant in the solution, because the order of $D + 2$ is 1. Substituting $x(t)$ and $y(t)$ into the system yields

$$(D - 2)(c_1 e^{-2t}) + (D - 4)\left(k_1 e^{-2t} - \frac{1}{3}e^t + \frac{1}{3}e^{4t}\right)$$

$$= -2c_1 e^{-2t} - 2c_1 e^{-2t} - 2k_1 e^{-2t} - \frac{1}{3}e^t + \frac{4}{3}e^{4t} - 4k_1 e^{-2t} + \frac{4}{3}e^t - \frac{4}{3}e^{4t}$$

$$= (-4c_1 - 6k_1)e^{-2t} + e^t = e^t.$$

Therefore, $-4c_1 - 6k_1 = 0$, so $k_1 = -\frac{2}{3}c_1$. Making this substitution, we have the solution

$$x(t) = c_1 e^{-2t}$$

$$y(t) = -\frac{2}{3}c_1 e^{-2t} - \frac{1}{3}e^t + \frac{1}{3}e^{4t}.$$

We now show how **dsolve** is used to find a general solution of the equation, which we name **Gen_Sol**.

```
> Gen_Sol:=dsolve({diff(x(t),t)+diff(y(t),t)=
  exp(t)+2*x(t)+4*y(t),
  diff(x(t),t)+diff(y(t),t)-y(t)=exp(4*t)},
  {x(t),y(t)});
```

$$Gen\_Sol := \left\{ x(t) = -\frac{3}{2}\_C1e^{-2t}, y(t) = \frac{1}{3}e^{4t} - \frac{1}{3}e^t + \_C1e^{-2t} \right\}$$

We then name $x(t)$ and $y(t)$ the result obtained in **Gen_Sol** with **assign**.

```
> assign(Gen_Sol):
```

To graph $(x(t), y(t))$ for various values of the arbitrary constant, we use **subs** to replace **_C1** by a number and then **plot** to graph the resulting parametric equations. Generally, to use **plot** to graph the parametric equations

$$\begin{cases} x = x(t) \\ y = y(t) \end{cases}, t_0 \leq t \leq t_1,$$

enter **plot([x(t),y(t),t=t0..t1])**. For example, entering

```
> to_plot_1:=subs(_C1=-4,[x(t),y(t),t=-2..2]);
```

$$to\_plot\_1 := \left[6e^{-2t}, \frac{1}{3}e^{4t} - \frac{1}{3}e^t - 4e^{-2t}, t = -2...2 \right]$$

defines **to_plot_1** to be the ordered triple consisting of $x(t)$ and $y(t)$ when **_C1** is replaced by -4 and **t=-2..2**. Thus, entering

> `plot(to_plot_1,view=[0..10,-10..10]);`

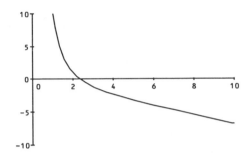

graphs

$$\begin{cases} x = 6e^{-2t} \\ y = \dfrac{1}{3}e^{4t} - \dfrac{1}{3}e^{t} - 4e^{-2t} \end{cases}, \quad -2 \le t \le 2.$$

The option **view=[0..10,-10..10]** instructs Maple to display the resulting graph on the rectangle $[0,10] \times [-10,10]$. Similarly, entering

> `to_plot:={seq(subs(_C1=i,[x(t),y(t),t=-2..2]),`
>   `i=-2..2)};`

$$to\_plot := \left\{ \left[3e^{-2t}, \frac{1}{3}e^{4t} - \frac{1}{3}e^{t} - 2e^{-2t}, t = -2...2 \right], \right.$$

$$\left[\frac{3}{2}e^{-2t}, \frac{1}{3}e^{4t} - \frac{1}{3}e^{t} - e^{-2t}, t = -2...2 \right],$$

$$\left[0, \frac{1}{3}e^{4t} - \frac{1}{3}e^{t}, t = -2...2 \right],$$

$$\left[-\frac{3}{2}e^{-2t}, \frac{1}{3}e^{4t} - \frac{1}{3}e^{t} + e^{-2t}, t = -2...2 \right],$$

$$\left. \left[-3e^{-2t}, \frac{1}{3}e^{4t} - \frac{1}{3}e^{t} + 2e^{-2t}, t = -2...2 \right] \right\}$$

and then

```
> plot(to_plot,view=[-10..10,-10..10]);
```

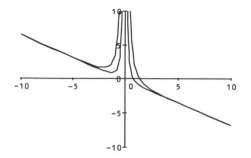

defines `to_plot` to be the set of ordered triples consisting of $x(t)$ and $y(t)$ when `_C1` is replaced by i for $i = -2, -1, 0, 1$, and 2 and `t=-2..2` and then graphs the parametric equations `to_plot`.

The determinant that we just discussed can be used to solve the system. From the earlier calculations, we obtained the formulas

$$(L_1L_4 - L_2L_3)x = L_4 f_1(t) - L_2 f_2(t)$$

and

$$(L_1L_4 - L_2L_3)y = L_1 f_2(t) - L_3 f_1(t).$$

By Cramer's rule, we have

$$\begin{vmatrix} L_1 & L_2 \\ L_3 & L_4 \end{vmatrix} x = \begin{vmatrix} f_1(t) & L_2 \\ f_2(t) & L_4 \end{vmatrix}$$

and

$$\begin{vmatrix} L_1 & L_2 \\ L_3 & L_4 \end{vmatrix} y = \begin{vmatrix} L_1 & f_1(t) \\ L_3 & f_2(t) \end{vmatrix}.$$

These two formulas can be used to solve the system of equations. Notice that

$$\begin{vmatrix} L_1 & f_1(t) \\ L_3 & f_2(t) \end{vmatrix} = L_1 f_2(t) - L_3 f_1(t)$$

and

$$\begin{vmatrix} f_1(t) & L_2 \\ f_2(t) & L_4 \end{vmatrix} = L_4 f_1(t) - L_2 f_2(t),$$

so we evaluate these determinants by applying the operators to the functions $f_1(t)$ and $f_2(t)$.

EXAMPLE: Solve the initial-value problem

$$\begin{cases} x' = x - 4y + 1 \\ y' = x + y \end{cases}, x(0) = 1, y(0) = 0.$$

SOLUTION: In operator notation, the system is

$$\begin{cases} (D - 1)x + 4y = 1 \\ -x + (D - 1)y = 0 \end{cases}.$$

Then, with Cramer's rule, we have

$$\begin{vmatrix} D - 1 & 4 \\ -1 & D - 1 \end{vmatrix} x = \begin{vmatrix} 1 & 4 \\ 0 & D - 1 \end{vmatrix}$$

and

$$\begin{vmatrix} D - 1 & 4 \\ -1 & D - 1 \end{vmatrix} y = \begin{vmatrix} D - 1 & 1 \\ -1 & 0 \end{vmatrix}.$$

Solving for x yields $[(D - 1)(D - 1) + 4]x = (D - 1)[1] - 0 = 0 - 1 = -1$, which can be written as $(D^2 - 2D + 5)x = -1$. The homogeneous equation $(D^2 - 2D + 5)x = 0$ has the characteristic equation $m^2 - 2m + 5 = 0$ with roots $m = 1 \pm 2i$. Thus, the homogeneous solution is $x_h(t) = e^t(c_1 \cos 2t + c_2 \sin 2t)$. By the method of undetermined coefficients, we assume that the particular solution is $x_p(t) = A$. Substitution into the equation shows that $5A = -1$, so $x_p(t) = -\frac{1}{5}$. Therefore, $x(t) = e^t(c_1 \cos 2t + c_2 \sin 2t) - \frac{1}{5}$.

Next, we solve for $y(t)$. Expanding $\begin{vmatrix} D - 1 & 4 \\ -1 & D - 1 \end{vmatrix} y = \begin{vmatrix} D - 1 & 1 \\ -1 & 0 \end{vmatrix}$, we obtain $(D^2 - 2D + 5)y = 1$. Following the same steps as those used in finding $x(t)$, we find that $y(t) = e^t(k_1 \cos 2t + k_2 \sin 2t) + \frac{1}{5}$. Then, because the operator $\begin{vmatrix} D - 1 & 4 \\ -1 & D - 1 \end{vmatrix} = D^2 - 2D + 5$ is of order 2, there are only two arbitrary constants in the solution. Substituting $x(t)$ and $y(t)$ into the first equation of the system yields

$$x'(t) - x + 4y = e^t(c_1 \cos 2t + c_2 \sin 2t) + e^t(-2c_1 \sin 2t + 2c_2 \cos 2t)$$

$$- e^t(c_1 \cos 2t + c_2 \sin 2t) + \frac{1}{5} + 4e^t(k_1 \cos 2t + k_2 \sin 2t) + \frac{4}{5}$$

$$= (2c_2 + 4k_1)e^t \cos 2t + (-2c_1 + 4k_2)e^t \sin 2t + 1.$$

Therefore, $2c_2 + 4k_1 = 0$ and $-2c_1 + 4k_2 = 0$, so $k_1 = -\frac{1}{2}c_2$ and $k_2 = \frac{1}{2}c_1$. This means that a general solution is

$$\begin{cases} x(t) = e^t(c_1 \cos 2t + c_2 \sin 2t) - \frac{1}{5} \\ y(t) = e^t\left(-\frac{1}{2}c_2 \cos 2t + \frac{1}{2}c_1 \sin 2t\right) + \frac{1}{5} \end{cases}.$$

A solution to the initial-value problem must satisfy $x(0) = 1$ and $y(0) = 0$. Therefore, because $x(0) = c_1 - \frac{1}{5} = 1$ and $y(0) = -\frac{1}{2}c_2 + \frac{1}{5} = 0$, $c_1 = \frac{6}{5}$ and $c_2 = \frac{2}{5}$. This means that the solution is

$$\begin{cases} x(t) = e^t\left(\frac{6}{5}\cos 2t + \frac{2}{5}\sin 2t\right) - \frac{1}{5} \\ y(t) = e^t\left(-\frac{1}{5}\cos 2t + \frac{3}{5}\sin 2t\right) + \frac{1}{5} \end{cases}.$$

In the same manner as in the previous example, we can use **dsolve** to solve this initial-value problem. For example, entering

```
> x:='x':y:='y':
  Sol:=dsolve({diff(x(t),t)=x(t)-4*y(t)+1,
  diff(y(t),t)=x(t)+y(t),x(0)=1,y(0)=0},{x(t),y(t)});
```

$$Sol := \left\{ y(t) = \frac{1}{5} + \frac{3}{5}e^t \sin(2t) - \frac{1}{5}e^t \cos(2t), x(t) = -\frac{1}{5} + \frac{2}{5}e^t \sin(2t) + \frac{6}{5}e^t \cos(2t) \right\}$$

solves the initial-value problem and names the resulting output **Sol**. Then entering

```
> assign(Sol):
  plot([x(t),y(t),t=0..10]);
```

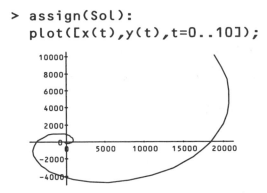

names $x(t)$ and $y(t)$ the result given in Sol and then graphs the parametric equations

$$\begin{cases} x = x(t) \\ y = y(t) \end{cases}, 0 \leq t \leq 10.$$

Note that if $\begin{vmatrix} L_1 & L_2 \\ L_3 & L_4 \end{vmatrix} = 0$, then we say the system is **degenerate**, which means that the system either has infinitely many solutions or no solutions. If $\begin{vmatrix} L_1 & L_2 \\ L_3 & L_4 \end{vmatrix} = \begin{vmatrix} f_1(t) & L_2 \\ f_2(t) & L_4 \end{vmatrix} = \begin{vmatrix} L_1 & f_1(t) \\ L_3 & f_2(t) \end{vmatrix} = 0$, then the system has **infinitely many solutions**. Otherwise, the system has **no solutions**.

Systems of three or more equations can be solved with the operator method as well.

EXAMPLE: Solve the system of differential equations

$$\begin{cases} x' = x + 2y + z \\ y' = 6x - y \\ z' = -x - 2y - z \end{cases}.$$

SOLUTION: In operator notation, this system is

$$\begin{cases} (D-1)x - 2y - z = 0 \\ -6x + (D+1)y = 0 \\ x + 2y + (D+1)z = 0 \end{cases}.$$

Because the second equation involves only x and y, we eliminate z from the first and third equations. Applying $(D + 1)$ to the first equation yields

$$(D + 1)(D - 1)x - 2(D + 1)y - (D + 1)z = 0.$$

Adding this equation and the third equation then gives us

$$[D^2(D - 6) - 12D]y = (D^3 - 6D^2 - 12D)y = 0.$$

This homogeneous equation has characteristic equation

$$m^3 - 6m^2 - 12m = m(m^2 - 6m - 12) = m(m + 4)(m - 3) = 0,$$

so

$$y(t) = c_1 + c_2 e^{-4t} + c_3 e^{3t}.$$

Instead of repeating the elimination process with x and z, we use the equation $y' = 6x - y$ or $x = \dfrac{1}{6}(y' + y)$ to find x. Therefore,

$$x(t) = \frac{1}{6}(-4c_2 e^{-4t} + 3c_3 e^{3t} + c_1 + c_2 e^{-4t} + c_3 e^{3t})$$

$$= \frac{1}{6}c_1 - \frac{1}{2}c_2 e^{-4t} + \frac{2}{3}c_3 e^{3t}.$$

Finally, we use $x' = x + 2y + z$ to find $z(t)$:

$$z(t) = x' - x - 2y$$

$$= 2c_2 e^{-4t} + 2c_3 e^{3t} - \left(\frac{1}{6}c_1 - \frac{1}{2}c_2 e^{-4t} + \frac{2}{3}c_3 e^{3t}\right) - 2\left(c_1 + c_2 e^{-4t} + c_3 e^{3t}\right)$$

$$= -\frac{13}{6}c_1 + \frac{1}{2}c_2 e^{-4t} - \frac{2}{3}c_3 e^{3t}$$

Note that we can use **dsolve** to solve a system of three equations in the same way that we use **dsolve** to solve a system of two equations.

```
> x:='x':y:='y':z:='z':
  Sol:=dsolve({diff(x(t),t)=x(t)+2*y(t)+z(t),
  diff(y(t),t)=6*x(t)-y(t),
  diff(z(t),t)=-x(t)-2*y(t)-z(t)},{x(t),y(t),z(t)});
```

$Sol := \Big\{ y(t) = 6\_C1 - \dfrac{3}{2}\_C2\, e^{3t} + 2\_C3\, e^{-4t},\ x(t) = \_C1 - \_C2\, e^{3t} - \_C3\, e^{-4t},$

$z(t) = -13\_C1 + \_C2\, e^{3t} + \_C3\, e^{-4t} \Big\}$

9.2 REVIEW OF MATRIX ALGEBRA AND CALCULUS

Because of their importance in the study of systems of linear equations, we now review matrices and the operations associated with them.

☰ Basic Operations

DEFINITION | $n \times m$ *Matrix*

An $n \times m$ **matrix** is an array of the form

$$\begin{pmatrix} a_{11} & a_{12} & \cdots & a_{1m} \\ a_{21} & a_{22} & \cdots & a_{2m} \\ \vdots & \vdots & \ddots & \vdots \\ a_{n1} & a_{n2} & \cdots & a_{nm} \end{pmatrix}$$

with n rows and m columns. This matrix can be denoted $\mathbf{A} = (a_{ij})$.

DEFINITION | *Transpose*

The **transpose** of the $n \times m$ matrix

$$\mathbf{A} = \begin{pmatrix} a_{11} & a_{12} & \cdots & a_{1m} \\ a_{21} & a_{22} & \cdots & a_{2m} \\ \vdots & \vdots & \ddots & \vdots \\ a_{n1} & a_{n2} & \cdots & a_{nm} \end{pmatrix}$$

is the $m \times n$ matrix

$$\mathbf{A}^T = \begin{pmatrix} a_{11} & a_{21} & \cdots & a_{n1} \\ a_{12} & a_{22} & \cdots & a_{n2} \\ \vdots & \vdots & \ddots & \vdots \\ a_{1m} & a_{2m} & \cdots & a_{nm} \end{pmatrix}.$$

Hence, $\mathbf{A}^T = (a_{ji})$.

DEFINITION | *Scalar Multiplication, Matrix Addition*

Let $\mathbf{A} = (a_{ij})$ be an $n \times m$ matrix and c a scalar. Then the **scalar multiple** of $\mathbf{A}$ by c is the $n \times m$ matrix given by $c\mathbf{A} = (ca_{ij})$. If $\mathbf{B} = (b_{ij})$ is also an $n \times m$ matrix, then the **sum** of matrices $\mathbf{A}$ and $\mathbf{B}$ is the $n \times m$ matrix $\mathbf{A} + \mathbf{B} = (a_{ij}) + (b_{ij}) = (a_{ij} + b_{ij})$.

Hence, $c\mathbf{A}$ is the matrix obtained by multiplying each element of $\mathbf{A}$ by c; $\mathbf{A} + \mathbf{B}$ is obtained by adding corresponding elements of $\mathbf{A}$ and $\mathbf{B}$.

EXAMPLE: Compute $3\mathbf{A} - 9\mathbf{B}$ if $\mathbf{A} = \begin{pmatrix} -1 & 4 & -2 \\ 6 & 2 & -10 \end{pmatrix}$ and $\mathbf{B} = \begin{pmatrix} 2 & -4 & 8 \\ 7 & 4 & 2 \end{pmatrix}$.

SOLUTION: Because $3\mathbf{A} = \begin{pmatrix} -3 & 12 & -6 \\ 18 & 6 & -30 \end{pmatrix}$ and $-9\mathbf{B} = \begin{pmatrix} -18 & 36 & -72 \\ -63 & -36 & -18 \end{pmatrix}$, $3\mathbf{A} - 9\mathbf{B} =$

$3\mathbf{A} + (-9\mathbf{B}) = \begin{pmatrix} -21 & 48 & -78 \\ -45 & -30 & -48 \end{pmatrix}$.

You should use `array` or `matrix`, which is contained in the `linalg` package, to define matrices. Elementary computations with matrices are carried out with `evalm` or commands contained in the `linalg` package.

For example, we now use `array` to define $\mathbf{A}$ and $\mathbf{B}$ to be the matrices $\mathbf{A} = \begin{pmatrix} -1 & 4 & -2 \\ 6 & 2 & -10 \end{pmatrix}$ and $\mathbf{B} = \begin{pmatrix} 2 & -4 & 8 \\ 7 & 4 & 2 \end{pmatrix}$.

```
> A:=array([[-1,4,-2],[6,2,-10]]);
  B:=array([[2,-4,8],[7,4,2]]);
```

$$A := \begin{bmatrix} -1 & 4 & -2 \\ 6 & 2 & -10 \end{bmatrix}$$

$$B := \begin{bmatrix} 2 & -4 & 8 \\ 7 & 4 & 2 \end{bmatrix}$$

We then use `evalm` to compute $3\mathbf{A} - 9\mathbf{B}$.

```
> evalm(3*A-9*B);
```

$$\begin{bmatrix} -21 & 48 & -78 \\ -45 & -30 & -48 \end{bmatrix}$$

DEFINITION | *Matrix Multiplication*

If

$$A = \begin{pmatrix} a_{11} & a_{12} & \cdots & a_{1j} \\ a_{21} & a_{22} & \cdots & a_{2j} \\ \vdots & \vdots & \ddots & \vdots \\ a_{n1} & a_{n2} & \cdots & a_{nj} \end{pmatrix}$$

is an $n \times j$ matrix and

$$B = \begin{pmatrix} b_{11} & b_{12} & \cdots & b_{1m} \\ b_{21} & b_{22} & \cdots & b_{2m} \\ \vdots & \vdots & \ddots & \vdots \\ b_{j1} & b_{j2} & \cdots & b_{jm} \end{pmatrix}$$

is a $j \times m$ matrix, then

$$AB = \begin{pmatrix} a_{11} & a_{12} & \cdots & a_{1j} \\ a_{21} & a_{22} & \cdots & a_{2j} \\ \vdots & \vdots & \ddots & \vdots \\ a_{n1} & a_{n2} & \cdots & a_{nj} \end{pmatrix} \begin{pmatrix} b_{11} & b_{12} & \cdots & b_{1m} \\ b_{21} & b_{22} & \cdots & b_{2m} \\ \vdots & \vdots & \ddots & \vdots \\ b_{j1} & b_{j2} & \cdots & b_{jm} \end{pmatrix}$$

is the unique matrix

$$C = \begin{pmatrix} c_{11} & c_{12} & \cdots & c_{1m} \\ c_{21} & c_{22} & \cdots & c_{2m} \\ \vdots & \vdots & \ddots & \vdots \\ c_{n1} & c_{n2} & \cdots & c_{nm} \end{pmatrix},$$

where

$$c_{11} = a_{11}b_{11} + a_{12}b_{21} + \cdots + a_{1j}b_{j1} = \sum_{k=1}^{j} a_{1k}b_{k1}, \quad c_{12}$$

$$= a_{11}b_{12} + a_{12}b_{22} + \cdots + a_{1j}b_{j2} = \sum_{k=1}^{j} a_{1k}b_{k2}$$

and

$$c_{uv} = a_{u1}b_{1v} + a_{u2}b_{2v} + \cdots + a_{uj}b_{jv} = \sum_{k=1}^{j} a_{uk}b_{kv}.$$

In other words, the element c_{uv} is obtained by multiplying each member of the uth row of **A** by the corresponding entry in the vth column of **B** and adding the result.

EXAMPLE: Compute **AB** if $\mathbf{A} = \begin{pmatrix} 0 & 4 & 5 \\ -5 & -1 & 5 \end{pmatrix}$ and $B = \begin{pmatrix} -3 & 4 \\ -5 & -4 \\ 1 & -4 \end{pmatrix}$.

SOLUTION: Because **A** is a 2×3 matrix and **B** is a 3×2 matrix, **AB** is the 2×2 matrix:

$$\mathbf{AB} = \begin{pmatrix} 0 & 4 & 5 \\ -5 & -1 & 5 \end{pmatrix} \begin{pmatrix} -3 & 4 \\ -5 & -4 \\ 1 & -4 \end{pmatrix}$$

$$= \begin{pmatrix} 0 \cdot -3 + 4 \cdot -5 + 5 \cdot 1 & 0 \cdot 4 + 4 \cdot -4 + 5 \cdot -4 \\ -5 \cdot -3 + -1 \cdot -5 + 5 \cdot 1 & -5 \cdot 4 + -1 \cdot -4 + 5 \cdot -4 \end{pmatrix} = \begin{pmatrix} -15 & -36 \\ 25 & -36 \end{pmatrix}$$

As in the previous example, we can use **evalm** to compute the matrix product. First, we define **A** and **B** to be

$$\mathbf{A} = \begin{pmatrix} 0 & 4 & 5 \\ -5 & -1 & 5 \end{pmatrix}$$

and

$$\mathbf{B} = \begin{pmatrix} -3 & 4 \\ -5 & -4 \\ 1 & -4 \end{pmatrix}$$

with **array**

```
> A:=array([[0,4,5],[-5,-1,5]]):
  B:=array([[-3,4],[-5,-4],[1,-4]]):
```

and then use **evalm** together with **&***, which represents noncommutative multiplication, to compute the matrix product **AB**.

```
> evalm(A &* B);
```

$$\begin{bmatrix} -15 & -36 \\ 25 & -36 \end{bmatrix}$$

DEFINITION | *Identity Matrix*

The $n \times n$ matrix

$$\begin{pmatrix} 1 & 0 & 0 & 0 \\ 0 & 1 & 0 & 0 \\ \vdots & \vdots & \ddots & \vdots \\ 0 & 0 & 0 & 1 \end{pmatrix}$$

is called the $n \times n$ **identity matrix**, denoted by $\mathbf{I}$ or $\mathbf{I}_n$.

We can define the $n \times n$ identity matrix, `Id`, using `alias` and `&*` by entering

```
> alias(Id=&*());
```

If $\mathbf{A}$ is an $n \times n$ matrix, then $\mathbf{IA} = \mathbf{AI} = \mathbf{A}$. More generally, an $n \times n$ matrix is called a **square matrix**.

☰ Determinants and Inverses

DEFINITION | *Determinant*

If $\mathbf{A} = (a_{11})$, the determinant of $\mathbf{A}$, denoted by $\det(\mathbf{A})$ or $|\mathbf{A}|$, is $\det(\mathbf{A}) = a_{11}$; if $\mathbf{A} = \begin{pmatrix} a_{11} & a_{12} \\ a_{21} & a_{22} \end{pmatrix}$, then

$$\det(\mathbf{A}) = \begin{vmatrix} a_{11} & a_{12} \\ a_{21} & a_{22} \end{vmatrix} = a_{11}a_{22} - a_{12}a_{21}.$$

More generally, if

$$\mathbf{A} = \begin{pmatrix} a_{11} & a_{12} & \cdots & a_{1n} \\ a_{21} & a_{22} & \cdots & a_{2n} \\ \vdots & \vdots & \ddots & \vdots \\ a_{n1} & a_{n2} & \cdots & a_{nn} \end{pmatrix}$$

(continued on p. 396)

(continued)

is an $n \times n$ matrix and $\mathbf{A}_{ij}$ is the $(n-1) \times (n-1)$ matrix obtained by deleting the ith row and jth column from $\mathbf{A}$, then

$$\det(\mathbf{A}) = \begin{vmatrix} a_{11} & a_{12} & \cdots & a_{1n} \\ a_{21} & a_{22} & \cdots & a_{2n} \\ \vdots & \vdots & \ddots & \vdots \\ a_{n1} & a_{n2} & \cdots & a_{nn} \end{vmatrix} = \sum_{j=1}^{n} (-1)^{i+j} a_{ij} \det(\mathbf{A}_{ij}) = \sum_{j=1}^{n} (-1)^{i+j} a_{ij} |\mathbf{A}_{ij}|.$$

The number $(-1)^{i+j} a_{ij} \det(\mathbf{A}_{ij}) = (-1)^{i+j} a_{ij} |\mathbf{A}_{ij}|$ is called the **cofactor** of a_{ij}. The **cofactor matrix**, $\mathbf{A}^c$, of $\mathbf{A}$ is the matrix obtained by replacing each element of $\mathbf{A}$ by its cofactor. Hence,

$$\mathbf{A}^c = \begin{pmatrix} |\mathbf{A}_{11}| & -|\mathbf{A}_{12}| & \cdots & (-1)^{n+1}|\mathbf{A}_{1n}| \\ -|\mathbf{A}_{21}| & |\mathbf{A}_{22}| & \cdots & (-1)^{n}|\mathbf{A}_{2n}| \\ \vdots & \vdots & \ddots & \vdots \\ (-1)^{n+1}|\mathbf{A}_{n1}| & (-1)^{n}|\mathbf{A}_{n2}| & \cdots & |\mathbf{A}_{nn}| \end{pmatrix}.$$

EXAMPLE: Calculate $|\mathbf{A}|$ if

$$\mathbf{A} = \begin{pmatrix} -4 & -2 & -1 \\ 5 & -4 & -3 \\ 5 & 1 & -2 \end{pmatrix}.$$

SOLUTION:

$$|\mathbf{A}| = \begin{vmatrix} -4 & -2 & -1 \\ 5 & -4 & -3 \\ 5 & 1 & -2 \end{vmatrix} = (-4) \begin{vmatrix} -4 & -3 \\ 1 & -2 \end{vmatrix} + (-1)^3(-2) \begin{vmatrix} 5 & -3 \\ 5 & -2 \end{vmatrix} + (-1) \begin{vmatrix} 5 & -4 \\ 5 & 1 \end{vmatrix}$$
$$= -4((-4)(-2) - (-3)(1)) + 2((5)(-2) - (-3)(5)) - ((5)(1) - (-4)(5)) = -59.$$

Determinants are computed using the **det** command, which is contained in the **linalg** package. We now load the **linalg** package. Note that the commands contained in the package are displayed; if a colon had been included instead of a semicolon, this output would be suppressed.

```
> with(linalg);
```

[*BlockDiagonal, GramSchmidt, JordanBlock, Wronskian, add, addcol,*
    addrow, adj, adjoint, angle, augment, backsub, band, basis, bezout,
    blockmatrix, charmat, charpoly, col, coldim, colspace, colspan,
    companion, concat, cond, copyinto, crossprod, curl, definite, delcols,
    delrows, det, diag, diverge, dotprod, eigenvals, eigenvects,
    entermatrix, equal, exponential, extend, ffgausselim, fibonacci,
    frobenius, gausselim, gaussjord, genmatrix, grad, hadamard,
    hermite, hessian, hilbert, htranspose, ihermite, indexfunc, innerprod,
    intbasis, inverse, ismith, iszero, jacobian, jordan, kernel, laplacian,
    leastsqrs, linsolve, matrix, minor, minpoly, mulcol, mulrow, multiply,
    norm, normalize, nullspace, orthog, permanent, pivot, potential,
    randmatrix, randvector, range, rank, ratform, row, rowdim,
    rowspace, rowspan, rref, scalarmul, singularvals, smith, stack,
    submatrix, subvector, sumbasis, swapcol, swaprow, sylvester,
    toeplitz, trace, transpose, vandermonde, vecpotent, vectdim, vector]

Then we define **A** and compute the determinant of **A** with `det`.

```
> A:=array([[-4,-2,-1],[5,-4,-3],[5,1,-2]]):
  det(A);
```

$$-59$$

DEFINITION

Adjoint and Inverse

B is an **inverse** of the $n \times n$ matrix **A** means that $\mathbf{AB} = \mathbf{BA} = \mathbf{I}$. The **adjoint**, $\mathbf{A}^a$, of an $n \times n$ matrix **A** is the transpose of the cofactor matrix: $\mathbf{A}^a = (\mathbf{A}^c)^T$. If $|\mathbf{A}| \neq 0$ and $\mathbf{B} = \dfrac{1}{|\mathbf{A}|}\mathbf{A}^a$, then $\mathbf{AB} = \mathbf{BA} = \mathbf{I}$. Therefore, if $|\mathbf{A}| \neq 0$, the inverse of **A** is given by

$$\mathbf{A}^{-1} = \frac{1}{|\mathbf{A}|}\mathbf{A}^a.$$

EXAMPLE: Find A^{-1} if $A = \begin{pmatrix} 2 & -1 \\ -3 & 1 \end{pmatrix}$.

SOLUTION: In this case, $|A| = \begin{vmatrix} 2 & -1 \\ -3 & 1 \end{vmatrix} = 2 - 3 = -1 \neq 0$, so A^{-1} exists. Moreover, $A^c = \begin{pmatrix} 1 & 3 \\ 1 & 2 \end{pmatrix}$, so $A^a = \begin{pmatrix} 1 & 1 \\ 3 & 2 \end{pmatrix}$ and $A^{-1} = \frac{1}{|A|} A^a = \begin{pmatrix} -1 & -1 \\ -3 & -2 \end{pmatrix}$. The commands `adjoint` and `inverse`, both of which are contained in the `linalg` package, can be used to compute the adjoint and inverse of an $n \times n$ matrix. For example, entering

```
> with(linalg):
  A:=array([[2,-1],[-3,1]]):
  adjoint(A);
  inverse(A);
```

$$\begin{bmatrix} 1 & 1 \\ 3 & 2 \end{bmatrix}$$

$$\begin{bmatrix} -1 & -1 \\ -3 & -2 \end{bmatrix}$$

loads the `linalg` package, defines $A = \begin{pmatrix} 2 & -1 \\ -3 & 1 \end{pmatrix}$, computes the adjoint of A, and then computes the inverse of A.

☰ Eigenvalues and Eigenvectors

DEFINITION

Eigenvalues and Eigenvectors

A nonzero vector **x** is an **eigenvector** of the square matrix **A** means there is a number λ, called an **eigenvalue** of **A**, so that

$$Ax = \lambda x.$$

EXAMPLE: Show that $\begin{pmatrix} -1 \\ 2 \end{pmatrix}$ and $\begin{pmatrix} 1 \\ 1 \end{pmatrix}$ are eigenvectors of $\begin{pmatrix} -1 & 2 \\ 4 & -3 \end{pmatrix}$ with corresponding eigenvalues -5 and 1, respectively.

SOLUTION: Since $\begin{pmatrix} -1 & 2 \\ 4 & -3 \end{pmatrix}\begin{pmatrix} -1 \\ 2 \end{pmatrix} = \begin{pmatrix} 5 \\ -10 \end{pmatrix} = -5\begin{pmatrix} -1 \\ 2 \end{pmatrix}$ and $\begin{pmatrix} -1 & 2 \\ 4 & -3 \end{pmatrix}\begin{pmatrix} 1 \\ 1 \end{pmatrix} = \begin{pmatrix} 1 \\ 1 \end{pmatrix} = 1\begin{pmatrix} 1 \\ 1 \end{pmatrix}$, $\begin{pmatrix} -1 \\ 2 \end{pmatrix}$ and $\begin{pmatrix} 1 \\ 1 \end{pmatrix}$ are eigenvectors of $\begin{pmatrix} -1 & 2 \\ 4 & -3 \end{pmatrix}$ with corresponding eigenvalues -5 and 1, respectively.

If **x** is an eigenvector of **A** with corresponding eigenvalue λ, then $\mathbf{Ax} = \lambda\mathbf{x}$, which is equivalent to the equation $(\mathbf{A} - \lambda\mathbf{I})\mathbf{x} = 0$. Note that λ is an eigenvalue if and only if $\det(\mathbf{A} - \lambda\mathbf{I}) = 0$.

DEFINITION

> *Characteristic Polynomial*
>
> The equation $\det(\mathbf{A} - \lambda\mathbf{I}) = 0$ is called the **characteristic equation** of **A**; $\det(\mathbf{A} - \lambda\mathbf{I})$ is called the **characteristic polynomial** of **A**; the roots of the characteristic polynomial of **A** are the eigenvalues of **A**.

Generally, to find the eigenvectors and corresponding eigenvalues of a square matrix **A**, we will begin by computing the eigenvalues.

EXAMPLE: Calculate the eigenvalues and corresponding eigenvectors of $\mathbf{A} = \begin{pmatrix} 4 & -6 \\ 3 & -7 \end{pmatrix}$.

SOLUTION: The characteristic polynomial of $\mathbf{A} = \begin{pmatrix} 4 & -6 \\ 3 & -7 \end{pmatrix}$ is

$$\begin{vmatrix} 4 - \lambda & -6 \\ 3 & -7 - \lambda \end{vmatrix} = \lambda^2 + 3\lambda - 10 = (\lambda + 5)(\lambda - 2).$$

Alternatively, we can compute the characteristic polynomial with `charpoly`, which is contained in the `linalg` package.

```
> with(linalg):
  A:=array([[4,-6],[3,-7]]):
  char_A:=charpoly(A,lambda);
```

$char\_A := \lambda^2 + 3\lambda - 10$

Since the eigenvalues are found by solving $(\lambda + 5)(\lambda - 2) = 0$,

```
> solve(char_A=0);
```

$2, -5$

the eigenvalues are $\lambda = -5$ and $\lambda = 2$. Let $\begin{pmatrix} x_1 \\ y_1 \end{pmatrix}$ denote the eigenvector corresponding to the eigenvalue $\lambda = -5$. Then

$$\left\{ \begin{pmatrix} 4 & -6 \\ 3 & -7 \end{pmatrix} - (-5) \begin{pmatrix} 1 & 0 \\ 0 & 1 \end{pmatrix} \right\} \begin{pmatrix} x_1 \\ y_1 \end{pmatrix} = 0.$$

Simplifying yields the system of equations $\begin{cases} 9x_1 - 6y_1 = 0 \\ 3x_1 - 2y_1 = 0 \end{cases}$, so $y_1 = \dfrac{3}{2} x_1$.

We perform the same steps with Maple. After using **alias** to define **Id** to be the identity matrix, we use **evalm** to compute $\left\{ \begin{pmatrix} 4 & -6 \\ 3 & -7 \end{pmatrix} - (-5) \begin{pmatrix} 1 & 0 \\ 0 & 1 \end{pmatrix} \right\} \begin{pmatrix} x_1 \\ y_1 \end{pmatrix}$ and name the result **step_1**.

```
> alias(Id=&*()):
  step_1:=evalm((A+5*Id) &* [[x1],[y1]]);
```

$$step\_1 := \begin{bmatrix} 9x1 - 6y1 \\ 3x1 - 2y1 \end{bmatrix}$$

To form the system of equations $\begin{cases} 9x_1 - 6y_1 = 0 \\ 3x_1 - 2y_1 = 0 \end{cases}$, we use **equate**, which is contained in the **student** package. Once the **student** package has been loaded, **equate(A,B)** returns the system of equations obtained by equating the entries of **A** and **B** where **A** and **B** are vectors, matrices, or lists (note that they must have compatible dimensions); **equate(A)** returns the system of equations obtained by setting all the entries of **A** equal to zero.

```
> with(student);
```

$[D, Doubleint, Int, Limit, Sum, Tripleint, changevar, combine,$
$\quad completesquare, distance, equate, extrema, integrand, intercept,$
$\quad intparts, isolate, leftbox, leftsum, makeproc, maximize, middlebox,$
$\quad middlesum, midpoint, minimize, powsubs, rightbox, rightsum,$
$\quad showtangent, simpson, slope, trapezoid, value]$

The resulting system of equations is named **eqs_1**, and it is then solved with **solve**.

```
> eqs_1:=equate(step_1);
```

$eqs\_1 := \{3x1 - 2y1 = 0, 9x1 - 6y1 = 0\}$

```
> solve(eqs_1);
```

$$\left\{ x1 = x1, y1 = \frac{3}{2}x1 \right\}$$

Therefore, if x_1 is any real number, then $\begin{pmatrix} x_1 \\ \frac{3}{2}x_1 \end{pmatrix}$ is an eigenvector. In particular, if $x_1 = 2$, then $\begin{pmatrix} 2 \\ 3 \end{pmatrix}$

is an eigenvector of $\mathbf{A} = \begin{pmatrix} 4 & -6 \\ 3 & -7 \end{pmatrix}$ with corresponding eigenvalue $\lambda = -5$. Similarly, if we let

$\begin{pmatrix} x_2 \\ y_2 \end{pmatrix}$ denote the eigenvector corresponding to $\lambda = 2$, then $\left\{ \begin{pmatrix} 4 & -6 \\ 3 & -7 \end{pmatrix} - 2 \begin{pmatrix} 1 & 0 \\ 0 & 1 \end{pmatrix} \right\} \begin{pmatrix} x_2 \\ y_2 \end{pmatrix} = 0,$

which yields the system $\begin{cases} 2x_2 - 6y_2 = 0 \\ 3x_2 - 9y_2 = 0 \end{cases},$

```
> step_1:=evalm((A-2*Id)&*[[x2],[y2]]):
  eqs_2:=equate(step_1);
```

$$eqs\_2 := \{2x2 - 6y2 = 0, 3x2 - 9y2 = 0\}$$

```
> solve(eqs_2);
```

$$\{x2 = 3y2, y2 = y2\}$$

Thus, $y_2 = \frac{1}{3}x_2$. If $x_2 = 3$, then $\begin{pmatrix} 3 \\ 1 \end{pmatrix}$ is an eigenvector of $\mathbf{A} = \begin{pmatrix} 4 & -6 \\ 3 & -7 \end{pmatrix}$ with corresponding

eigenvalue $\lambda = 2$.

Note that the commands **eigenvals** and **eigenvects**, both contained in the **linalg**
package, can be used to compute the exact value of the eigenvalues and corresponding eigenvectors, when possible. Thus, entering

```
> eigenvals(A);
```

$$2, -5$$

returns the eigenvalues of $\mathbf{A}$, and entering

```
> eigenvects(A);
```

$$\left[-5, 1, \left\{ \left[1, \frac{3}{2} \right] \right\}, [2, 1, \{[3, 1]\}] \right]$$

returns the eigenvalues and corresponding eigenvectors.

EXAMPLE: Find the eigenvalues and corresponding eigenvectors of $\mathbf{A} = \begin{pmatrix} 0 & 1 \\ -1 & 0 \end{pmatrix}$.

SOLUTION: In this case, the characteristic polynomial is $\begin{vmatrix} -\lambda & 1 \\ -1 & -\lambda \end{vmatrix} = \lambda^2 + 1$, so the eigenvalues are the roots of the equation $\lambda^2 + 1 = 0$. These are the imaginary numbers $\lambda = i$ and $\lambda = -i$, where $i = \sqrt{-1}$. The corresponding eigenvectors are found by substituting the eigenvalues into the equation $(\mathbf{A} - \lambda\mathbf{I})\mathbf{x} = 0$ and solving for $\mathbf{x}$. For $\lambda = i$, this equation is $\begin{pmatrix} -i & 1 \\ -1 & -i \end{pmatrix} \begin{pmatrix} x_1 \\ y_1 \end{pmatrix} = \begin{pmatrix} 0 \\ 0 \end{pmatrix}$, which is equivalent to the system $\begin{cases} -ix_1 + y_1 = 0 \\ -x_1 - iy_1 = 0 \end{cases}$. Notice that the second equation of this system is a constant multiple of the first equation. Hence, an eigenvector $\begin{pmatrix} x_1 \\ y_1 \end{pmatrix}$ must satisfy $y_1 = ix_1$. Therefore, $\begin{pmatrix} x_1 \\ ix_1 \end{pmatrix} = \begin{pmatrix} 1 \\ i \end{pmatrix} x_1$ is an eigenvector for any value of x_1. For example, if $x_1 = 1$, then $\begin{pmatrix} 1 \\ i \end{pmatrix}$ is an eigenvector. For $\lambda = -i$, the equation is $\begin{pmatrix} i & 1 \\ -1 & i \end{pmatrix} \begin{pmatrix} x_2 \\ y_2 \end{pmatrix} = \begin{pmatrix} 0 \\ 0 \end{pmatrix}$, which is equivalent to $\begin{cases} ix_2 + y_2 = 0 \\ -x_2 + iy_2 = 0 \end{cases}$. Because the second equation equals i times the first equation, the eigenvector $\begin{pmatrix} x_2 \\ y_2 \end{pmatrix}$ must satisfy $y_2 = -ix_2$. Hence, $\begin{pmatrix} x_2 \\ -ix_2 \end{pmatrix} = \begin{pmatrix} 1 \\ -i \end{pmatrix} x_2$ is an eigenvector for any value of x_2. Therefore, if $x_2 = 1$, then $\begin{pmatrix} 1 \\ -i \end{pmatrix}$ is an eigenvector. Equivalent results are obtained using `eigenvals` and `eigenvects`. After loading the `linalg` package and defining **A**, we use `eigenvals` to compute the eigenvalues of **A** and `eigenvects` to compute the eigenvalues and corresponding eigenvectors of **A**. Note that the results of the `eigenvects` command are given in terms of `RootOf`.

```
> with(linalg):
  A:=array([[0,1],[-1,0]]):
  eigenvals(A);
  step_1:=eigenvects(A);
  I, -I
```

$$step\_1 := [\text{RootOf}(\_Z^2 + 1), 1, \{1\,[\text{RootOf}(\_Z^2 + 1)]\}]$$

To compute the exact value of the eigenvalues and corresponding eigenvectors we use **map** and `allvalues`.

```
> map(allvalues,step_1);
```

$$[I, -I, 1, \{[1\ I]\}, \{[1 - I]\}]$$

Note: Recall that the complex conjugate of the complex number $z = a + bi$ is $\bar{z} = a - bi$. Similarly, the complex conjugate of the vector

$$\mathbf{x} = \begin{pmatrix} a_1 + b_1 i \\ a_2 + b_2 i \\ \vdots \\ a_n + b_n i \end{pmatrix}$$

is the vector

$$\bar{\mathbf{x}} = \begin{pmatrix} a_1 - b_1 i \\ a_2 - b_2 i \\ \vdots \\ a_n - b_n i \end{pmatrix}.$$

Notice that the eigenvectors corresponding to the complex conjugate eigenvalues $\lambda = i$ and $\lambda = -i$ in the previous example are $\begin{pmatrix} 1 \\ i \end{pmatrix}$ and $\begin{pmatrix} 1 \\ -i \end{pmatrix}$, which are complex conjugates. We will see that the eigenvectors that correspond to complex conjugate eigenvalues are themselves complex conjugates.

DEFINITION

> *Eigenvalue of Multiplicity m*
>
> Suppose that $(\lambda - \lambda_1)^m$, where m is a positive integer, is a factor of the characteristic polynomial of the $n \times n$ matrix $\mathbf{A}$, while $(\lambda - \lambda_1)^{m+1}$ is not a factor of this polynomial. Then $\lambda = \lambda_1$ is an **eigenvalue of multiplicity m**.

We often say that the eigenvalue of an $n \times n$ matrix $\mathbf{A}$ is repeated if it is of multiplicity m, where $m \geq 2$ and $m \leq n$. When trying to find the eigenvector(s) corresponding to an eigenvalue of multiplicity m, two situations may be encountered: (1) m linearly independent eigenvectors can be found that correspond to λ or (2) only one eigenvector can be found that corresponds to λ.

EXAMPLE: Find the eigenvector(s) corresponding to (a) the eigenvalue $\lambda = -2$ of

$$A = \begin{pmatrix} 2 & -2 & -2 \\ 4 & -4 & -2 \\ -2 & 1 & -1 \end{pmatrix}$$

and (b) the eigenvalue $\lambda = 5$ of

$$B = \begin{pmatrix} 5 & -4 & 0 \\ 1 & 0 & 2 \\ 0 & 2 & 5 \end{pmatrix}.$$

SOLUTION: (a) After loading the `linalg` and `student` packages and defining **A**, we see that the characteristic polynomial of **A** is

```
> with(linalg):
  with(student):
  A:=array([[2,-2,-2],[4,-4,-2],[-2,1,-1]]):
  char_A:=charpoly(A,lambda);
```

$char\_A := \lambda^3 + 3\lambda^2 - 4$

and then we find the eigenvalues with `solve`.

```
> solve(char_A=0);
```

$1, -2, -2$

Hence, $\lambda = -2$ is an eigenvalue of multiplicity 2. To find the eigenvector(s)

$$v = \begin{pmatrix} x_1 \\ y_1 \\ z_1 \end{pmatrix}$$

corresponding to $\lambda = -2$, we see that we must solve the equation

$$\left\{ \begin{pmatrix} 2 & -2 & -2 \\ 4 & -4 & -2 \\ -2 & 1 & -1 \end{pmatrix} - (-2) \begin{pmatrix} 1 & 0 & 0 \\ 0 & 1 & 0 \\ 0 & 0 & 1 \end{pmatrix} \right\} \begin{pmatrix} x_1 \\ y_1 \\ z_1 \end{pmatrix} = 0.$$

```
> alias(Id=&*()):
  step_1:=evalm((A+2*Id)*[[x1],[y1],[z1]]);
```

$$step\_1 := \begin{bmatrix} 4x1 - 2y1 - 2z1 \\ 4x1 - 2y1 - 2z1 \\ -2x1 + y1 + z1 \end{bmatrix}$$

```
> eqs:=equate(step_1);
```

$$eqs := \{4x1 - 2y1 - 2z1 = 0, -2x1 + y1 + z1 = 0\}$$

```
> solve(eqs);
```

$$\{y1 = 2x1 - z1, x1 = x1, z1 = z1\}$$

This indicates that $y_1 = 2x_1 - z_1$. Thus, if $z_1 = 0$, then $y_1 = 2x_1$. Hence,

$$\mathbf{v}_1 = \begin{pmatrix} x_1 \\ 2x_1 \\ 0 \end{pmatrix} = \begin{pmatrix} 1 \\ 2 \\ 0 \end{pmatrix} x_1$$

is one eigenvalue. On the other hand, if in $y_1 = 2x_1 - z_1$, $x_1 = 0$, then $y_1 = -z_1$. Therefore,

$$\mathbf{v}_2 = \begin{pmatrix} 0 \\ -z_1 \\ z_1 \end{pmatrix} = \begin{pmatrix} 0 \\ -1 \\ 1 \end{pmatrix} z_1$$

is another eigenvector corresponding to $\lambda = -2$ that is not a constant multiple of $\mathbf{v}_1$. Hence, we have found two linearly independent eigenvectors that correspond to the eigenvalue $\lambda = -2$ of multiplicity 2.

Note that the result of using **eigenvects** indicates that $\lambda = -2$ is an eigenvalue of multiplicity 2; Maple returns two linearly independent eigenvectors corresponding to the eigenvalue $\lambda = -2$.

```
> eigenvects(A);
```

$$[-2, 2, \{[1 \quad 2 \quad 0], [0 \quad -1 \quad 1]\}], \left[1, 1, \left\{\left[1 \quad 1 \quad -\frac{1}{2}\right]\right\}\right]$$

(b) In the same manner as in (a), the eigenvalues of **B** are determined by finding the zeros of the characteristic polynomial.

```
> B:=array([[5,-4,0],[1,0,2],[0,2,5]]):
  char_B:=charpoly(B,lambda);
  solve(char_B=0);
```

$$char\_B := \lambda^3 - 10\lambda^2 + 25\lambda$$

$$0, 5, 5$$

Hence, $\lambda = 5$ is an eigenvalue of multiplicity 2. In this case, when we find the corresponding eigenvector

$$\mathbf{v} = \begin{pmatrix} x_1 \\ y_1 \\ z_1 \end{pmatrix},$$

we solve

$$\left\{ \begin{pmatrix} 5 & -4 & 0 \\ 1 & 0 & 2 \\ 0 & 2 & 5 \end{pmatrix} - 5 \begin{pmatrix} 1 & 0 & 0 \\ 0 & 1 & 0 \\ 0 & 0 & 1 \end{pmatrix} \right\} \begin{pmatrix} x_1 \\ y_1 \\ z_1 \end{pmatrix} = 0.$$

```
> step_1:=evalm((B-5*Id)&*[[x1],[y1],[z1]]):
  eqs:=equate(step_1):
  solve(eqs);
```

$$\{z1 = z1, x1 = -2z1, y1 = 0\}$$

Thus, the components of $\mathbf{v}$ must satisfy $x_1 = -2z_1$ and $y_1 = 0$. Hence, we find only one eigenvector

$$\mathbf{v} = \begin{pmatrix} -2z_1 \\ 0 \\ z_1 \end{pmatrix} = \begin{pmatrix} -2 \\ 0 \\ 1 \end{pmatrix} z_1$$

that corresponds to the eigenvalue $\lambda = 5$ of multiplicity 2. Equivalent results are returned by `eigenvects`.

```
> eigenvects(B);
```

$$[5, 2, \{[-2 \quad 0 \quad 1]\}], \left[0, 1, \left\{\left[-2 \quad -\frac{5}{2} \quad 1\right]\right\}\right]$$

☰ *Matrix Calculus*

DEFINITION | *Derivative and Integral of a Matrix*

The **derivative** of the $n \times m$ matrix

$$\mathbf{A}(t) = \begin{pmatrix} a_{11}(t) & a_{12}(t) & \cdots & a_{1m}(t) \\ a_{21}(t) & a_{22}(t) & \cdots & a_{2m}(t) \\ \vdots & \vdots & \ddots & \vdots \\ a_{n1}(t) & a_{n2}(t) & \cdots & a_{nm}(t) \end{pmatrix},$$

where $a_{ij}(t)$ is differentiable for all values of i and j, is

$$\frac{d}{dt}\mathbf{A}(t) = \begin{pmatrix} \dfrac{d}{dt}a_{11}(t) & \dfrac{d}{dt}a_{12}(t) & \cdots & \dfrac{d}{dt}a_{1m}(t) \\ \dfrac{d}{dt}a_{21}(t) & \dfrac{d}{dt}a_{22}(t) & \cdots & \dfrac{d}{dt}a_{2m}(t) \\ \vdots & \vdots & \ddots & \vdots \\ \dfrac{d}{dt}a_{n1}(t) & \dfrac{d}{dt}a_{n2}(t) & \cdots & \dfrac{d}{dt}a_{nm}(t) \end{pmatrix}.$$

The **integral** of $\mathbf{A}(t)$, where $a_{ij}(t)$ is integrable for all values of i and j, is

$$\int \mathbf{A}(t)\,dt = \begin{pmatrix} \displaystyle\int a_{11}(t)\,dt & \displaystyle\int a_{12}(t)\,dt & \cdots & \displaystyle\int a_{1m}(t)\,dt \\ \displaystyle\int a_{21}(t)\,dt & \displaystyle\int a_{22}(t)\,dt & \cdots & \displaystyle\int a_{2m}(t)\,dt \\ \vdots & \vdots & \ddots & \vdots \\ \displaystyle\int a_{n1}(t)\,dt & \displaystyle\int a_{n2}(t)\,dt & \cdots & \displaystyle\int a_{nm}(t)\,dt \end{pmatrix}.$$

EXAMPLE: Find $\dfrac{d}{dt}A(t)$ and $\displaystyle\int A(t)\,dt$ if $A(t) = \begin{pmatrix} \cos 3t & \sin 3t & e^{-t} \\ t & t\sin t^2 & e^t \end{pmatrix}$.

SOLUTION: We find $\dfrac{d}{dt}A(t)$ by differentiating each element of $\mathbf{A}(t)$. This yields

$$\frac{d}{dt}A(t) = \begin{pmatrix} -3\sin 3t & 3\cos 3t & -e^{-t} \\ 1 & \sin t^2 + 2t^2\cos t^2 & e^t \end{pmatrix}.$$

Similarly, we find $\int \mathbf{A}(t)\, dt$ by integrating each element of $\mathbf{A}(t)$. Hence,

$$\int \mathbf{A}(t)\, dt = \begin{pmatrix} \dfrac{1}{3}\sin(3t) + c_{11} & \dfrac{-1}{3}\cos(3t) + c_{12} & -e^{-t} + c_{13} \\[2mm] \dfrac{1}{2}t^2 + c_{21} & \dfrac{-1}{2}\cos(t^2) + c_{22} & e^t + c_{23} \end{pmatrix},$$

where each c_{ij} represents an arbitrary constant.

After defining $\mathbf{A}$ with `array`,

```
> A:=array([[cos(3*t),sin(3*t),exp(-t)],[t,t*sin(t^2),
  exp(t)]])
```

$$A := \begin{bmatrix} \cos(3t) & \sin(3t) & e^{-t} \\ t & t\sin(t^2) & e^t \end{bmatrix}$$

we can use `map` and `diff` to differentiate each entry of $\mathbf{A}$ with respect to t

```
> map(diff,A,t);
```

$$\begin{bmatrix} -3\sin(3t) & 3\cos(3t) & -e^{-t} \\ 1 & \sin(t^2) + 2t^2\cos(t^2) & e^t \end{bmatrix}$$

and use `map` and `int` to differentiate each entry of $\mathbf{A}$ with respect to t.

```
> map(int,A,t);
```

$$\begin{bmatrix} \dfrac{1}{3}\sin(3t) & -\dfrac{1}{3}\cos(3t) & -e^{-t} \\[2mm] \dfrac{1}{2}t^2 & -\dfrac{1}{2}\cos(t^2) & e^t \end{bmatrix}$$

9.3 PRELIMINARY DEFINITIONS AND NOTATION
• •

We begin our study of systems of ordinary differential equations by introducing several definitions along with some convenient notation.

Let

$$\mathbf{X} = \mathbf{X}(t) = \begin{pmatrix} x_1(t) \\ x_2(t) \\ \vdots \\ x_n(t) \end{pmatrix},$$

$$\mathbf{A}(t) = \begin{pmatrix} a_{11}(t) & a_{12}(t) & \cdots & a_{1n}(t) \\ a_{21}(t) & a_{22}(t) & \cdots & a_{2n}(t) \\ \vdots & \vdots & \ddots & \vdots \\ a_{n1}(t) & a_{n2}(t) & \cdots & a_{nn}(t) \end{pmatrix},$$

and

$$\mathbf{F}(t) = \begin{pmatrix} f_1(t) \\ f_2(t) \\ \vdots \\ f_n(t) \end{pmatrix}.$$

Then the homogeneous system of first-order linear differential equations

$$\begin{cases} x_1'(t) = a_{11}(t)x_1(t) + a_{12}(t)x_2(t) + \cdots + a_{1n}(t)x_n(t) \\ x_2'(t) = a_{21}(t)x_1(t) + a_{22}(t)x_2(t) + \cdots + a_{2n}(t)x_n(t) \\ \qquad\qquad\qquad\qquad \vdots \\ x_n'(t) = a_{n1}(t)x_1(t) + a_{n2}(t)x_2(t) + \cdots + a_{nn}(t)x_n(t) \end{cases}$$

is equivalent to $\mathbf{X}'(t) = \mathbf{A}(t)\mathbf{X}(t)$ and the nonhomogeneous system

$$\begin{cases} x_1'(t) = a_{11}(t)x_1(t) + a_{12}(t)x_2(t) + \cdots + a_{1n}(t)x_n(t) + f_1(t) \\ x_2'(t) = a_{21}(t)x_1(t) + a_{22}(t)x_2(t) + \cdots + a_{2n}(t)x_n(t) + f_2(t) \\ \qquad\qquad\qquad\qquad \vdots \\ x_n'(t) = a_{n1}(t)x_1(t) + a_{n2}(t)x_2(t) + \cdots + a_{nn}(t)x_n(t) + f_n(t) \end{cases}$$

is equivalent to $\mathbf{X}'(t) = \mathbf{A}(t)\mathbf{X}(t) + \mathbf{F}(t)$.

EXAMPLE: (a) Write the homogeneous system $\begin{cases} x' = -5x + 5y \\ y' = -5x + y \end{cases}$ in matrix form. (b) Write the nonhomogeneous system $\begin{cases} x' = x + 2y - \sin t \\ y' = 4x - 3y + t^2 \end{cases}$ in matrix form.

SOLUTION: (a) The homogeneous system $\begin{cases} x' = -5x + 5y \\ y' = -5x + y \end{cases}$ is equivalent to the system $\begin{pmatrix} x' \\ y \end{pmatrix} = \begin{pmatrix} -5 & 5 \\ -5 & 1 \end{pmatrix} \begin{pmatrix} x \\ y \end{pmatrix}$. (b) The nonhomogeneous system $\begin{cases} x' = x + 2y - \sin t \\ y' = 4x - 3y + t^2 \end{cases}$ is equivalent to $\begin{pmatrix} x' \\ y' \end{pmatrix} = \begin{pmatrix} 1 & 2 \\ 4 & -3 \end{pmatrix} \begin{pmatrix} x \\ y \end{pmatrix} + \begin{pmatrix} -\sin t \\ t^2 \end{pmatrix}$.

The higher-order equations that we solved earlier in the text can be written as a system of first-order equations. Consider the nth-order differential equation with constant coefficients

$$y^{(n)}(t) + a_{n-1}y^{(n-1)}(t) + \cdots + a_2 y''(t) + a_1 y'(t) + a_0 y(t) = f(t).$$

Let $x_1 = y, x_2 = \dfrac{dx_1}{dt} = y'(t), x_3 = \dfrac{dx_2}{dt} = y''(t), \ldots, x_{n-1} = \dfrac{dx_{n-2}}{dt} = y^{(n-2)}, x_n = \dfrac{dx_{n-1}}{dt} = y^{(n-1)}$. Then the equation $y^{(n)}(t) + a_{n-1}y^{(n-1)}(t) + \cdots + a_2 y''(t) + a_1 y'(t) + a_0 y(t) = f(t)$ is equivalent to the system

$$\begin{cases} x_1' = x_2 \\ x_2' = x_3 \\ \quad\vdots \\ x_{n-1}' = x_n \\ x_n' = -a_{n-1}y^{(n-1)} - \cdots - a_2 y''(t) - a_1 y'(t) - a_0 y + f(t) \\ \quad\; = -a_{n-1}x_n - \cdots - a_2 x_3 - a_1 x_2 - a_0 x_1 + f(t) \end{cases},$$

which can be written in matrix form as

$$\begin{pmatrix} x_1' \\ x_2' \\ \vdots \\ x_{n-1}' \\ x_n' \end{pmatrix} = \begin{pmatrix} 0 & 1 & 0 & 0 & 0 \\ 0 & 0 & 1 & 0 & 0 \\ \vdots & \vdots & \vdots & \ddots & \vdots \\ 0 & 0 & 0 & 0 & 1 \\ -a_0 & -a_1 & -a_2 & \cdots & -a_n \end{pmatrix} \begin{pmatrix} x_1 \\ x_2 \\ \vdots \\ x_{n-1} \\ x_n \end{pmatrix} + \begin{pmatrix} 0 \\ 0 \\ 0 \\ 0 \\ f(t) \end{pmatrix},$$

EXAMPLE: Write the equation $y'' + 5y' + 6y = \cos t$ as a system.

SOLUTION: We begin by letting $x_1 = y$, $x_2 = x_1' = y'$, and $x_2' = y'' = \cos t - 6y - 5y' = \cos t - 6x_1 - 5x_2$. Hence, $\begin{cases} x_1' = x_2 \\ x_2' = \cos t - 6x_1 - 5x_2 \end{cases}$, and this can be written as $\begin{pmatrix} x_1' \\ x_2' \end{pmatrix} = \begin{pmatrix} 0 & 1 \\ -6 & -5 \end{pmatrix} + \begin{pmatrix} 0 \\ \cos t \end{pmatrix}$.

Nonlinear differential equations can be written as a system of equations as well, as we show in the following example.

EXAMPLE: The **Van der Pol equation** is the nonlinear ordinary differential equation

$$w'' - \mu(1 - w^2)w' + w = 0.$$

Write the Van der Pol equation as a system.

SOLUTION: Let $x_1 = w$ and $x_2 = w' = x_1'$. Then, $x_2' = w'' = \mu(1 - w^2)w' - w = \mu(1 - x_1^2)x_2 - x_1$, so the Van der Pol equation is equivalent to the nonlinear system $\begin{cases} x_1' = x_2 \\ x_2' = \mu(1 - x_1^2)x_2 - x_1 \end{cases}$.

At this point, given a system of ordinary differential equations, our goal will be to construct either an explicit or numerical solution of the system of equations.

We now state the following theorems and terminology that are used in establishing the fundamentals of solving systems of differential equations. All proofs are omitted. In each case, we assume that the matrix $\mathbf{A}(t)$ in the systems $\mathbf{X}'(t) = \mathbf{A}(t)\mathbf{X}(t) + \mathbf{F}(t)$ and $\mathbf{X}'(t) = \mathbf{A}(t)\mathbf{X}(t)$ is an $n \times n$ matrix.

DEFINITION | *Solution Vector*

A **solution vector** of the system $X'(t) = A(t)X(t) + F(t)$ on the interval I is a matrix of the form

$$X(t) = \begin{pmatrix} x_1(t) \\ x_2(t) \\ \vdots \\ x_n(t) \end{pmatrix},$$

where the $x_i(t)$ are differentiable functions that satisfy $X'(t) = A(t)X(t) + F(t)$ on I.

Let $X'(t) = A(t)X(t)$, where

$$X(t) = \begin{pmatrix} x_1(t) \\ x_2(t) \\ \vdots \\ x_n(t) \end{pmatrix},$$

and

$$A(t) = \begin{pmatrix} a_{11}(t) & a_{12}(t) & \cdots & a_{1n}(t) \\ a_{21}(t) & a_{22}(t) & \cdots & a_{2n}(t) \\ \vdots & \vdots & \ddots & \vdots \\ a_{n1}(t) & a_{n2}(t) & \cdots & a_{nn}(t) \end{pmatrix},$$

where $a_{ij}(t)$ is continuous for all $1 \le j \le n$ and $1 \le i \le n$. Let

$$\{\Phi_i\}_{i=1}^m = \left\{ \begin{pmatrix} \Phi_{1i} \\ \Phi_{2i} \\ \vdots \\ \Phi_{ni} \end{pmatrix} \right\}_{i=1}^m$$

be a set of m solutions of $X'(t) = A(t)X(t)$. We define linear dependence and independence of the set of vectors

$$\{\Phi_i\}_{i=1}^m = \left\{ \begin{pmatrix} \Phi_{1i} \\ \Phi_{2i} \\ \vdots \\ \Phi_{ni} \end{pmatrix} \right\}_{i=1}^m$$

in the same ways as we define linear dependence and independence of sets of functions. The set

$$\{\Phi_i\}_{i=1}^m = \left\{ \begin{pmatrix} \Phi_{1i} \\ \Phi_{2i} \\ \vdots \\ \Phi_{ni} \end{pmatrix} \right\}_{i=1}^m$$

is **linearly dependent** on an interval $\mathbf{I}$ means that there is a set of constants $\{c_i\}_{i=1}^m$ not all zero such that $\sum_{i=1}^m c_i \Phi_i = 0$; otherwise, the set is **linearly independent**.

DEFINITION

Fundamental Set of Solutions

Any set

$$\{\Phi_i\}_{i=1}^n = \left\{ \begin{pmatrix} \Phi_{1i} \\ \Phi_{2i} \\ \vdots \\ \Phi_{ni} \end{pmatrix} \right\}_{i=1}^n$$

of n linearly independent solution vectors of $\mathbf{X}'(t) = \mathbf{A}(t)\mathbf{X}(t)$ on an interval $\mathbf{I}$ is called a **fundamental set of solutions** on $\mathbf{I}$.

We can determine if a set of vectors is linearly independent or linearly dependent by computing the Wronskian.

. . . .

THEOREM   The set

$$\{\Phi_i\}_{i=1}^n = \left\{ \begin{pmatrix} \Phi_{1i} \\ \Phi_{2i} \\ \vdots \\ \Phi_{ni} \end{pmatrix} \right\}_{i=1}^n$$

is linearly independent if and only if the **Wronskian**

$$W(\Phi_1, \Phi_2, \ldots, \Phi_n) = \det \begin{pmatrix} \Phi_{11} & \Phi_{12} & \cdots & \Phi_{1n} \\ \Phi_{21} & \Phi_{22} & \cdots & \Phi_{2n} \\ \vdots & \vdots & \ddots & \vdots \\ \Phi_{n1} & \Phi_{n2} & \cdots & \Phi_{nn} \end{pmatrix} \neq 0.$$

. . . .

The following theorem implies that a fundamental set of solutions cannot contain more than n vectors because the solutions could not be linearly independent.

. . . .

THEOREM   Any $n + 1$ nontrivial solutions of $\mathbf{X}'(t) = \mathbf{A}(t)\mathbf{X}(t)$ are linearly dependent.

. . . .

Finally, we state the following theorems that indicate that a fundamental set of solutions can always be found and a general solution can be constructed.

. . . .

THEOREM   There is a set of n nontrivial linearly independent solutions of $\mathbf{X}'(t) = \mathbf{A}(t)\mathbf{X}(t)$.

. . . .

THEOREM

Let

$$\{\Phi_i\}_{i=1}^n = \left\{ \begin{pmatrix} \Phi_{1i} \\ \Phi_{2i} \\ \vdots \\ \Phi_{ni} \end{pmatrix} \right\}_{i=1}^n$$

be a set of n linearly independent solutions of $X'(t) = A(t)X(t)$. Then every solution of $X'(t) = A(t)X(t)$ is a linear combination of these solutions. Hence, a **general solution** of $X'(t) = A(t)X(t)$ is

$$X(t) = c_1\Phi_1(t) + c_2\Phi_2(t) + \cdots + c_n\Phi_n(t).$$

DEFINITION

Fundamental Matrix

Let

$$\{\Phi_i\}_{i=1}^n = \left\{ \begin{pmatrix} \Phi_{1i} \\ \Phi_{2i} \\ \vdots \\ \Phi_{ni} \end{pmatrix} \right\}_{i=1}^n$$

be a set of n linearly independent solutions of $X'(t) = A(t)X(t)$. Then

$$\Phi(t) = \begin{pmatrix} \Phi_1 & \Phi_2 & \cdots & \Phi_n \end{pmatrix} = \begin{pmatrix} \Phi_{11} & \Phi_{12} & \cdots & \Phi_{1n} \\ \Phi_{21} & \Phi_{22} & \cdots & \Phi_{2n} \\ \vdots & \vdots & \ddots & \vdots \\ \Phi_{n1} & \Phi_{n2} & \cdots & \Phi_{nn} \end{pmatrix}$$

is called a **fundamental matrix** of the system $X'(t) = A(t)X(t)$. Thus, a **general solution** can be written as $X(t) = \Phi(t)C$, where $C = \begin{pmatrix} c_1 \\ c_2 \\ \vdots \\ c_n \end{pmatrix}$.

Notice that the form of this general solution, as $X(t) = \Phi(t)C$, is equivalent to the form expressed earlier,

$$X(t) = c_1\Phi_1(t) + c_2\Phi_2(t) + \cdots + c_n\Phi_n(t).$$

EXAMPLE: Show that $\Phi(t) = \begin{pmatrix} e^{-2t} & -3e^{5t} \\ 2e^{-2t} & e^{5t} \end{pmatrix}$ is a fundamental matrix for the system $X'(t) = \begin{pmatrix} 4 & -3 \\ -2 & -1 \end{pmatrix} X(t)$. Use the matrix to find a general solution of $X'(t) = \begin{pmatrix} 4 & -3 \\ -2 & -1 \end{pmatrix} X(t)$.

SOLUTION: Since

$$\begin{pmatrix} e^{-2t} \\ 2e^{-2t} \end{pmatrix}' = \begin{pmatrix} -2e^{-2t} \\ -4e^{-2t} \end{pmatrix} = \begin{pmatrix} 4 & -3 \\ -2 & -1 \end{pmatrix} \begin{pmatrix} e^{-2t} \\ 2e^{-2t} \end{pmatrix}$$

and

$$\begin{pmatrix} -3e^{5t} \\ e^{5t} \end{pmatrix}' = \begin{pmatrix} -15e^{5t} \\ 5e^{5t} \end{pmatrix} = \begin{pmatrix} 4 & -3 \\ -2 & -1 \end{pmatrix} \begin{pmatrix} -3e^{5t} \\ e^{5t} \end{pmatrix},$$

both $\begin{pmatrix} e^{-2t} \\ 2e^{-2t} \end{pmatrix}$ and $\begin{pmatrix} -3e^{5t} \\ e^{5t} \end{pmatrix}$ are solutions of the system $X'(t) = \begin{pmatrix} 4 & -3 \\ -2 & -1 \end{pmatrix} X(t)$. The solutions are linearly independent because

$$W\left(\begin{pmatrix} e^{-2t} \\ 2e^{-2t} \end{pmatrix}, \begin{pmatrix} -3e^{5t} \\ e^{5t} \end{pmatrix}\right) = \begin{vmatrix} e^{-2t} & -3e^{5t} \\ 2e^{-2t} & e^{5t} \end{vmatrix} = 7e^{3t} \neq 0.$$

A general solution is given by

$$X(t) = \Phi(t)C = \begin{pmatrix} e^{-2t} & -3e^{5t} \\ 2e^{-2t} & e^{5t} \end{pmatrix} \begin{pmatrix} c_1 \\ c_2 \end{pmatrix} = \begin{pmatrix} c_1e^{-2t} - 3c_2e^{5t} \\ 2c_1e^{-2t} + c_2e^{5t} \end{pmatrix} = c_1 \begin{pmatrix} e^{-2t} \\ 2e^{-2t} \end{pmatrix} + c_2 \begin{pmatrix} -3e^{5t} \\ e^{5t} \end{pmatrix}.$$

The same steps are carried out with Maple. After defining **A** and **vec_1** to be the vector $\begin{pmatrix} e^{-2t} \\ 2e^{-2t} \end{pmatrix}$,

```
> A:=array([[4,-3],[-2,-1]]):
  vec_1:=array([[exp(-2*t)],[2*exp(-2*t)]])
```

$$vec\_1 := \begin{bmatrix} e^{-2t} \\ 2e^{-2t} \end{bmatrix}$$

we use **map** and **diff** to compute $\left(\begin{array}{c} e^{-2t} \\ 2e^{-2t} \end{array}\right)'$

```
> map(diff,vec_1,t);
```

$$\begin{bmatrix} -2e^{-2t} \\ -4e^{-2t} \end{bmatrix}$$

and **evalm** to compute $\begin{pmatrix} 4 & -3 \\ -2 & -1 \end{pmatrix}\begin{pmatrix} e^{-2t} \\ 2e^{-2t} \end{pmatrix}$. As expected, the results are the same.

```
> evalm(A &* vec_1);
```

$$\begin{bmatrix} -2e^{-2t} \\ -4e^{-2t} \end{bmatrix}$$

Similarly, after defining **vec_2** to be $\begin{pmatrix} -3e^{5t} \\ e^{5t} \end{pmatrix}$, we use **map**, **diff**, and **evalm** to see that

$$\begin{pmatrix} -3e^{5t} \\ e^{5t} \end{pmatrix}' = \begin{pmatrix} -15e^{5t} \\ 5e^{5t} \end{pmatrix} = \begin{pmatrix} 4 & -3 \\ -2 & -1 \end{pmatrix}\begin{pmatrix} -3e^{5t} \\ e^{5t} \end{pmatrix}.$$

```
> vec_2:=array([[-3*exp(5*t)],[exp(5*t)]]):
  map(diff,vec_2,t);
  evalm(A &* vec_2);
```

$$\begin{bmatrix} -15e^{5t} \\ 5e^{5t} \end{bmatrix}$$

$$\begin{bmatrix} -15e^{5t} \\ 5e^{5t} \end{bmatrix}$$

To compute $W\left(\begin{pmatrix} e^{-2t} \\ 2e^{-2t} \end{pmatrix}, \begin{pmatrix} -3e^{5t} \\ e^{5t} \end{pmatrix}\right)$, we load the **linalg** package and then use **augment**,

which is contained in the **linalg** package, to form the matrix $\Phi(t) = \begin{pmatrix} e^{-2t} & -3e^{5t} \\ 2e^{-2t} & e^{5t} \end{pmatrix}$.

```
> with(linalg):
  Phi:=augment(vec_1,vec_2);
```

$$\Phi := \begin{bmatrix} e^{-2t} & -3e^{5t} \\ 2e^{-2t} & e^{5t} \end{bmatrix}$$

Then, $W\left(\begin{pmatrix} e^{-2t} \\ 2e^{-2t} \end{pmatrix}, \begin{pmatrix} -3e^{5t} \\ e^{5t} \end{pmatrix}\right) = \begin{vmatrix} e^{-2t} & -3e^{5t} \\ 2e^{-2t} & e^{5t} \end{vmatrix} = 7e^{3t}$ is computed and simplified with `det`, which is contained in the `linalg` package, and `simplify`.

```
> simplify(det(Phi));
```

$$7e^{3t}$$

9.4 HOMOGENEOUS LINEAR SYSTEMS WITH CONSTANT COEFFICIENTS

Now that we have covered the necessary terminology, we can turn our attention to solving linear systems with constant coefficients. Let

$$A = \begin{pmatrix} a_{11} & a_{12} & \cdots & a_{1n} \\ a_{21} & a_{22} & \cdots & a_{2n} \\ \vdots & \vdots & \ddots & \vdots \\ a_{n1} & a_{n2} & \cdots & a_{nn} \end{pmatrix}$$

be an $n \times n$ real matrix, and let $\{\lambda_k\}_{k=1}^n$ be the eigenvalues and $\{v_k\}_{k=1}^n$ the corresponding eigenvectors of A. Then the general solution of the system $X' = AX$ is determined by the eigenvalues of A. For the moment, we consider the cases when the eigenvalues of A are distinct and real or the eigenvalues of A are distinct and complex. We will consider the case when A has repeated eigenvalues (eigenvalues of multiplicity greater than 1) separately.

≡ Distinct Real Eigenvalues

We may expect a general solution of any $n \times n$ system to be written as

$$X(t) = c_1 v_1 e^{\lambda_1 t} + c_2 v_2 e^{\lambda_2 t} + \cdots + c_n v_n e^{\lambda_n t},$$

where $\lambda_1, \lambda_2, \ldots, \lambda_n$ are n distinct real eigenvalues of A with corresponding eigenvectors $v_1, v_2, \ldots, v_n$, respectively. We investigate this claim by assuming that $X = ve^{\lambda t}$ is a solution of $X' = AX$. Then, $X' = \lambda ve^{\lambda t}$ must satisfy the differential equation which implies that

$$\lambda ve^{\lambda t} = Ave^{\lambda t}.$$

Now, since $\mathbf{Iv} = \mathbf{v}$, we make the following substitution so that the expression can be simplified.

$$\lambda \mathbf{I} v e^{\lambda t} = \mathbf{A} v e^{\lambda t}$$
$$(\mathbf{A} - \lambda \mathbf{I}) v e^{\lambda t} = \mathbf{0}.$$

Then, because $e^{\lambda t} \neq 0$, we have

$$(\mathbf{A} - \lambda \mathbf{I}) \mathbf{v} = \mathbf{0}.$$

In order for this system of equations to have a solution other than $\mathbf{v} = \mathbf{0}$ (which is required for an eigenvector),

$$|\mathbf{A} - \lambda \mathbf{I}| = 0.$$

We recall that a solution λ to this equation is an eigenvalue of $\mathbf{A}$, while the vector $\mathbf{v}$ satisfying $(\mathbf{A} - \lambda \mathbf{I})\mathbf{v} = \mathbf{0}$ is the eigenvector that corresponds to λ. Hence, if $\mathbf{A}$ has n distinct eigenvalues $\{\lambda_1, \lambda_2, \dots, \lambda_n\}$, then we can find a set of n linearly independent eigenvectors $\{\mathbf{v}_1, \mathbf{v}_2, \dots, \mathbf{v}_n\}$. From these eigenvalues and corresponding eigenvectors, we form the n linearly independent solutions

$$\mathbf{X}_1 = \mathbf{v}_1 e^{\lambda_1 t}, \mathbf{X}_2 = \mathbf{v}_2 e^{\lambda_2 t}, \dots, \mathbf{X}_n = \mathbf{v}_n e^{\lambda_n t}.$$

Therefore, if $\mathbf{A}$ is an $n \times n$ matrix with n distinct real eigenvalues $\{\lambda_k\}_{k=1}^n$, then a general solution of $\mathbf{X}' = \mathbf{A}\mathbf{X}$ is the linear combination of the set of solutions $\{\mathbf{X}_1, \mathbf{X}_2, \dots, \mathbf{X}_n\}$,

$$\mathbf{X}(t) = c_1 \mathbf{v}_1 e^{\lambda_1 t} + c_2 \mathbf{v}_2 e^{\lambda_2 t} + \cdots + c_n \mathbf{v}_n e^{\lambda_n t} = \sum_{i=1}^n c_i \mathbf{v}_i e^{\lambda_i t}.$$

EXAMPLE: Find a general solution of $\mathbf{X}' = \begin{pmatrix} 5 & -1 \\ 0 & 3 \end{pmatrix} \mathbf{X}$.

SOLUTION: We find the eigenvalues and corresponding eigenvectors of $\mathbf{A} = \begin{pmatrix} 5 & -1 \\ 0 & 3 \end{pmatrix}$ with `eigenvects`.

```
> with(linalg):
  A:=array([[5,-1],[0,3]]):
  vals_and_vecs:=eigenvects(A);
```

$$vals\_and\_vecs := [[5, 1, \{[1 \quad 0]\}], [3, 1, \{[1 \quad 2]\}]]$$

Hence, the eigenvalues of $\begin{pmatrix} 5 & -1 \\ 0 & 3 \end{pmatrix}$ are $\lambda_1 = 3$ and $\lambda_2 = 5$ with corresponding eigenvectors $\begin{pmatrix} 1 \\ 2 \end{pmatrix}$

and $\begin{pmatrix} 1 \\ 0 \end{pmatrix}$, respectively. Therefore, a general solution of the system $X' = \begin{pmatrix} 5 & -1 \\ 0 & 3 \end{pmatrix} X$ is

$$X = c_1 v_1 e^{\lambda_1 t} + c_2 v_2 e^{\lambda_2 t} = c_1 \begin{pmatrix} 1 \\ 2 \end{pmatrix} e^{3t} + c_2 \begin{pmatrix} 1 \\ 0 \end{pmatrix} e^{5t}.$$

Remember that the system $X' = \begin{pmatrix} 5 & -1 \\ 0 & 3 \end{pmatrix} X$ is the same as the system $\begin{cases} x' = 5x - y \\ y' = 3y \end{cases}$, which we form as follows using **equate**, which is contained in the **student** package.

```
> with(student):
  Eqs:=equate([[diff(x(t),t)],[diff(y(t),t)]],
   A &* [[x(t)],[y(t)]]);
```

$$Eqs := \left\{ \frac{\partial}{\partial t} x(t) = 5x(t) - y(t), \frac{\partial}{\partial t} y(t) = 3y(t) \right\}$$

Thus, we can write the general solution we obtained as $\begin{cases} x = c_1 e^{3t} + c_2 e^{5t} \\ y = 2c_1 e^{3t} \end{cases}$.

The same general solution is obtained with **dsolve**.

```
> Gen_Sol:=dsolve(Eqs,{x(t),y(t)});
```

$$Gen\_Sol := \{ y(t) = 2\_C2 e^{3t}, x(t) = \_C1 e^{5t} + \_C2 e^{3t} \}$$

We then assign $x(t)$ and $y(t)$ the results obtained in **Gen_Sol** with **assign**.

```
> assign(Gen_Sol):
```

We then graph $x(t)$ and $y(t)$ for various values of the arbitrary constants using the same techniques that we used in Section 9.1. For example, entering

```
> to_plot:={seq(seq(subs({_C1=i,_C2=j},[x(t),y(t),t=-2..2]),
   i=-1..1),j=-1..1)};
```

$$to\_plot := \left\{ [-e^{5t} - e^{3t}, -2e^{3t}, t = -2..2], [-e^{3t}, -2e^{3t}, t = -2..2], \right.$$
$$[e^{5t} - e^{3t}, -2e^{3t}, t = -2..2], [-e^{5t}, 0, t = -2..2],$$
$$[0, 0, t = -2..2], [e^{5t}, 0, t = -2..2], [-e^{5t} + e^{3t}, 2e^{3t}, t = -2..2],$$
$$\left. [e^{3t}, 2e^{3t}, t = -2..2], [e^{5t} + e^{3t}, 2e^{3t}, t = -2..2] \right\}$$

defines **to_plot** to be the set of ordered triples $x(t)$, $y(t)$, **t=-2..2** obtained by replacing each occurrence of **_C1** and **_C2** in $x(t)$ and $y(t)$ by i and j, respectively, for $i = -1, 0,$ and 1 and $j = -1, 0,$ and 1. Then entering

```
> plot(to_plot,view=[-2..2,-2..2]);
```

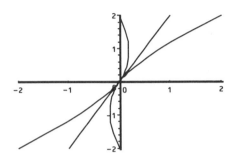

graphs the set of parametric functions in `to_plot`. Alternatively, we can take advantage of the `DEplot2` command, which is contained in the `DEtools` package. First, we load the `DEtools` package.

```
> with(DEtools);
```

  [*DEplot, DEplot1, DEplot2, Dchangevar, PDEplot, dfieldplot, phaseportrait*]

Then entering

```
> DEplot2(A,[x,y],t=-2..2,x=-2..2,y=-2..2);
```

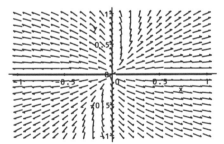

graphs the direction field associated with the system $\begin{pmatrix} x' \\ y' \end{pmatrix} = A \begin{pmatrix} x \\ y \end{pmatrix}$ on the rectangle $[-2, 2] \times [-2, 2]$. Conversely, entering

```
> DEplot2(A,[x,y],t=-2..2,{[0,1,.25],[0,0,.25],
   [0,-.5,-.5],[0,0,-.75]},
   x=-2..2,y=-2..2);
```

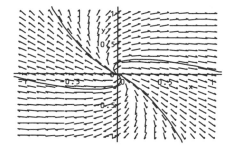

graphs the direction field associated with the system on the rectangle $[-2, 2] \times [-2, 2]$ along with graphs of the solutions that satisfy $\begin{cases} x(0) = 1 \\ y(0) = .25 \end{cases}$, $\begin{cases} x(0) = 0 \\ y(0) = .25 \end{cases}$, $\begin{cases} x(0) = -.5 \\ y(0) = -.5 \end{cases}$, and $\begin{cases} x(0) = 0 \\ y(0) = -.75 \end{cases}$.

☰ Complex Conjugate Eigenvalues

Next, consider the system $X' = AX$ with complex conjugate eigenvalues $\lambda_1 = \alpha + \beta i$ and $\lambda_2 = \alpha - \beta i$ and corresponding eigenvectors $v_1 = a + bi$ and $v_2 = a - bi$. Then, one solution of $X' = AX$ is

$$X = v_1 e^{\lambda t} = (a + bi)e^{(\alpha + \beta i)t} = e^{\alpha t}(a + bi)e^{i\beta t} = e^{\alpha t}(a + bi)(\cos \beta t + \sin)$$
$$= e^{\alpha t}(a \cos \beta t - b \sin \beta t) + ie^{\alpha t}(a \sin \beta t + b \cos \beta t)$$
$$= x_1(t) + ix_2(t).$$

Now, because X is a solution of the system $X' = AX$, we have $x_1'(t) + ix_2'(t) = Ax_1(t) + iAx_2(t)$. Equating the real and imaginary parts of this equation yields $x_1'(t) = Ax_1(t)$ and $x_2'(t) = Ax_2(t)$. Therefore, $x_1(t)$ and $x_2(t)$ are solutions of $X' = AX$, so any linear combination of $x_1(t)$ and $x_2(t)$ is also a solution. We can show that $x_1(t)$ and $x_2(t)$ are linearly independent, so this linear combination forms a portion of a general solution of $X' = AX$. Hence, for a square matrix A of any size, we can state the following.

THEOREM

· ·

If A has complex conjugate eigenvalues $\lambda_1 = \alpha + \beta i$ and $\lambda_2 = \alpha - \beta i$ and corresponding eigenvectors $v_1 = a + bi$ and $v_2 = a - bi$, then two linearly independent solutions of $X' = AX$ are $x_1(t) = e^{\alpha t}(a \cos \beta t - b \sin \beta t)$ and $x_2(t) = e^{\alpha t}(a \sin \beta t + b \cos \beta t)$.

· ·

Notice that in the case of complex conjugate eigenvalues, we are able to obtain two linearly independent solutions from knowing one eigenvalue and an eigenvector that corresponds to it.

EXAMPLE: Find a general solution of $X' = \begin{pmatrix} 3 & -2 \\ 4 & -1 \end{pmatrix} X$.

SOLUTION: In this case, $A = \begin{pmatrix} 3 & -2 \\ 4 & -1 \end{pmatrix}$; the eigenvalues and corresponding eigenvectors are found with **eigenvects**.

```
> with(linalg):
  A:=array([[3,-2],[4,-1]]):
  vals_and_vecs:=eigenvects(A);
```

$$vals\_and\_vecs := \left[\text{RootOf}(\_Z^2 - 2\_Z + 5), 1, \right.$$

$$\left\{ \left[1, -\frac{1}{2}\text{RootOf}(\_Z^2 - 2\_Z + 5) + \frac{3}{2} \right] \right\} \right]$$

We see that the results are given in terms of **RootOf** so we use **map** and **allvalaues** to determine the exact value of the eigenvalues and corresponding eigenvectors.

```
> map(allvalues,vals_and_vecs);
```

$$[1 + 2I, 1 - 2I, 1, \{[1 \quad 1 - I]\}, \{[1 \quad 1 + I]\}]$$

Hence, $\lambda_1 = 1 + 2i$ and $\lambda_2 = 1 - 2i$ with corresponding eigenvectors $\begin{pmatrix} 1 \\ 1 - i \end{pmatrix}$ and $\begin{pmatrix} 1 \\ 1 + i \end{pmatrix}$, respectively. Therefore, in the notation we used,

$$a = \begin{pmatrix} 1 \\ 1 \end{pmatrix} \text{ and } b = \begin{pmatrix} 0 \\ -1 \end{pmatrix}.$$

Hence, with $\alpha = 1$ and $\beta = 2$, a general solution is

$$X(t) = c_1 e^t \left[\begin{pmatrix} 1 \\ 1 \end{pmatrix} \cos 2t - \begin{pmatrix} 0 \\ -1 \end{pmatrix} \sin 2t \right] + c_2 e^t \left[\begin{pmatrix} 1 \\ 1 \end{pmatrix} \sin 2t + \begin{pmatrix} 0 \\ -1 \end{pmatrix} \cos 2t \right]$$

$$= \begin{pmatrix} c_1 e^t \cos 2t + c_2 \sin 2t \\ c_1 e^t (\cos 2t + \sin 2t) + c_2 (\sin 2t - \cos 2t) \end{pmatrix}.$$

As in the previous example, we observe that the system $X' = \begin{pmatrix} 3 & -2 \\ 4 & -1 \end{pmatrix} X$ is the same as the system $\begin{cases} x' = 3x - 2y \\ y' = 4x - y \end{cases}$,

```
> with(student):
  Eqs:=equate([[diff(x(t),t)],[diff(y(t),t)]],
  A &* [[x(t)],[y(t)]]);
```

$$Eqs := \left\{ \frac{\partial}{\partial t} x(t) = 3x(t) - 2y(t), \frac{\partial}{\partial t} y(t) = 4x(t) - y(t) \right\}$$

so that we may write the solution as

$$\begin{cases} x = c_1 e^t \cos 2t + c_2 \sin 2t \\ y = c_1 e^t (\cos 2t + \sin 2t) + c_2 (\sin 2t - \cos 2t) \end{cases}.$$

We obtain the same result with **dsolve**.

```
> dsolve(Eqs,{x(t),y(t)});
```

$\{y(t) = (\_C1 + \_C2)e^t \sin(2t) + (-\_C1 + \_C2)e^t \cos(2t),$
$x(t) = \_C1e^t \sin(2t) + \_C2e^t \cos(2t)\}$

We can graph the solution for various values of the arbitrary constants using **plot** or we can take advantage of the **DEplot2** command, which is contained in the **DEtools** package, as in the previous example. After defining **init_conds** to be the set of ordered triples $(0, .25i, .25j)$ for $i = -1, 0,$ and 1 and $j = -1, 0,$ and 1, loading the **DEtools** package, we use **DEplot2** to graph the direction field associated with the system and then use **DEplot2** to graph the solutions that satisfy the initial conditions specified in **init_conds**. In the second **DEplot2** command, the option **stepsize=0.1** instructs Maple to use a step size of 0.1, which is smaller than the default value of $\dfrac{3 - (-3)}{20} = \dfrac{3}{10}$, and helps assure that the resulting graphs appear smooth, while the option **arrows=NONE** instructs Maple not to display the direction field associated with the system.

```
> init_conds:={seq(seq([0,.25*i,.25*j],i=-1..1),j=-1..1)}:
  with(DEtools):
  DEplot2(A,[x,y],t=-3..3,x=-5..5,y=-5..5);
  DEplot2(A,[x,y],t=-3..3,init_conds,x=-5..5,y=-5..5,
   stepsize=0.1,arrows=NONE);
```

Initial-value problems can be solved as well.

EXAMPLE: Solve $\mathbf{X}' = \begin{pmatrix} 5 & 5 & 2 \\ -6 & -6 & -5 \\ 6 & 6 & 5 \end{pmatrix} \mathbf{X}$ subject to $\mathbf{X}(0) = \begin{pmatrix} 0 \\ 0 \\ 2 \end{pmatrix}$.

SOLUTION: The eigenvalues and corresponding eigenvectors of

$$\mathbf{A} = \begin{pmatrix} 5 & 5 & 2 \\ -6 & -6 & -5 \\ 6 & 6 & 5 \end{pmatrix}$$

are determined with **eigenvects**.

```
> with(linalg):
  A:=array([[5,5,2],[-6,-6,-5],[6,6,5]]):
  vals_and_vecs:=eigenvects(A);
```

$$vals\_and\_vecs := [0, 1, \{[1 \quad -1 \quad 0]\}], \left[\text{RootOf}\left(\_Z^2 - 4\_Z + 13\right), 1,\right.$$
$$\left.\left\{\left[\frac{1}{6} + \frac{1}{6}\text{RootOf}\left(\_Z^2 - 4\_Z + 13\right) \quad -1 \quad 1\right]\right\}\right]$$

We see that the second list in **vals_and_vecs**, which is extracted from **vals_and_vecs** as follows, involves **RootOf**.

```
> vals_and_vecs[2];
```

$$\left[\text{RootOf}(\_Z^2 - 4\_Z + 13), 1, \left\{ \left[\frac{1}{6} + \frac{1}{6} \text{RootOf}(\_Z^2 - 4\_Z + 13) \quad -1 \quad 1 \right] \right\} \right]$$

Therefore, we use **map** and **allvalues** to compute these values exactly.

```
> map(allvalues,vals_and_vecs[2]);
```

$$\left[2 + 3I, 2 - 3I, 1, \left\{ \left[\frac{1}{2} + \frac{1}{2}I \quad -1 \quad 1 \right] \right\}, \left\{ \left[\frac{1}{2} - \frac{1}{2}I \quad -1 \quad 1 \right] \right\} \right]$$

Hence, $\lambda_1 = 0$, $\lambda_2 = 2 + 3i$, and $\lambda_3 = 2 - 3i$ with corresponding eigenvectors

$$\begin{pmatrix} 1 \\ -1 \\ 0 \end{pmatrix}, \begin{pmatrix} \frac{1}{2}(1 + i) \\ -1 \\ 1 \end{pmatrix}, \text{ and } \begin{pmatrix} \frac{1}{2}(1 - i) \\ -1 \\ 1 \end{pmatrix}.$$

Hence, one solution of the system of differential equations is

$$\mathbf{X}_1 = \mathbf{v}_1 e^{\lambda_1 t} = \begin{pmatrix} 1 \\ -1 \\ 0 \end{pmatrix} e^{(0)t} = \begin{pmatrix} 1 \\ -1 \\ 0 \end{pmatrix}.$$

Multiplying the vectors

$$\begin{pmatrix} \frac{1}{2}(1 + i) \\ -1 \\ 1 \end{pmatrix} \text{ and } \begin{pmatrix} \frac{1}{2}(1 - i) \\ -1 \\ 1 \end{pmatrix}$$

by 2 yields

$$\begin{pmatrix} 1 + i \\ -2 \\ 2 \end{pmatrix} \text{ and } \begin{pmatrix} 1 - i \\ -2 \\ 2 \end{pmatrix}.$$

Thus, the two linearly independent solutions that correspond to the eigenvalues $\lambda = 2 \pm 3i$ are

$$\mathbf{X}_2 = e^{2t} \left[\begin{pmatrix} 1 \\ -2 \\ 2 \end{pmatrix} \cos 3t - \begin{pmatrix} 1 \\ 0 \\ 0 \end{pmatrix} \sin 3t \right] = \begin{pmatrix} e^{2t}(\cos 3t - \sin 3t) \\ -2e^{2t} \cos 3t \\ 2e^{2t} \cos 3t \end{pmatrix}$$

and

$$\mathbf{X}_3 = e^{2t}\left[\begin{pmatrix}1\\-2\\2\end{pmatrix}\sin 3t + \begin{pmatrix}1\\0\\0\end{pmatrix}\cos 3t\right] = \begin{pmatrix}e^{2t}(\sin 3t + \cos 3t)\\-2e^{2t}\sin 3t\\2e^{2t}\sin 3t\end{pmatrix}.$$

Hence, a general solution is

$$\mathbf{X} = c_1\mathbf{X}_1 + c_2\mathbf{X}_2 + c_3\mathbf{X}_3 = \mathbf{v}_1 e^{\lambda_1 t}$$

$$= c_1\begin{pmatrix}1\\-1\\0\end{pmatrix} + c_2\begin{pmatrix}e^{2t}(\cos 3t - \sin 3t)\\2e^{2t}\cos 3t\\2e^{2t}\cos 3t\end{pmatrix} + c_3\begin{pmatrix}e^{2t}(\sin 3t + \cos 3t)\\-2e^{2t}\sin 3t\\2e^{2t}\sin 3t\end{pmatrix}$$

$$= \begin{pmatrix}c_1 + c_2 e^{2t}(\cos 3t - \sin 3t) + c_3 e^{2t}(\sin 3t + \cos 3t)\\-c_1 - 2c_2 e^{2t}\cos 3t - 2c_3 e^{2t}\sin 3t\\2c_2 e^{2t}\cos 3t + 2c_3 e^{2t}\sin 3t\end{pmatrix}.$$

Application of the initial condition

$$\mathbf{X}(0) = \begin{pmatrix}0\\0\\2\end{pmatrix}$$

yields the system of equations

$$\mathbf{X}(0) = \begin{pmatrix}c_1 + c_2 + c_3\\-c_1 - 2c_2\\2c_2\end{pmatrix} = \begin{pmatrix}0\\0\\2\end{pmatrix}$$

with solution $c_2 = 1$, $c_1 = -2$, and $c_3 = 1$. Therefore, the solution of this initial-value problem is

$$\mathbf{X} = \begin{pmatrix}-2 + e^{2t}(\cos 3t - \sin 3t) + e^{2t}(\sin 3t + \cos 3t)\\2 - 2e^{2t}\cos 3t - 2e^{2t}\sin 3t\\2e^{2t}\cos 3t + 2e^{2t}\sin 3t\end{pmatrix}.$$

We see that **dsolve** is also successful in solving the initial-value problem. After loading the **student** package, we use **equate**, which is contained in the **student** package, to form the set of equations

$$\begin{cases}\dfrac{dx}{dt} = 5x + 5y + 2z\\[2mm]\dfrac{dy}{dt} = -6x - 6y - 5z \;.\\[2mm]\dfrac{dz}{dt} = 6x + 6y + 5z\end{cases}$$

```
> with(student):
  Eqs:=equate([[diff(x(t),t)],[diff(y(t),t)],
   [diff(z(t),t)]],A &* [[x(t)],[y(t)],[z(t)]]);
```

$$Eqs := \left\{ \frac{\partial}{\partial t}x(t) = 5x(t) + 5y(t) + 2z(t), \frac{\partial}{\partial t}y(t) = -6x(t) - 6y(t) - 5z(t), \right.$$

$$\left. \frac{\partial}{\partial t}z(t) = 6x(t) + 6y(t) + 5z(t) \right\}$$

We then use **union** to add the initial conditions to **Eqs** and name the resulting set of equations **init_val_prob**, which is then solved with **dsolve**.

```
> init_val_prob:=Eqs union {x(0)=0,y(0)=0,z(0)=2};
```

$$init\_val\_prob := \left\{ \frac{\partial}{\partial t}x(t) = 5x(t) + 5y(t) + 2z(t), \right.$$

$$\frac{\partial}{\partial t}y(t) = -6x(t) - 6y(t) - 5z(t),$$

$$\frac{\partial}{\partial t}z(t) = 6x(t) + 6y(t) + 5z(t),$$

$$\left. x(0) = 0, y(0) = 0, z(0) = 2 \right\}$$

```
> Sol:=dsolve(init_val_prob,{x(t),y(t),z(t)});
```

$$Sol := \left\{ y(t) = 2 - 2e^{2t}\sin(3t) - 2e^{2t}\cos(3t), \right.$$

$$\left. z(t) = 2e^{2t}\sin(3t) + 2e^{2t}\cos(3t), x(t) = -2 + 2e^{2t}\cos(3t) \right\}$$

To graph the solution, we use **assign** to name $x(t)$, $y(t)$, and $z(t)$ the results obtained in **Sol**. Then, after loading the **plots** package, we use **spacecurve**, which is contained in the **plots** package, to graph $(x(t), y(t), z(t))$ for $0 \leq t \leq 2\pi$. The option **axes=BOXED** instructs Maple to enclose a box around the resulting three-dimensional graphics object.

```
> assign(Sol):
  with(plots):
  spacecurve([x(t),y(t),z(t)],t=0..Pi,axes=BOXED);
```

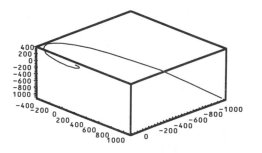

To graph the direction field associated with the system, we use the `fieldplot3d` command, which like the `spacecurve` command is also contained in the `plots` package.

```
> fieldplot3d([5*x+5*y-2*z,-6*x-6*y-5*z,
  6*x+6*y+5*z],x=-10..10,y=-10..10,z=-10..10,
  axes=BOXED);
```

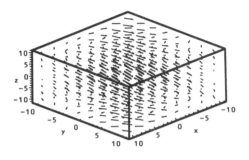

An alternate method can be used to solve initial-value problems. Let $\Phi(t)$ be a fundamental matrix for the system of equations $\mathbf{X}' = \mathbf{A}\mathbf{X}$. Then a general solution is $\mathbf{X}(t) = \Phi(t)\mathbf{C}$, where $\mathbf{C}$ is a constant vector. If the initial condition $\mathbf{X}(0) = \mathbf{X}_0$ is given, then $\mathbf{X}(t) = \Phi(t)\mathbf{C}$. Hence,

$$\mathbf{X}(0) = \Phi(0)\mathbf{C}$$
$$\mathbf{X}_0 = \Phi(0)\mathbf{C}$$
$$\mathbf{C} = \Phi^{-1}(0)\mathbf{X}_0.$$

Therefore, the solution to the initial-value problem is $\mathbf{X}(t) = \Phi(t)\Phi^{-1}(0)\mathbf{X}_0$.

EXAMPLE: Use a fundamental matrix to solve the initial-value problem $\mathbf{X}' = \begin{pmatrix} 1 & 1 \\ 4 & -2 \end{pmatrix} \mathbf{X}$ subject to $\mathbf{X}(0) = \begin{pmatrix} 1 \\ -2 \end{pmatrix}$.

SOLUTION: The eigenvalues of $\mathbf{A} = \begin{pmatrix} 1 & 1 \\ 4 & -2 \end{pmatrix}$ are found with `eigenvects`.

```
> with(linalg):
  A:=array([[1,1],[4,-2]]):
  vals_and_vecs:=eigenvects(A);
```

$vals\_and\_vecs := [-3, 1, \{[1 \quad -4]\}], [2, 1, \{[1 \quad 1]\}]$

Hence, $\lambda_1 = 2$ and $\lambda_2 = -3$ with corresponding eigenvectors $\mathbf{v}_1 = \begin{pmatrix} 1 \\ 1 \end{pmatrix}$ and $\mathbf{v}_2 = \begin{pmatrix} 1 \\ -4 \end{pmatrix}$, respec-

tively. A fundamental matrix is then given by $\Phi(t) = \begin{pmatrix} e^{2t} & e^{-3t} \\ e^{2t} & -4e^{-3t} \end{pmatrix}$. We calculate $\Phi^{-1}(0)$ with

`inverse`, which is contained in the `linalg` package, after defining `Phi`.

```
> Phi:=t->array([[exp(2*t),exp(-3*t)],[exp(2*t),
  -4*exp(-3*t)]]):
  Phi(0);
```

$$\begin{bmatrix} 1 & 1 \\ 1 & -4 \end{bmatrix}$$

```
> inverse(Phi(0));
```

$$\begin{bmatrix} \dfrac{4}{5} & \dfrac{1}{5} \\ \dfrac{1}{5} & -\dfrac{1}{5} \end{bmatrix}$$

Hence, $\mathbf{X}(t) = \Phi(t)\Phi^{-1}(0)\mathbf{X}_0$:

```
> X:=t->evalm(Phi(t) &* inverse(Phi(0)) &* [[1],[-2]]):
  X(t);
```

$$\begin{bmatrix} \dfrac{2}{5}e^{2t} + \dfrac{3}{5}e^{-3t} \\ \dfrac{2}{5}e^{2t} - \dfrac{12}{5}e^{-3t} \end{bmatrix}$$

☰ Repeated Eigenvalues

We now consider the case of repeated eigenvalues. As we have already seen, an eigenvalue of multiplicity m can have m corresponding linearly independent eigenvectors or it can have less than m corresponding eigenvectors. In the case of m linearly independent eigenvectors, a general solution is found in the same manner as in the case of n distinct eigenvalues.

EXAMPLE: Solve

$$X' = \begin{pmatrix} -1 & 1 & 2 \\ 6 & 4 & 12 \\ -2 & -2 & -6 \end{pmatrix} X.$$

SOLUTION: We find the eigenvalues and corresponding eigenvectors of

$$A = \begin{pmatrix} -1 & 1 & 2 \\ 6 & 4 & 12 \\ -2 & -2 & -6 \end{pmatrix}$$

using **eigenvects**.

```
> with(linalg):
  A:=array([[-1,1,2],[6,4,12],[-2,-2,-6]]):
  vals_and_vecs:=eigenvects(A);
```

$vals\_and\_vecs := [-2, 2, \{[-2 \quad 0 \quad 1], [-1 \quad 1 \quad 0]\}], [1, 1, \{[1 \quad 6 \quad \leftarrow 2]\}]$

From the result, we see that the eigenvalue $\lambda_1 = \lambda_2 = -2$ of multiplicity 2 has two corresponding linearly independent eigenvectors,

$$v_1 = \begin{pmatrix} -2 \\ 0 \\ 1 \end{pmatrix}$$

and

$$v_2 = \begin{pmatrix} -1 \\ 1 \\ 0 \end{pmatrix},$$

while the eigenvalue $\lambda_3 = 1$ of multiplicity 1 has corresponding eigenvector

$$v_2 = \begin{pmatrix} 1 \\ 6 \\ -2 \end{pmatrix}.$$

A general solution is then

$$
\mathbf{X} = c_1\mathbf{v}_1 e^{\lambda_1 t} + c_2\mathbf{v}_2 e^{\lambda_2 t} + c_3\mathbf{v}_3 e^{\lambda_3 t} = c_1 \begin{pmatrix} -2 \\ 0 \\ 1 \end{pmatrix} e^{-2t} + c_2 \begin{pmatrix} -1 \\ 1 \\ 0 \end{pmatrix} e^{-2t} + c_3 \begin{pmatrix} 1 \\ 6 \\ -2 \end{pmatrix} e^{t}
$$

$$
= \begin{pmatrix} (-2c_1 - c_2)e^{-2t} + c_3 e^{t} \\ c_2 e^{-2t} + 6c_3 e^{t} \\ c_1 e^{-2t} - 2c_3 e^{t} \end{pmatrix}.
$$

We now use **dsolve** to find a general solution and name the result **Gen_Sol**.

```
> with(student):
  Eqs:=equate([[diff(x(t),t)],[diff(y(t),t)],[diff(z(t),t)]],
   A &* [[x(t)],[y(t)],[z(t)]]);
  Gen_Sol:=dsolve(Eqs,{x(t),y(t),z(t)});
```

$$
Eqs := \left\{ \frac{\partial}{\partial t}x(t) = -x(t) + y(t) + 2z(t), \frac{\partial}{\partial t}y(t) = 6x(t) + 4y(t) + 12z(t), \right.
$$

$$
\left. \frac{\partial}{\partial t}z(t) = -2x(t) - 2y(t) - 6z(t) \right\}
$$

$$
Gen\_Sol := \left\{ y(t) = -3e^{t}\_C2 + e^{-2t}\_C1, z(t) = e^{t}\_C2 + \_C3 e^{-2t}, \right.
$$

$$
\left. x(t) = -\frac{1}{2}e^{t}\_C2 + (-2\_C3 - \_C1)e^{-2t} \right\}
$$

$x(t)$, $y(t)$, and $z(t)$ are then named the result obtained in **Gen_Sol** with **assign**. After defining **vals** to be the list consisting of -1 and 1, we use **seq** and **subs** to define **to_plot** to be the set of eight ordered triples $(x(t), y(t), z(t))$ obtained by replacing _C1, _C2, and _C3 in $x(t)$, $y(t)$ and $z(t)$ by i, j, and k, where i, j, and k take on the values in **vals**.

```
> assign(Gen_Sol):
  vals:=[-1,1]:
  to_plot:={seq(seq(seq(subs({_C1=i,_C2=j,_C3=k},
  [x(t),y(t),z(t)]),i=vals),
  j=vals),k=vals)};
```

$$to\_plot := \left\{\left[-\frac{1}{2}e^t + e^{-2t}, -3e^t + e^{-2t}, e^t - e^{-2t}\right],\right.$$

$$\left[\frac{1}{2}e^t - e^{-2t}, 3e^t - e^{-2t}, -e^t + e^{-2t}\right],$$

$$\left[\frac{1}{2}e^t + 3e^{-2t}, 3e^t - e^{-2t}, -e^t - e^{-2t}\right],$$

$$\left[\frac{1}{2}e^t + e^{-2t}, 3e^t + e^{-2t}, -e^t - e^{-2t}\right],$$

$$\left[-\frac{1}{2}e^t + 3e^{-2t}, -3e^t - e^{-2t}, e^t - e^{-2t}\right],$$

$$\left[\frac{1}{2}e^t - 3e^{-2t}, 3e^t + e^{-2t}, -e^t + e^{-2t}\right],$$

$$\left[-\frac{1}{2}e^t - e^{-2t}, -3e^t - e^{-2t}, e^t + e^{-2t}\right],$$

$$\left.\left[-\frac{1}{2}e^t - 3e^{-2t}, -3e^t + e^{-2t}, e^t + e^{-2t}\right]\right\}$$

Then, after loading the **plots** package, the set of eight functions in **to_plot** is graphed for $-1 \leq t \leq 1$ with **spacecurve**, which is contained in the **plots** package.

```
> with(plots):
  spacecurve(to_plot,t=-1..1,axes=BOXED);
```

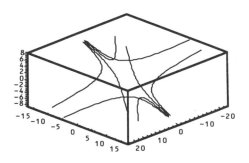

Let us restrict our attention to a system with the repeated eigenvalue $\lambda_1 = \lambda_2$ that has only one corresponding eigenvector v_1. (Assume that all other eigenvalues are distinct.) Hence, we obtain one solution to the system $X_1 = v_1 e^{\lambda_1 t}$ that corresponds to λ_1. We now seek a second solution corresponding to λ_1 in a manner similar to that considered in the case of repeated characteristic values of higher-order equations. In this case, however, suppose that the second linearly independent solution corresponding to λ_1 is of the form

$$X_2 = (v_2 t + w_2)e^{\lambda_1 t}.$$

In order to find v_2 and w_2, we substitute X_2 into $X' = AX$. Since $X_2' = \lambda_1(v_2 t + w_2)e^{\lambda_1 t} + v_2 e^{\lambda_1 t}$, we have

$$X_2' = AX_2$$
$$\lambda_1(v_2 t + w_2)e^{\lambda_1 t} + v_2 e^{\lambda_1 t} = A(v_2 t + w_2)e^{\lambda_1 t}$$
$$\lambda_1 v_2 t + (\lambda_1 w_2 + v_2) = Av_2 t + Aw_2.$$

Equating coefficients yields $\lambda_1 v_2 = Av_2$ and $\lambda_1 w_2 + v_2 = Aw_2$. The equation $\lambda_1 v_2 = Av_2$ indicates that v_2 is an eigenvector that corresponds to λ_1, so $v_2 = v_1$. We simplify the equation $\lambda_1 w_2 + v_2 = Aw_2$ as follows:

$$\lambda_1 w_2 + v_2 = Aw_2$$
$$v_2 = Aw_2 - \lambda_1 w_2$$
$$v_2 = (A - \lambda_1 I)w_2.$$

Hence, w_2 satisfies the equation

$$(A - \lambda_1 I)w_2 = v_1.$$

EXAMPLE: Find a general solution of $X'(t) = \begin{pmatrix} -8 & -1 \\ 16 & 0 \end{pmatrix} X(t)$.

SOLUTION: We find the eigenvalues of $A = \begin{pmatrix} -8 & -1 \\ 16 & 0 \end{pmatrix}$ with **eigenvects**.

```
> with(linalg):
  A:=array([[-8,-1],[16,0]]):
  vals_and_vecs:=eigenvects(A);latex([vals_and_vecs]);
  vals_and_vecs := [-4, 2, {[1  -4]}]
```

Hence, $\lambda_1 = \lambda_2 = -4$ with corresponding eigenvector $v_1 = \begin{pmatrix} 1 \\ -4 \end{pmatrix}$ so one solution to the system is $X_1 = \begin{pmatrix} 1 \\ -4 \end{pmatrix} e^{-4t}$. Therefore, to find $w_2 = \begin{pmatrix} x_2 \\ y_2 \end{pmatrix}$ in the second linearly independent solution

$X_2 = (v_1 t + w_2)e^{\lambda_1 t}$, we solve $(A - \lambda_1 I)w_2 = v_1$ that in this case, is

$$\begin{pmatrix} -4 & -1 \\ 16 & 4 \end{pmatrix} \begin{pmatrix} x_2 \\ y_2 \end{pmatrix} = \begin{pmatrix} 1 \\ -4 \end{pmatrix}.$$

After using **alias** to define **Id** to represent the identity matrix and **array** to define the vector $\begin{pmatrix} 1 \\ -4 \end{pmatrix}$, we use **linsolve**, which is contained in the **linalg** package, to solve this matrix equation.

```
> alias(Id=&*()):
  b:=array([1,-4]):
  sol_vec:=linsolve(A+4*Id,b);
```

$$sol\_vec := [\_t_{[1]}, -4\_t_{[1]} - 1]$$

The result means that if x_2 is any number, then $y_2 = -4x_2 - 1$. For example, if we choose $x_2 = 0$, then $y_2 = -1$.

```
> subs(_t[1]=0,eval(sol_vec));
```

$$[0, -1]$$

With $w_2 = \begin{pmatrix} 0 \\ -1 \end{pmatrix}$, the second solution is

$$X_2 = \left(\begin{pmatrix} 1 \\ -4 \end{pmatrix} t + \begin{pmatrix} 0 \\ -1 \end{pmatrix} \right) e^{-4t}.$$

Hence, a general solution is

$$X = c_1 \begin{pmatrix} 1 \\ -4 \end{pmatrix} e^{-4t} + c_2 \left(\begin{pmatrix} 1 \\ -4 \end{pmatrix} t + \begin{pmatrix} 0 \\ -1 \end{pmatrix} \right) e^{-4t} = \begin{pmatrix} c_1 + c_2 t \\ (-4c_1 - c_2) - 4tc_2 \end{pmatrix} e^{-4t}.$$

As in the previous examples in this section, we see that **dsolve** returns the same general solution as that found earlier.

```
> with(student):
  Eqs:=equate([[diff(x(t),t)],[diff(y(t),t)]],
    A &* [[x(t)],[y(t)]]);
  Gen_Sol:=dsolve(Eqs,{x(t),y(t)});
```

$$Eqs := \left\{ \frac{\partial}{\partial t} x(t) = -8x(t) - y(t), \frac{\partial}{\partial t} y(t) = 16x(t) \right\}$$

$$Gen\_Sol := \left\{ x(t) = \_C1 e^{-4t} + \_C2 e^{-4t} t, \right.$$
$$\left. y(t) = (-4\_C1 - \_C2)e^{-4t} - 4\_C2 e^{-4t} t \right\}$$

After loading the `DEtools` package, we use `DEplot2` to graph the direction field associated with the system on the rectangle $[-1,1] \times [-1,1]$. The second `DEplot2` command graphs the solutions that satisfy the initial conditions $\begin{cases} x(0) = 1 \\ y(0) = 1 \end{cases} \begin{cases} x(0) = 1 \\ y(0) = -1 \end{cases} \begin{cases} x(0) = -1 \\ y(0) = 1 \end{cases}$, and $\begin{cases} x(0) = -1 \\ y(0) = -1 \end{cases}$ on the rectangle $[-2,2] \times [-2,2]$. The option `stepsize=0.1` is included to help assure that the resulting graphs are smooth; the direction field is not displayed because the command `arrows=NONE` is included.

```
> with(DEtools):
  DEplot2(A,[x,y],t=-4..4,x=-1..1,y=-1..1);
  DEplot2(A,[x,y],t=-4..4,{[0,1,1],[0,1,-1],
  [0,-1,1],[0,-1,-1]},x=-2..2,y=-2..2,arrows=NONE,
  stepsize=0.1);
```

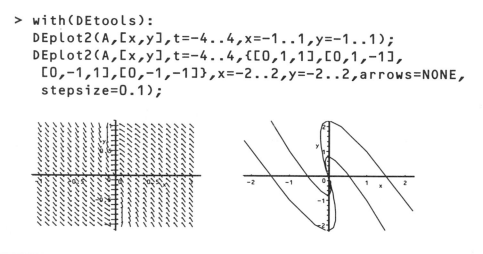

A similar method is carried out in the case of three equal eigenvalues $\lambda_1 = \lambda_2 = \lambda_3$. When we encounter this situation, we assume that

$$X_1 = v_1 e^{\lambda_1 t}, X_2 = (v_2 t + w_2)e^{\lambda_1 t}, \text{ and } X_3 = \left(v_3 \frac{t^2}{2} + w_3 t + u_3\right) e^{\lambda_1 t}.$$

Substitution of these solutions into the system of differential equations yields the following system of equations that is solved for the unknown vectors $v_2, w_2, v_3, w_3,$ and u_3:

$$\lambda_1 v_2 = Av_2, (A - \lambda_1 I)w_2 = v_2, \lambda_1 v3 = Av3, (A - \lambda_1 I)w3 = v3, \text{ and } (A - \lambda_1 I)u_3 = w_3.$$

Similar to the previous case, $v_3 = v_2 = v_1$, $w_2 = w_3$, and the vector u_3 is found by solving the system

$$(A - \lambda_1 I)u_3 = w_2.$$

Hence, the three solutions have the form

$$X_1 = v_1 e^{\lambda_1 t}, X_2 = (v_1 t + w_2)e^{\lambda_1 t}, \text{ and } X_3 = \left(v_1 \frac{t^2}{2} + w_2 t + u_3\right) e^{\lambda_1 t}.$$

Notice that this method is easily generalized for instances when the multiplicity of the repeated eigenvalue is greater than 3. We demonstrate the case of an eigenvalue of multiplicity 3 in the following example.

EXAMPLE: Solve

$$\mathbf{X}' = \begin{pmatrix} 1 & 1 & 1 \\ 2 & 1 & -1 \\ -3 & 2 & 4 \end{pmatrix} \mathbf{X}.$$

SOLUTION: The eigenvalues and corresponding eigenvector(s) are found with **eigenvects**.

```
> with(linalg):
  A:=array([[1,1,1],[2,1,-1],[-3,2,4]]):
  vals_and_vecs:=eigenvects(A);
```

$$vals\_and\_vecs := [2, 3, \{[0 \quad -1 \quad 1]\}]$$

Hence, $\lambda_1 = \lambda_2 = \lambda_3 = 2$ with corresponding eigenvector

$$\mathbf{v}_1 = \begin{pmatrix} 0 \\ -1 \\ 1 \end{pmatrix},$$

so one solution is

$$\mathbf{X}_1 = \mathbf{v}_1 e^{2t} = \begin{pmatrix} 0 \\ -1 \\ 1 \end{pmatrix} e^{2t}.$$

The vector

$$\mathbf{w}_2 = \begin{pmatrix} x_2 \\ y_2 \\ z_2 \end{pmatrix}$$

in the second linearly independent solution of the form $\mathbf{X}_2 = (\mathbf{v}_1 t + \mathbf{w}_2)e^{2t}$ is found by solving the system $(\mathbf{A} - \lambda\mathbf{I})\mathbf{w}_2 = \mathbf{v}_1$. This system is

$$\begin{pmatrix} -1 & 1 & 1 \\ 2 & -1 & -1 \\ -3 & 2 & 2 \end{pmatrix} \begin{pmatrix} x_2 \\ y_2 \\ z_2 \end{pmatrix} = \begin{pmatrix} 0 \\ -1 \\ 1 \end{pmatrix},$$

and it is solved as follows with **linsolve**.

```
> alias(Id=&*()):
  b:=array([0,-1,1]):
  sol_1:=linsolve(A-2*Id,b);
```

$$sol\_1 := [-1 \quad \_t_{[1]} \quad -1 - \_t_{[1]}]$$

This result means that $x_2 = -1$ and $z_2 = -y_2 - 1$. If we choose $y_2 = -1$,

```
> w2:=subs(_t[1]=-1,eval(sol_1));
```

$$w2 := [-1 \quad -1 \quad 0]$$

then $z_2 = 0$, so

$$\mathbf{w}_2 = \begin{pmatrix} -1 \\ -1 \\ 0 \end{pmatrix} \text{ and } \mathbf{X}_2 = \left(\begin{pmatrix} 0 \\ -1 \\ 1 \end{pmatrix} t + \begin{pmatrix} -1 \\ -1 \\ 0 \end{pmatrix} \right) e^{2t}.$$

Finally, we must determine the vector

$$\mathbf{u}_3 = \begin{pmatrix} x_3 \\ y_3 \\ z_3 \end{pmatrix}$$

in the third linearly independent solution $\mathbf{X}_3 = \left(\mathbf{v}_1 \dfrac{t^2}{2} + \mathbf{w}_2 t + \mathbf{u}_3 \right) e^{\lambda_1 t}$ by solving the system $(\mathbf{A} - \lambda \mathbf{I})\mathbf{u}_3 = \mathbf{w}_2$.

```
> sol_2:=linsolve(A-2*Id,w2);
```

$$sol\_2 := [-2 \quad \_t_{[1]} \quad -3 - \_t_{[1]}]$$

Therefore, $x_3 = -2$ and $z_3 = -y_3 - 3$. If we select $y_3 = -3$, then $z_3 = 0$.

```
> u3:=subs(_t[1]=-3,eval(sol_2));
```

$$u3 := [-2 \quad -3 \quad 0]$$

Hence,

$$\mathbf{u}_3 = \begin{pmatrix} -2 \\ -3 \\ 0 \end{pmatrix},$$

so the third linearly independent solution is

$$X_3 = \left(\begin{pmatrix} 0 \\ -1 \\ 1 \end{pmatrix} \frac{t^2}{2} + \begin{pmatrix} -1 \\ -1 \\ 0 \end{pmatrix} t + \begin{pmatrix} -2 \\ -3 \\ 0 \end{pmatrix} \right) e^{2t}.$$

A general solution is then given by

$$X = c_1 X_1 + c_2 X_2 + c_3 X_3 = c_1 \begin{pmatrix} 0 \\ -1 \\ 1 \end{pmatrix} e^{2t} + c_2 \left(\begin{pmatrix} 0 \\ -1 \\ 1 \end{pmatrix} t + \begin{pmatrix} -1 \\ -1 \\ 0 \end{pmatrix} \right) e^{2t} + c_3 \left(\begin{pmatrix} 0 \\ -1 \\ 1 \end{pmatrix} \frac{t^2}{2} \right.$$

$$\left. + \begin{pmatrix} -1 \\ -1 \\ 0 \end{pmatrix} t + \begin{pmatrix} -2 \\ -3 \\ 0 \end{pmatrix} \right) e^{2t}$$

$$= \begin{pmatrix} -c_2 e^{2t} + c_3(-t - 2)e^{2t} \\ -c_1 e^{2t} + c_2(-t - 1)e^{2t} + c_3 \left(-\frac{t^2}{2} - t - 3 \right) e^{2t} \\ c_1 e^{2t} + c_2 t e^{2t} + c_3 \left(\frac{t^2}{2} \right) e^{2t} \end{pmatrix}$$

As we have seen in previous examples, we obtain the same result with **dsolve**.

```
> with(student):
  Eqs:=equate([[diff(x(t),t)],[diff(y(t),t)],[diff(z(t),t)]],
    A &* [[x(t)],[y(t)],[z(t)]]);
  Gen_Sol:=dsolve(Eqs,{x(t),y(t),z(t)});
```

$$Eqs := \left\{ \frac{\partial}{\partial t} y(t) = 2x(t) + y(t) - z(t), \frac{\partial}{\partial t} z(t) = -3x(t) + 2y(t) + 4z(t), \right.$$

$$\left. \frac{\partial}{\partial t} x(t) = x(t) + y(t) + z(t) \right\}$$

$$Gen\_Sol := \left\{ y(t) = (\_C1 - 2\_C3 - \_C2)e^{2t} + (2\_C3 + \_C1)e^{2t}t - \_C3e^{2t}t^2, \right.$$

$$z(t) = \_C2e^{2t} + (-\_C1 - 4\_C3)e^{2t}t + \_C3e^{2t}t^2,$$

$$\left. x(t) = \_C1e^{2t} - 2\_C3e^{2t}t \right\}$$

To graph the direction field associated with the equation, we use **fieldplot3d**, which is contained in the **plots** package.

```
> with(plots):
  fieldplot3d([x+y+z,2*x+y-z,-3*x+2*y+4*z],x=-2..2,
   y=-2..2,z=-2..2,axes=BOXED);
```

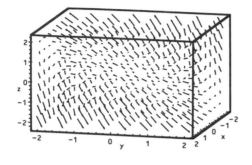

9.5 VARIATION OF PARAMETERS

In Chapter 4, we learned how to solve nonhomogeneous differential equations through the use of undetermined coefficients and variation of parameters. Here we approach the solution of systems of nonhomogeneous equations using variation of parameters.

Let $\mathbf{X} = \mathbf{X}(t) = \begin{pmatrix} x_1(t) \\ x_2(t) \\ \vdots \\ x_n(t) \end{pmatrix}$, $\mathbf{A} = \begin{pmatrix} a_{11} & a_{12} & \cdots & a_{1n} \\ a_{21} & a_{22} & \cdots & a_{2n} \\ \vdots & \vdots & \ddots & \vdots \\ a_{n1} & a_{n2} & \cdots & a_{nn} \end{pmatrix}$, $\mathbf{F}(t) = \begin{pmatrix} f_1(t) \\ f_2(t) \\ \vdots \\ f_n(t) \end{pmatrix}$, and $\Phi(t)$ be a fundamental

matrix of the system $\mathbf{X}' = \mathbf{AX}$. Then a general solution of the homogeneous system $\mathbf{X}' = \mathbf{AX}$ is $\mathbf{X} = \Phi(t)\mathbf{C}$, where

$$\mathbf{C} = \begin{pmatrix} c_1 \\ c_2 \\ \vdots \\ c_n \end{pmatrix}$$

is an $n \times 1$ constant vector.

In much the same way we derived the method of variation of parameters for solving higher-order differential equations, we assume that the particular solution of the nonhomogeneous system can be expressed in the form $X_p = \Phi(t)V(t)$, where

$$V(t) = \begin{pmatrix} v_1(t) \\ v_2(t) \\ \vdots \\ v_n(t) \end{pmatrix}.$$

Notice that $X'_p = \Phi(t)V'(t) + \Phi'(t)V(t)$. Then if X_p satisfies $X' = AX + F(t)$, we have

$$\Phi(t)V'(t) + \Phi'(t)V(t) = A\Phi(t)V(t) + F(t).$$

However, the fundamental matrix $\Phi(t)$ satisfies $X' = AX$, so $\Phi'(t) = A\Phi(t)$. Hence,

$$\Phi(t)V'(t) + A\Phi(t)V(t) = A\Phi(t)V(t) + F(t)$$
$$\Phi(t)V'(t) = F(t).$$

Multiplying both sides of this equation by $\Phi^{-1}(t)$ yields

$$\Phi^{-1}(t)\Phi(t)V'(t) = \Phi^{-1}(t)F(t)$$
$$V'(t) = \Phi^{-1}(t)F(t).$$

Therefore, $V(t) = \int \Phi^{-1}(t)F(t)\,dt$, so the particular solution is

$$X_p = \Phi(t)\int \Phi^{-1}(t)F(t)\,dt,$$

and a general solution of the system is

$$X = \Phi(t)C + X_p = \Phi(t)C + \Phi(t)\int \Phi^{-1}(t)F(t)\,dt.$$

EXAMPLE: Solve $X' = \begin{pmatrix} -5 & 3 \\ 2 & -10 \end{pmatrix} X + \begin{pmatrix} e^{-2t} \\ 1 \end{pmatrix}$.

SOLUTION: In order to apply variation of parameters, we first calculate a fundamental matrix for the associated homogeneous system $X' = \begin{pmatrix} -5 & 3 \\ 2 & -10 \end{pmatrix} X$. We compute the eigenvalues and corresponding eigenvectors of $A = \begin{pmatrix} -5 & 3 \\ 2 & -10 \end{pmatrix}$ with **eigenvects**.

```
> with(linalg):
  A:=array([[-5,3],[2,-10]]):
  vals_and_vecs:=eigenvects(A);
```

$$vals\_and\_vecs := [-11, 1, \{[1 \quad -2]\}], [-4, 1, \{[3 \quad 1]\}]$$

Hence, $\lambda_1 = -4$ and $\lambda_2 = -11$ with corresponding eigenvectors $\mathbf{v}_1 = \begin{pmatrix} 3 \\ 1 \end{pmatrix}$ and $\mathbf{v}_2 = \begin{pmatrix} 1 \\ -2 \end{pmatrix}$,

respectively. Thus, a general solution of this system is $\mathbf{X}(t) = c_1 \begin{pmatrix} 1 \\ -2 \end{pmatrix} e^{-11t} + c_2 \begin{pmatrix} 3 \\ 1 \end{pmatrix} e^{-4t}$, so a

fundamental matrix is then given by $\Phi(t) = \begin{pmatrix} e^{-11t} & 3e^{-4t} \\ -2e^{-11t} & e^{-4t} \end{pmatrix}$ and $\Phi^{-1}(t) = \begin{pmatrix} \frac{1}{7}e^{11t} & \frac{-3}{7}e^{11t} \\ \frac{2}{7}e^{4t} & \frac{1}{7}e^{4t} \end{pmatrix}$.

```
> Phi:=t->array([[exp(-11*t),3*exp(-4*t)],
  [-2*exp(-11*t),exp(-4*t)]]):
  inverse(Phi(t));
```

$$\begin{bmatrix} \dfrac{1}{7}\dfrac{1}{e^{-11t}} & -\dfrac{3}{7}\dfrac{1}{e^{-11t}} \\ \dfrac{2}{7}\dfrac{1}{e^{-4t}} & \dfrac{1}{7}\dfrac{1}{e^{-4t}} \end{bmatrix}$$

Thus,

$$\Phi^{-1}(t)\mathbf{F}(t) = \begin{pmatrix} \dfrac{1}{7}e^{9t} - \dfrac{3}{7}e^{11t} \\ \dfrac{2}{7}e^{2t} + \dfrac{1}{7}e^{4t} \end{pmatrix}$$

```
> F:=t->array([[exp(-2*t)],[1]]):
  step_1:=evalm(inverse(Phi(t)) &* F(t));
```

$$step_1 := \begin{bmatrix} \dfrac{1}{7}\dfrac{e^{-2t}-3}{e^{-11t}} \\ \dfrac{1}{7}\dfrac{2e^{-2t}+1}{7e^{-4t}} \end{bmatrix}$$

and

$$\mathbf{V}(t) = \int \Phi^{-1}(t)\mathbf{F}(t)\,dt = \int \begin{pmatrix} \dfrac{-2}{7}e^{9t} + \dfrac{6}{7}e^{11t} \\ \dfrac{2}{7}e^{2t} + \dfrac{1}{7}e^{4t} \end{pmatrix} dt = \begin{pmatrix} \dfrac{-6}{23}e^{9t} + \dfrac{6}{77}e^{11t} \\ \dfrac{1}{7}e^{2t} + \dfrac{1}{28}e^{4t} \end{pmatrix}.$$

```
> V:=t->map(integrate,step_1,t):
  map(simplify,V(t));
```

$$\begin{bmatrix} \dfrac{1}{63}e^{9t} - \dfrac{3}{77}e^{11t} \\[2ex] \dfrac{1}{7}e^{2t} + \dfrac{1}{28}e^{4t} \end{bmatrix}$$

By variation of parameters, we have the particular solution

$$X_p(t) = \Phi(t) \int \Phi^{-1}(t)F(t)\, dt = \begin{pmatrix} \dfrac{3}{44} + \dfrac{4}{9}e^{-2t} \\[2ex] \dfrac{5}{44} + \dfrac{1}{9}e^{-2t} \end{pmatrix}.$$

```
> map(simplify,evalm(Phi(t) &* V(t)));
```

$$\begin{bmatrix} \dfrac{4}{9}e^{-2t} + \dfrac{3}{44} \\[2ex] \dfrac{1}{9}e^{-2t} + \dfrac{5}{44} \end{bmatrix}$$

Therefore, a general solution is given by

$$X(t) = \Phi(t)C + X_p(t) = \begin{pmatrix} \dfrac{-1}{2}e^{-11t} & 3e^{-4t} \\[2ex] e^{-11t} & e^{-4t} \end{pmatrix}\begin{pmatrix} c_1 \\[1ex] c_2 \end{pmatrix} + \begin{pmatrix} \dfrac{3}{44} + \dfrac{4}{9}e^{-2t} \\[2ex] \dfrac{5}{44} + \dfrac{1}{9}e^{-2t} \end{pmatrix}.$$

We obtain the same results with **dsolve**.

```
> Sol:=dsolve({diff(x(t),t)=-5*x(t)+3*y(t)+exp(-2*t),
  diff(y(t),t)=2*x(t)-10*y(t)+1},
  {x(t),y(t)});
```

$$Sol := \left\{ x(t) = \frac{3}{44} + \frac{4}{9}e^{-2t} + 3e^{-4t}\_C2 + e^{-11t}\_C1, \right.$$

$$\left. y(t) = \frac{1}{9}e^{-2t} + \frac{5}{44} + e^{-4t}\_C2 - 2e^{-11t}\_C1 \right\}$$

To graph the solution for various values of the constants _C1 and _C2, we first name $x(t)$ and $y(t)$ the result obtained in **Sol** with **assign** and then use **seq** and **subs** to define **to_plot** to be the set of ordered triples consisting of $x(t)$, $y(t)$, and **t=-2..2**. Note that **to_plot** is not displayed because a colon is included at the end of the command instead of a semicolon. We then graph the set of functions in **to_plot** with **plot**. The option **view=[-3..3,-3..3]** instructs Maple to display the resulting graphs on the rectangle $[-3, 3] \times [-3, 3]$, while the option `numpoints=200` increases the number of sample points Maple uses in generating the graphs to help the resulting graphs appear smooth.

```
> assign(Sol):
  to_plot:={seq(seq(subs({_C1=i,_C2=j},[x(t),y(t),t=-2..2]),
    i=-1..1),j=-1..1)}:
  plot(to_plot,view=[-3..3,-3..3],numpoints=200);
```

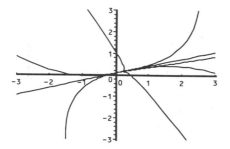

EXAMPLE: Solve $X'(t) = \begin{pmatrix} 2 & 5 \\ -4 & -2 \end{pmatrix} X(t) + \begin{pmatrix} \cos 4t \\ \sin 4t \end{pmatrix}$.

SOLUTION: The eigenvalues and corresponding eigenvectors of $A = \begin{pmatrix} 2 & 5 \\ -4 & -2 \end{pmatrix}$ are obtained using `eigenvects` followed by `map` and `allvalues`.

```
> with(linalg):
  A:=array([[2,5],[-4,-2]]):
  vals_and_vecs:=eigenvects(A);
```

$$vals\_and\_vecs := \left[\text{RootOf}(\_Z^2 + 16), 1, \left\{ \left[-\frac{1}{4}\text{RootOf}(\_Z^2 + 16) - \frac{1}{2} \quad 1 \right] \right\} \right]$$

```
> map(allvalues,vals_and_vecs);
```

$$\left[4I, -4I, 1, \left\{ \left[-\frac{1}{2} - I \quad 1 \right] \right\}, \left\{ \left[-\frac{1}{2} + I \quad 1 \right] \right\} \right]$$

This result means that the eigenvalues are $\lambda_1 = 4i$, with corresponding eigenvector $\mathbf{v}_1 = \begin{pmatrix} -\dfrac{1}{2} - i \\ 1 \end{pmatrix}$

and $\lambda_2 = -4i$ with corresponding eigenvector $\mathbf{v}_2 = \begin{pmatrix} -\dfrac{1}{2} + i \\ 1 \end{pmatrix}$. Using the notation used in Sec-

tion 9.3, $\mathbf{a} = \begin{pmatrix} -\dfrac{1}{2} \\ 1 \end{pmatrix}$ and $\mathbf{b} = \begin{pmatrix} -1 \\ 0 \end{pmatrix}$. Hence, a general solution is

$$X = c_1 \left[\begin{pmatrix} -\dfrac{1}{2} \\ 1 \end{pmatrix} \cos 4t - \begin{pmatrix} -1 \\ 0 \end{pmatrix} \sin 4t \right] + c_2 \left[\begin{pmatrix} -\dfrac{1}{2} \\ 1 \end{pmatrix} \sin 4t + \begin{pmatrix} -1 \\ 0 \end{pmatrix} \cos 4t \right]$$

$$= \begin{pmatrix} -\dfrac{1}{2} \cos 4t - \sin 4t & -\dfrac{1}{2} \sin 4t - \cos 4t \\ \cos 4t & \sin 4t \end{pmatrix} \begin{pmatrix} c_1 \\ c_2 \end{pmatrix}.$$

Multiplying by 2 to eliminate the fractions, a fundamental matrix is given by

$$\Phi(t) = \begin{pmatrix} -\cos 4t + 2\sin 4t & -\sin 4t - 2\cos 4t \\ 2\cos 4t & 2\sin 4t \end{pmatrix}$$

```
> Phi:=t->array([[-cos(4*t)+2*sin(4*t),-sin(4*t)-2*cos(4*t)],
  [2*cos(4*t),2*sin(4*t)]]):
  Phi(t);
```

$$\begin{bmatrix} -\cos(4t) + 2\sin(4t) & -\sin(4t) - 2\cos(4t) \\ 2\cos(4t) & 2\sin(4t) \end{bmatrix}$$

with inverse

$$\Phi^{-1}(t) = \begin{pmatrix} \dfrac{\sin 4t}{2} & \dfrac{2\cos 4t + \sin 4t}{4} \\ -\dfrac{\cos 4t}{2} & \dfrac{-(\cos 4t + 2\sin 4t)}{4} \end{pmatrix}.$$

```
> inv_Phi:=map(simplify,inverse(Phi(t)),trig);
```

$$inv\_Phi := \begin{bmatrix} \dfrac{1}{2}\sin(4t) & \dfrac{1}{4}\sin(4t) + \dfrac{1}{2}\cos(4t) \\ -\dfrac{1}{2}\cos(4t) & -\dfrac{1}{4}\cos(4t) + \dfrac{1}{2}\sin(4t) \end{bmatrix}$$

Therefore,

$$\Phi^{-1}(t)\begin{pmatrix}\cos 4t\\ \sin 4t\end{pmatrix}=\begin{pmatrix}\dfrac{\sin 8t}{2}+\dfrac{1}{8}-\dfrac{\cos 8t}{8}\\[2mm] -\dfrac{\sin 8t}{8}-\dfrac{1}{2}\cos 8t\end{pmatrix}.$$

```
> F:=t->array([[cos(4*t)],[sin(4*t)]]):
  step_1:=evalm(inv_Phi &* F(t));latex("");
```

$$step\_1:=\begin{bmatrix}\sin(4t)\cos(4t)+\dfrac{1}{4}\sin(4t)^2\\[2mm] -\dfrac{1}{2}\cos(4t)^2-\dfrac{1}{4}\sin(4t)\cos(4t)+\dfrac{1}{2}\sin(4t)^2\end{bmatrix}$$

```
> step_2:=map(combine,step_1,trig);
```

$$step\_2:=\begin{bmatrix}\dfrac{1}{2}\sin(8t)+\dfrac{1}{8}-\dfrac{1}{8}\cos(8t)\\[2mm] -\dfrac{1}{2}\cos(8t)-\dfrac{1}{8}\sin(8t)\end{bmatrix}$$

Because

$$\int\Phi^{-1}(t)\begin{pmatrix}\cos(4t)\\ \sin(4t)\end{pmatrix}dt=\begin{pmatrix}-\dfrac{\cos 8t}{16}+\dfrac{t}{8}-\dfrac{\sin 8t}{64}\\[2mm] \dfrac{\cos 8t}{64}-\dfrac{\sin 8t}{16}\end{pmatrix},$$

```
> step_3:=map(int,step_2,t);
```

$$step\_3:=\begin{bmatrix}-\dfrac{1}{16}\cos(8t)+\dfrac{1}{8}t-\dfrac{1}{64}\sin(8t)\\[2mm] -\dfrac{1}{16}\sin(8t)16+\dfrac{1}{64}\cos(8t)\end{bmatrix}$$

the particular solution is

$$\mathbf{X}_p(t) = \Phi(t) \int \Phi^{-1}(t) \begin{pmatrix} \cos(4t) \\ \sin(4t) \end{pmatrix} dt$$

$$= \begin{pmatrix} -\cos 4t + 2\sin 4t & -\sin 4t - 2\cos 4t \\ 2\cos 4t & 2\sin 4t \end{pmatrix} \begin{pmatrix} -\dfrac{\cos 8t}{16} + \dfrac{t}{8} - \dfrac{\sin 8t}{64} \\ \dfrac{\cos 8t}{64} - \dfrac{\sin 8t}{16} \end{pmatrix}$$

$$= \begin{pmatrix} \dfrac{3\cos 4t \cos 8t}{32} + \dfrac{7t\cos 4t}{8} + \dfrac{\cos 4t \sin 8t}{64} - \dfrac{7\sin 4t \cos 8t}{64} + \dfrac{3t\sin 4t}{4} - \dfrac{\sin 4t \sin 8t}{32} \\ -\dfrac{\cos 4t \cos 8t}{8} + \dfrac{t\cos 4t}{4} - \dfrac{\cos 4t \sin 8t}{32} + \dfrac{\sin 4t \cos 8t}{32} - \dfrac{\sin 4t \sin 8t}{8} \end{pmatrix}.$$

```
> evalm(Phi(t) &* step_3);
```

$$\left[\frac{1}{32} \cos(4t)\cos(8t) - \frac{1}{8}\cos(4t)t + \frac{9}{64}\cos(4t)\sin(8t) \right.$$

$$\left. -\frac{9}{64}\sin(4t)\cos(8t) + \frac{1}{4}\sin(4t)t + \frac{1}{32}\sin(4t)\sin(8t) \right]$$

$$\left[-\frac{1}{8}\cos(4t)\cos(8t) + \frac{1}{4}\cos(4t)t - \frac{1}{32}\cos(4t)\sin(8t) \right.$$

$$\left. -\frac{1}{8}\sin(4t)\sin(8t) + \frac{1}{32}\sin(4t)\cos(8t) \right]$$

Therefore, a general solution is

$$\mathbf{X}(t) = \Phi(t)\mathbf{C} + \mathbf{X}_p(t) = \begin{pmatrix} -\cos 4t + 2\sin 4t & -\sin 4t - 2\cos 4t \\ 2\cos 4t & 2\sin 4t \end{pmatrix} \begin{pmatrix} c_1 \\ c_2 \end{pmatrix}$$

$$+ \begin{pmatrix} \dfrac{3\cos 4t \cos 8t}{32} + \dfrac{7t\cos 4t}{8} + \dfrac{\cos 4t \sin 8t}{64} - \dfrac{7\sin 4t \cos 8t}{64} + \dfrac{3t\sin 4t}{4} - \dfrac{\sin 4t \sin 8t}{32} \\ -\dfrac{\cos 4t \cos 8t}{8} + \dfrac{t\cos 4t}{4} - \dfrac{\cos 4t \sin 8t}{32} + \dfrac{\sin 4t \cos 8t}{32} - \dfrac{\sin 4t \sin 8t}{8} \end{pmatrix}.$$

After forming the system using `equate`, which is contained in the `student` package,

```
> with(student):
  Eqs:=equate([[diff(x(t),t)],[diff(y(t),t)]],
   A &* [[x(t)],[y(t)]] + F(t));
```

$$Eqs := \left\{ \frac{\partial}{\partial t} y(t) = -4x(t) - 2y(t) + \sin(4t), \frac{\partial}{\partial t} x(t) = 2x(t) + 5y(t) + \cos(4t) \right\}$$

we obtain the same solution as that obtained previously using **dsolve** together with the **laplace** option. Note that if we forget to include the **laplace** option, the result returned by **dsolve** is incorrect.

```
> x:='x':y:='y':
  Gen_Sol:=dsolve(Eqs,{x(t),y(t)},laplace);
```

$$Gen\_Sol := \left\{ x(t) = -\frac{1}{8}\cos(4t)t + \frac{1}{4}\sin(4t)t + \frac{9}{32}\sin(4t) + x(0)\cos(4t) \right.$$

$$+ \frac{5}{4}y(0)\sin(4t) + \frac{1}{2}x(0)\sin(4t), y(t) = \frac{1}{4}\cos(4t)t - \frac{1}{16}\sin(4t)$$

$$\left. + y(0)\cos(4t) - \frac{1}{2}y(0)\sin(4t) - x(0)\sin(4t) \right\}$$

To graph the solution obtained with **dsolve** for various initial conditions, we first name $x(t)$ and $y(t)$ the results given in **Gen_Sol** with **assign** and then use **seq** to form the list of ordered pairs, named **c_vals**, consisting of (i, j) for $i = -1, 0$, and 1 and $j = -1, 0$, and 1.

```
> assign(Gen_Sol):
  c_vals:=[seq(seq([i,j],i=-1..1),j=-1..1)];
```

$$c\_vals := [[-1,-1],[0,-1],[1,-1],[-1,0],[0,0],[1,0],[-1,1],[0,1],[1,1]]$$

The command

```
> subs({x(0)=c_vals[k][1],y(0)=c_vals[k][2]},[x(t),y(t),
  t=-4..4])
```

substitutes $x(0)$ by the first coordinate of the kth ordered pair in the list **c_vals** and $y(0)$ by the second coordinate of the kth ordered pair in the list **c_vals**.

We now use a **for** loop to graph the nine solutions obtained by replacing $x(0)$ by the first coordinate of the kth ordered pair in the list **c_vals** and $y(0)$ by the second coordinate of the kth ordered pair in the list **c_vals** for $k = 1, 2, \ldots, 9$ and for $-4 \leq t \leq 4$. The option **numpoints=200** in the **plot** command instructs Maple to use 200 sample points when generating each graph, thus helping to assure that the resulting graphs appear smooth.

```
>  k:='k':
   for k to 9 do
     plot(subs({x(0)=c_vals[k][1],y(0)=c_vals[k][2]},
     [x(t),y(t),t=-4..4]),numpoints=200) od;
```

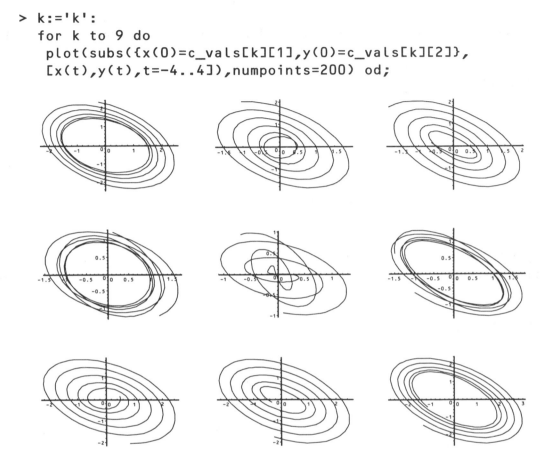

9.6 LAPLACE TRANSFORMS

In many cases, Laplace transforms can be used to solve initial-value problems that involve a system of linear differential equations. This method is applied in much the same way that it was in solving initial-value problems involving higher-order differential equations. In this case, however, a system of algebraic equations is obtained after taking the Laplace transform of each equation. After solving for the Laplace transform of each of the unknown functions, the inverse Laplace transform is used to find each unknown function in the solution of the system. We illustrate this technique in the following example.

EXAMPLE: Solve $X' = \begin{pmatrix} 0 & 1 \\ 1 & 0 \end{pmatrix} X + \begin{pmatrix} \sin t \\ 2\cos t \end{pmatrix}$ subject to $X(0) = \begin{pmatrix} 2 \\ 0 \end{pmatrix}$.

SOLUTION: Let $X(t) = \begin{pmatrix} x(t) \\ y(t) \end{pmatrix}$. Then we can rewrite this problem as

$$\begin{cases} x' = y + \sin t \\ y' = x + 2\cos t \end{cases}, x(0) = 2, y(0) = 0.$$

```
> Eqs:={diff(x(t),t)=y(t)+sin(t),
    diff(y(t),t)=x(t)+2*cos(t)};
```

$$Eqs := \left\{ \frac{\partial}{\partial t} x(t) = y(t) + \sin(t), \frac{\partial}{\partial t} y(t) = x(t) + 2\cos(t) \right\}$$

Taking the Laplace transform of both sides of each equation yields the system

$$\begin{cases} sX(s) - x(0) = Y(s) + \dfrac{1}{s^2 + 1} \\ sY(s) - y(0) = X(s) + \dfrac{2s}{s^2 + 1} \end{cases},$$

```
> step_1:=map(laplace,Eqs,t,s);
```

$$step\_1 := \left\{ \text{laplace}(x(t), t, s)s - x(0) = \text{laplace}(y(t), t, s) + \frac{1}{s^2 + 1}, \right.$$

$$\left. \text{laplace}(y(t), t, s)s - y(0) = \text{laplace}(x(t), t, s) + 2\frac{s}{s^2 + 1} \right\}$$

and applying the initial conditions results in

$$\begin{cases} sX(s) - Y(s) = \dfrac{1}{s^2 + 1} + 2 \\ -X(s) + sY(s) = \dfrac{2s}{s^2 + 1} \end{cases}.$$

```
> to_solve:=subs({x(0)=2,y(0)=0},step_1);
```

$$to\_solve := \left\{ \text{laplace}(x(t), t, s)s - 2 = \text{laplace}(y(t), t, s) + \frac{1}{s^2 + 1}, \right.$$

$$\left. \text{laplace}(y(t), t, s)s = \text{laplace}(x(t), t, s) + 2\frac{s}{s^2 + 1} \right\}$$

Solving the system of equations `to_solve` for `laplace(x(t),t,s)`, and `laplace(y(t),t,s)`, which represent the Laplace transform of $x(t)$, $X(s)$, and the Laplace transform of $y(t)$, $Y(s)$, respectively, with `solve` yields

```
> step_2:=solve(to_solve,
    {laplace(x(t),t,s),laplace(y(t),t,s)});
```

$$step\_2 := \left\{ \text{laplace}(y(t), t, s) = \frac{4s^2 + 3}{-1 + s^4}, \text{laplace}(x(t), t, s) = \frac{(2s^2 + 5)s}{-1 + s^4} \right\}$$

We then use `invlaplace` to compute the inverse Laplace transform to see that

$$x(t) = \frac{7}{4}e^t + \frac{7}{4}e^{-t} - \frac{3}{2}\cos t \qquad \text{and} \qquad y(t) = \frac{7}{4}e^t - \frac{7}{4}e^{-t} + \frac{1}{2}\sin t.$$

```
> readlib(laplace):
  Sol:=map(invlaplace,step_2,s,t);
```

$$Sol := \left\{ y(t) = \frac{7}{4}e^t - \frac{7}{4}e^{-t} + \frac{1}{2}\sin(t), x(t) = \frac{7}{4}e^t + \frac{7}{4}e^{-t} - \frac{3}{2}\cos(t) \right\}$$

To graph the solution, we first use `assign` to name $x(t)$ and $y(t)$ the result obtained in `Sol` and then use `plot` to graph the solution for $-3 \le t \le 3$.

```
> assign(Sol):
  plot([x(t),y(t),t=-3..3]);
```

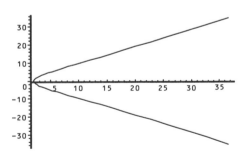

We obtain the same results with `dsolve`. First, we use `union` to add the initial conditions to the set of equations `Eqs`, naming the resulting set of equations `Sys_of_Eqs`,

```
> x:='x':y:='y':
  Sys_of_Eqs:=Eqs union {x(0)=2,y(0)=0};
```

$$Sys\_of\_Eqs := \left\{ \frac{\partial}{\partial t}x(t) = y(t) + \sin(t), \frac{\partial}{\partial t}y(t) = x(t) + 2\cos(t), x(0) = 2, y(0) = 0 \right\}$$

and then use `dsolve` to solve the initial-value problem.

```
> dsolve(Sys_of_Eqs,{x(t),y(t)});
```

$$\left\{ y(t) = \frac{7}{4}e^t - \frac{7}{4}e^{-t} + \frac{1}{2}\sin(t), x(t) = \frac{7}{4}e^t + \frac{7}{4}e^{-t} - \frac{3}{2}\cos(t) \right\}$$

In Section 9.1, we solved systems that involve higher-order differential equations with differential operators. In many cases, these systems can be solved with Laplace transforms, as illustrated in the following example.

EXAMPLE: Solve $\begin{cases} x'' = 3x' - y' - 2x + y \\ x' + y' = 2x - y \end{cases}$, $x(0) = 0, x'(0) = 0, y(0) = -1$.

SOLUTION: After defining `Eqs` to be the system of equations $\begin{cases} x'' = 3x' - y' - 2x + y \\ x' + y' = 2x - y \end{cases}$, we use `laplace` to take the Laplace transform of both equations, naming the resulting set of equations `step_1`,

```
> x:='x':y:='y':
  Eqs:={diff(x(t),t$2)=3*diff(x(t),t)-diff(y(t),t)-2*x(t)+y(t),
  diff(x(t),t)+diff(y(t),t)=2*x(t)-y(t)};
```

$$Eqs := \left\{ \left(\frac{\partial}{\partial t}x(t)\right) + \left(\frac{\partial}{\partial t}y(t)\right) = 2x(t) - y(t), \right.$$

$$\left. \frac{\partial^2}{\partial t^2}x(t) = 3\left(\frac{\partial}{\partial t}x(t)\right) - \left(\frac{\partial}{\partial t}y(t)\right) - 2x(t) + y(t) \right\}$$

```
> step_1:=map(laplace,Eqs,t,s);
```

$$step\_1 := \{(\text{laplace}\,(x(t), t, s)s - x(0))s - D(x)(0) = 3\,\text{laplace}\,(x(t), t, s)s - 3x(0)$$
$$- \text{laplace}(y(t), t, s)s + y(0) - 2\,\text{laplace}\,(x(t), t, s) + \text{laplace}\,(y(t), t, s),$$
$$\text{laplace}\,(x(t), t, s)s - x(0) + \text{laplace}\,(y(t), t, s)s - y(0)$$
$$= 2\,\text{laplace}\,(x(t), t, s) - \text{laplace}\,(y(t), t, s)\}$$

and then use `subs` to apply the initial conditions, naming the resulting set of equations `to_solve`.

```
> to_solve:=subs({x(0)=0,D(x)(0)=0,y(0)=-1},step_1);
```

$$to\_solve := \{\text{laplace}\,(x(t), t, s)s^2 = 3\,\text{laplace}\,(x(t), t, s)s - \text{laplace}\,(y(t), t, s)s - 1$$
$$-2\,\text{laplace}\,(x(t), t, s) + \text{laplace}\,(y(t), t, s),$$
$$\text{laplace}\,(x(t), t, s)s + \text{laplace}\,(y(t), t, s)s + 1$$
$$= 2\,\text{laplace}\,(x(t), t, s) - \text{laplace}\,(y(t), t, s)\}$$

We then use **solve** to solve the system of equations **to_solve** for **laplace(x(t),t,s)**, which represents the Laplace transform of $x(t)$, and **laplace(y(t),t,s)**, which represents the Laplace transform of $y(t)$.

```
> step_2:=solve(to_solve,
    {laplace(x(t),t,s),laplace(y(t),t,s)});
```

$$step\_2 := \left\{\text{laplace}\,(y(t), t, s) = -\frac{-2 + s}{(s - 1)s}, \text{laplace}\,(x(t), t, s) = -2\frac{1}{(-2 + s)s(s - 1)}\right\}$$

The solution is obtained by computing the inverse Laplace transform with **invlaplace**. The resulting set of equations is named **Sol**.

```
> readlib(laplace):
    Sol:=map(invlaplace,step_2,s,t);
```

$$Sol := \{y(t) = -2 + e^t, x(t) = -e^{2t} - 1 + 2e^t\}$$

After naming $x(t)$ and $y(t)$ the result obtained in **Sol** with **assign**, we graph the solution for $-1 \le t \le 1$ with **plot**.

```
> assign(Sol):
    plot([x(t),y(t),t=-1..1]);
```

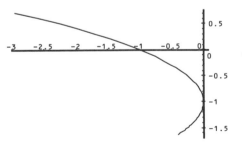

In the same way as in the previous example, we see that we obtain the same result with **dsolve** after using **union** to add the initial conditions to the set of equations **Eqs**.

```
> x:='x':y:='y':
  Sys_of_Eqs:=Eqs union {x(0)=0,D(x)(0)=0,y(0)=-1}:
  dsolve(Sys_of_Eqs,{x(t),y(t)});
```

$$\{y(t) = -2 + e^t, x(t) = -e^{2t} - 1 + 2e^t\}$$

9.7 NONLINEAR SYSTEMS, LINEARIZATION, AND CLASSIFICATION OF EQUILIBRIUM POINTS

We now turn our attention to the systems of equations of the form

$$\begin{cases} x' = f(x, y) \\ y' = g(x, y) \end{cases}.$$

This system is called **autonomous**, because f and g do not depend explicitly on the independent variable t. To understand problems of this type, however, we must first introduce some terminology.

DEFINITION

> *Equilibrium Point*
>
> A point (x_0, y_0) is an **equilibrium point** of the system $\begin{cases} x' = f(x, y) \\ y' = g(x, y) \end{cases}$ if $f(x_0, y_0) = 0$ and $g(x_0, y_0) = 0$.

EXAMPLE: Find the equilibrium points of the system $\begin{cases} x' = 2x - y \\ y' = -x + 3y \end{cases}$.

SOLUTION: The equilibrium points of the system $\begin{cases} x' = 2x - y \\ y' = -x + 3y \end{cases}$ are the solutions of the system of equations $\begin{cases} 2x - y = 0 \\ -x + 3y = 0 \end{cases}$. The only solution of this system is $(0,0)$, so the only equilibrium point is $(0, 0)$.

Before we move on to nonlinear systems, we will first investigate the properties of systems of the form

$$\begin{cases} x' = ax + by \\ y' = cx + dy \end{cases},$$

where $\begin{vmatrix} a & b \\ c & d \end{vmatrix} = ad - bc \neq 0$. The only equilibrium point of this system is $(0, 0)$. We have solved many

systems of this type by using the eigenvalues and corresponding eigenvectors of $A = \begin{pmatrix} a & b \\ c & d \end{pmatrix}$.

The behavior of the solutions of a system of differential equations and the classification of the equilibrium point depend on the eigenvalues and corresponding eigenvectors of the system. We now investigate more thoroughly the cases that can arise in solving the system

$$\begin{cases} x' = ax + by \\ y' = cx + dy \end{cases}.$$

☰ Real Distinct Eigenvalues

Suppose that λ_1 and λ_2 are real eigenvalues of the system where $\lambda_2 < \lambda_1$. Then a general solution is

$$X(t) = \begin{pmatrix} x(t) \\ y(t) \end{pmatrix} = c_1 v_1 e^{\lambda_1 t} + c_2 v_2 e^{\lambda_2 t} = e^{\lambda_1 t} \left[c_1 v_1 + c_2 v_2 e^{(\lambda_2 - \lambda_1)t} \right].$$

(a) Suppose that both eigenvalues are negative. If we assume that $\lambda_2 < \lambda_1 < 0$, then $\lambda_2 - \lambda_1 < 0$. This means that $e^{(\lambda_2 - \lambda_1)t}$ is small for large values of t, so $X(t) \approx c_1 v_1 e^{\lambda_1 t}$ is small for large values of t. If $c_1 \neq 0$, then $\lim_{t \to \infty} X(t) = 0$ in one of the directions determined by v_1 or v_1. If $c_1 = 0$, then $X(t) = c_2 v_2 e^{\lambda_2 t}$. Again, because $\lambda_2 < 0$, $\lim_{t \to \infty} X(t) = 0$ in the directions determined by v_2 and $-v_2$. In this case, $(0, 0)$ is a **stable node**.

(b) Suppose that both eigenvalues are positive. If $0 < \lambda_2 < \lambda_1$, then $e^{\lambda_1 t}$ and $e^{\lambda_2 t}$ both become unbounded as t increases. If $c_1 \neq 0$, then $X(t)$ becomes unbounded in either the direction of v_1 or $-v_1$. If $c_1 = 0$, then $X(t)$ becomes unbounded in the directions given by v_2 and $-v_2$. In this case, $(0, 0)$ is an **unstable node**.

(c) Suppose that the eigenvalues have opposite sign. Then, if $\lambda_2 < 0 < \lambda_1$ and $c_1 \neq 0$, $X(t)$ becomes unbounded in either the direction of v_1 or $-v_1$ as it did in (b). However, if $c_1 = 0$, then due to the fact that $\lambda_2 < 0$, $\lim_{t \to \infty} X(t) = 0$ along the line determined by v_2. If the initial point $X(0)$ is not on the line determined by v_2, then the line given by v_1 is an asymptote for the solution. We say that $(0, 0)$ is a **saddle point** in this case.

EXAMPLE: Classify the equilibrium point $(0, 0)$ in the systems (a) $\begin{cases} x' = 2x - 7y \\ y' = -3y \end{cases}$,

(b) $\begin{cases} x' = -4x + 8y \\ y' = x - 6y \end{cases}$, and (c) $\begin{cases} x' = -5x + 6y \\ y' = -7x + 8y \end{cases}$.

SOLUTION: (a) After loading the **linalg** package and defining $A = \begin{pmatrix} 2 & -7 \\ 0 & -3 \end{pmatrix}$, the eigenvalues and corresponding eigenvectors are found with **eigenvects**.

```
> with(linalg):
  A:=array([[2,-7],[0,-3]]):
  eigenvects(A);
```

$$\left[-3, 1, \left\{\begin{bmatrix} \frac{7}{5} & 1 \end{bmatrix}\right\}\right], [2, 1, \{[1 \quad 0]\}]$$

Hence, $\lambda_1 = -3$ and $\lambda_2 = 2$ with corresponding eigenvectors $v_1 = \begin{pmatrix} 7/5 \\ 1 \end{pmatrix}$ and $v_2 = \begin{pmatrix} 1 \\ 0 \end{pmatrix}$, respectively. Because these eigenvalues have opposite sign, $(0, 0)$ is a saddle point. The solution becomes unbounded in the directions associated with the positive eigenvalue, $v_2 = \begin{pmatrix} 1 \\ 0 \end{pmatrix}$ and $-v_2 = \begin{pmatrix} -1 \\ 0 \end{pmatrix}$. Along the line determined by $v_1 = \begin{pmatrix} 7/5 \\ 1 \end{pmatrix}$, the solution approaches $(0, 0)$. After loading the DEtools package, the direction field and graphs of several solutions to the system are generated with DEplot2.

```
> with(DEtools):
  DEplot2(A,[x,y],t=-1..1,x=-1..1,y=-1..1);
  DEplot2(A,[x,y],t=-1..1,{[0,.5,.25],[0,-.5,-.25],
     [0,.5,-.25],[0,-.5,.25]},x=-1..1,y=-1..1,stepsize=0.05);
```

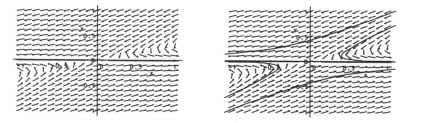

(b) As in (a), the eigenvalues and corresponding eigenvectors of $A = \begin{pmatrix} -4 & 8 \\ 1 & -6 \end{pmatrix}$ are determined with eigenvects. The eigenvalues $\lambda_1 = -1$ and $\lambda_2 = -2$ are both negative.

```
> A:=array([[-4,8],[1,-6]]):
  eigenvects(A);
```

$$[-8, 1, \{[-2 \quad 1]\}], [-2, 1, \{[4 \quad 1]\}]$$

The result means that the eigenvalues are $\lambda_1 = -8$ and $\lambda_2 = -2$ with corresponding eigenvectors $v_1 = \begin{pmatrix} -2 \\ 1 \end{pmatrix}$ and $v_2 = \begin{pmatrix} 4 \\ 1 \end{pmatrix}$, respectively. Hence, $(0, 0)$ is a stable node. Therefore, the solutions approach $(0, 0)$ along the lines given by these vectors. The direction field and graphs of several solutions to the system are generated here with DEplot2.

```
> DEplot2(A,[x,y],t=-1..1,x=-1..1,y=-1..1);
  DEplot2(A,[x,y],t=-1..1,{[0,.5,.25],[0,-.5,-.25],
  [0,.5,-.25],[0,-.5,.25]},x=-1..1,y=-1..1,stepsize=0.05);
```

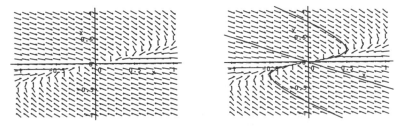

(c) After determining the eigenvalues and corresponding eigenvectors of $\mathbf{A} = \begin{pmatrix} -5 & 6 \\ -7 & 8 \end{pmatrix}$ with

`eigenvects`,

```
> A:=array([[-5,6],[-7,8]]);
  eigenvects(A);
```

$$\left[1, 1, \{[1 \quad 1]\}\right], \left[2, 1, \left\{\left[1 \quad \frac{7}{6}\right]\right\}\right]$$

we see that both eigenvalues are positive, so $(0,0)$ is an unstable node. Note that the corresponding eigenvectors are $\mathbf{v}_1 = \begin{pmatrix} 1 \\ 1 \end{pmatrix}$ and $\mathbf{v}_2 = \begin{pmatrix} 1 \\ 7/6 \end{pmatrix}$, respectively. Hence, the solutions become unbounded along the lines determined by these vectors. As in (a) and (b), the direction field and graphs of several solutions to the system are generated with `DEplot2`.

```
> DEplot2(A,[x,y],t=-1..1,x=-1..1,y=-1..1);
  DEplot2(A,[x,y],t=-1..1,{[0,.5,.25],[0,-.5,-.25],
  [0,.5,-.35],[0,-.5,.35]},x=-1..1,y=-1..1,stepsize=0.05);
```

≡ Repeated Eigenvalues

We recall from our previous experience with repeated eigenvalues that the eigenvalue can have either two linearly independent eigenvectors or only one eigenvector associated with it. Hence, we investigate the behavior of solutions in this case by considering both of these possibilities.

(1) Suppose that the eigenvalue $\lambda = \lambda_1 = \lambda_2$ has two corresponding linearly independent eigenvectors $\mathbf{v}_1$ and $\mathbf{v}_2$. Then a general solution is

$$\mathbf{X}(t) = c_1\mathbf{v}_1 e^{\lambda t} + c_2\mathbf{v}_2 e^{\lambda t} = (c_1\mathbf{v}_1 + c_2\mathbf{v}_2)e^{\lambda t}.$$

Hence, if $\lambda > 0$, then $\mathbf{X}(t)$ becomes unbounded along the line determined by the vector $c_1\mathbf{v}_1 + c_2\mathbf{v}_2$, where c_1 and c_2 are arbitrary constants. In this case, we call the equilibrium point a **degenerate unstable node** (or an **unstable star**). Conversely, if $\lambda < 0$, then $\mathbf{X}(t)$ approaches $(0,0)$ along these lines, and we call $(0,0)$ a **degenerate stable node** (or **stable star**). Note that the name "star" was selected due to the shape of the solutions.

(2) Suppose that $\lambda = \lambda_1 = \lambda_2$ has only one corresponding eigenvector $\mathbf{v}_1$. Hence, a general solution is

$$\mathbf{X}(t) = c_1\mathbf{v}_1 e^{\lambda t} + c_2[\mathbf{v}_1 t + \mathbf{w}_2]e^{\lambda t} = (c_1\mathbf{v}_1 + c_2\mathbf{w}_2)e^{\lambda t} + c_2\mathbf{v}_1 te^{\lambda t},$$

where $(\mathbf{A} - \lambda\mathbf{I})\mathbf{w}_2 = \mathbf{v}_1$. If we write this solution as $\mathbf{X}(t) = te^{\lambda t}\left[\frac{1}{t}(c_1\mathbf{v}_1 + c_2\mathbf{w}_2) + c_2\mathbf{v}_1\right]$, then we can more easily investigate the behavior of this solution. If $\lambda < 0$, then $\lim_{t\to\infty} te^{\lambda t} = 0$ and $\lim_{t\to\infty}\left[\frac{1}{t}(c_1\mathbf{v}_1 + c_2\mathbf{w}_2) + c_2\mathbf{v}_1\right] = c_2\mathbf{v}_1$. Hence, the solutions approach $(0,0)$ along the line determined by $\mathbf{v}_1$, and we call $(0,0)$ a **degenerate stable node**. If $\lambda > 0$, then the solutions become unbounded along this line, and we say that $(0,0)$ is a **degenerate unstable node**.

EXAMPLE: Classify the equilibrium point $(0,0)$ in the systems (a) $\begin{cases} x' = x + 9y \\ y' = -x - 5y \end{cases}$ and (b) $\begin{cases} x' = 2x \\ y' = 2y \end{cases}$.

SOLUTION: (a) After loading the `linalg` package, the eigenvalues and corresponding eigenvector(s) of $\mathbf{A} = \begin{pmatrix} 1 & 9 \\ -1 & -5 \end{pmatrix}$ are found with `eigenvects`.

```
> with(linalg):
  A:=array([[1,9],[-1,-5]]):
  eigenvects(A);

  [-2,2,{[-3  1]}]
```

Hence, $\lambda_1 = \lambda_2 = -2$ with corresponding eigenvector $v_1 = \begin{pmatrix} -3 \\ 1 \end{pmatrix}$. Therefore, because $\lambda = -2 < 0$, $(0,0)$ is a degenerate stable node. Notice that in the graph of several members of the family of solutions of this system along with the direction field generated with DEplot2, the solutions approach $(0,0)$ along the line in the direction of $v_1 = \begin{pmatrix} -3 \\ 1 \end{pmatrix}$ and $y = -\dfrac{x}{3}$.

```
> with(DEtools):
  DEplot2(A,[x,y],t=-1..1,x=-1..1,y=-1..1);
  DEplot2(A,[x,y],t=-1..1,{[0,.75,.5],[0,.5,-.75],
  [0,-.5,.75],[0,-.75,-.5]},x=-1..1,y=-1..1);
```

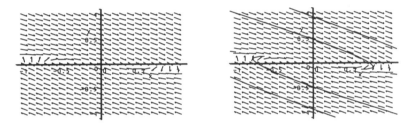

(b) As in (a), we use **eigenvects** to find the eigenvalues and corresponding eigenvector(s) of $\mathbf{A} = \begin{pmatrix} 2 & 0 \\ 0 & 2 \end{pmatrix}$.

```
> A:=array([[2,0],[0,2]]):
  eigenvects(A);
```

$$[2, 2, \{[1 \quad 0], [0 \quad 1]\}]$$

We have $\lambda_1 = \lambda_2 = 2$ with corresponding eigenvectors $v_1 = \begin{pmatrix} 1 \\ 0 \end{pmatrix}$ and $v_2 = \begin{pmatrix} 0 \\ 1 \end{pmatrix}$. Since $\lambda = 2 > 0$, we classify $(0,0)$ as a degenerate unstable node (or star). Some of these solutions along with the direction field are graphed here with DEplot2. Notice that they become unbounded in the direction of any vector in the xy-plane since $v_1 = \begin{pmatrix} 1 \\ 0 \end{pmatrix}$ and $v_2 = \begin{pmatrix} 0 \\ 1 \end{pmatrix}$.

```
> DEplot2(A,[x,y],t=-1..1,x=-1..1,y=-1..1);
  DEplot2(A,[x,y],t=-1..1,{[0,.75,.5],[0,.5,-.75],
  [0,-.5,.75],[0,-.75,-.5]},x=-1..1,y=-1..1);
```

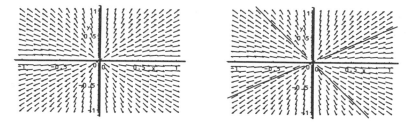

☰ Complex Conjugate Eigenvalues

We have seen that if the eigenvalues of the system of differential equations are $\lambda_1 = \alpha + \beta i$ and $\lambda_2 = \alpha - \beta i$ with corresponding eigenvectors $\mathbf{v}_1 = \mathbf{a} + \mathbf{b}i$ and $\mathbf{v}_2 = \mathbf{a} - \mathbf{b}i$, then two linearly independent solutions of the system are $\mathbf{X}_1(t) = e^{\alpha t}(\mathbf{a}\cos\beta t - \mathbf{b}\sin\beta t)$ and $\mathbf{X}_2(t) = e^{\alpha t}(\mathbf{b}\cos\beta t + \mathbf{a}\sin\beta t)$.

Hence, a general solution is $\mathbf{X}(t) = c_1\mathbf{X}_1(t) + c_2\mathbf{X}_2(t)$, so there are constants $\mathbf{A}_1$, $\mathbf{A}_2$, $\mathbf{B}_1$, and $\mathbf{B}_2$ such that x and y are given by

$$\mathbf{X}(t) = \begin{pmatrix} x(t) \\ y(t) \end{pmatrix} = \begin{pmatrix} \mathbf{A}_1 e^{\alpha t}\cos\beta t + \mathbf{A}_2 e^{\alpha t}\sin\beta t \\ \mathbf{B}_1 e^{\alpha t}\cos\beta t + \mathbf{B}_2 e^{\alpha t}\sin\beta t \end{pmatrix}.$$

(a) If $\alpha = 0$, then the solution is

$$\mathbf{X}(t) = \begin{pmatrix} x(t) \\ y(t) \end{pmatrix} = \begin{pmatrix} \mathbf{A}_1\cos\beta t + \mathbf{A}_2\sin\beta t \\ \mathbf{B}_1\cos\beta t + \mathbf{B}_2\sin\beta t \end{pmatrix}.$$

Hence, both x and y are periodic. If fact, if $\mathbf{A}_2 = \mathbf{B}_1 = 0$, then

$$\mathbf{X}(t) = \begin{pmatrix} x(t) \\ y(t) \end{pmatrix} = \begin{pmatrix} \mathbf{A}_1\cos\beta t \\ \mathbf{B}_2\sin\beta t \end{pmatrix}.$$

In rectangular coordinates this solution is

$$\frac{x^2}{\mathbf{A}_1^2} + \frac{y^2}{\mathbf{B}_2^2} = 1,$$

where the graph is either a circle or an ellipse centered at $(0,0)$ depending on the value of A_1 and B_2. Hence, $(0,0)$ is classified as a **center**. Note that the motion around these circles or ellipses is either clockwise or counterclockwise for all solutions.

(b) If $\alpha \neq 0$, then $e^{\alpha t}$ is present in the solution. This term causes the solution to spiral around the equilibrium point. If $\alpha > 0$, then the solution spirals away from $(0,0)$, so we classify $(0,0)$ as an **unstable spiral**. Otherwise, if $\alpha < 0$, the solution spirals toward $(0,0)$, so we say that $(0,0)$ is a **stable spiral**.

EXAMPLE: Classify the equilibrium point $(0,0)$ in each of the systems (a) $\begin{cases} x' = 10x - 8y \\ y' = 4x + 2y \end{cases}$,

(b) $\begin{cases} x' = -2x - y \\ y' = 10x \end{cases}$, and (c) $\begin{cases} x' = -5x + 10y \\ y' = -5x + 5y \end{cases}$.

SOLUTION: (a) After loading the `linalg` package, the eigenvalues of $\mathbf{A} = \begin{pmatrix} 10 & -8 \\ 4 & 2 \end{pmatrix}$ are found with `eigenvals`.

```
> with(linalg):
  A:=array([[10,-8],[4,2]]);
  eigenvals(A);
```

$$A := \begin{bmatrix} 10 & -8 \\ 4 & 2 \end{bmatrix}$$

$$6 + 4I, 6 - 4I$$

Thus, $(0,0)$ is an unstable spiral because $\alpha = 6 > 0$. Several solutions along with the direction field are graphed with `DEplot2`.

```
> with(DEtools):
  DEplot2(A,[x,y],t=-1..1,x=-1..1,y=-1..1);
  DEplot2(A,[x,y],t=-1..1,{[0,.5,.25],[0,-.5,-.25],[0,.5,-.35],
  [0,-.5,.35]},x=-1..1,y=-1..1,stepsize=0.05);
```

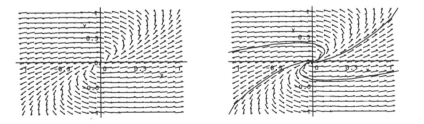

(b) As in (a), the eigenvalues of $\mathbf{A} = \begin{pmatrix} -2 & -1 \\ 10 & 0 \end{pmatrix}$ are found with **eigenvals**.

```
> A:=array([[-2,-1],[10,0]]);
  eigenvals(A);
```

$$A := \begin{bmatrix} -2 & -1 \\ 10 & 0 \end{bmatrix}$$

$$-1 + 3 \quad I, -1 - 3 \quad I$$

Thus, $(0,0)$ is a stable spiral because $\alpha = -1 < 0$. Several solutions along with the direction field are graphed now with **DEplot2**.

```
> DEplot2(A,[x,y],t=-2..2,x=-1..1,y=-1..1);
  DEplot2(A,[x,y],t=-2..2,{[0,.5,.25],[0,-.5,-.25],[0,.5,-.35],
  [0,-.5,.35]},x=-1..1,y=-1..1,stepsize=0.05);
```

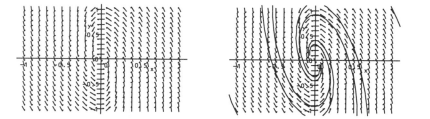

(c) As in (a) and (b), we find the eigenvalues of $\mathbf{A} = \begin{pmatrix} -5 & 10 \\ -5 & 5 \end{pmatrix}$ with **eigenvals**.

```
> A:=array([[-5,10],[-5,5]]);
  eigenvals(A);
```

$$A := \begin{bmatrix} -5 & 10 \\ -5 & 5 \end{bmatrix}$$

$$5 \quad I, -5 \quad I$$

Thus, $(0,0)$ is a center because $\alpha = 0$. Several solutions along with the direction field are graphed here with **DEplot2**.

```
> DEplot2(A,[x,y],t=-2..2,x=-1..1,y=-1..1);
  DEplot2(A,[x,y],t=-2..2,{[0,.2,.2],[0,.4,.4],[0,.6,.6],
  [0,.8,.8]},x=-1..1,y=-1..1,stepsize=0.05);
```

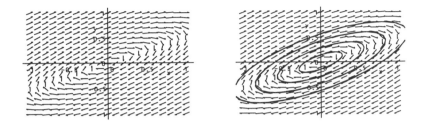

☰ Nonlinear Systems

When working with nonlinear systems, we can often gain a great deal of information concerning the system by making a linear approximation near each equilibrium point of the nonlinear system and solving the linear system. Although the solution to the linearized system only approximates the actual solution to the nonlinear system, the general behavior of solutions to the nonlinear system near each equilibrium is the same as that of the corresponding linear system in most cases. The first step toward approximating a nonlinear system near each equilibrium point is to find the equilibrium points of the system and the matrix for linearization near each point, defined as follows:

Near each equilibrium point (x_0, y_0) of the nonlinear system

$$\begin{cases} x' = f(x, y) \\ y' = g(x, y) \end{cases},$$

the system can be approximated with

$$\begin{cases} x' = f_x(x_0, y_0)(x - x_0) + f_y(x_0, y_0)(y - y_0) + f(x_0, y_0) \\ y' = g_x(x_0, y_0)(x - x_0) + g_y(x_0, y_0)(y - y_0) + g(x_0, y_0) \end{cases}.$$

Then, since $f(x_0, y_0) = 0$ and $g(x_0, y_0) = 0$, the approximate system is

$$\begin{cases} x' = f_x(x_0, y_0)(x - x_0) + f_y(x_0, y_0)(y - y_0) \\ y' = g_x(x_0, y_0)(x - x_0) + g_y(x_0, y_0)(y - y_0) \end{cases},$$

which can be written in matrix form as

$$\begin{pmatrix} x' \\ y' \end{pmatrix} = \begin{pmatrix} f_x(x_0, y_0) & f_y(x_0, y_0) \\ g_x(x_0, y_0) & g_y(x_0, y_0) \end{pmatrix} \begin{pmatrix} x - x_0 \\ y - y_0 \end{pmatrix}.$$

Note that we often call this system the **linearized system** corresponding to the nonlinear system because we have removed the nonlinear terms from the original system. Now that the system is approximated by a system of the form $\begin{cases} x' = ax + by \\ y' = cx + dy \end{cases}$, an equilibrium point (x_0, y_0) of the system

$\begin{cases} x' = f(x, y) \\ y' = g(x, y) \end{cases}$ is classified by the eigenvalues of the matrix $\begin{pmatrix} f_x(x_0, y_0) & f_y(x_0, y_0) \\ g_x(x_0, y_0) & g_y(x_0, y_0) \end{pmatrix}$, which is called

the **Jacobian matrix** and is symbolized as $J(x_0, y_0)$. Of course, this linearization must be carried out for each equilibrium point. After determining the matrix for linearization for each equilibrium point, the eigenvalues for the matrix must be found. Then we classify each equilibrium point according to the following criteria.

Classification of Equilibrium Points

Let (x_0, y_0) be an equilibrium point of the system, and let λ_1 and λ_2 be the eigenvalues of the matrix

$$\begin{pmatrix} f_x(x_0, y_0) & f_y(x_0, y_0) \\ g_x(x_0, y_0) & g_y(x_0, y_0) \end{pmatrix}.$$

(a) Suppose that λ_1 and λ_2 are real. If $\lambda_1 > \lambda_2 > 0$, then (x_0, y_0) is an **unstable node**; if $\lambda_2 < \lambda_1 < 0$, then (x_0, y_0) is a **stable node**; and if $\lambda_2 < 0 < \lambda_1$, then (x_0, y_0) is a **saddle**.

(b) Suppose that $\lambda_1 = \alpha + \beta i$ and $\lambda_2 = \alpha - \beta i$, where $\beta \neq 0$. If $\alpha < 0$, then (x_0, y_0) is a **stable spiral**; if $\alpha > 0$, then (x_0, y_0) is an **unstable spiral**; and if $\alpha = 0$, then (x_0, y_0) may be a center, unstable spiral, or stable spiral: we can draw no conclusion. We will not discuss the case when the eigenvalues are the same or one eigenvalue is zero.

EXAMPLE: Find and classify the equilibrium points of $\begin{cases} x' = 5x + 3y - 4 \\ y' = -4x - 3y + 2 \end{cases}$.

SOLUTION: Notice that the system of equations is linear. However, it is not homogeneous, so we demonstrate how the equilibrium points are located and classified using the Jacobian matrix. In this case, $f(x, y) = 5x + 3y - 4$ and $g(x, y) = -4x - 3y + 2$, so we proceed by defining **A**.

```
> with(linalg):
  A:=array([5*x+3*y-4,-4*x-3*y+2]):
```

The equilibrium point is found by solving $\begin{cases} 5x + 3y - 4 = 0 \\ -4x - 3y + 2 = 0 \end{cases}$ that has solution $(2, -2)$.

```
> with(student):
  to_solve:=equate(A);
```

$\qquad to\_solve := \{5x + 3y - 4 = 0, -4x - 3y + 2 = 0\}$

```
> solve(to_solve);
```

$\qquad \{x = 2, y = -2\}$

The Jacobian matrix is found with `jacobian`, which is contained in the `linalg` package, is named `jac_A`, and then the eigenvalues of `jac_A` are found with `eignvals`.

```
> jac_A:=jacobian(A,[x,y]);
```

$$jac\_A := \begin{bmatrix} 5 & 3 \\ -4 & -3 \end{bmatrix}$$

```
> eigenvals(jac_A);
```

$$3, -1$$

Hence, the eigenvalues are $\lambda_1 = 3$ and $\lambda_2 = -1$. Since these values are of opposite sign, the equilibrium point $(2, -2)$ is a saddle. Note that the corresponding eigenvectors are found to be $\mathbf{v}_1 = \begin{pmatrix} -3/2 \\ 1 \end{pmatrix}$ and $\mathbf{v}_2 = \begin{pmatrix} 1 \\ -2 \end{pmatrix}$, respectively.

```
> eigenvects(jac_A);
```

$$\left[3, 1, \left\{ \begin{bmatrix} -\dfrac{3}{2} & 1 \end{bmatrix} \right\} \right], [-1, 1, \{[1 \quad -2]\}]$$

Therefore, the solutions approach $(2, -2)$ in the direction of the line determined by $\mathbf{v}_2 = \begin{pmatrix} 1 \\ -2 \end{pmatrix}$ that passes through $(2, -2)$. Similarly, it becomes unbounded in the direction of the line that passes through $(2, -2)$ that is determined by $\mathbf{v}_1 = \begin{pmatrix} -3/2 \\ 1 \end{pmatrix}$. After using `convert` together with the `list` option to convert **A** from an array to a list, we graph several members of the family of solutions of this nonhomogeneous linear system and the direction field with `DEplot2`.

```
> Sys:=convert(A,list);
```

$$Sys := [5x + 3y - 4, -4x - 3y + 2]$$

```
> with(DEtools):
  DEplot2(Sys,[x,y],t=-2..2,x=0..4,y=-4..0);
  DEplot2(Sys,[x,y],t=-2..2,{[0,2,-1],[0,2,-3],
    [0,1,-1],[0,3,-3]},x=0..4,y=-4..0,stepsize=0.1);
```

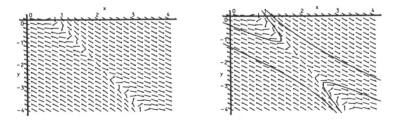

For analyzing nonlinear systems, we state the following theorem.

· ·

THEOREM

Suppose that (x_0, y_0) is an equilibrium point of the nonlinear system $\begin{cases} x' = f(x, y) \\ y' = g(x, y) \end{cases}$.
Then the relationships shown in the following table hold for the classification of (x_0, y_0) in the nonlinear system and that in the associated linearized system.

Associated Linearized System	Nonlinear System
Stable node	Stable node
Unstable node	Unstable node
Stable spiral	Stable spiral
Unstable spiral	Unstable spiral
Saddle	Saddle
Center	No conclusion

· ·

EXAMPLE: Find and classify the equilibrium points of $\begin{cases} x' = 1 - y \\ y' = x^2 - y^2 \end{cases}$.

SOLUTION: We begin by finding the equilibrium points of this nonlinear system by solving
$\begin{cases} 1 - y = 0 \\ x^2 - y^2 = 0 \end{cases}$ with **solve**.

```
>  f:=(x,y)->1-y:
   g:=(x,y)->x^2-y^2:
   Eq_pts:=solve({f(x,y)=0,g(x,y)=0});
```

$$Eq\_pts := \{y = 1, x = 1\}, \{y = 1, x = -1\}$$

Thus, the two equilibrium points are $(1, 1)$ and $(-1, 1)$. The Jacobian matrix is $J(x, y) = \begin{pmatrix} 0 & -1 \\ 2x & -2y \end{pmatrix}$
as computed with **jacobian**.

```
>  with(linalg):
   jac:=jacobian([f(x,y),g(x,y)],[x,y]);
```

$$jac := \begin{bmatrix} 0 & -1 \\ 2x & -2y \end{bmatrix}$$

Next, we obtain a linearized system about each equilibrium point.

For $(1, 1)$, we obtain the Jacobian matrix $J(1, 1) = \begin{pmatrix} 0 & -1 \\ 2 & -2 \end{pmatrix}$, while with $(-1, 1)$ we obtain

$J(-1, 1) = \begin{pmatrix} 0 & -1 \\ -2 & -2 \end{pmatrix}$. The eigenvalues of each matrix are computed with **eigenvals**.

```
> eigenvals(subs(Eq_pts[1],eval(jac)));
  eigenvals(subs(Eq_pts[2],eval(jac)));
```

$-1 + I, -1 - I$

$-1 + \sqrt{3}, -1 - \sqrt{3}$

Thus, $(1, 1)$ is a stable spiral and $(-1, 1)$ is a saddle.

The direction field for the equation along with the graphs of several solutions are generated with **DEplot2**.

```
> with(DEtools):
  DEplot2([f(x,y),g(x,y)],[x,y],t=-2..2,x=-2..2,y=0..2);
  DEplot2([f(x,y),g(x,y)],[x,y],t=-3..3,{[0,-1,0],
  [0,-1,1.5],[0,0,1],[0,-.5,1.5],[0,-1.5,1],[0,1,0],
  [0,1.5,1],[0,.5,1.5],[0,1.5,1.5],[0,-.75,.5],
  [0,-1.25,.75]},x=-2..2,y=0..2,
  stepsize=0.1);
```

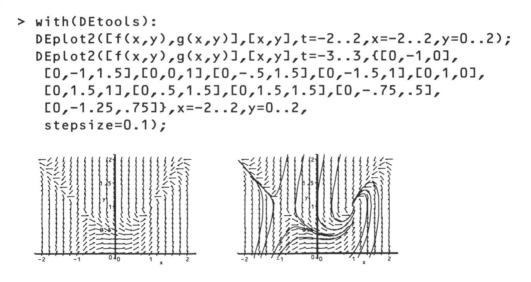

EXAMPLE: Find and classify the equilibrium points of $\begin{cases} x' = x(7 - x - 2y) \\ y' = y(5 - x - y) \end{cases}$.

SOLUTION: The equilibrium points of this system satisfy $\begin{cases} x(7 - x - 2y) = 0 \\ y(5 - x - y) = 0 \end{cases}$ and are found with **solve** and named **Eq_pts**.

```
> f:=(x,y)->x*(7-x-2*y):
  g:=(x,y)->y*(5-x-y):
  Eq_pts:=solve({f(x,y)=0,g(x,y)=0});
```

$$Eq\_pts := \{y = 0, x = 0\}, \{x = 0, y = 5\}, \{x = 7, y = 0\}, \{y = 2, x = 3\}$$

The Jacobian matrix and the eigenvalues of the Jacobian matrix are found with **jacobian** and **eigenvals**, respectively, and named **jac** and **eigs**, respectively.

```
> with(linalg):
  jac:=jacobian([f(x,y),g(x,y)],[x,y]);
  eigs:=[eigenvals(jac)];
```

$$jac := \begin{bmatrix} 7 - 2x - 2y & -2x \\ -y & 5 - x - 2y \end{bmatrix}$$

$$eigs := \left[6 - \frac{3}{2}x - 2y + \frac{1}{2}\sqrt{4 - 4x + x^2 + 8xy}, \right.$$

$$\left. 6 - \frac{3}{2}x - 2y - \frac{1}{2}\sqrt{4 - 4x + x^2 + 8xy} \right]$$

To classify the equilibrium points of the system, we evaluate the eigenvalues of the Jacobian matrix, **eigs**, for each of the equilibrium points of the system, **Eq_pts**, using a **for** loop.

```
> for i from 1 to 4 do
    simplify(subs(Eq_pts[i],eigs)) od;
```

$$[7, 5]$$
$$[-3, -5]$$
$$[-2, -7]$$
$$[1, -6]$$

The equilibrium points are classified using the eigenvalues of $J(x_0, y_0)$ obtained earlier.

$$J(0, 0) = \begin{pmatrix} 7 & 0 \\ 0 & 5 \end{pmatrix}; \lambda_1 = 7, \lambda_2 = 5; (0, 0) \text{ is an unstable node.}$$

$$J(0, 5) = \begin{pmatrix} -3 & 0 \\ -5 & -5 \end{pmatrix}; \lambda_1 = -3, \lambda_2 = -5; (0, 5) \text{ is a stable node.}$$

$$J(7, 0) = \begin{pmatrix} -7 & -14 \\ 0 & -2 \end{pmatrix}; \lambda_1 = -2, \lambda_2 = -7; (7, 0) \text{ is a stable node.}$$

$$J(3, 2) = \begin{pmatrix} -3 & -6 \\ -2 & -2 \end{pmatrix}; \lambda_1 = 1, \lambda_2 = -6; (3, 2) \text{ is a saddle.}$$

The direction field associated with the system is generated with `DEplot2`. In the second `DEplot2` command, solutions satisfying different initial conditions are graphed. Note that in the second `DEplot2` command the direction field is not displayed because the option `arrows=NONE` is included.

```
> with(DEtools):
  DEplot2([f(x,y),g(x,y)],[x,y],t=-2..2,x=-3..10,y=-3..8);
  DEplot2([f(x,y),g(x,y)],[x,y],t=-3..3,{[0,1,2],[0,8,4],
    [0,7,1.5],[0,4.5,-1],[0,1.2,-1.7],[0,-.6,3],
    [0,-2.5,5],[0,1,7.5],[0,2.4,4.5],[0,3.7,1.5],
    [0,6.2,-2],[0,1.9,2.4],[0,7.1,5],[0,-2.7,4.5],
    [0,-1.6,4.1],[0,-2.3,.5],[0,2.3,6.7],[0,8.3,7.2],
    [0,8.4,2.9]},
    x=-3..10,y=-3..8,stepsize=0.05,arrows=NONE);
```

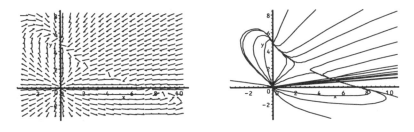

9.8 NUMERICAL METHODS

≡ Built-In Methods

As we saw in Section 9.7, we can use several of the commands contained in the `DEtools` package to provide us with numerical information about solutions to systems of differential equations. The following examples illustrate the use of some of the other commands contained in the `DEtools` package that can help us determine information about a system of differential equations in addition to showing how to use `dsolve` together with the `numeric` option to generate a numerical solution of a system of differential equations.

▼ **A P P L I C A T I O N**

The FitzHugh-Nagumo Equation

EXAMPLE: Under certain assumptions, the **Fitzhugh-Nagumo equation**, which arises in the study of the impulses in a nerve fiber, can be written as the system of ordinary differential equations

$$
\begin{cases}
\dfrac{dV}{d\xi} = W \\[2mm]
\dfrac{dW}{d\xi} = F(V) + R - uW, \\[2mm]
\dfrac{dR}{d\xi} = \dfrac{\varepsilon}{u}(bR - V - a)
\end{cases}
$$

where $F(V) = \dfrac{1}{3}V^3 - V$. (a) Sketch the graph of the solution to the FitzHugh-Nagumo equation that satisfies the initial conditions $V(0) = 1, W(0) = 0, R(0) = 1$ if $\varepsilon = 0.08, a = 0.7, b = 0$, and $u = 1$. (b) Sketch the graph of the solution that satisfies the initial conditions $V(0) = 1, W(0) = 0.5, R(0) = 0.5$ if $\varepsilon = 0.08, a = 0.7, b = 0.8$, and $u = 0.6$.

SOLUTION: We begin by defining the FitzHugh-Nagumo system in **FHN** and then the constants $\varepsilon = 0.08, a = 0.7, b = 0$, and $u = 1$. In this case, we use lowercase letters to avoid any ambiguity with built-in Maple functions like **W**, which represents the omega function.

```
> FHN:=[diff(v(xi),xi)=w,
    diff(w(xi),xi)=1/3*v^3-v+r-u*w,
    diff(r(xi),xi)=epsilon/u*(b*r-v-a)];
```

$$
FHN := \left[\frac{\partial}{\partial \xi} v(\xi) = w, \ \frac{\partial}{\partial \xi} w(\xi) = \frac{1}{3} v^3 - v + r - w, \ \frac{\partial}{\partial \xi} r(\xi) = -0.08v - 0.056 \right]
$$

```
> epsilon:=0.08:
  b:=0:
  a:=0.7:
  u:=1:
  FHN;
```

$$
\left[\frac{\partial}{\partial \xi} v(\xi) = w, \ \frac{\partial}{\partial \xi} w(\xi) = \frac{1}{3} v^3 - v + r - w, \ \frac{\partial}{\partial \xi} r(\xi) = -.08v - .056 \right]
$$

Next, we use **DEplot** to graph the solution for $0 \leq \xi \leq 30$. Note that the ordered quadruple **[0,1,0,1]** corresponds to the initial conditions $V(0) = 1$, $W(0) = 0$, $R(0) = 1$. Several solutions could be graphed together by including additional ordered quadruples corresponding to various initial conditions. The option **stepsize** is included to help assure that the resulting graph is smooth.

```
> with(DEtools):
  DEplot(FHN,[v,w,r],0..30,[0,1,0,1],stepsize=0.1);
```

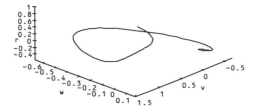

In the same manner as in Section 2.7, we can obtain a numerical solution that can be evaluated for particular values of ζ using **dsolve** together with the **numeric** option. For example, entering

```
> Sys_of_Eqs:={diff(v(xi),xi)=w(xi),
  diff(w(xi),xi)=1/3*v(xi)^3-v(xi)+r(xi)-u*w(xi),
  diff(r(xi),xi)=epsilon/u*(b*r(xi)-v(xi)-a),
  v(0)=1,w(0)=0,r(0)=1}:
  Num_Sol:=dsolve(Sys_of_Eqs,{v(xi),w(xi),r(xi)},numeric);
  Num_Sol := proc(xi) `dsolve/numeric/result2`(xi,2845256,
  [1,1,1])
  end
```

returns a numerical solution to the system that satisfies $V(0) = 1$, $W(0) = 0$, $R(0) = 1$ and names the result **Num_Sol**. The advantage of using **dsolve** together with the **numeric** option is that we obtain a function that can be evaluated for particular numbers, as shown with **seq** and **array**.

```
> array([seq(Num_Sol(i/10),i=0..10)]);
```

$[\{\xi = 0, w(\xi) = 0, v(\xi) = 1., r(\xi) = 1.\},$

$\{\xi = .1000000000, r(\xi) = .9863957089, v(\xi) = 1.001590360, w(\xi) = .03106291647\},$

$\{\xi = .2000000000, r(\xi) = .9727668507, v(\xi) = 1.006070952, w(\xi) = .05787555787\},$

$\{\xi = .3000000000, r(\xi) = .9590919482, v(\xi) = 1.013037403, w(\xi) = .08084504770\},$

$\{v(\xi) = 1.022124357, w(\xi) = .1003454787, \xi = .4000000000, r(\xi) = .9453526008\},$

$\{\xi = .5000000000, r(\xi) = .9315331852, v(\xi) = 1.033002419, w(\xi) = .1167226905\},$

$\{\xi = .6000000000, r(\xi) = .9176205781, v(\xi) = 1.045375506, w(\xi) = .1302974590\},$

$\{\xi = .7000000000, r(\xi) = .9036039000, v(\xi) = 1.058978447, w(\xi) = .1413675974\},$

$\{\xi = .8000000000, r(\xi) = .8894742763, v(\xi) = 1.073574768, w(\xi) = .1502093411\},$

$\{w(\xi) = .1570782838, r(\xi) = .8752246166, v(\xi) = 1.088954578, \xi = .9000000000\},$

$\{\xi = 1.0, r(\xi) = .8608494100, v(\xi) = 1.104932541, w(\xi) = .1622100611\}]$

The result we obtain can also be graphed in either two or three dimensions using **odeplot**, which is contained in the **plots** package.

Two-dimensional graphs are generated by including the **scene** option. We use now **DEplot** together with the **scene** option to graph V as a function of ξ, W as a function of ξ, and V as a function of W.

```
> DEplot(FHN,[v,w,r],0..30,{[0,1,0,1]},stepsize=0.1,
    scene=[xi,v]);
  DEplot(FHN,[v,w,r],0..30,{[0,1,0,1]},stepsize=0.1,
    scene=[xi,w]);
  DEplot(FHN,[v,w,r],0..30,{[0,1,0,1]},stepsize=0.1,
    scene=[w,v]);
```

For (b), we redefine the constants $\varepsilon = 0.08$, $a = 0.7$, $b = 0.8$, and $u = 0.6$ and then use **DEplot** in the same manner as in (a) to generate various graphs of the solutions that satisfy $V(0) = 1$, $W(0) = 0.5$, $R(0) = 0.5$.

```
> epsilon:=0.08:
  b:=0.8:
  a:=0.7:
  u:=0.6:
  FHN;
```

$$\left[\frac{\partial}{\partial \xi}v(\xi) = w, \frac{\partial}{\partial \xi}w(\xi) = \frac{1}{3}v^3 - v + r - w, \frac{\partial}{\partial \xi}r(\xi) = -0.08v - 0.056\right]$$

```
> DEplot(FHN,[v,w,r],0..15,{[0,1,.5,.5]},
  stepsize=0.1);
  DEplot(FHN,[v,w,r],0..15,{[0,1,.5,.5]},
  stepsize=0.1,scene=[xi,v]);
  DEplot(FHN,[v,w,r],0..15,{[0,1,.5,.5]},
  stepsize=0.1,scene=[xi,w]);
  DEplot(FHN,[v,w,r],0..15,{[0,1,.5,.5]},
  stepsize=0.1,scene=[w,v]);
```

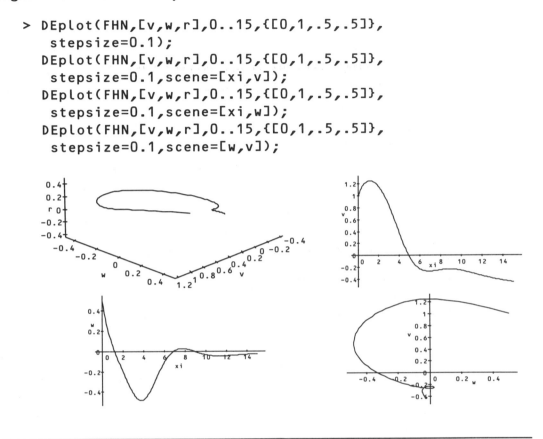

▲

▼ A P P L I C A T I O N

The SIR Model with Vital Dynamics

EXAMPLE: The initial-value problem for the **SIR model with vital dynamics** is

$$\begin{cases} s'(t) = -asi + b - bs \\ i'(t) = asi - ci - bi \quad , \\ s(0) = s_0, i(0) = i_0 \end{cases}$$

where a, b, c, and d represent positive constants corresponding to the **contact rate**, **death rate**, and **removal rate**, respectively, and $s(t)$ and $i(t)$ represent the percentage of the population susceptible to the disease and infected by the disease, respectively. Sketch the graph of the solution to the SIR model using various initial conditions if $a = 8$, $b = 1$, and $c = 1$.

SOLUTION: After defining a, b, and c, we define **SIR** to be the list consisting of the two differential equations in the SIR model.

```
> a:=8:
  b:=1:
  c:=1:
```

```
> SIR:=[diff(s(t),t)=-a*s*i+b-b*s,
           diff(i(t),t)=a*s*i-c*i-b*i];
```

$$SIR := \left[\frac{\partial}{\partial t}s(t) = -8si + 1 - s, \frac{\partial}{\partial t}i(t) = 8si - 2i \right]$$

We will graph the solutions satisfying $s(0) = \dfrac{j}{10}$, $i(0) = 1 - \dfrac{j}{10}$ for $j = 1, 2, \ldots, 9$. We use **seq** to define the set of ordered triples corresponding to these initial conditions. For each ordered triple in **inits**, the first coordinate will correspond to the value of t and the second and third coordinates will correspond to the values of s and i for that particular value of t.

```
> inits:={seq([0,j/10,1-j/10],j=1..9)};
```

$$inits := \left\{ \left[0, \frac{2}{5}, \frac{3}{5} \right], \left[0, \frac{1}{5}, \frac{4}{5} \right], \left[0, \frac{1}{2}, \frac{1}{2} \right], \left[0, \frac{3}{5}, \frac{2}{5} \right], \left[0, \frac{7}{10}, \frac{3}{10} \right], \left[0, \frac{1}{10}, \frac{9}{10} \right], \right.$$
$$\left. \left[0, \frac{3}{10}, \frac{7}{10} \right], \left[0, \frac{4}{5}, \frac{1}{5} \right], \left[0, \frac{9}{10}, \frac{1}{10} \right] \right\}$$

Last, we use **phaseportrait** to generate the graphs satisfying these initial conditions. The option **stepsize=0.5** is included to help assure that the resulting graphs appear smooth; the option **arrows=THIN** instructs Maple to display the direction field, using thin arrows, for the equation; while **s=0..1,i=0..1** instructs Maple that the s- and i-axes displayed correspond to the interval $[0, 1]$.

```
> with(DEtools):
  phaseportrait(SIR,[s,i],0..10,inits,stepsize=0.05,
   arrows=THIN,s=0..1,i=0..1);
```

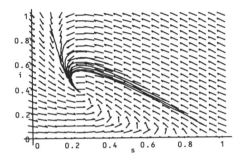

Next, we graph the solutions satisfying $s(0) = \dfrac{j}{10}$, $i(0) = 0.1$ for $j = 1, 2, \ldots, 9$. As before, we define `inits_2` to correspond to these initial conditions and then use `phaseportrait` to generate the graphs. In this case, the direction field is not displayed as the default option for `arrows` for this command is `arrows=NONE`.

```
> inits_2:={seq([0,j/10,.1],j=1..9)}:
  phaseportrait(SIR,[s,i],0..10,inits_2,stepsize=0.05,
   s=0..1,i=0..1);
```

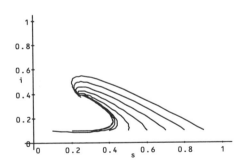

In each of these cases, the exact same results are obtained by using `DEplot` instead of `phaseportrait`.

☰ *Euler's Method*

Euler's method for approximation, which was discussed for first-order equations, may be extended to include systems of first-order equations. Therefore, the initial-value problem

$$
\begin{cases}
\dfrac{dx}{dt} = f(t, x, y) \\[2mm]
\dfrac{dy}{dt} = g(t, x, y) \\[2mm]
x(t_0) = x_0, \ y(t_0) = y_0
\end{cases}
$$

is approximated at each step by the recursive relationship based on the Taylor expansion of x and y up to order h,

$$
x_{n+1} = x_n + hf(t_n, x_n, y_n)
$$

and

$$
y_{n+1} = y_n + hg(t_n, x_n, y_n),
$$

where $t_n = t_0 + nh, n = 0, 1, 2, \ldots$.

EXAMPLE: Use Euler's method with $h = 0.1$ to approximate the solution of the initial-value problem

$$
\begin{cases}
\dfrac{dx}{dt} = x - y + 1 \\[2mm]
\dfrac{dy}{dt} = x + 3y + e^{-t} \\[2mm]
x(0) = 0, \ y(0) = 1
\end{cases}
$$

Compare these results to those of the exact solution of the system of equations.

SOLUTION: In this case, $f(x, y) = x - y + 1$, $g(x, y) = x + 3y + e^{-t}$, $t_0 = 0$, $x_0 = 0$, and $y_0 = 1$, so we use the formulas

$$
x_{n+1} = x_n + h(x_n - y_n + 1)
$$

and

$$
y_{n+1} = y_n + h(x_n + 3y_n + e^{-t_n}),
$$

where $t_n = (0.1)n, n = 0, 1, 2, \ldots$.

If $n = 0$, then

$$x_1 = x_0 + h(x_0 - y_0 + 1) = 0$$

and

$$y_1 = y_0 + h(x_0 + 3y_0 + e^{-t_0}) = 1.4.$$

To use Maple to implement Euler's method, we define the functions f and g as well as the increment size, the formula for incrementing t, the recursive formulas for determining subsequent values of x and y, and the initial values of x and y. Notice that using the option `remember` in defining these functions causes previous values to be retained so that subsequent values may be based on them.

```
> f:='f':g:='g':t:='t':h:='h':x:='x':y:='y':
  f:=(t,x,y)->x-y+1:
  g:=(t,x,y)->x+3*y+exp(-t):
  h:=0.1:
  t:=n->n*h:
  x:=proc(n) option remember;
    x(n-1)+h*f(t(n-1),x(n-1),y(n-1))
    end:
  x(0):=0:
  y:=proc(n) option remember;
    y(n-1)+h*g(t(n-1),x(n-1),y(n-1))
    end:
  y(0):=1:
```

The actual solution of this problem is determined with `dsolve`. Note that we use capital letters to avoid conflict with the definitions of x and y entered previously.

```
> Sol:=dsolve({diff(X(t),t)=X(t)-Y(t)+1,
    diff(Y(t),t)=X(t)+3*Y(t)+exp(-t),X(0)=0,
    Y(0)=1},{X(t),Y(t)});
  assign(Sol):
```

$$Sol := \left\{ Y(t) = \frac{1}{4} - \frac{2}{9}e^{-t} + \frac{35}{36}e^{2t} + \frac{11}{6}e^{2t}t, X(t) = -\frac{1}{9}e^{-t} - \frac{3}{4} + \frac{31}{36}e^{2t} - \frac{11}{6}e^{2t}t \right\}$$

We then use `array`, `seq`, `evalv`, and `subs` to display the results obtained with this method and compare them to the actual function values. The first column gives the value of t, the second column represents the approximate value of x and should be compared to the third column that gives the exact value of x, and the fourth and fifth columns give the approximate and exact values of y, respectively.

```
> array([seq([t(i),x(i),evalf(subs(t=t(i),X(t))),y(i),
  evalf(subs(t=t(i),Y(t)))],i=0..10)]);
```

$$
\begin{bmatrix}
0 & 0 & 0 & 1 & 1.000000000 \\
.1 & 0 & -.0226978438 & 1.4 & 1.460323761 \\
.2 & -.04 & -.1033456496 & 1.910483742 & 2.065447345 \\
.3 & -.1350483742 & -.2654317310 & 2.561501940 & 2.859043457 \\
.4 & -.3047034056 & -.5401053307 & 3.390529507 & 3.896823683 \\
.5 & -.5742266969 & -.9684079522 & 4.444250023 & 5.249747751 \\
.6 & -.9760743689 & -1.604118114 & 5.780755426 & 7.008061926 \\
.7 & -1.551757348 & -2.517371685 & 7.472255781 & 9.286376522 \\
.8 & -2.354158661 & -3.799261739 & 9.608415310 & 12.23004486 \\
.9 & -3.450416058 & -5.567674073 & 12.30045693 & 16.02317121 \\
1.0 & -4.925503357 & -7.974680032 & 15.68620937 & 20.89865639
\end{bmatrix}
$$

Because the accuracy of this approximation diminishes as t increases, we attempt to improve the approximation by decreasing the increment size. We do this by entering the value $h = 0.05$ and repeating the procedure that we followed before. Notice that the approximations are more accurate with the smaller value of h.

```
> h:='h':x:='x':y:='y':
  h:=0.05:
  t:=n->n*h:
  x:=proc(n) option remember;
    x(n-1)+h*f(t(n-1),x(n-1),y(n-1))
    end:
  x(0):=0:
  y:=proc(n) option remember;
    y(n-1)+h*g(t(n-1),x(n-1),y(n-1))
    end:
  y(0):=1:
  array([seq([t(i),x(i),evalf(subs(t=t(i),X(t))),y(i),
   evalf(subs(t=t(i),Y(t)))],i=0..20)]);latex(");
```

$$
\begin{bmatrix}
0 & 0 & 0 & 1 & 1.000000000 \\
.05 & 0 & -.0053245352 & 1.20 & 1.214394744 \\
.10 & -.0100 & -.0226978438 & 1.427561471 & 1.460323761 \\
.15 & -.03187807355 & -.0544669738 & 1.686437563 & 1.742305463 \\
.20 & -.06779385540 & -.1033456496 & 1.980844693 & 2.065447345 \\
.25 & -.1202257828 & -.1724651306 & 2.315518242 & 2.435520534 \\
.30 & -.1920129841 & -.2654317310 & 2.695774728 & 2.859043457 \\
.35 & -.2864023697 & -.3863918325 & 3.127581199 & 3.343375766 \\
.40 & -.4071015482 & -.5401053307 & 3.617632665 & 3.896823683 \\
.45 & -.5583382589 & -.7320285710 & 4.173438490 & 4.528758223 \\
.50 & -.7449270964 & -.9684079522 & 4.803418758 & 5.249747751 \\
.55 & -.9723443891 & -1.256385532 & 5.517011750 & 6.071706638 \\
.60 & -1.246812196 & -1.604118114 & 6.324793784 & 7.008061926 \\
.65 & -1.575392495 & -2.020911482 & 7.238612824 & 8.073940114 \\
.70 & -1.966092761 & -2.517371685 & 8.271737412 & 9.286376522 \\
.75 & -2.427984270 & -3.105575390 & 9.439022651 & 10.66454984 \\
.80 & -2.971334616 & -3.799261739 & 10.75709516 & 12.23004486 \\
.85 & -3.607756105 & -4.614048311 & 12.24455915 & 14.00714688 \\
.90 & -4.350371868 & -5.567674073 & 13.92222596 & 16.02317121 \\
.95 & -5.214001760 & -6.680272715 & 15.81336974 & 18.30883218 \\
1.00 & -6.215370335 & -7.974680032 & 17.94401216 & 20.89865639
\end{bmatrix}
$$

☰ *Runge-Kutta Method*

We would like to be able to improve the approximation without using such a small value for h. As with first-order equations, the Runge-Kutta method can be extended to systems. In this case, the recursive formulas at each step are

$$
x_{n+1} = x_n + \frac{h}{6}(k_1 + 2k_2 + 2k_3 + k_4)
$$

and

$$y_{n+1} = y_n + \frac{h}{6}(m_1 + 2m_2 + 2m_3 + m_4),$$

where

$$k_1 = f(t_n, x_n, y_n), \qquad\qquad m_1 = g(t_n, x_n, y_n),$$

$$k_2 = f\left(t_n + \frac{h}{2}, x_n + \frac{hk_1}{2}, y_n + \frac{hm_1}{2}\right), \qquad m_2 = g\left(t_n + \frac{h}{2}, x_n + \frac{hk_1}{2}, y_n + \frac{hm_1}{2}\right),$$

$$k_3 = f\left(t_n + \frac{h}{2}, x_n + \frac{hk_2}{2}, y_n + \frac{hm_2}{2}\right), \qquad m_3 = g\left(t_n + \frac{h}{2}, x_n + \frac{hk_2}{2}, y_n + \frac{hm_2}{2}\right),$$

$$k_4 = f(t_n + h, x_n + hk_3, y_n + hm_3), \qquad m_4 = g(t_n + h, x_n + hk_3, y_n + hm_3).$$

EXAMPLE: Use the Runge-Kutta method to approximate the solution of the initial value problem from the previous example

$$\begin{cases} \dfrac{dx}{dt} = x - y + 1 \\[2mm] \dfrac{dy}{dt} = x + 3y + e^{-t} \\[2mm] x(0) = 0, y(0) = 1 \end{cases}$$

for $h = 0.1$. Compare these results to those of the exact solution of the system of equations as well as those obtained with Euler's method.

SOLUTION: Again, because $f(x, y) = x - y + 1$, $g(x, y) = x + 3y + e^{-t}$, $t_0 = 0$, $x_0 = 0$, and $y_0 = 1$, we use the formulas

$$x_{n+1} = x_n + \frac{h}{6}(k_1 + 2k_2 + 2k_3 + k_4)$$

and

$$y_{n+1} = y_n + \frac{h}{6}(m_1 + 2m_2 + 2m_3 + m_4),$$

where

$$k_1 = f(t_n, x_n, y_n) = x_n - y_n + 1, \qquad m_1 = g(t_n, x_n, y_n) = x_n + 3y_n + e^{-t_n},$$

$$k_2 = \left(x_n + \frac{hk_1}{2}\right) - \left(y_n + \frac{hm_1}{2}\right) + 1, \qquad m_2 = \left(x_n + \frac{hk_1}{2}\right) + 3\left(y_n + \frac{hm_1}{2}\right) + e^{-\left(t_n + \frac{h}{2}\right)},$$

$$k_3 = \left(x_n + \frac{hk_2}{2}\right) - \left(y_n + \frac{hm_2}{2}\right) + 1, \qquad m_3 = \left(x_n + \frac{hk_2}{2}\right) + 3\left(y_n + \frac{hm_2}{2}\right) + e^{-\left(t_n + \frac{h}{2}\right)},$$

$$k_4 = (x_n + hk_3) - (y_n + hm_3) + 1, \qquad m_4 = (x_n + hk_3) + 3(y_n + hm_3) + e^{-(t_n + h)}.$$

To use Maple to implement the Runge-Kutta method, we begin by defining the appropriate functions.

```
> f:=(t,x,y)->x-y+1:
  g:=(t,x,y)->x+3*y+exp(-t):
  h:=0.1:
  t:=n->n*h:
```

The recursive formulas for xrk and yrk are defined using the option remember so that Maple "remembers" the values of xrk and yrk computed. Thus, previously computed values are retained and subsequent values may be based on them.

```
> xrk:='xrk':yrk:='yrk':
  xrk:=proc(n)
    local k1,m1,k2,m2,k3,m3,k4,m4;
    option remember;
    k1:=f(t(n-1),xrk(n-1),yrk(n-1));
    m1:=g(t(n-1),xrk(n-1),yrk(n-1));
    k2:=f(t(n-1)+h/2,xrk(n-1)+h*k1/2,yrk(n-1)+h*m1/2);
    m2:=g(t(n-1)+h/2,xrk(n-1)+h*k1/2,yrk(n-1)+h*m1/2);
    k3:=f(t(n-1)+h/2,xrk(n-1)+h*k2/2,yrk(n-1)+h*m2/2);
    m3:=g(t(n-1)+h/2,xrk(n-1)+h*k2/2,yrk(n-1)+h*m2/2);
    k4:=f(t(n-1)+h,xrk(n-1)+h*k3,yrk(n-1)+h*m3);
    m4:=g(t(n-1)+h,xrk(n-1)+h*k3,yrk(n-1)+h*m3);
    xrk(n-1)+h/6*(k1+2*k2+2*k3+k4)
  end:
  xrk(0):=0:
```

```
> yrk:=proc(n)
   local k1,m1,k2,m2,k3,m3,k4,m4;
   option remember;
   k1:=f(t(n-1),xrk(n-1),yrk(n-1));
   m1:=g(t(n-1),xrk(n-1),yrk(n-1));
   k2:=f(t(n-1)+h/2,xrk(n-1)+h*k1/2,yrk(n-1)+h*m1/2);
   m2:=g(t(n-1)+h/2,xrk(n-1)+h*k1/2,yrk(n-1)+h*m1/2);
   k3:=f(t(n-1)+h/2,xrk(n-1)+h*k2/2,yrk(n-1)+h*m2/2);
   m3:=g(t(n-1)+h/2,xrk(n-1)+h*k2/2,yrk(n-1)+h*m2/2);
   k4:=f(t(n-1)+h,xrk(n-1)+h*k3,yrk(n-1)+h*m3);
   m4:=g(t(n-1)+h,xrk(n-1)+h*k3,yrk(n-1)+h*m3);
   yrk(n-1)+h/6*(m1+2*m2+2*m3+m4)
   end:
yrk(0):=1:
```

In the same way as in the previous example, the actual solution of this problem is determined with **dsolve**. Note that we use capital letters to avoid conflict with the definitions of x and y entered previously.

```
> Y:='Y':
  Sol:=dsolve({diff(X(t),t)=X(t)-Y(t)+1,
   diff(Y(t),t)=X(t)+3*Y(t)+exp(-t),X(0)=0,
   Y(0)=1},X(t),Y(t)):
  assign(Sol):
```

We show the results obtained with this method and compare them to the exact values. This table has the same format as the two previous tables in the example using Euler's method.

```
> array([seq([t(i),xrk(i),evalf(subs(t=t(i),X(t))),yrk(i),
  evalf(subs(t=t(i),Y(t)))],i=0..10)]);
```

$$\begin{bmatrix}
0 & 0 & 0 & 1 & 1.000000000 \\
.1 & -.02268784122 & -.0226978438 & 1.460309033 & 1.460323761 \\
.2 & -.1033199102 & -.1033456496 & 2.065409956 & 2.065447345 \\
.3 & -.2653822036 & -.2654317310 & 2.858972430 & 2.859043457 \\
.4 & -.5400208435 & -.5401053307 & 3.896703967 & 3.896823683 \\
.5 & -.9682731501 & -.9684079522 & 5.249558887 & 5.249747751 \\
.6 & -1.603912061 & -1.604118114 & 7.007776299 & 7.008061926 \\
.7 & -2.517066042 & -2.517371685 & 9.285957081 & 9.286376522 \\
.8 & -3.798818367 & -3.799261739 & 12.22944216 & 12.23004486 \\
.9 & -5.567041919 & -5.567674073 & 16.02231955 & 16.02317121 \\
1.0 & -7.973791094 & -7.974680032 & 20.89746891 & 20.89865639
\end{bmatrix}$$

The Runge-Kutta method is much more accurate than Euler's method. In fact, the Runge-Kutta method with $h = 0.1$ is more accurate than Euler's method with $h = 0.05$.

Since the Runge-Kutta method can be extended to systems of first-order equations, this method can be used to solve higher-order differential equations. This is accomplished by transforming the higher-order equation into a system of first-order equations. We illustrate this now with an equation, the **pendulum equation**, that we have solved in several situations.

EXAMPLE: Use the Runge-Kutta method with $h = 0.1$ to approximate the solution of the non-linear initial-value problem $x'' + \sin x = 0, x(0) = 0, x'(0) = 1$.

SOLUTION: We begin by transforming the second-order equation into a system of first-order equations. We do this by letting $x' = y$, so $y' = x'' = -\sin x$. Hence, $f(t, x, y) = y$ and $g(t, x, y) = -\sin x$. After defining f, g, h, and t, we again use xrk and yrk to represent the approximate values.

```
> f:=(t,x,y)->y:
  g:=(t,x,y)->-sin(x):
  h:=0.1:
  t:=n->n*h:
```

```
> xrk:='xrk':yrk:='yrk':
  xrk:=proc(n)
  local k1,m1,k2,m2,k3,m3,k4,m4;
  option remember;
  k1:=f(t(n-1),xrk(n-1),yrk(n-1));
  m1:=g(t(n-1),xrk(n-1),yrk(n-1));
  k2:=f(t(n-1)+h/2,xrk(n-1)+h*k1/2,yrk(n-1)+h*m1/2);
  m2:=g(t(n-1)+h/2,xrk(n-1)+h*k1/2,yrk(n-1)+h*m1/2);
  k3:=f(t(n-1)+h/2,xrk(n-1)+h*k2/2,yrk(n-1)+h*m2/2);
  m3:=g(t(n-1)+h/2,xrk(n-1)+h*k2/2,yrk(n-1)+h*m2/2);
  k4:=f(t(n-1)+h,xrk(n-1)+h*k3,yrk(n-1)+h*m3);
  m4:=g(t(n-1)+h,xrk(n-1)+h*k3,yrk(n-1)+h*m3);
  xrk(n-1)+h/6*(k1+2*k2+2*k3+k4)
  end:
  xrk(0):=0:

> yrk:=proc(n)
  local k1,m1,k2,m2,k3,m3,k4,m4;
  option remember;
  k1:=f(t(n-1),xrk(n-1),yrk(n-1));
  m1:=g(t(n-1),xrk(n-1),yrk(n-1));
  k2:=f(t(n-1)+h/2,xrk(n-1)+h*k1/2,yrk(n-1)+h*m1/2);
  m2:=g(t(n-1)+h/2,xrk(n-1)+h*k1/2,yrk(n-1)+h*m1/2);
  k3:=f(t(n-1)+h/2,xrk(n-1)+h*k2/2,yrk(n-1)+h*m2/2);
  m3:=g(t(n-1)+h/2,xrk(n-1)+h*k2/2,yrk(n-1)+h*m2/2);
  k4:=f(t(n-1)+h,xrk(n-1)+h*k3,yrk(n-1)+h*m3);
  m4:=g(t(n-1)+h,xrk(n-1)+h*k3,yrk(n-1)+h*m3);
  yrk(n-1)+h/6*(m1+2*m2+2*m3+m4)
  end:
  yrk(0):=1:
```

Recall that we approximate the nonlinear equation $x'' + \sin x = 0$ with $x'' + x = 0$ because $\sin x \approx x$ for small values of x. The solution of $x'' + x = 0, x(0) = 0, x'(0) = 1$ is found to be $x = \sin t$, so $y = x' = \cos t$. Because the use of the approximation $\sin x \approx x$ is linear, we expect the approximations obtained with the Runge-Kutta method, which is a fourth-order method, to be more accurate than those obtained by solving $x'' + x = 0, x(0) = 0, x'(0) = 1$. In the same manner as in the previous two examples, we compare the values we obtained.

```
> array([seq([t(i),xrk(i),evalf(sin(t(i))),yrk(i),
  evalf(cos(t(i)))],i=0..10)]);
```

0	0	0	1	1.
.1	.09983340279	.09983341665	.9950083093	.9950041653
.2	.1986717763	.1986693308	.9801321815	.9800665778
.3	.2955396666	.2955202067	.9556620941	.9553364891
.4	.3894995326	.3894183423	.9220619210	.9210609940
.5	.4796677609	.4794255386	.8799410570	.8775825619
.6	.5652275343	.5646424734	.8300199467	.8253356149
.7	.6454380735	.6442176872	.7730926195	.7648421873
.8	.7196401313	.7173560909	.7099897691	.6967067093
.9	.7872579516	.7833269096	.6415454001	.6216099683
1.0	.8477981672	.8414709848	.5685692634	.5403023059

We generate a set of 70 approximate values in `approx_pts`. Notice that we use enough values of t so that the t-values eventually become larger than 2π.

```
> approx_pts:={seq([xrk(i),yrk(i)],i=0..70)}:
```

These points are plotted using `plot` together with the option `style=POINT`. Similarly, we use `plot` to generate a graph of $\begin{cases} x = \sin t \\ y = \cos t \end{cases}$, $0 \leq t \leq 2\pi$. These two graphs are then shown together with `display`, which is contained in the `plots` package, to reveal that they yield similar approximations.

```
> Plot_1:=plot(approx_pts,style=POINT):
  Plot_2:=plot([sin(t),cos(t),t=0..2*Pi]):
  with(plots):
  display({Plot_1,Plot_2});
```

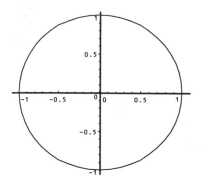

Applications of Systems of Ordinary Differential Equations

10.1 L-R-C CIRCUITS WITH LOOPS

As indicated in Chapter 5, an electrical circuit can be modeled with an ordinary differential equation with constant coefficients. In this section, we illustrate how a circuit involving loops can be described as a system of ordinary linear differential equations with constant coefficients. This derivation is based on the following principles:

DEFINITION | *Kirchhoff's Current Law*
The current entering a point of the circuit equals the current leaving the point.

DEFINITION | *Kirchhoff's Voltage Law*
The sum of the changes in voltage around each loop in the circuit is zero.

As was the case in Chapter 5, we use the following standard symbols for the components of the circuit: $I(t)$ = current, where $I(t) = \dfrac{dQ}{dt}(t)$; $Q(t)$ = charge; R = resistance; C = capacitance; V = voltage; and L = inductance.

The relationships corresponding to the drops in voltage in the various components of the circuit that were stated in Chapter 5 are also given in the following table.

Circuit Element	Voltage Drop
Inductor	$L\dfrac{dI}{dt}$
Resistor	RI
Capacitor	$\dfrac{1}{C}Q$
Voltage Source	$-V(t)$

≡ L-R-C Circuit with One Loop

In determining the drops in voltage around the circuit, we consistently add the voltages in the clockwise direction. The positive direction is directed from the negative symbol toward the positive symbol associated with the voltage source. In summing the voltage drops encountered in the circuit, a drop across a component is added to the sum if the positive direction through the component agrees with the clockwise direction. Otherwise, this drop is subtracted. In the case of the following L-R-C circuit with one loop involving each type of component, the current is equal around the circuit by Kirchhoff's current law, as illustrated in the following diagram.

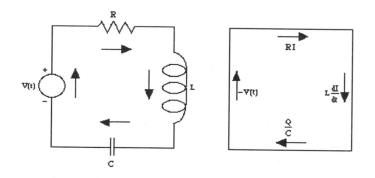

Also, by Kirchhoff's voltage law, we have the following sum:

$$RI + L\frac{dI}{dt} + \frac{1}{C}Q - V(t) = 0.$$

Solving this equation for $\dfrac{dI}{dt}$ and using the relationship between I and Q, $\dfrac{dQ}{dt} = I$, we have the following system of differential equations with indicated initial conditions:

$$\begin{cases} \dfrac{dQ}{dt} = I \\ \dfrac{dI}{dt} = -\dfrac{1}{LC}Q - \dfrac{R}{L}I + \dfrac{V(t)}{L} \end{cases}, Q(0) = Q_0, I(0) = I_0$$

The method of variation of parameters can be used to solve problems of this type.

EXAMPLE: Consider the L-R-C circuit with $L = 1, R = 2, C = 4$, and $V(t) = e^{-t}$. Determine that charge and current using the values given in the list $[0, 0]$, $[1, 1]$, $[-1, 1]$, $[1, -1]$, $[1, 2]$, $[1, -2]$, $[-1, 2]$, $[-1, -2]$, $[2, 1]$, $[2, -1]$, $[3, 1]$, $[3, -1]$, $[4, 1]$, $[4, -1]$, $[5, 1]$, and $[5, -1]$, where each ordered pair represents [_C1,_C2] in the general solution. Plot the phase plane for this system using the solution obtained with this set of constants.

SOLUTION: We begin by modeling the circuit with a system of differential equations. In this case, we have

$$\begin{cases} \dfrac{dQ}{dt} = I \\[2mm] \dfrac{dI}{dt} = -\dfrac{1}{4}Q - 2I + e^{-t} \end{cases}'$$

which we solve with $dsolve$, naming the resulting general solution $Dsys$.

```
> q:='q':i:='i':
  Dsys:=dsolve({diff(q(t),t)=i(t),
  diff(i(t),t)=-1/4*q(t)-2*i(t)+exp(-t)},{q(t),i(t)});
```

$$Dsys := \left\{ i(t) = \frac{4}{3}e^{-t} + \left(-\_C1 + \frac{1}{2}\_C1\sqrt{3} \right) e^{\frac{1}{2}(-2+\sqrt{3})t} + \_C2 e^{-\frac{1}{2}(2+\sqrt{3})t} \right.$$

$$\left. q(t) = -\frac{4}{3}e^{-t} + \_C1 e^{\frac{1}{2}(-2+\sqrt{3})t} + (-4\_C2 + 2\_C2\sqrt{3})e^{-\frac{1}{2}(2+\sqrt{3})t} \right\}$$

We then name $q(t)$ and $i(t)$ the result obtained in $Dsys$ with $assign$ and define the function to_plot. Given an ordered pair $pair$, $to\_plot(pair)$ returns the ordered triple consisting of $q(t)$, $i(t)$, and $t=-2..2$, where each occurrence of $\_C1$ in $q(t)$ and $i(t)$ is replaced by the first coordinate of $pair$, while each occurrence of $\_C2$ in $q(t)$ and $i(t)$ is replaced by the second coordinate of $pair$.

```
> assign(Dsys):
  to_plot:=proc(pair)
    subs({_C1=pair[1],_C2=pair[2]},[q(t),i(t),t=-2..2])
  end:
```

We then define the list $pairs$

```
> pairs:=[[0,0],[1,1],[-1,1],[1,-1],[1,2],[1,-2],
   [-1,2],[-1,-2],[2,1],[2,-1],[3,1],[3,-1],[4,1],
   [4,-1],[5,1],[5,-1]];
```

$pairs := [[0, 0], [1, 1], [-1, 1], [1, -1], [1, 2], [1, -2], [-1, 2], [-1, -2], [2, 1], [2, -1], [3, 1],$
$[3, -1], [4, 1], [4, -1], [5, 1], [5, -1]]$

and use **map** to apply **to_plot** to each ordered pair in **pairs**, naming the resulting set of ordered triples **to_graph**. Note that **to_graph** is not displayed because a colon is included at the end of the command. The resulting set of functions is graphed with **plot**. The option **view=[-10..10,-30..30]** instructs Maple to display the resulting graphs on the rectangle $[-10, 10] \times [-30, 30]$.

```
> to_graph:=map(to_plot,pairs):
   plot(to_graph,view=[-10..10,-30..30]);
```

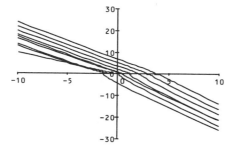

Using **nops**, we see that **to_graph** consists of 16 ordered triples.

```
> nops(to_graph);
```

16

The first element of each component of **to_graph** represents $q(t)$. We now use **seq** to extract the first component of each ordered triple of **to_graph** and name the resulting set **qs**.

```
> qs:={seq(to_graph[i][1],i=1..16)};
```

$$qs := \left\{ -\frac{4}{3}e^{-t} + \%3 + \%2, \; -\frac{4}{3}e^{-t} + \%3 + (-8 + 4\sqrt{3})\%1, \; -\frac{4}{3}e^{-t} + \%3, +\%4, \right.$$

$$-\frac{4}{3}e^{-t} - \%3 + \%2, \; -\frac{4}{3}e^{-t} - \%3 + (-8 + 4\sqrt{3})\%1, \; -\frac{4}{3}e^{-t} + \%3 + (8 - 4\sqrt{3})\%1,$$

$$-\frac{4}{3}e^{-t}, \; -\frac{4}{3}e^{-t} + 2\%3 + \%2, \; -\frac{4}{3}e^{-t} - \%3 + (8 - 4\sqrt{3})\%1, \; -\frac{4}{3}e^{-t} + 3\%3 + \%4,$$

$$-\frac{4}{3}e^{-t} + 3\%3 + \%2, \; -\frac{4}{3}e^{-t} - 2\%3 + \%4, \; -\frac{4}{3}e^{-t} + 4\%3 + \%4$$

$$\left. -\frac{4}{3}e^{-t} + 4\%3 + \%2, \; -\frac{4}{3}e^{-t} - 5\%3 + \%4, \; -\frac{4}{3}e^{-t} + 5\%3 + \%2 \right\}$$

$$\%1 := e^{-\frac{1}{2}(2+\sqrt{3})t}$$

$$\%2 := (-4 + 2\sqrt{3})\%1$$

$$\%3 := e^{\frac{1}{2}(-2+\sqrt{3})t}$$

$$\%4 := (4 - 2\sqrt{3})\%1$$

Next, we use **plot** to graph the set of functions **qs** on the interval $[0, 6]$.

```
> plot(qs,t=0..6);
```

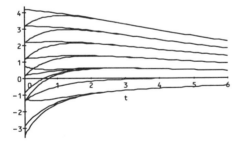

Similarly, the second element of each component of **to_graph** represents $i(t)$. These sixteen functions are extracted from **to_graph** in the same manner as before and then graphed with **plot**.

```
> is:={seq(to_graph[i][2],i=1..16)}:
  plot(is,t=0..6);
```

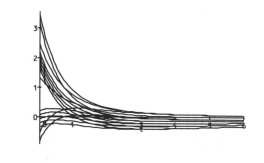

≡ *L-R-C Circuit with Two Loops*

The differential equation that models the circuit becomes more difficult to derive as the number of loops in the circuit is increased. For example, consider the following circuit that contains two loops.

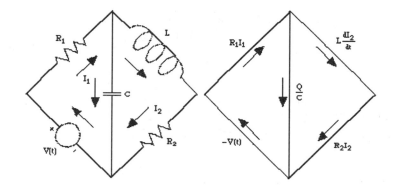

In this case, the current through the capacitor is equivalent to $I_1 - I_2$. Summing the voltage drops around each loop, we have

$$\begin{cases} R_1 I_1 + \dfrac{1}{C}Q - V(t) = 0 \\ L\dfrac{dI_2}{dt} + R_2 I_2 - \dfrac{1}{C}Q = 0 \end{cases}.$$

Solving the first equation for I_1 and using the relationship

$$\frac{dQ}{dt} = I = I_1 - I_2,$$

we have the following system:

$$
\begin{cases}
\dfrac{dQ}{dt} = -\dfrac{1}{R_1 C}Q - I_2 + \dfrac{V(t)}{R_1} \\[2mm]
\dfrac{dI_2}{dt} = \dfrac{1}{LC}Q - \dfrac{R_2}{L}I_2
\end{cases}.
$$

EXAMPLE: Consider the L-R-C circuit with two loops given that $R_1 = 16$, $L = R_2 = 4$, $C = 1$, and $V(t) = e^{-t}$. (a) Plot the phase plane using the solutions obtained with the following pairs of constants indicating [_C1,_C2]: [0,0], [1,1], [−1,−1], [2,2], [−2,−2], [2,−1], [−3,−3], [3,3], [4,4], [−4,−4], [5,5], [−5,−5], [6,6], [−6,−6]. (b) Plot $q(t)$ and $i_2(t)$ separately to compare the behavior of the solutions. (c) Determine the current $i(t)$ and plot the corresponding solutions using these constants.

SOLUTION: The nonhomogeneous system that models this circuit is

$$
\begin{cases}
\dfrac{dQ}{dt} = -\dfrac{1}{16}Q - I_2 + \dfrac{1}{16}e^{-t} \\[2mm]
\dfrac{dI_2}{dt} = \dfrac{1}{4}Q - I_2
\end{cases}.
$$

In the same way as in the previous example, a general solution is found with **dsolve** in **Dsys**.

```
> Dsys:='Dsys':q:='q':i:='i':
  Dsys:=dsolve({diff(q(t),t)=
  -1/16*q(t)-itwo(t)+1/16*exp(-t),
  diff(itwo(t),t)=1/4*q(t)-itwo(t)},
  {q(t),itwo(t)});
```

$$
Dsys := \left\{ itwo(t) = \frac{1}{16}e^{-t} + \_C1 e^{-\frac{17}{32}t}\sin\left(\frac{1}{32}\sqrt{31}t\right) + \_C2 e^{-\frac{17}{32}t}\cos\left(\frac{1}{32}\sqrt{31}t\right), \right.
$$

$$
q(t) = \left(\frac{15}{8}\_C1 - \frac{1}{8}\_C2\sqrt{31}\right)e^{-\frac{17}{32}t}\sin\left(\frac{1}{32}\sqrt{31}t\right)
$$

$$
\left. + \left(\frac{1}{8}\_C1\sqrt{31} + \frac{15}{8}\_C2\right)e^{-\frac{17}{32}t}\cos\left(\frac{1}{32}\sqrt{31}t\right)\right\}
$$

In the same way as in the previous example, we name the solutions obtained with **dsolve** with **assign**, define the function **to_plot**, and enter the constants in the list **pairs**.

```
> assign(Dsys);
  to_plot:=proc(pair)
    subs({_C1=pair[1],_C2=pair[2]},[q(t),itwo(t),t=-1..20])
    end:
  pairs:=[[0,0],[1,1],[-1,-1],[2,2],[-2,-2],[2,-1],
    [-3,-3],[3,3],[4,4],[-4,-4],[5,5],[-5,-5],
    [6,6],[-6,-6]];
```

$$pairs := [[0,0],[1,1],[-1,-1],[2,2],[-2,-2],[2,-1],[-3,-3],$$
$$[3,3],[4,4],[-4,-4],[5,5],[-5,-5],[6,6],[-6,-6]]$$

We then use **map** to apply **to_plot** to the list of ordered pairs **pairs** and graph the resulting set of functions with **plot**.

```
> to_graph:=map(to_plot,pairs):
  plot(to_graph,view=[-0.02..0.02,-0.02..0.02]);
```

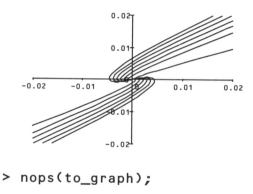

```
> nops(to_graph);
```

14

The first element of each component of **to_graph** represents $q(t)$. We now use **seq** to extract the first component of the first five ordered triples of **to_graph** and name the resulting set **q_table_1**.

```
> q_table_1:={seq(to_graph[i][1],i=1..5)};
```

$$q\_table\_1 := \left\{ 0, \left(\frac{15}{8} - \frac{1}{8}\sqrt{31} \right) e^{-\frac{17}{32}t} \%2 + \left(\frac{1}{8}\sqrt{31} + \frac{15}{8} \right) e^{-\frac{17}{32}t} \%1, \right.$$

$$\left(-\frac{15}{8} + \frac{1}{8}\sqrt{31} \right) e^{-\frac{17}{32}t} \%2 + \left(-\frac{1}{8}\sqrt{31} - \frac{15}{8} \right) e^{-\frac{17}{32}t} \%1,$$

$$\left(\frac{15}{4} - \frac{1}{4}\sqrt{31} \right) e^{-\frac{17}{32}t} \%2 + \left(\frac{1}{4}\sqrt{31} + \frac{15}{4} \right) e^{-\frac{17}{32}t} \%1,$$

$$\left. \left(-\frac{15}{4} + \frac{1}{4}\sqrt{31} \right) e^{-\frac{17}{32}t} \%2 + \left(-\frac{1}{4}\sqrt{31} - \frac{15}{4} \right) e^{-\frac{17}{32}t} \%1 \right\}$$

$$\%1 := \cos \left(\frac{1}{32}\sqrt{31}\, t \right)$$

$$\%2 := \sin \left(\frac{1}{32}\sqrt{31}\, t \right)$$

We then graph **q_table_1** with **plot**.

```
> plot(q_table_1,t=0..5);
```

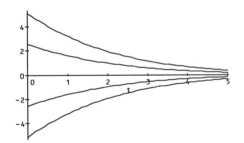

Similarly, we define **q_table_2** and **q_table_3** and graph the resulting sets with **plot**.

```
> q_table_2:={seq(to_graph[i][1],i=6..10)}:
  plot(q_table_2,t=0..5);
  q_table_3:={seq(to_graph[i][1],i=11..14)}:
  plot(q_table_3,t=0..5);
```

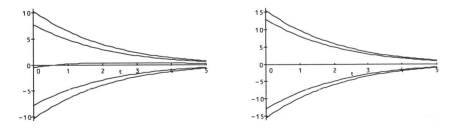

We graph the sets `i_table_1`, `i_table_2`, and `i_table_3`, representing the $i_2(t)$'s, in the same way.

```
> i_table_1:={seq(to_graph[i][2],i=1..5)}:
  plot(i_table_1,t=0..5);
  i_table_2:={seq(to_graph[i][2],i=6..10)}:
  plot(i_table_2,t=0..5);
  i_table_3:={seq(to_graph[i][2],i=11..14)}:
  plot(i_table_3,t=0..5);
```

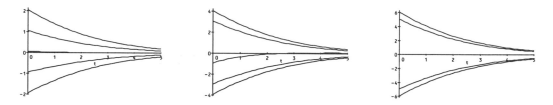

The overall current, $i(t)$, is defined as the derivative of the charge, $q(t)$. We now use `union` to define `qs` to be the union of `q_table_1`, `q_table_2`, and `q_table_3` and then use `map` and `diff` to differentiate each of the functions in `qs` with respect to t, naming the resulting set of functions `current`. The 14 functions in `current` are then graphed with `plot`.

```
> qs:=`union`(q_table_1,q_table_2,q_table_3):
  current:=map(diff,qs,t):
  plot(current,t=0..5);
```

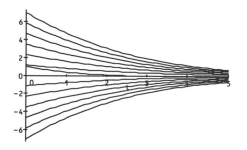

≡ *L-R-C Circuit with Three Loops*

Next, we consider the following circuit that is made up of three loops and illustrated in the following diagram.

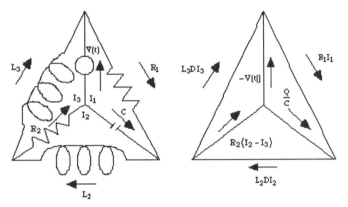

In this circuit, the current through the resistor R_2 is $I_2 - I_3$, and the current through the capacitor is $I - I_2$. Using these quantities in the voltage drop sum equations yields the following three-dimensional system:

$$\begin{cases} -V(t) + R_1 I_1 + \dfrac{1}{C}Q = 0 \\[2mm] -\dfrac{1}{C}Q + L_2\dfrac{dI_2}{dt} + R_2(I_2 - I_3) = 0 \\[2mm] V(t) - R_2(I_2 - I_1) + L_3\dfrac{dI_3}{dt} = 0 \end{cases}.$$

Using the relationship

$$\frac{dQ}{dt} = I_1 - I_2$$

and solving the first equation for I_1, we have the system

$$\begin{cases} \dfrac{dQ}{dt} = -\dfrac{1}{R_1 C}Q - I_2 + \dfrac{V(t)}{R_1} \\[2mm] \dfrac{dI_2}{dt} = \dfrac{1}{L_2 C}Q - \dfrac{R_2}{L_2}I_2 + \dfrac{R_2}{L_2}I_3 \\[2mm] \dfrac{dI_3}{dt} = \dfrac{R_2}{L_3}I_2 - \dfrac{R_2}{L_3}I_3 - \dfrac{V(t)}{L_3} \end{cases}.$$

EXAMPLE: Solve the three-loop circuit if $R_1 = R_2 = L_1 = L_2 = L_3 = 1$ and $V(t) = 10$ and the initial conditions are $Q(0) = 10$ and $I_2(0) = I_3(0) = 0$.

SOLUTION: In this case, the nonhomogeneous system is

$$\begin{cases} \dfrac{dQ}{dt} = -Q - I_2 + 10 \\[2mm] \dfrac{dI_2}{dt} = -Q - I_2 + I_3 \\[2mm] \dfrac{dI_3}{dt} = I_2 - I_3 - 10 \end{cases},$$

which we define as `Sys_of_Eqs`.

```
> Sys_of_Eqs:={diff(Q(t),t)=-Q(t)-I2(t)+10,
    diff(I2(t),t)=-Q(t)-I2(t)+I3(t),
    diff(I3(t),t)=I2(t)-I3(t)-10,
    Q(0)=10,I2(0)=0,I3(0)=0};
```

$$Sys\_of\_Eqs := \left\{ \frac{\partial}{\partial t} I3(t) = I2(t) - I3(t) - 10, Q(0) = 10, \right.$$

$$\frac{\partial}{\partial t} Q(t) = -Q(t) - I2(t) + 10, \frac{\partial}{\partial t} I2(t) = -Q(t) - I2(t) + I3(t),$$

$$\left. I2(0) = 0, I3(0) = 0 \right\}$$

We then solve `Sys_of_Eqs` with `dsolve` and name the result `Sol`.

```
> Sol:=dsolve(Sys_of_Eqs,{Q(t),I2(t),I3(t)});
```

$$Sol := \left\{ Q(t) = \left(-10 + 5e^{-t} - \left(-\frac{15}{2} - 5\sqrt{2} \right) e^{(\sqrt{2}-1)t} \right. \right.$$

$$\left. - \left(5\sqrt{2} - \frac{15}{2} \right) e^{-(\sqrt{2}+1)t} \right) \Big/ \left((\sqrt{2} - 1)(\sqrt{2} + 1) \right),$$

$$I2(t) = \left(20 + \left(-\frac{15}{2} - 5\sqrt{2} \right) e^{(\sqrt{2}-1)t} \sqrt{2} - \left(5\sqrt{2} - \frac{15}{2} \right) e^{-(\sqrt{2}+1)t} \sqrt{2} \right)$$

$$\Big/ \left((\sqrt{2} - 1)(\sqrt{2} + 1) \right),$$

$$I3(t) = \left(10 + 5e^{-t} + \left(-\frac{15}{2} - 5\sqrt{2} \right) e^{(\sqrt{2}-1)t} + \left(5\sqrt{2} - \frac{15}{2} \right) e^{-(\sqrt{2}+1)t} \right)$$

$$\Big/ \left((\sqrt{2} - 1)(\sqrt{2} + 1) \right) \right\}$$

After naming the solutions obtained in `Sol` with `assign`, we graph these three functions with `plot`.

```
> assign(Sol):
  plot(Q(t),t=0..10);
  plot(I2(t),t=0..10);
  plot(I3(t),t=0..10);
```

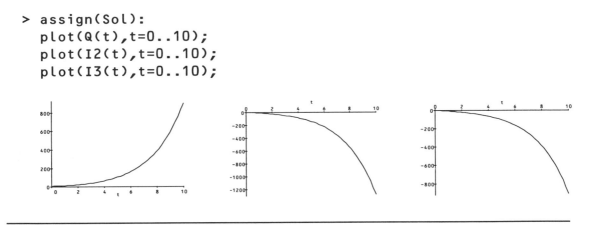

10.2 DIFFUSION PROBLEMS

≡ Diffusion through a Membrane

Solving problems to determine the diffusion of a material in a medium also leads to first-order systems of ordinary linear differential equations. For example, consider the situation in which two solutions of a substance are separated by a membrane of permeability P. Assume that the amount of substance that passes through the membrane at any particular time is proportional to the difference between the concentrations of the two solutions. Therefore, if we let x_1 and x_2 represent the two concentrations, and V_1 and V_2 represent the corresponding volume of each solution, the system of differential equations is given by

$$\begin{cases} \dfrac{dx_1}{dt} = \dfrac{P}{V_1}(x_2 - x_1) \\ \dfrac{dx_2}{dt} = \dfrac{P}{V_2}(x_1 - x_2) \end{cases},$$

where the initial amounts of x_1 and x_2 are given.

EXAMPLE: Suppose that two salt concentrations of equal volume V are separated by a membrane of permeability P. Given that $P = V$, determine the amount of salt in each concentration at time t if $x_1(0) = 2$ and $x_2(0) = 10$.

SOLUTION: In this case, the initial-value problem that models the situation is

$$\begin{cases} \dfrac{dx_1}{dt} = x_2 - x_1 \\[2mm] \dfrac{dx_2}{dt} = x_1 - x_2 \end{cases}, x_1(0) = 2, x_2(0) = 10,$$

which is defined as follows in `Init_val_prob` and then solved with `dsolve`.

```
> Init_val_prob:={diff(x1(t),t)=x2(t)-x1(t),
    diff(x2(t),t)=x1(t)-x2(t),x1(0)=2,x2(0)=10};
```

$$Init\_val\_prob := \left\{ \frac{\partial}{\partial t}x1(t) = x2(t) - x1(t), \ \frac{\partial}{\partial t}x2(t) = x1(t) - x2(t), \ x2(0) = 10, x1(0) = 2 \right\}$$

```
> Sol:=dsolve(Init_val_prob,{x1(t),x2(t)});
```

$$Sol := \left\{ x1(t) = 6 - 4e^{-2t}, x2(t) = 6 + 4e^{-2t} \right\}$$

We graph this solution parametrically with **plot** and then graph $x_1(t)$ and $x_2(t)$ simultaneously. Notice that the amount of salt in each concentration approaches 6, which is the average value of the two initial amounts.

```
> assign(Sol):
  plot([x1(t),x2(t),t=0..2]);
  plot({x1(t),x2(t)},t=0..2);
```

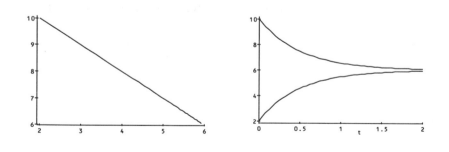

≡ Diffusion through a Double-Walled Membrane

Next, consider the situation in which two solutions are separated by a double-walled membrane, where the inner wall has permeability P_1 and the outer wall has permeability P_2 with $0 < P_1 < P_2$. Suppose that the volume of solution within the inner wall is V_1 and that between the two walls, V_2. Let x represent the concentration of the solution within the inner wall and y, the concentration between the two walls. Assuming that the concentration of the solution outside the outer wall is constantly C, we have the following system of first-order ordinary differential equations:

$$
\begin{cases}
\dfrac{dx}{dt} = \dfrac{P_1}{V_1}(y - x) \\[2mm]
\dfrac{dy}{dt} = \dfrac{1}{V_2}(P_2(C - y) + P_1(x - y)) = \dfrac{-(P_1 + P_2)y}{V_2} + \dfrac{P_1 x}{V_2} + \dfrac{P_2 C}{V_2} \\[2mm]
x(0) = x_0, \, y(0) = y_0
\end{cases}
$$

EXAMPLE: Given that $P_1 = 3$, $P_2 = 8$, $V_1 = 2$, and $V_2 = 10$. Determine and graph the solution of this system for various initial conditions and values of C.

SOLUTION: To graph the solution for various initial conditions, we must solve the initial-value problem

$$
\begin{cases}
\dfrac{dx}{dt} = \dfrac{3}{2}(y - x) \\[2mm]
\dfrac{dy}{dt} = \dfrac{-11y}{10} + \dfrac{3x}{10} + \dfrac{4}{5}C \\[2mm]
x(0) = x_0, \, y(0) = y_0
\end{cases}
$$

We find the solution without the use of **dsolve**. Note that depending on the speed of the computer you are using, when working with nonhomogeneous systems, this sometimes decreases the computation time when compared to that of **dsolve**. The eigenvalues and corresponding eigenvectors of the corresponding homogeneous system are determined with **eigenvects**.

```
> p1:=3:p2:=8:vol1:=2:vol2:=10:
  A:=array([[-p1/vol1,p1/vol1],[p1/vol2,-(p1+p2)/vol2]]);
```

$$
A := \begin{bmatrix} -\dfrac{3}{2} & \dfrac{3}{2} \\[2mm] \dfrac{3}{10} & -\dfrac{11}{10} \end{bmatrix}
$$

```
> with(linalg):
  Esys:=eigenvects(A);
```

$$
Esys := \left[-2, 1, \{[-3 \quad 1]\}\right], \left[\dfrac{-3}{5}, 1, \left\{\left[\dfrac{5}{3} \quad 1\right]\right\}\right]
$$

Hence, the eigenvalues are $\lambda_1 = -2$ and $\lambda_2 = -\dfrac{3}{5}$ with corresponding eigenvectors $v_1 = \begin{pmatrix} -3 \\ 1 \end{pmatrix}$

and $v_2 = \begin{pmatrix} 5/3 \\ 1 \end{pmatrix}$, respectively. Therefore, a fundamental matrix for the homogeneous system is

$$\Phi(t) = \begin{pmatrix} -3e^{-2t} & \dfrac{5}{3}e^{-\frac{3}{5}t} \\ e^{-2t} & e^{-\frac{3}{5}t} \end{pmatrix}$$

```
> phi:=t->array([[-3*exp(-2*t),5/3*exp(-3*t/5)],
  [exp(-2*t),exp(-3*t/5)]]):
  phi(t);
```

$$\begin{bmatrix} -3e^{-2t} & \dfrac{5}{3}e^{-\frac{3}{5}t} \\ e^{-2t} & e^{-\frac{3}{5}t} \end{bmatrix}$$

with inverse

$$\Phi^{-1}(t) = \begin{pmatrix} -\dfrac{3}{14}e^{2t} & \dfrac{5}{14}e^{2t} \\ \dfrac{3}{14}e^{\frac{3}{5}t} & \dfrac{9}{14}e^{\frac{3}{5}t} \end{pmatrix}.$$

```
> invphi:=t->inverse(phi(t)):
  invphi(t);
```

$$\begin{bmatrix} -\dfrac{3}{14}\dfrac{1}{e^{-2t}} & \dfrac{5}{14}\dfrac{1}{e^{-2t}} \\ \dfrac{3}{14}\dfrac{1}{e^{-\frac{3}{5}t}} & \dfrac{9}{14}\dfrac{1}{e^{-\frac{3}{5}t}} \end{bmatrix}$$

To compute a general solution, we first define **F**.

```
> f:=t->array([0,p2*c/vol2]):
```

Then we compute $\Phi^{-1}(u)\,\mathbf{F}(u)$, naming the result `step_1`;

```
> step_1:=evalm(invphi(u) &* f(u));
```

$$step\_1 := \begin{bmatrix} \dfrac{2}{7}\dfrac{c}{e^{-2u}} & \dfrac{18}{35}\dfrac{c}{e^{-\frac{3}{5}u}} \end{bmatrix}$$

compute $\int_0^t \Phi^{-1}(u)\, F(u)\, du$, naming the result s t e p_2;

```
> step_2:=map(int,step_1,u=0..t);
```

$$step\_2 := \left[\frac{1}{7}\frac{c}{e^{-2t}} - \frac{1}{7}c\frac{6}{7}\frac{c}{e^{-\frac{3}{5}t}} - \frac{6}{7}c \right]$$

and then compute $\Phi(t) \int_0^t \Phi^{-1}(u)\, F(u)\, du$, naming the result s o l p. Note that s o l p is a particular solution of the nonhomogeneous equation.

```
> solp:=t->evalm(phi(t) &* step_2):
  solp(t);
```

$$\left[\frac{1}{7}c\left(7 + 3e^{-2t} - 10e^{-\frac{3}{5}t}\right) - \frac{1}{7}c\left(-7 + e^{-2t} + 6e^{-\frac{3}{5}t}\right) \right]$$

A general solution of the corresponding homogeneous problem is found by computing $\Phi(t)\, \Phi^{-1}(0)\, X(0)$, naming the result s o l h.

```
> solh:=t->evalm(phi(t) &* invphi(0) &* [x0,y0]):
  solh(t);
```

$$\left[\frac{9}{14}x0\,e^{-2t} + \frac{5}{14}x0\,e^{-\frac{3}{5}t} - \frac{15}{14}y0\,e^{-2t} + \frac{15}{14}y0\,e^{-\frac{3}{5}t} \right.$$

$$\left. -\frac{3}{14}x0\,e^{-2t} + \frac{3}{14}x0\,e^{-\frac{3}{5}t} + \frac{5}{14}y0\,e^{-2t} + \frac{9}{14}y0\,e^{-\frac{3}{5}t} \right]$$

Finally, a general solution of the nonhomogeneous system $X(t) = \Phi(t)\, \Phi^{-1}(0)\, X(0) + \Phi(t) \int_0^t \Phi^{-1}(u)\, F(u)\, du$ is obtained by adding s o l h and s o l p, naming the result s o l u t i o n.

```
> solution:=evalm(solh(t)+solp(t));
```

$$solution := \left[\vphantom{\frac{9}{14}} \right.$$

$$\frac{9}{14}x0\,e^{-2t} + \frac{5}{14}x0\,e^{-\frac{3}{5}t} - \frac{15}{14}y0\,e^{-2t} + \frac{15}{14}y0\,e^{-\frac{3}{5}t} - \frac{1}{7}c\left(7 + 3e^{-2t} - 10e^{-\frac{3}{5}t}\right)$$

$$-\frac{3}{14}x0\,e^{-2t} + \frac{3}{14}x0\,e^{-\frac{3}{5}t} + \frac{5}{14}y0\,e^{-2t} + \frac{9}{14}y0\,e^{-\frac{3}{5}t} - \frac{1}{7}c\left(-7 + e^{-2t} + 6e^{-\frac{3}{5}t}\right)$$

$$\left. \vphantom{\frac{9}{14}} \right]$$

Note that $x(t)$ corresponds to the first part of the list **solution**, while $y(t)$ corresponds to the second part of **solution**. To graph $x(t)$ and $y(t)$ for various initial conditions and values of C, we use **proc** to define x and y.

```
> x:=proc(t0,x00,y00,c0)
    subs({t=t0,x0=x00,y0=y00,c=c0},solution[1])
    end:
  y:=proc(t0,x00,y00,c0)
    subs({t=t0,x0=x00,y0=y00,c=c0},solution[2])
    end:
```

For example, entering

```
> x(t,2,1,10);
```

$$\frac{9}{2}e^{-2t} - \frac{25}{2}e^{-\frac{3}{5}t} + 10$$

and

```
> y(t,2,1,10);
```

$$-\frac{3}{2}e^{-2t} - \frac{15}{2}e^{-\frac{3}{5}t} + 10$$

returns the solution to the initial-value problem

$$\begin{cases} \dfrac{dx}{dt} = \dfrac{3}{2}(y - x) \\[2mm] \dfrac{dy}{dt} = \dfrac{-11y}{10} + \dfrac{3x}{10} + 8 \\[2mm] x(0) = 2, \, y(0) = 1 \end{cases}.$$

Similarly, entering

```
> plot({x(t,2,1,10),y(t,2,1,10)},t=0..8);
```

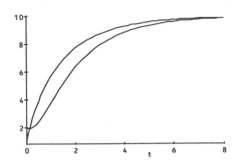

graphs the solution for $0 \le t \le 8$.

We now investigate other initial conditions. For example, entering

```
> to_graph:=[seq({x(t,2,i,10),y(t,2,i,10)},i=1..4)];
```

$$to\_graph := \left[\left\{ \frac{9}{2}e^{-2t} - \frac{25}{2}e^{-\frac{3}{5}t} + 10, -\frac{3}{2}e^{-2t} - \frac{15}{2}e^{-\frac{3}{5}t} + 10 \right\}, \right.$$

$$\left\{ \frac{24}{7}e^{-2t} - \frac{80}{7}e^{-\frac{3}{5}t} + 10, -\frac{8}{7}e^{-2t} - \frac{48}{7}e^{-\frac{3}{5}t} + 10 \right\},$$

$$\left\{ -\frac{11}{14}e^{-2t} - \frac{87}{14}e^{-\frac{3}{5}t} + 10, \frac{33}{14}e^{-2t} - \frac{145}{14}e^{-\frac{3}{5}t} + 10 \right\},$$

$$\left. \left\{ -\frac{3}{7}e^{-2t} - \frac{39}{7}e^{-\frac{3}{5}t} + 10, \frac{9}{7}e^{-2t} - \frac{65}{7}e^{-\frac{3}{5}t} + 10 \right\} \right]$$

returns a list of solutions to

$$\begin{cases} \dfrac{dx}{dt} = \dfrac{3}{2}(y - x) \\[2mm] \dfrac{dy}{dt} = \dfrac{-11y}{10} + \dfrac{3x}{10} + 8 \\[2mm] x(0) = 2, y(0) = i \end{cases}$$

for $i = 1, 2, 3$, and 4. After loading the **plots** package, we use **map** and **plot** to graph each of the pairs of solutions in **to_plot** together for $0 \le t \le 8$. The four graphs are then displayed using **display**, which is contained in the **plots** package.

Notice that all the graphs approach 10, the value of C.

```
> with(plots):
  graphs:=map(plot,to_graph,t=0..8):
  for i from 1 to 4 do display({graphs[i]}) od;
```

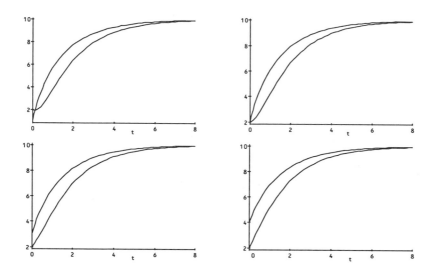

Similarly, entering

```
> to_graph:=[seq({x(t,i,3,10),y(t,i,3,10)},i=1..4)];
  graphs:=map(plot,to_graph,t=0..8):
  for i from 1 to 4 do display({graphs[i]}) od;
```

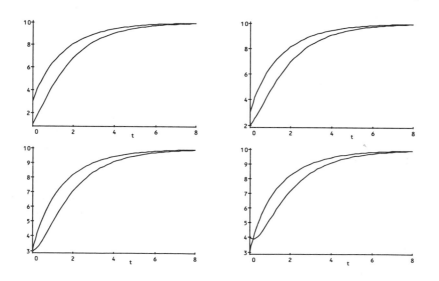

solves

$$
\begin{cases}
\dfrac{dx}{dt} = \dfrac{3}{2}(y - x) \\[2mm]
\dfrac{dy}{dt} = \dfrac{-11y}{10} + \dfrac{3x}{10} + 8 \\[2mm]
x(0) = i, y(0) = 3
\end{cases}
$$

for $i = 1, 2, 3$, and 4; graphs the solutions together; and displays the resulting four graphs.

To investigate the effects of different concentrations outside the outer wall, we determine and plot the solution for other parameter values using $x(0) = 2$ and $y(0) = 5$ and $C = i$ for i in the list of numbers `i_vals`. We generate four plots corresponding to $i = 3, 6, 9$, and 12.

```
>  i_vals:=[3,6,9,12]:
   to_graph:=[seq({x(t,2,5,i),y(t,2,5,i)},i=i_vals)];
   graphs:=map(plot,to_graph,t=0..8):
   for i from 1 to 4 do display({graphs[i]}) od;
```

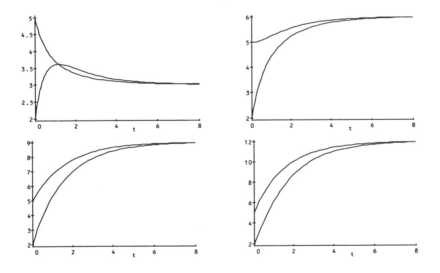

We also generate four plots for $i = 2, 4, 6$, and 8. Notice that in each case the solutions approach the value of C.

```
> i_vals:=[2,4,6,8]:
  to_graph:=[seq({x(t,2,5,i),y(t,2,5,i)},i=i_vals)];
  graphs:=map(plot,to_graph,t=0..8):
  for i from 1 to 4 do display({graphs[i]}) od;
```

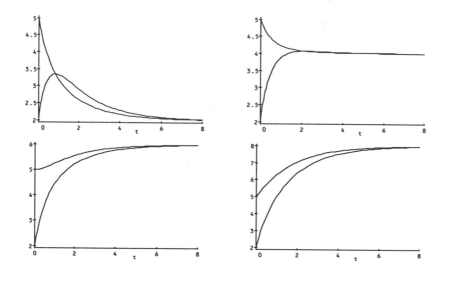

10.3 SPRING-MASS SYSTEMS

The motion of a mass attached to the end of a spring was modeled with a second-order linear differential equation with constant coefficients in Chapter 5. This situation can then be expressed as a system of first-order ordinary differential equations as well. Recall that if there is no external forcing function, then the second-order differential equation that models this situation is

$$m\frac{d^2x}{dt^2} + c\frac{dx}{dt} + kx = 0,$$

where m is the mass attached to the end of the spring, c is the damping coefficient, k is the spring constant found with Hooke's law, and $f(t)$ is the forcing function. This equation is easily transformed into a system of equations with the following substitution. Let $\frac{dx}{dt} = y$. Hence,

$\dfrac{dy}{dt} = \dfrac{d^2x}{dt^2} = -\dfrac{k}{m}x - \dfrac{c}{m}\dfrac{dx}{dt}$. Therefore, after substitution, we have the system

$$\begin{cases} \dfrac{dx}{dt} = y \\[2mm] \dfrac{dy}{dt} = -\dfrac{k}{m}x - \dfrac{c}{m}y \end{cases}$$

In the previous examples, the motion of each spring was illustrated as a function of time. However, problems of this type may also be investigated using the phase plane. In the following example, the phase plane corresponding to the various situations encountered by spring-mass systems discussed in previous sections (undamped, damped, overdamped, and critically damped) are determined.

EXAMPLE: Determine the phase portrait for each of the following situations: (a) $m = 1, c = 0$, and $k = 1$; (b) $m = 1, c = 1, k = \dfrac{1}{2}$; (c) $m = 1, c = \sqrt{5}, k = 1$; and (d) $m = 1, c = 2, k = 1$.

SOLUTION: For (a), the system is

$$\begin{cases} \dfrac{dx}{dt} = y \\[2mm] \dfrac{dy}{dt} = -x \end{cases}$$

We find a general solution of this system with **dsolve** and name the result **Spr_1**.

> ```
  x:='x':y:='y':
  Spr_1:=dsolve({diff(x(t),t)=y(t),
  diff(y(t),t)=-x(t)},{x(t),y(t)});
  ```

$Spr\_1 := \{y(t) = \_C1\sin(t) + \_C2\cos(t), x(t) = \_C2\sin(t) - \_C1\cos(t)\}$

We remark that **Spr_1** is a set: in this particular case, $y(t)$ is the first element and $x(t)$ is the second. Depending on the version of Maple and the computer you are using, your results may be the reverse of those obtained here. In this case, however, we obtain the formula for $y(t)$ by using **rhs** to extract the right-hand side of the first part of **Spr_1**.

> ```
  rhs(Spr_1[1]);
  ```

$\_C1\sin(t) + \_C2\cos(t)$

Similarly, the formula for $x(t)$ is obtained by using **rhs** to extract the right-hand side of the second part of **Spr_1**.

```
> rhs(Spr_1[2]);
```

$$\_C2 \sin(t) - \_C1 \cos(t)$$

A list of constants is entered in **pairs** where each represents **[_C1,_C2]**.

```
> pairs:=[[0,0],[1,1],[-1,1],[1,-1],[1,2],[1,-2],
    [-1,2],[-1,-2],[2,1],[2,-1],[3,1],[3,-1],
    [4,1],[4,-1],[5,1],[5,-1]];
```

$$pairs := [[0, 0], [1, 1], [-1, 1], [1, -1], [1, 2], [1, -2], [-1, 2], [-1, -2], [2, 1], [2, -1], [3, 1],$$
$$[3, -1], [4, 1], [4, -1], [5, 1], [5, -1]]$$

We then define **to_plot**. Given an ordered pair **pair**, **to_plot(pair)** returns the ordered triple corresponding to $y(t)$, $x(t)$, and **t=0..2*Pi**, where each occurrence of **_C1** in $x(t)$ and $y(t)$ is replaced by the first coordinate of **pair** and each occurrence of **_C2** is $x(t)$ and $y(t)$ is replaced by the second coordinate of **pair**. We then use **map** to apply **to_plot** to the list **pairs**, naming the resulting list **to_graph**.

```
> to_plot:=proc(pair)
    subs({_C1=pair[1],_C2=pair[2]},
     [rhs(Spr_1[2]),rhs(Spr_1[1]),t=0..2*Pi])
    end:
  to_graph:=map(to_plot,pairs);
```

$$to\_graph := [[0, 0, \%1], [\sin(t) - \cos(t) \sin(t) + \cos(t), \%1],$$
$$[\sin(t) + \cos(t), -\sin(t) + \cos(t), \%1],$$
$$[-\sin(t) - \cos(t), \sin(t) - \cos(t), \%1],$$
$$[2 \sin(t) - \cos(t), \sin(t) + 2 \cos(t), \%1],$$
$$[-2 \sin(t) - \cos(t), \sin(t) - 2 \cos(t), \%1],$$
$$[2 \sin(t) + \cos(t), -\sin(t) + 2 \cos(t), \%1],$$
$$[-2 \sin(t) + \cos(t), -\sin(t) - 2 \cos(t), \%1],$$
$$[\sin(t) - 2 \cos(t), 2 \sin(t) + \cos(t), \%1],$$
$$[-\sin(t) - 2 \cos(t), 2 \sin(t) - \cos(t), \%1],$$
$$[\sin(t) - 3 \cos(t), 3 \sin(t) + \cos(t), \%1],$$
$$[-\sin(t) - 3 \cos(t), 3 \sin(t) - \cos(t), \%1],$$
$$[\sin(t) - 4 \cos(t), 4 \sin(t) + \cos(t), \%1],$$
$$[-\sin(t) - 4 \cos(t), 4 \sin(t) - \cos(t), \%1],$$
$$[\sin(t) - 5 \cos(t), 5 \sin(t) + \cos(t), \%1],$$
$$[-\sin(t) - 5 \cos(t), 5 \sin(t) - \cos(t), \%1]]$$
$$\%1 := t = 0..2_\pi$$

We then graph the functions in to_graph by first using convert together with the set option to convert to_graph to a set object and then using plot. Note that the direction associated with the trajectories can be determined by using the first differential equation of the system, $x' = y$. Hence, x increases in the first and second quadrant where $y > 0$ and decreases in the third and fourth quadrants where $y < 0$. Alternatively, you can use DEplot2 to graph the direction field associated with the system.

```
> plot(convert(to_graph,set));
```

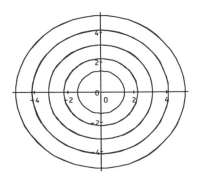

(b) A general solution for this undamped system is determined with dsolve in Spr_2.

```
> Spr_2:=dsolve({diff(x(t),t)=y(t),
    diff(y(t),t)=-1/2*x(t)-y(t)},{x(t),y(t)});
```

$$Spr\_2 := \left\{ y(t) = \_C1\,e^{-\frac{1}{2}t} \sin\left(\frac{1}{2}t\right) + \_C2\,e^{-\frac{1}{2}t} \cos\left(\frac{1}{2}t\right),\right.$$
$$\left. x(t) = (-\_C1 + \_C2)e^{-\frac{1}{2}t} \sin\left(\frac{1}{2}t\right) + (-\_C1 - \_C2)e^{-\frac{1}{2}t} \cos\left(\frac{1}{2}t\right) \right\}$$

As in (a), values for [_C1,_C2] are taken from the list pairs defined as follows

```
> to_plot:=proc(pair)
    subs({_C1=pair[1],_C2=pair[2]},
    [rhs(Spr_2[2]),rhs(Spr_2[1]),t=-Pi/2..2*Pi])
    end:
  pairs:=[[0,0],[1,1],[-1,1],[1,-1],[1,2],[1,-2],
    [-1,2],[-1,-2],[2,1],[2,-1],[3,1],[3,-1],
    [3,2],[-2,-3],[-2,3],[-2,-1],[-3,1],[-3,-1],
    [-4,1],[-4,-2],[-4,3],[-4,-4],[-4,5]];
```

$$pairs := [[0,0],[1,1],[-1,1],[1,-1],[1,2],[1,-2],[-1,2],[-1,-2],$$
$$[2,1],[2,-1],[3,1],[3,-1],[3,2],[-2,-3],[-2,3],[-2,-1],$$
$$[-3,1],[-3,-1],[-4,1],[-4,-2],[-4,3],[-4,-4],[-4,5]]$$

by applying `to_plot` to `pairs` with `map`. `plot` is then used to graph the phase plane for this system. Using $x' = y$, we see that the trajectories are directed inward. This is also determined by observing the exponential terms in the solution. Hence, both x and y approach a value of zero as t increases.

```
> to_graph:=map(to_plot,pairs);
  plot(convert(to_graph,set),view=[-5..5,-5..5]);
```

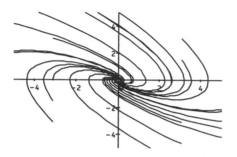

(c) Following the same method, we determine a general solution in `Spr_3` with `dsolve`. Note that this system is overdamped.

```
> Spr_3:=dsolve({diff(x(t),t)=y(t),
  diff(y(t),t)=-x(t)-sqrt(5)*y(t)},{x(t),y(t)});
```

$$Spr\_3 := \left\{ x(t) = \_C1\,e^{-\frac{1}{2}(\sqrt{5}-1)t} + \left(-\frac{1}{2}\_C2\sqrt{5} + \frac{1}{2}\_C2\right)e^{-\frac{1}{2}(\sqrt{5}+1)t}, \right.$$

$$\left. y(t) = \left(-\frac{1}{2}\_C1\sqrt{5} + \frac{1}{2}\_C1\right)e^{-\frac{1}{2}(\sqrt{5}-1)t} + \_C2\,e^{-\frac{1}{2}(\sqrt{5}+1)t} \right\}$$

Entering nearly identical code as in (a) and (b), we define `to_plot`, `pairs`, and `to_graph` and then use `plot` to graph the phase plane. The trajectories are directed toward the origin because both x and y approach zero as t increases.

```
> to_plot:=proc(pair)
    subs({_C1=pair[1],_C2=pair[2]},
      [rhs(Spr_3[1]),rhs(Spr_3[2]),t=-Pi..Pi])
    end:
  pairs:=[[0,0],[1,1],[-1,1],[1,-1],
    [1,2],[1,-2],[-1,2],[-1,-2],
    [2,1],[2,-1],[3,1],[3,-1],
    [4,1],[4,-1],[5,1],[5,-1],
    [6,1],[6,2],[6,3],[6,4],[6,5],
    [-6,1],[-6,2],[-6,3],[-6,4],[-6,5],
    [-6,-1],[-6,-2],[-6,-3],[-6,-4],[-6,-5],
    [6,-1],[6,-2],[6,-3],[6,-4],[6,-5]]:
  to_graph:=map(to_plot,pairs):
  plot(convert(to_graph,set),view=[-10..10,-10..10]);
```

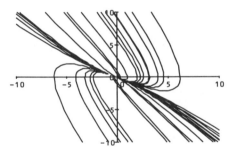

(d) Once again, dsolve is used to find a general solution of this system. Note that this system is critically damped.

```
> Spr_4:=dsolve({diff(x(t),t)=y(t),
    diff(y(t),t)=-x(t)-2*y(t)},{x(t),y(t)});
```

$$Spr\_4 := \{y(t) = \_C1\,e^{-t} + \_C2\,e^{-t}t, x(t) = (-\_C1 - \_C2)e^{-t} - \_C2\,e^{-t}\}$$

After defining to_plot, pairs, and to_graph, the phase plane is produced with plot. Again, the trajectories are directed toward the origin.

```
> to_plot:=proc(pair)
   subs({_C1=pair[1],_C2=pair[2]},
    [rhs(Spr_4[2]),rhs(Spr_4[1]),t=-Pi..Pi])
   end:
  pairs:=[[0,0],[1,1],[-1,1],[1,-1],[1,2],[1,-2],
    [-1,2],[-1,-2],[2,1],[2,-1],[3,1],[3,-1],
    [3,2],[-2,-3],[-2,3],[-2,-1],[-3,1],[-3,-1],
    [-4,1],[-4,-2],[-4,3],[-4,-4],[-4,5]]:
  to_graph:=map(to_plot,pairs);
  plot(convert(to_graph,set),view=[-5..5,-5..5]);
```

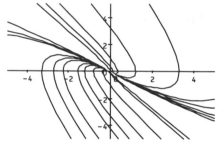

10.4 POPULATION PROBLEMS

In Chapter 3, the population problems discussed were based on the simple principle that the rate at which a population grows (or decays) is proportional to the number present in the population at any time t. Hence, if $x(t)$ represents the population at time t, then

$$\frac{dx}{dt} = kx$$

for some constant k. These ideas can be extended to other examples involving more than one population that lead to systems of ordinary differential equations. We illustrate several situations through the following examples. In each problem, however, we determine the rate at which a population P changes with the equation

$$\frac{dP}{dt} = \text{(rate entering)} - \text{(rate leaving)}.$$

We begin by determining the population in two neighboring territories. Suppose that the population x and y of two neighboring territories depends on several factors. The birth rate of x is a_1, while that of y is b_1. The rate at which citizens of x move to y is a_2, while that at which citizens move from y to x is b_2. Finally, the mortality rate of each territory is disregarded. Determine the respective populations of these two territories for any time t.

Using the simple principles of previous examples, we have that the rate at which population x changes is

$$\frac{dx}{dt} = a_1 x - a_2 x + b_1 y = (a_1 - a_2)x + b_1 y,$$

while the rate at which population y changes is

$$\frac{dy}{dt} = b_1 y - b_2 y + a_2 x = (b_1 - b_2)y + a_2 x.$$

Therefore, the system of equations that must be solved is

$$\begin{cases} \dfrac{dx}{dt} = (a_1 - a_2)x + b_1 y \\[2mm] \dfrac{dy}{dt} = a_2 x + (b_1 - b_2)y \end{cases}$$

where the initial populations of the two territories $x(0) = x_0$ and $y(0) = y_0$ are given.

EXAMPLE: Determine the populations $x(t)$ and $y(t)$ in each territory if $a_1 = 2$, $a_2 = 3$, $b_1 = 2$, and $b_2 = 1$, given that $x(0) = 2000$ and $y(0) = 400$.

SOLUTION: Using these values, we have the following initial-value problem:

$$\begin{cases} \dfrac{dx}{dt} = x + y \\[2mm] \dfrac{dy}{dt} = x + 4y \end{cases}, x(0) = 1000, y(0) = 2000.$$

By observing the positive coefficients in both of the differential equations in this system, we expect both populations to increase.

```
> Sol:=dsolve({diff(x(t),t)=x(t)+y(t),
    diff(y(t),t)=x(t)+4*y(t),x(0)=1000,y(0)=2000},
    {x(t),y(t)});
```

$$Sol := \left\{ y(t) = \left(750 + \frac{750}{13}\sqrt{13} + \frac{1}{2}\left(500 + \frac{500}{13}\sqrt{13} \right)\sqrt{13} \right) e^{\frac{1}{2}(5+\sqrt{13})t} \right.$$

$$+ \left(1000 - \frac{4000}{13}\sqrt{13} \right) e^{-\frac{1}{2}(-5+\sqrt{13})t},$$

$$x(t) = \left(500 + \frac{500}{13}\sqrt{13} \right) e^{\frac{1}{2}(5+\sqrt{13})t}$$

$$\left. + \left(-1500 + \frac{6000}{13}\sqrt{13} - \frac{1}{2}\left(1000 - \frac{4000}{13}\sqrt{13} \right)\sqrt{13} \right) e^{-\frac{1}{2}(-5+\sqrt{13})t} \right\}$$

After naming $x(t)$ and $y(t)$ the result obtained in **Sol** with **assign**, we graph these two population functions with **plot**.

```
> assign(Sol):
  plot({x(t),y(t)},t=0..2);
```

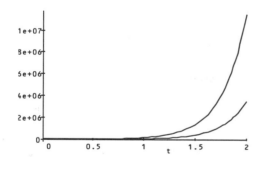

EXAMPLE: Determine the populations $x(t)$ and $y(t)$ in each territory if $a_1 = 5$, $a_2 = 4$, $b_1 = 2$, and $b_2 = 3$, given that $x(0) = 60$ and $y(0) = 10$.

SOLUTION: In this case, the initial-value problem that models the situation is

$$\begin{cases} \dfrac{dx}{dt} = x + y \\[2mm] \dfrac{dy}{dt} = 4x - 2y \end{cases}, x(0) = 60, y(0) = 10.$$

After defining $\mathbf{A} = \begin{pmatrix} 1 & 1 \\ 4 & -2 \end{pmatrix}$ and loading the **student** package, we use **equate**, which is contained in the **student** package to form the system of equations

$$\begin{cases} \dfrac{dx}{dt} = x + y \\[2mm] \dfrac{dy}{dt} = 4x - 2y \end{cases}$$

and then use **union** to form the initial-value problem

$$\begin{cases} \dfrac{dx}{dt} = x + y \\[2mm] \dfrac{dy}{dt} = 4x - 2y \end{cases}, x(0) = 60, y(0) = 10,$$

naming the resulting set of equations **Sys_of_Eqs**.

```
>  x:='x':y:='y':
   A:=array([[1,1],[4,-2]]):
   with(student):
   Eqs:=equate([[diff(x(t),t)],[diff(y(t),t)]],
    A &* [[x(t)],[y(t)]]);
   Sys_of_Eqs:=Eqs union x(0)=60,y(0)=10;
```

$$Eqs := \left\{ \frac{\partial}{\partial t}x(t) = x(t) + y(t), \frac{\partial}{\partial t}y(t) = 4x(t) - 2y(t) \right\}$$

$$Sys\_of\_Eqs := \left\{ \frac{\partial}{\partial t}x(t) = x(t) + y(t), \frac{\partial}{\partial t}y(t) = 4x(t) - 2y(t), x(0) = 60, y(0) = 10 \right\}$$

We then use **dsolve** to solve the initial-value problem and name the result **Sol**.

```
>  Sol:=dsolve(Sys_of_Eqs,{x(t),y(t)});
```

$$Sol := \{x(t) = 50e^{2t} + 10e^{-3t}, y(t) = 50e^{2t} - 40e^{-3t}\}$$

After naming $x(t)$ and $y(t)$ the result obtained in **Sol** with **assign**, we graph these two population functions with **plot**. Notice that as t increases, the two populations are approximately the same.

```
> assign(Sol):
  plot({x(t),y(t)},t=0..1.2);
```

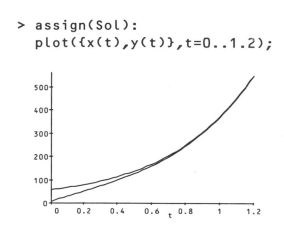

After loading the **DEtools** package, we graph the direction field for the system along with several solutions of the system of differential equations for various initial conditions with **DEplot2**. Notice that in each case, the values of $x(t)$ and $y(t)$ approach the same value as t increases.

```
> with(DEtools):
  DEplot2(A,[x,y],t=0..4,x=0..400,y=0..400);
  Inits:={seq([0,50*i,0],i=1..7)} union
    {seq([0,0,50*i],i=1..7)}:
  DEplot2(A,[x,y],t=0..4,Inits,x=0..400,y=0..400);
```

Population problems that involve more than two neighboring populations can be solved with a system of differential equations as well. Suppose that the population of three neighboring territories x, y, and z depends on several factors. The birth rates of x, y, and z are a_1, b_1, and c_1, respectively. The rate at which citizens of x move to y is a_2, while that at which citizens move from x to z is a_3. Similarly, the rate at which citizens of y move to x is b_2, while that at which citizens move from y to z is b_3. Also, the rate at which citizens of z move to x is c_2, while that at which citizens move from z to y is c_3. Suppose that the mortality rate of each territory is ignored in the model. Then we can determine the respective populations of the three territories for any time t using differential equations with the parameters given here.

The system of equations in this case is similar to that derived in the previous example. The rate at which population x changes is

$$\frac{dx}{dt} = a_1 x - a_2 x - a_3 x + b_2 y + c_2 z = (a_1 - a_2 - a_3)x + b_2 y + c_2 z,$$

while the rate at which population y changes is

$$\frac{dy}{dt} = b_1 y - b_2 y - b_3 y + a_2 x + c_3 z = (b_1 - b_2 - b_3)y + a_2 x + c_3 z,$$

and that of z is

$$\frac{dz}{dt} = c_1 z - c_2 z - c_3 z + a_3 x + b_3 y = (c_1 - c_2 - c_3)z + a_3 x + b_3 y.$$

Hence, we must solve the 3×3 system

$$\begin{cases} \dfrac{dx}{dt} = (a_1 - a_2 - a_3)x + b_2 y + c_2 z \\[2mm] \dfrac{dy}{dt} = a_2 x + (b_1 - b_2 - b_3)y + c_3 z , \\[2mm] \dfrac{dz}{dt} = a_3 x + b_3 y + (c_1 - c_2 - c_3)z \end{cases}$$

where the initial populations $x(0) = x_0$, $y(0) = y_0$, and $z(0) = z_0$ are given.

EXAMPLE: Determine the population of the three territories if $a_1 = 3$, $a_2 = 0$, $a_3 = 2$, $b_1 = 4$, $b_2 = 2$, $b_3 = 1$, $c_1 = 5$, $c_2 = 3$, and $c_3 = 0$ if $x(0) = 50$, $y(0) = 60$, and $z(0) = 25$.

SOLUTION: In this case, the system of differential equations is

$$\begin{cases} \dfrac{dx}{dt} = x + 2y + 3z \\[2mm] \dfrac{dy}{dt} = y \\[2mm] \dfrac{dz}{dt} = 2x + y + 2z \end{cases}$$

After defining the initial-value problem in $\mathtt{Sys\_of\_Eqs}$,

```
> x:='x':y:='y':z:='z':
  Sys_of_Eqs:={diff(x(t),t)=x(t)+2*y(t)+3*z(t),
   diff(y(t),t)=y(t),diff(z(t),t)=2*x(t)+y(t)+2*z(t),
   x(0)=50,y(0)=60,z(0)=25};
```

$$Sys\_of\_Eqs := \left\{ z(0) = 25, y(0) = 60, \frac{\partial}{\partial t}z(t) = 2x(t) + y(t) + 2z(t), \right.$$

$$\left. x(0) = 50, \frac{\partial}{\partial t}x(t) = x(t) + 2y(t) + 3z(t), \frac{\partial}{\partial t}y(t) = y(t) \right\}$$

we solve the initial-value problem with **dsolve** and name the result **Sol**.

```
> Sol:=dsolve(Sys_of_Eqs,{x(t),y(t),z(t)});
```

$$Sol := \{y(t) = 60e^t, z(t) = -40e^t + 63e^{4t} + 2e^{-t}, x(t) = -10e^t + 63e^{4t} - 3e^{-t}\}$$

After naming $x(t)$, $y(t)$, and $z(t)$ the results obtained in **Sol** with **assign**, the graph of these three population functions is generated with **plot**. We notice that although y was initially greater than x and z, these populations increase at a much higher rate.

```
> assign(Sol);
  plot({x(t),y(t),z(t)},t=0..1/2);
```

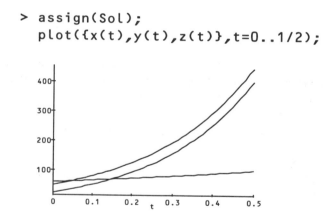

10.5 APPLICATIONS USING LAPLACE TRANSFORMS

☰ Coupled Spring-Mass Systems

The motion of a mass attached to the end of a spring was modeled with a second-order linear differential equation with constant coefficients in Chapter 5. Similarly, if a second spring and mass are attached to the end of the first mass, then the model becomes that of a system of second-order equations. To state the problem more precisely, let masses m_1 and m_2 be attached to the ends of springs S_1 and S_2 having spring constants k_1 and k_2, respectively. Then spring S_2 is attached to the base of mass m_1. Suppose that $x(t)$ and $y(t)$ represent the vertical displacement from the equilibrium position of springs S_1 and S_2, respectively. Since spring S_2 undergoes both elongation and compression when the system is in motion (due to the spring S_1 and the mass m_2), then according to Hooke's law, S_2 exerts the force $k_2(y - x)$ on m_2 while S_1 exerts the force $-k_1x$ on m_2. Therefore, the force acting on mass m_1 is the sum $-k_1x + k_2(y - x)$, and that acting on m_2 is $-k_2(y - x)$. Hence, using Newton's second law ($F = ma$) with each mass, we have the system

$$\begin{cases} m_1 \dfrac{d^2x}{dt^2} = -k_1x + k_2(y - x) \\[2mm] m_2 \dfrac{d^2y}{dt^2} = -k_2(y - x) \end{cases}$$

The initial position and velocity of the two masses m_1 and m_2 are given by $x(0)$, $x'(0)$, $y(0)$, and $y'(0)$, respectively. Hence, the method of Laplace transforms can be used to solve problems of this type. Recall the following property of the Laplace transform: $L\{f(t)\} = s^2F(s) - sf(0) - f'(0)$, where $F(s)$ is the Laplace transform of $f(t)$. This property is of great use in solving this problem because both equations involve second derivatives. We illustrate the solution of higher-order systems of differential equations with Laplace transforms in the following examples.

EXAMPLE: Consider the spring-mass system with $m_1 = m_2 = 1$, $k_1 = 3$, and $k_2 = 2$. Find the position functions $x(t)$ and $y(t)$ if $x(0) = 0$, $x'(0) = 1$, $y(0) = 1$, and $y'(0) = 0$.

SOLUTION: In order to find $x(t)$ and $y(t)$, we must solve the initial-value problem

$$\begin{cases} \dfrac{d^2x}{dt^2} = -5x + 2y \\[2mm] \dfrac{d^2y}{dt^2} = 2x - 2y \end{cases}, x(0) = 0, x'(0) = 1, y(0) = 1, y'(0) = 0.$$

After defining the equations

$$\begin{cases} \dfrac{d^2x}{dt^2} = -5x + 2y \\[2mm] \dfrac{d^2y}{dt^2} = 2x - 2y \end{cases}$$

in **Eqs**, we use **Laplace** to take the Laplace transform of both sides of each equation.

```
> x:='x':y:='y':
  Eqs:={diff(x(t),t$2)=-5*x(t)+2*y(t),
  diff(y(t),t$2)=2*x(t)-2*y(t)};
```

$$Eqs := \left\{ \frac{\partial^2}{\partial t^2} x(t) = -5x(t) + 2y(t), \frac{\partial^2}{\partial t^2} y(t) = 2x(t) - 2y(t) \right\}$$

```
> step_1:=map(laplace,Eqs,t,s);
```

$$step\_1 := \{(\text{laplace}(x(t), t, s)s - x(0))s - D(x)(0)$$
$$= -5 \,\text{laplace}\,(x(t), t, s) + 2 \,\text{laplace}\,(y(t), t, s),$$
$$(\text{laplace}(y(t), t, s)s - y(0))s - D(y)(0)$$
$$= 2 \,\text{laplace}\,(x(t), t, s) - 2 \,\text{laplace}\,(y(t), t, s)\}$$

Applying the initial conditions results in

```
> step_2:=subs({x(0)=0,D(x)(0)=1,y(0)=1,D(y)(0)=0},
  step_1);
```

$$step\_2 := \{\text{laplace}(x(t), t, s)s^2 - 1 = -5 \,\text{laplace}\,(x(t), t, s) + 2 \,\text{laplace}\,(y(t), t, s),$$
$$(\text{laplace}(y(t), t, s)s - 1)s = 2 \,\text{laplace}\,(x(t), t, s) - 2 \,\text{laplace}\,(y(t), t, s)\}$$

Solving for `laplace(x(t),t,s)`, which represents $X(s)$, and `laplace(y(t), t,s)`, which represents $Y(s)$, with `solve`, we find that

```
> step_3:=solve(step_2,{laplace(x(t),t,s),laplace(y(t),t,s)});
```

$$step\_3 := \left\{ \text{laplace}\,(y(t), t, s) = \frac{s^3 + 5s + 2}{6 + s^4 + 7s^2}, \right.$$

$$\left. \text{laplace}\,(x(t), t, s) = \frac{2s + 2 + s^2}{6 + s^4 + 7s^2} \right\}$$

and, finally, using `invlaplace` yields the solution, which we name `Sol`.

```
> readlib(laplace):
  Sol:=map(invlaplace,step_3,s,t);
```

$$Sol := \left\{ y(t) = -\frac{1}{15} \sin(\sqrt{2}\sqrt{3}\,t)\sqrt{2}\sqrt{3} + \frac{1}{5} \cos(\sqrt{2}\sqrt{3}\,t) + \frac{2}{5} \sin(t) + \frac{4}{5} \cos(t), \right.$$

$$\left. x(t) = \frac{2}{15} \sin(\sqrt{2}\sqrt{3}\,t)\sqrt{2}\sqrt{3} - \frac{2}{5} \cos(\sqrt{2}\sqrt{3}\,t) + \frac{1}{5} \sin(t) + \frac{2}{5} \cos(t) \right\}$$

Note that **dsolve** is equally successful in solving the initial-value problem.

```
> Init_Val_Prob:=Eqs union {x(0)=0,D(x)(0)=1,y(0)=1,D(y)(0)=0}:
  dsolve(Init_Val_Prob,{x(t),y(t)});
```

$$\left\{ y(t) = -\frac{1}{15}\sin(\sqrt{2}\sqrt{3}\,t)\sqrt{2}\sqrt{3} + \frac{1}{5}\cos(\sqrt{2}\sqrt{3}\,t) + \frac{2}{5}\sin(t) + \frac{4}{5}\cos(t), \right.$$

$$\left. x(t) = \frac{2}{15}\sin(\sqrt{2}\sqrt{3}\,t)\sqrt{2}\sqrt{3} - \frac{2}{5}\cos(\sqrt{2}\sqrt{3}\,t) + \frac{1}{5}\sin(t) + \frac{2}{5}\cos(t) \right\}$$

After naming $x(t)$ and $y(t)$ the result obtained in **Sol** with **assign**, we graph $x(t)$ and $y(t)$ simultaneously with **plot**. Note that $y(t)$ starts at $(0,1)$, while $x(t)$ has initial point $(0,0)$. Of course, these functions can be graphed parametrically in the xy-plane, as is done in the second **plot** command. Notice that this phase plane is different from those discussed in previous sections. One of the reasons for this is that the equations in the system of differential equations are second-order instead of first-order.

```
> assign(Sol):
  plot({x(t),y(t)},t=0..2*Pi);
  plot([x(t),y(t),t=0..2*Pi],numpoints=200);
```

We can take advantage of Maple's animation capabilities to view the motion of the spring. We begin by defining the function **spring** in the same way that we did in Section 5.1.

```
> spring:='spring':
  n:=15:
  eps:=0.1:
  L1:=1:
  L2:=1:
  spring:=proc(t0)
   local xt0,yt0,pts;
   xt0:=evalf(subs(t=t0,x(t)));
   yt0:=evalf(subs(t=t0,y(t)));
   pts:=[[0,yt0-L2],
     seq([eps*(-1)^m,yt0-L2+m*(xt0-yt0+L2)/n],m=1..n-1),
     [0,xt0],
     seq([eps*(-1)^m,xt0+m*(L1-xt0)/n],m=1..n-1),
     [0,L1]];
   plot(pts,view=[-1..1,-2.5..1.5],xtickmarks=2,ytickmarks=2);
   end:
```

We then use **seq** to generate the list of graphs **spring(k)** for 60 equally spaced values of k between 0 and 16, naming the resulting list of 60 graphs **to_animate**.

```
> k_vals:=seq(k*16/59,k=0..59):
  to_animate:=[seq(spring(k),k=k_vals)]:
```

The list of graphs **to_animate** is animated using **display**, which is contained in the **plots** package, together with the **insequence** option. Two frames from the resulting animation are shown here.

```
> with(plots):
  display(to_animate,insequence=true);
```

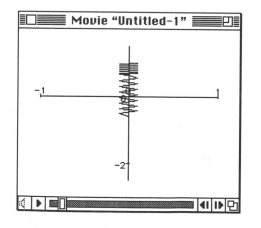 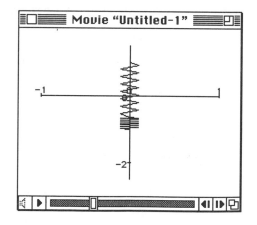

Suppose that external forces $F_1(t)$ and $F_2(t)$ are applied to the masses. This causes the system of equations to become

$$\begin{cases} m_1 \dfrac{d^2x}{dt^2} = -k_1 x + k_2(y - x) + F_1(t) \\[2mm] m_2 \dfrac{d^2y}{dt^2} = -k_2(y - x) + F_2(t) \end{cases}$$

We investigate the effects of these external forcing functions in the following example that is again solved through the method of Laplace transforms.

EXAMPLE: Solve the problem with $m_1 = m_2 = 1$, $k_1 = 3$, and $k_2 = 2$ if the forcing functions are $F_1(t) = 1$ and $F_2(t) = \sin t$ and the initial conditions are $x(0) = 0$, $x'(0) = 1$, $y(0) = 1$, and $y'(0) = 0$.

SOLUTION: In this case, the initial-value problem to be solved is

$$\begin{cases} \dfrac{d^2x}{dt^2} = -5x + 2y + 1 \\[2mm] \dfrac{d^2y}{dt^2} = 2x - 2y + \sin t \end{cases} \qquad x(0) = 0, x'(0) = 1, y(0) = 1, y'(0) = 0.$$

As in the previous example, we can use either **dsolve** to solve the initial-value problem directly or Maple to implement the method of Laplace transforms.

Taking the Laplace transform of both sides of each equation with **laplace** yields

```
> Eqs:={diff(x(t),t$2)=-5*x(t)+2*y(t)+1,
    diff(y(t),t$2)=2*x(t)-2*y(t)+sin(t)}:
  step_1:=map(laplace,Eqs,t,s);
```

$step\_1 := \Big\{ (\text{laplace}(x(t), t, s)s - x(0))s - D(x)(0)$

$= -5\,\text{laplace}\,(x(t), t, s) + 2\,\text{laplace}\,(y(t), t, s) + \dfrac{1}{s},$

$(\text{laplace}(y(t), t, s)s - y(0))s - D(y)(0)$

$= 2\,\text{laplace}\,(x(t), t, s) - 2\,\text{laplace}\,(y(t), t, s) + \dfrac{1}{s^2 + 1}\Big\}$

and then applying the initial conditions with **subs** results in

```
> step_2:=subs({x(0)=0,D(x)(0)=1,y(0)=1,D(y)(0)=0},step_1);
```

$step\_2 := \Big\{ \text{laplace}\,(x(t), t, s)s^2 - 1$

$$= -5\,\text{laplace}\,(x(t), t, s) + 2\,\text{laplace}\,(y(t), t, s) + \frac{1}{s},$$

$(\text{laplace}(y(t), t, s)s - 1)s$

$$= 2\,\text{laplace}\,(x(t), t, s) - 2\,\text{laplace}\,(y(t), t, s) + \frac{1}{s^2 + 1}\Big\}$$

Solving the system of equations in $step\_2$ for $laplace(x(t), t, s)$ and $laplace(y(t), t, s)$ with $solve$ gives us

```
> step_3:=solve(step_2,{laplace(x(t),t,s),laplace(y(t),t,s)});
```

$step\_3 := \Big\{ \text{laplace}\,(x(t), t, s) = \dfrac{3s^4 + 5s^2 + s^5 + 4s + 2 + 3s^3}{s(13s^2 + 6 + s^6 + 8s^4)},$

$$\text{laplace}\,(y(t), t, s) = \dfrac{6s^4 + 7s^2 + 7s + 2 + 3s^3 + s^6}{s(13s^2 + 6 + s^6 + 8s^4)}\Big\}$$

Finally, the solution, named Sol, is obtained with $invlaplace$.

```
> readlib(laplace):
  Sol:=map(invlaplace,step_3,s,t);
```

$Sol := \Big\{ x(t) = \dfrac{1}{3} + \dfrac{11}{75}\sin(\sqrt{2}\sqrt{3}\,t)\sqrt{2}\sqrt{3} - \dfrac{8}{15}\cos(\sqrt{2}\sqrt{3}\,t) + \dfrac{8}{25}\sin(t)$

$$+ \dfrac{1}{5}\cos(t) - \dfrac{1}{5}t\cos(t), y(t) = \dfrac{1}{3} - \dfrac{11}{150}\sin(\sqrt{2}\sqrt{3}\,t)\sqrt{2}\sqrt{3}$$

$$+ \dfrac{4}{15}\cos(\sqrt{2}\sqrt{3}\,t) + \dfrac{21}{25}\sin(t) + \dfrac{2}{5}\cos(t) - \dfrac{2}{5}t\cos(t)\Big\}$$

We obtain equivalent results with $dsolve$. However, implementing the method of Laplace transforms as before may take less time than using $dsolve$.

```
> Init_Val_Prob:=Eqs union {x(0)=0,D(x)(0)=1,y(0)=1,D(y)(0)=0}:
  dsolve(Init_Val_Prob,{x(t),y(t)});
```

$$\Big\{ x(t) = \frac{1}{25}\%4\cos(\%3) - \frac{1}{25}\%2\sin(\%3) + \frac{11}{75}\%4\sqrt{2}\sqrt{3} + \frac{1}{3} + \frac{1}{5}\cos(t)$$

$$+ \frac{2}{5}\sin(t) - \frac{8}{15}\%2 - \frac{1}{150}\%4\cos(\%1)\sqrt{2}\sqrt{3}$$

$$- \frac{1}{150}\%4\cos(\%3)\sqrt{2}\sqrt{3} + \frac{1}{150}\%2\sin(\%1)\sqrt{2}\sqrt{3}$$

$$+ \frac{1}{25}\%2\sin(\%1) + \frac{1}{150}\%2\sin(\%3)\sqrt{2}\sqrt{3} - \frac{1}{5}t\cos(t) - \frac{1}{25}\%4\cos(\%1),$$

$$y(t) = -\frac{2}{25}\%4\cos(\%1) + \frac{1}{75}\%2\sin(\%1)\sqrt{2}\sqrt{3} + \frac{4}{15}\%2 + \sin(t) + \frac{2}{5}\cos(t)$$

$$- \frac{11}{150}\%4\sqrt{2}\sqrt{3} + \frac{1}{3} - \frac{1}{75}\%4\cos(\%3)\sqrt{2}\sqrt{3}$$

$$- \frac{1}{75}\%4\cos(\%1)\sqrt{2}\sqrt{3} + \frac{1}{75}\%2\sin(\%3)\sqrt{2}\sqrt{3} - \frac{2}{5}t\cos(t)$$

$$+ \frac{2}{25}\%4\cos(\%3) - \frac{2}{25}\%2\sin(\%3) + \frac{2}{25}\%2(\%1)\Big\}$$

$$\%1 := \sqrt{2}\sqrt{3}\,t - t$$

$$\%2 := \cos(\sqrt{2}\sqrt{3}\,t)$$

$$\%3 := \sqrt{2}\sqrt{3}\,t + t$$

$$\%4 := \sin(\sqrt{2}\sqrt{3}\,t)$$

After naming $x(t)$ and $y(t)$ the result obtained in **Sol** with **assign**, we graph $x(t)$ and $y(t)$ simultaneously and then parametrically to illustrate the solution in the phase plane. Note that $x(t)$ and $y(t)$ have initial points $(0,0)$ and $(0,1)$, respectively.

```
> assign(Sol):
  plot({x(t),y(t)},t=0..2*Pi);
  plot([x(t),y(t),t=0..2*Pi],numpoints=200);
```

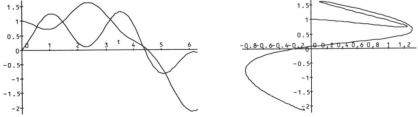

As in the previous example, we can also use Maple to help us visualize the motion of the spring. As before, we begin by defining `spring`.

```
> spring:='spring':
  n:=15:
  eps:=0.1:
  L1:=1:
  L2:=1:
  spring:=proc(t0)
   local xt0,yt0,pts;
   xt0:=evalf(subs(t=t0,x(t)));
   yt0:=evalf(subs(t=t0,y(t)));
   pts:=[[0,yt0-L2],
     seq([eps*(-1)^m,yt0-L2+m*(xt0-yt0+L2)/n],m=1..n-1),
     [0,xt0],
     seq([eps*(-1)^m,xt0+m*(L1-xt0)/n],m=1..n-1),
     [0,L1]];
   plot(pts,view=[-1..1,-3..1.5],xtickmarks=2,ytickmarks=2);
   end:
```

One way to visualize the motion is to generate several graphs of the spring, as in the following `for` loop:

```
> for k from 0 to 8 do spring(k) od;
```

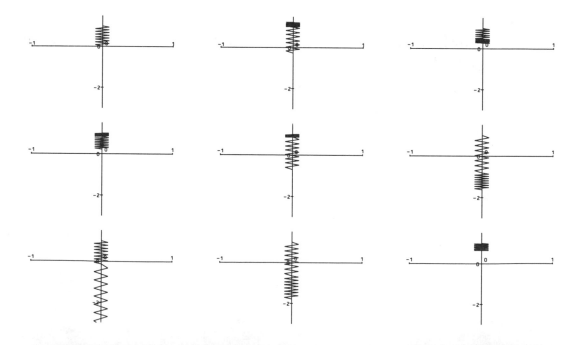

Alternatively, we can generate several graphs and then animate the result using display, which is contained in the plots package, together with the insequence option.

```
> with(plots):
  k_vals:=seq(k*8/59,k=0..59):
  to_animate:=[seq(spring(k),k=k_vals)]:
  display(to_animate,insequence=true);
```

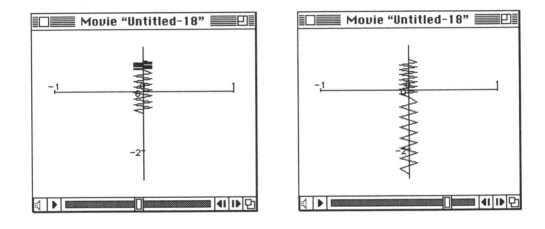

≡ The Double Pendulum

In a method similar to that of the simple pendulum in Chapter 5 and that of the coupled spring system in the previous section, the motion of a double pendulum is modeled by the following system of equations using the approximation $\sin \theta \approx \theta$ for small displacements:

$$\begin{cases} (m_1 + m_2)\ell_1^2\theta_1'' + m_2\ell_1\ell_2\theta_2'' + (m_1 + m_2)\ell_1 g\theta_1 = 0 \\ m_2\ell_2^2\theta_2'' + m_2\ell_1\ell_2\theta_1'' + m_2\ell_2 g\theta_2 = 0 \end{cases}$$

where θ_1 represents the displacement of the upper pendulum and θ_2 that of the lower pendulum. Also, m_1 and m_2 represent the mass attached to the upper and lower pendulums, respectively, while the length of each is given by ℓ_1 and ℓ_2.

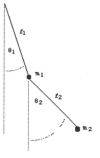

EXAMPLE: Suppose that $m_1 = 3$, $m_2 = 1$, and each pendulum has length 16. If $g = 32$, then determine $\theta_1(t)$ and $\theta_2(t)$ if $\theta_1(0) = 1$, $\theta_1'(0) = 0$, $\theta_2(0) = 0$, and $\theta_2'(0) = -1$.

SOLUTION: In this case, the system is

$$\begin{cases} 4(16)^2 \theta_1'' + 16^2 \theta_2'' + 4(16)(32)\theta_1 = 0 \\ 16^2 \theta_1'' + 16^2 \theta_2'' + (16)(32)\theta_2 = 0 \end{cases}$$

which can be simplified to obtain

$$\begin{cases} 4\theta_1'' + \theta_2'' + 8\theta_1 = 0 \\ \theta_1'' + \theta_2'' + 2\theta_2 = 0 \end{cases}$$

After defining **Eqs**, we use **laplace** to compute the Laplace transform of each side of both equations, naming the result **step_1**, and then apply the initial conditions to **step_1** with **subs**, naming the resulting output **step_2**.

```
> Eqs:={4*diff(theta[1](t),t$2)+diff(theta[2](t),t$2)+
  8*theta[1](t)=0,
  diff(theta[1](t),t$2)+diff(theta[2](t),t$2)+
  2*theta[2](t)=0}:
  step_1:=map(laplace,Eqs,t,s):
  step_2:=subs({theta[1](0)=1,D(theta[1])(0)=0,
  theta[2](0)=0,D(theta[2])(0)=-1},step_1);
```

$step\_2 := \{4(\mathrm{laplace}(\theta_{[1]}(t), t, s)s - 1)s + \mathrm{laplace}(\theta_{[2]}(t), t, s)s^2 + 1$
$\qquad + 8\,\mathrm{laplace}(\theta_{[1]}(t), t, s) = 0, (\mathrm{laplace}(\theta_{[1]}(t), t, s)s - 1)s$
$\qquad + \mathrm{laplace}(\theta_{[2]}(t), t, s)s^2 + 1 + 2\,\mathrm{laplace}(\theta_{[2]}(t), t, s) = 0\}$

Solving the equations in **step_2** for **laplace(theta[1](t),t,s)**, which represents the Laplace transform of $\theta_1(t)$, and **laplace(theta[2](t),t,s)**, which represents the Laplace transform of $\theta_2(t)$, with **solve** results in

```
> step_3:=solve(step_2,
    {laplace(theta[1](t),t,s),laplace(theta[2](t),t,s)});
```

$$
step\_3 := \left\{ \text{laplace}\,(\theta_{[2]}(t), t, s) = -\frac{-8s + 3s^2 + 8}{3s^4 + 16s^2 + 16}, \right.
$$

$$
\left. \text{laplace}\,(\theta_{[1]}(t), t, s) = \frac{8s - 2 + 3s^3}{3s^4 + 16s^2 + 16} \right\}
$$

and then computing the inverse Laplace transform with **invlaplace** yields the solution, named **Sol**.

```
> readlib(laplace):
  Sol:=map(invlaplace,step_3,s,t);
```

$$
Sol := \left\{ \theta_{[2]}(t) = -\frac{1}{4}\sqrt{3}\sin\left(\frac{2}{3}\sqrt{3}\,t\right) + \cos\left(\frac{2}{3}\sqrt{3}\,t\right) - \frac{1}{4}\sin(2t) - \cos(2t), \right.
$$

$$
\left. \theta_{[1]}(t) = -\frac{1}{8}\sqrt{3}\sin\left(\frac{2}{3}\sqrt{3}\,t\right) + \frac{1}{2}\cos\left(\frac{2}{3}\sqrt{3}\,t\right) + \frac{1}{8}\sin(2t) + \frac{1}{2}\cos(2t) \right\}
$$

As in many previous examples, we see that we obtain the same results with **dsolve**.

```
> Init_Val_Prob:=Eqs union {theta[1](0)=1,D(theta[1])(0)=0,
    theta[2](0)=0,D(theta[2])(0)=-1}:
  dsolve(Init_Val_Prob,{theta[1](t),theta[2](t)});
```

$$
\left\{ \theta_{[2]}(t) = -\frac{1}{4}\sqrt{3}\sin\left(\frac{2}{3}\sqrt{3}\,t\right) + \cos\left(\frac{2}{3}\sqrt{3}\,t\right) - \frac{1}{4}\sin(2t) - \cos(2t), \right.
$$

$$
\left. \theta_{[1]}(t) = -\frac{1}{8}\sqrt{3}\sin\left(\frac{2}{3}\sqrt{3}\,t\right) + \frac{1}{2}\cos\left(\frac{2}{3}\sqrt{3}\,t\right) + \frac{1}{8}\sin(2t) + \frac{1}{2}\cos(2t) \right\}
$$

These two functions are graphed together and then parametrically with **plot** to show the solution in the phase plane.

```
> assign(Sol):
  plot({theta[1](t),theta[2](t)},t=0..15);
  plot([theta[1](t),theta[2](t),t=0..15],numpoints=200);
```

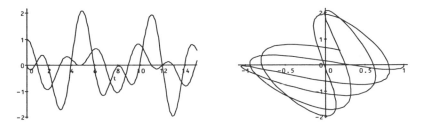

In the same way as in Section 5.5, we can animate the motion of the pendulum with Maple's animation capabilities. First, we define the function `pen2`. The result of entering `pen2(t,l1,l2)` looks like the motion of the pendulum with lengths `l1` and `l2` at time `t`.

```
> pen2:=proc(t0,len1,len2)
    local pt1,pt2,xt0,yt0;
    xt0:=evalf(subs(t=t0,theta[1](t)));
    yt0:=evalf(subs(t=t0,theta[2](t)));
    pt1:=[len1*cos(3*Pi/2+xt0),len1*sin(3*Pi/2+xt0)];
    pt2:=[len1*cos(3*Pi/2+xt0)+len2*cos(3*Pi/2+yt0),
     len1*sin(3*Pi/2+xt0)+len2*sin(3*Pi/2+yt0)];
    plot([[0,0],pt1,pt2],xtickmarks=2,ytickmarks=2,
     view=[-32..32,-32..0]);
    end:
```

As in the previous example, we can generate several graphs with a `for` loop:

```
> for k from 0 to 4 by 0.5 do pen2(k,16,16) od;
```

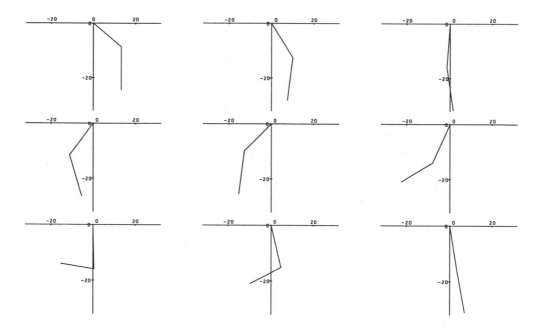

Alternatively, we can generate a list of graphs and then animate the result using the `display` function together with the `insequence` option.

```
> k_vals:=seq(k*10/119,k=0..119):
  to_animate:=[seq(pen2(k,16,16),k=k_vals)]:
  with(plots):
  display(to_animate,insequence=true);
```

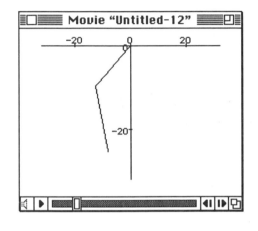 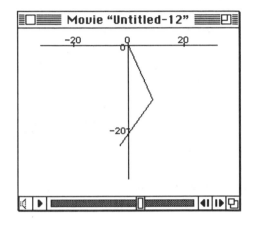

10.6 SPECIAL NONLINEAR EQUATIONS AND
SYSTEMS OF EQUATIONS

Several special equations and systems that arise in the study of many areas of applied mathematics can be solved and/or investigated using the techniques of Chapter 9. These include the predator-prey population dynamics problem, the Van der Pol equation that models variable damping in a spring-mass system, and the Bonhoeffer-Van der Pol (BVP) oscillator. We begin by considering the Lotka-Volterra system that models the interaction between two populations.

☰ Biological Systems: Predator-Prey Interaction

Let $x(t)$ and $y(t)$ represent the population at time t of the prey and predator, respectively. Suppose that a is the birth rate of $x(t)$ and b is the rate of interaction between populations $x(t)$ and $y(t)$. Because an interaction between the two populations indicates that a predator overtakes a member of the prey population, the rate at which $x(t)$ changes with respect to time is

$$\frac{dx}{dt} = ax - bxy.$$

Similarly, if the growth rate of the predator population is d and the death rate is c, then the rate at which $y(t)$ changes with respect to time is

$$\frac{dy}{dt} = -cy + dxy.$$

Therefore, we must solve the **Lotka-Volterra system**

$$\begin{cases} \dfrac{dx}{dt} = ax - bxy \\[2mm] \dfrac{dy}{dt} = -cy + dxy \end{cases}$$

subject to the initial populations $x(0) = x_0$ and $y(0) = y_0$.

EXAMPLE: Find and classify the equilibrium points of the Lotka-Volterra equations.

SOLUTION: After defining f and g, we use **solve** to solve the system of equations $\begin{cases} ax - bxy = 0 \\ -cy + dxy = 0 \end{cases}$ for x and y.

```
> f:='f':g:='g':
  f:=(x,y)->a*x-b*x*y:
  g:=(x,y)->-c*y+d*x*y:
  Eq_pts:=solve({f(x,y)=0,g(x,y)=0},{x,y});
```

$$Eq\_pts := \left\{ y = \frac{a}{b}, x = \frac{c}{d} \right\}, \{y = 0, x = 0\}$$

This shows that the equilibrium points are $(0, 0)$ and $\left(\dfrac{c}{d}, \dfrac{a}{b} \right)$. After loading the **linalg** package, the Jacobian matrix of the nonlinear system is found with **jacobian** and named **jac**.

```
> with(linalg):
  jac:=jacobian([f(x,y),g(x,y)],[x,y]);
```

$$jac := \begin{bmatrix} a - by & -bx \\ dy & -c + dx \end{bmatrix}$$

Near $\left(\dfrac{c}{d}, \dfrac{a}{b} \right)$, we have $J\left(\dfrac{c}{d}, \dfrac{a}{b} \right) = \begin{pmatrix} 0 & -\dfrac{bc}{d} \\ \dfrac{da}{b} & 0 \end{pmatrix}$

```
> m1:=subs(Eq_pts[1],eval(jac));
```

$$m1 := \begin{bmatrix} 0 & -\dfrac{bc}{d} \\ \dfrac{da}{b} & 0 \end{bmatrix}$$

with eigenvalues

```
> eigenvals(m1);
```

$$\sqrt{-ca}, -\sqrt{-ca}$$

so the point $\left(\dfrac{c}{d}, \dfrac{a}{b}\right)$ is classified as a center.

Similarly, near $(0, 0)$, we have $J(0, 0) = \begin{pmatrix} a & 0 \\ 0 & -c \end{pmatrix}$ with eigenvalues $\lambda_1 = c$ and $\lambda_2 = -a$. Because these eigenvalues are real with opposite sign, we classify $(0, 0)$ as a saddle.

```
> m2:=subs(Eq_pts[2],eval(jac)):
  eigenvals(m2);
```

$$a, -c$$

We graph the solution of the Lotka-Volterra system for various initial conditions for the particular case of $a = 2, b = 1, c = 3$, and $d = 1$. After defining Eq_1 and Eq_2,

```
> x:='x':y:='y':t:='t':
  Eq_1:=diff(x(t),t)=2*x(t)-x(t)*y(t):
  Eq_2:=diff(y(t),t)=-3*y(t)+x(t)*y(t):
```

we use **dsolve** together with the **numeric** option to generate a numerical solution to the initial-value problem

$$\begin{cases} x' = 2x - xy \\ y' = -3y + xy \end{cases}, x(0) = 1, y(0) = 1$$

and name the resulting procedure Sol_1.

```
> Sol_1:=dsolve({Eq_1,Eq_2,x(0)=1,y(0)=1},{x(t),y(t)},numeric);

  Sol_1 := proc(t) 'dsolve/numeric/result2'(t,5089212,[1,1])
  end
```

We can graph the solution for various values of t using **odeplot**, which is contained in the **plots** package. For example, entering

```
> with(plots):
  odeplot(Sol_1,[x(t),y(t)],0..3,view=[0..8,0..6]);
```

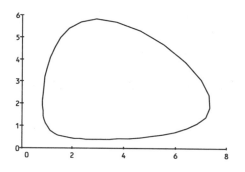

generates a graph of y versus x for $0 \le t \le 3$. The resulting graph is displayed on the rectangle $[0, 8] \times [0, 6]$ because the option **view=[0..8,0..6]** is included in the **odeplot** command. In addition, entering

```
> odeplot(Sol_1,[t,x(t),y(t)],0..6,axes=BOXED);
  odeplot(Sol_1,[t,y(t)],0..6);
  odeplot(Sol_1,[t,y(t)],0..6);
```

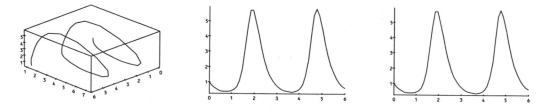

generates a three-dimensional graph of y versus x versus t for $0 \le t \le 6$, a two-dimensional graph of y versus t for $0 \le t \le 6$, and a two-dimensional graph of x versus t for $0 \le t \le 6$.

To graph the solution for various initial conditions, we proceed by defining **sol**. Given a number s, **sol** declares the variables **Num_Sol, x, y, t, Eq_1**, and **Eq_2** local to the procedure **sol**; defines **Eq_1** to be the equation $x' = 2x - xy$, **Eq_2** to be the equation $y' = -3y + xy$, and **Num_Sol** to be a numerical solution of the initial-value problem

$$\begin{cases} x' = 2x - xy \\ y' = -3y + xy \end{cases}, x(0) = 3s, y(0) = 2s;$$

and then graphs **Num_Sol** for $0 \le t \le 4$. The resulting graph is displayed on the rectangle $[0, 15] \times [0, 15]$ because the option **view=[0..15,0..15]** is included in the **odeplot** command.

```
> x:='x':y:='y':t:='t':Num_Sol:='Num_Sol':sol:='sol':
  sol:=proc(s)
    local Num_Sol,x,y,t,Eq_1,Eq_2;
    Eq_1:=diff(x(t),t)=2*x(t)-x(t)*y(t);
    Eq_2:=diff(y(t),t)=-3*y(t)+x(t)*y(t);
    Num_Sol:=dsolve(Eq_1,Eq_2,x(0)=3*s,y(0)=2*s,
     x(t),y(t),numeric);
    odeplot(Num_Sol,[x(t),y(t)],0..4,view=[0..15,0..15])
  end:
```

For example, entering

```
> sol(11/40);
```

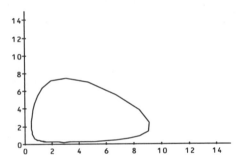

graphs the solution to the initial-value problem

$$\begin{cases} x' = 2x - xy \\ y' = -3y + xy \end{cases}, x(0) = \frac{33}{40}, y(0) = \frac{11}{20}.$$

To graph the solution for several initial conditions, we first define t_vals to be a set of numbers, which will correspond to various s-values,

```
> t_vals:={seq(1/8+3/40*i,i=0..10)};
```

$$t\_vals := \left\{ \frac{1}{2}, \frac{11}{40}, \frac{1}{5}, \frac{1}{8}, \frac{7}{20}, \frac{17}{40}, \frac{23}{40}, \frac{13}{20}, \frac{29}{40}, \frac{4}{5}, \frac{7}{8} \right\}$$

and then use map to apply sol to each number in the set of numbers t_vals. No output is displayed because a colon is included at the end of the map command. The resulting set of graphs, named to_display, is displayed simultaneously using display, which is contained in the plots package.

```
> to_display:=map(sol,t_vals):
  display(to_display);
```

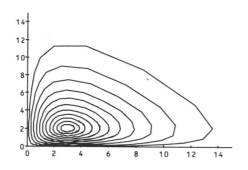

Notice that all of the solutions oscillate about the center. These solutions reveal the relationship between the two populations: prey, $x(t)$, and predator, $y(t)$. As we follow one cycle counterclockwise beginning, for example, near the point $\left(2, \dfrac{1}{2}\right)$, we notice that as $x(t)$ increases, then $y(t)$ increases until $y(t)$ becomes overpopulated. Then, because the prey population is too small to supply the predator population, $y(t)$ decreases, and this leads to an increase in the population of $x(t)$. At this point, because the number of predators becomes too small to control the number in the prey population, $x(t)$ becomes overpopulated and the cycle repeats itself.

☰ Physical Systems: Variable Damping

In some physical systems, energy is fed into the system when there are small oscillations, while energy is taken from the system when there are large oscillations. This indicates that the system undergoes **negative damping** for small oscillations and **positive damping** for large oscillations. A differential equation that models this situation is **Van der Pol's equation,**

$$x'' + \mu(x^2 - 1)x' + x = 0,$$

where μ is a positive constant. We can transform this second-order differential equation into a system of first-order differential equations with the substitution $x' = y$. Hence, $y' = x'' = -x - \mu(x^2 - 1)x' = -x - \mu(x^2 - 1)y$, so the system of equations is

$$\begin{cases} x' = y \\ y' = -x - \mu(x^2 - 1)y \end{cases}.$$

Notice that $\mu(x^2 - 1)$ represents the damping coefficient. This system models variable damping because $\mu(x^2 - 1) < 0$ when $-1 < x < 1$ and $\mu(x^2 - 1) > 0$ when $|y| > 1$. Therefore, damping is negative for small oscillations, $-1 < x < 1$, and positive for large oscillations, $|y| > 1$.

EXAMPLE: Find and classify the equilibrium points of the system of differential equations that is equivalent to Van der Pol's equation.

SOLUTION: We find these equilibrium points by solving

$$\begin{cases} y = 0 \\ -x - \mu(x^2 - 1)y = 0 \end{cases}.$$

From the first equation, we see that $y = 0$. Then substitution of $y = 0$ into the second equation yields $x = 0$ as well. Therefore, the equilibrium point is $(0, 0)$.

The Jacobian matrix for this system is

$$J(x, y) = \begin{pmatrix} 0 & 1 \\ -1 - 2\mu xy & -\mu(x^2 - 1) \end{pmatrix}.$$

Then, at $(0, 0)$, we have the matrix

$$J(0, 0) = \begin{pmatrix} 0 & 1 \\ -1 & \mu \end{pmatrix}.$$

We find the eigenvalues of $J(0, 0)$ by solving

$$\begin{vmatrix} -\lambda & 1 \\ -1 & \mu - \lambda \end{vmatrix} = \lambda^2 - \mu\lambda + 1 = 0$$

that has roots $\lambda_{1,2} = \dfrac{\mu \pm \sqrt{\mu^2 - 4}}{2}$.

```
> f:='f':g:='g':
  f:=(x,y)->y:
  g:=(x,y)->-x-mu*(x^2-1)*y:
  Eq_pts:=solve({f(x,y)=0,g(x,y)=0},{x,y});
```

$Eq\_pts := \{y = 0, x = 0\}$

```
> with(linalg):
  jac:=jacobian([f(x,y),g(x,y)],[x,y]);
```

$$jac := \begin{bmatrix} 0 & 1 \\ -1 - 2\mu xy & -\mu(x^2 - 1) \end{bmatrix}$$

```
> m:=subs(Eq_pts,eval(jac));
```

$$m := \begin{bmatrix} 0 & 1 \\ -1 & \mu \end{bmatrix}$$

```
> eigenvals(m);
```

$$\frac{1}{2}\mu + \frac{1}{2}\sqrt{\mu^2 - 4}, \frac{1}{2}\mu - \frac{1}{2}\sqrt{\mu^2 - 4}$$

Notice that if $\mu > 2$, then both eigenvalues are positive and real. Hence, we classify $(0,0)$ as an **unstable node**. Conversely, if $0 < \lambda < 2$, then the eigenvalues are a complex conjugate pair with a positive real part. Hence, $(0,0)$ is an **unstable spiral**. (We omit the case when $\mu = 2$ since the eigenvalues are repeated.)

In the previous example, we illustrated how to use `dsolve` together with the `numeric` option to generate a numerical solution to an initial-value problem and then how to use `odeplot` to graph the resulting numerical solution. In cases when a graph but not a numerical solution is needed, DEplot2, which is contained in the `DEtools` package, can be used to graph various solutions.

After loading the `DEtools` package, we use `DEplot2` to graph the direction field associated with the system

$$\begin{cases} x' = y \\ y' = (1 - x^2)y - x \end{cases}$$

on the rectangle $[-3, 3] \times [-3, 3]$. The second `DEplot2` command graphs the solution to the initial-value problem

$$\begin{cases} x' = y \\ y' = (1 - x^2)y - x \end{cases}, x(0) = 1, y(0) = 1$$

for $0 \leq t \leq 10$. In this case, the option `stepsize=0.05` is used to help assure that the resulting graph appears smooth; the option `arrows=NONE` instructs Maple not to display the direction field associated with the system.

```
> with(DEtools):
  x:='x':y:='y':t:='t':
  DEplot2([y,(1-x^2)*y-x],
    [x,y],t=0..10,x=-3..3,y=-3..3);
  DEplot2([y,(1-x^2)*y-x],
    [x,y],t=0..10,{[0,1,1]},x=-3..3,y=-3..3,
    stepsize=0.05,arrows=NONE);
```

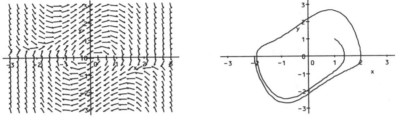

Notice that the curve moves away from the unstable spiral $(0, 0)$. Also, the curve does not move an infinite distance from the origin. Instead, it approaches a curve that we call a **limit cycle**.

We can also use DEplot2 to graph $(t, x(t), y(t))$, $x(t)$, and $y(t)$. For example, entering

```
> DEplot2([y,(1-x^2)*y-x],
    [t,x,y],t=0..15,[0,1,1],
    stepsize=0.05);
  DEplot2([y,(1-x^2)*y-x],
    [t,x,y],t=0..15,{[0,1,1]},scene=[t,x],
    stepsize=0.05);
  DEplot2([y,(1-x^2)*y-x],
    [t,x,y],t=0..15,{[0,1,1]},scene=[t,y],
    stepsize=0.05);
```

first graphs the solution $(t, x(t), y(t))$ that satisfies the initial conditions $x(0) = 1$ and $y(0) = 1$ for $0 \le t \le 15$, then graph $x(t)$ for $0 \le t \le 15$, and finally graphs $y(t)$ for $0 \le t \le 15$.

To graph the solution for various initial conditions, we begin by defining the set of numbers `t_vals`

```
> t_vals:=seq(t/4,t=1..16);
```

$$t\_vals := \left\{1, 2, 3, 4, \frac{1}{2}, \frac{13}{4}, \frac{7}{2}, \frac{15}{4}, \frac{5}{2}, \frac{11}{4}, \frac{3}{2}, \frac{7}{4}, \frac{1}{4}, \frac{3}{4}, \frac{5}{4}, \frac{9}{4}\right\}$$

and then defining `init_conds` to be a set of ordered triples that will correspond to various initial conditions in the subsequent command.

```
> init_conds:={seq(
  [0,exp(-t/4)*cos(2*Pi*t),exp(-t/4)*sin(2*Pi*t)],
  t=t_vals)};
```

$$init\_conds := \left\{[0, -e^{-\frac{7}{8}}, 0], [0, e^{-\frac{1}{4}}, 0], [0, e^{-\frac{3}{4}}, 0], [0, e^{-\frac{1}{2}}, 0], [0, e^{-1}, 0], [0, -e^{-\frac{1}{8}}, 0], \right.$$
$$[0, 0, e^{-\frac{1}{16}}], [0, 0, e^{-\frac{5}{16}}], [0, 0, e^{-\frac{9}{16}}], [0, 0, -e^{-\frac{3}{16}}], [0, -e^{-\frac{3}{8}}, 0], [0, 0, -e^{-\frac{7}{16}}],$$
$$\left. [0, 0, e^{-\frac{13}{16}}], [0, 0, -e^{-\frac{15}{16}}], [0, 0, -e^{-\frac{11}{16}}], [0, -e^{-\frac{5}{8}}, 0]\right\}$$

We then use `DEplot2` to graph the solutions that satisfy the initial conditions specified in `init_conds`.

```
> t:='t':
  DEplot2([y,(1-x^2)*y-x],
    [x,y],t=0..10,init_conds,x=-3..3,y=-3..3,
    stepsize=0.05,arrows=NONE);
```

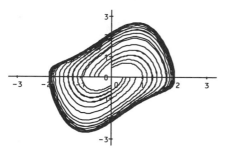

If instead we had entered

```
> init_conds:={seq(
  [0,exp(t/4)*cos(2*Pi*t),exp(t/4)*sin(2*Pi*t)],
  t=t_vals)}:
```

the resulting graph would look like

```
> t:='t':
  DEplot2([y,(1-x^2)*y-x],
   [x,y],t=0..10,init_conds,x=-3..3,y=-3..3,
   stepsize=0.05,arrows=NONE);
```

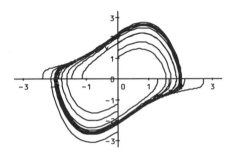

Eigenvalue Problems and Fourier Series

In previous chapters, we have seen that many physical situations can be modeled by either ordinary differential equations or systems of ordinary differential equations. However, to understand the motion of a spring at a particular location and at a particular time, the heat of a thin wire at a particular location and a particular time, or the electrostatic potential at a point on a plate, we must solve partial differential equations as each of these quantities depends on (at least) two changing conditions.

Wave equation: $c^2 u_{xx} = u_{tt}$

Heat equation: $u_t = c^2 u_{xx}$

Laplace's equation: $u_{xx} + u_{yy} = 0$

We now turn our attention to partial differential equations. As you may suspect, obtaining solutions of partial differential equations is connected with solving ordinary differential equations.

11.1 BOUNDARY VALUE, EIGENVALUE, AND STURM-LIOUVILLE PROBLEMS

☰ *Boundary Value Problems*

In previous sections, we have solved initial-value problems. However, at this time we will consider boundary value problems that are solved in much the same way as initial-value problems except that the value of the function and its derivatives are given at two values of x instead of one. We then consider boundary value problems that include a parameter in the differential equation.

EXAMPLE: Solve $\begin{cases} y'' + y = 0, 0 < x < \pi \\ y'(0) = 0, y'(\pi) = 0 \end{cases}$.

SOLUTION: Because the characteristic equation is $m^2 + 1 = 0$ with roots $m = \pm i$, a general solution is $y(x) = c_1 \cos x + c_2 \sin x$, where $y'(x) = -c_1 \sin x + c_2 \cos x$. Applying the boundary conditions, we have $y'(0) = c_2 = 0$. Then $y(x) = c_1 \cos x$. With this solution, we have $y'(\pi) = -c_1 \sin \pi = 0$ for any value of c_1. Therefore, there are infinitely many solutions of the boundary value problem, depending on the choice of c_1.

In this case, we are able to use **dsolve** to solve the boundary value problem.

```
> Sol:=dsolve({diff(y(x),x$2)+y(x)=0,
    D(y)(0)=0,D(y)(Pi)=0},y(x));
```

$Sol := y(x) = \_C2 \cos(x)$

After naming $y(x)$ the solution obtained in **Sol**, we define **to_plot** to be the set of functions obtained by replacing **_C2** in $y(x)$ by i for $i = -3, -2, -1, 0, 1, 2,$ and 3. The set of functions **to_plot** is then graphed on the interval $[0, \pi]$ with **plot**.

```
> assign(Sol):
    to_plot:={seq(subs(_C2=i,y(x)),i=-3..3)}:
    plot(to_plot,x=0..Pi);
```

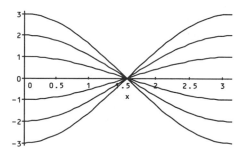

We can notice a difference between **initial-value problems** and **boundary value problems** from the previous example. This difference is that an initial-value problem has a unique solution while a boundary value problem may have more than one solution or no solution.

EXAMPLE: Solve $\begin{cases} y'' + y = 0, 0 < x < \pi \\ y'(0) = 0, y'(\pi) = 1 \end{cases}$.

SOLUTION: Using the general solution obtained in the previous example, we have $y(x) = c_1 \cos x + c_2 \sin x$. As before, $y'(0) = c_2 = 0$, so $y(x) = c_1 \cos x$. However, because $y'(\pi) = -c_1 \sin \pi = 0 \neq 1$, the boundary conditions cannot be satisfied with any choice of c_1. Therefore, there is no solution to the boundary value problem.

The boundary conditions in these problems can involve the function and its derivative. However, this modification to the problem does not affect the method of solution, as shown in the following example.

EXAMPLE: Solve $\begin{cases} y'' - y = 0, 0 < x < 1 \\ y'(0) + 3y(0) = 0, y'(1) + y(1) = 1 \end{cases}$.

SOLUTION: The characteristic equation is $m^2 - 1 = 0$ with roots $m = \pm 1$. Hence, a general solution is $y(x) = c_1 e^x + c_2 e^{-x}$ with derivative $y'(x) = c_1 e^x - c_2 e^{-x}$. Applying $y'(0) + 3y(0) = 0$ yields $y'(0) + 3y(0) = c_1 - c_2 + 3(c_1 + c_2) = 4c_1 + 2c_2 = 0$. Then

$$y'(1) + y(1) = c_1 e^1 - c_2 e^{-1} + c_1 e^1 + c_2 e^{-1} = 2c_1 e = 1,$$

so $c_1 = \dfrac{1}{2e}$ and $c_2 = -\dfrac{1}{e}$. Thus, the boundary value problem has the unique solution

$$y(x) = \frac{1}{2e} e^x - \frac{1}{e} e^{-x} = \frac{1}{2} e^{x-1} - e^{-x-1}.$$

We are not able to use **dsolve** to solve the boundary value problem directly as in the first example. However, we are able to find a general solution of the equation with **dsolve**.

```
> y:='y':
  Sol:=dsolve(diff(y(x),x$2)-y(x)=0,y(x));
```

$$Sol := y(x) = \_C1\, e^x + \_C2\, e^{-x}$$

We then name $y(x)$ the result obtained in **Sol** with **assign**. Note that $y(0)$ does not return the value of $y(x)$ if $x = 0$.

```
> assign(Sol):
  y(0);
```

$$y(0)$$

However, we are able to evaluate $y'(x) + 3y(x)$ if $x = 0$ and $y'(x) + y(x)$ if $x = 1$ with `eval` and `subs`.

```
> Eq_1:=eval(subs(x=0,diff(y(x),x)+3*y(x)=0));
  Eq_2:=eval(subs(x=1,diff(y(x),x)+y(x)=1));
```

$Eq\_1 := 4\_C1 + 2\_C2 = 0$

$Eq\_2 := 2\_C1\,e^1 = 1$

We then solve these two equations for $\_C1$ and $\_C2$ with `solve`.

```
> c_vals:=solve({Eq_1,Eq_2});
```

$$c\_vals := \left\{ \_C1 = \frac{1}{2}\frac{1}{e^1}, \_C2 = -\frac{1}{e} \right\}$$

$\_C1$ and $\_C2$ are then named the values obtained in `c_vals` with `assign`, and we determine $y(x)$ using `eval`.

```
> assign(c_vals):
  eval(y(x));
```

$$\frac{1}{2}\frac{e^x}{e} - \frac{e^{-x}}{e}$$

☰ Eigenvalue Problems

We now consider **eigenvalue problems**, boundary value problems that include a parameter. Values of the parameter that satisfy the differential equation are called **eigenvalues** of the problem, while for each eigenvalue, the nontrivial solution y that satisfies the problem is called the corresponding **eigenfunction**. We now show how the eigenvalues and eigenfunctions are found.

EXAMPLE: Solve $y'' + \lambda y = 0, 0 < x < p$, subject to $y(0) = 0$ and $y(p) = 0$.

SOLUTION: We recall from our study of higher-order differential equations with constant coefficients that the solution of this equation is determined by solving the characteristic or auxiliary equation $m^2 + \lambda = 0$, where we assume that solutions have the form $y = e^{mx}$. Of course, the values of m depend on the value of the parameter λ. Hence, we consider the three following cases.

Case I: ($\lambda = 0$)
In this case, the characteristic equation is $m^2 = 0$, so the repeated roots are $m_1 = m_2 = 0$. This indicates that a general solution is $y(x) = c_1 x + c_2$. Application of the boundary condition $y(0) = 0$ yields $y(0) = c_1(0) + c_2 = 0$, so $c_2 = 0$. Then $y(p) = c_1 p = 0$, so $c_1 = 0$. Therefore, $y(x) = 0$ in this case. Since we obtain the trivial solution, $\lambda = 0$ is **not** an eigenvalue.

Case II: ($\lambda < 0$)
To represent λ as a negative value, we let $\lambda = -k^2 < 0$. Then the characteristic equation is $m^2 - k^2 = 0$, so $m_1 = k$ and $m_2 = -k$. A general solution is, therefore, $y(x) = c_1 e^{kx} + c_2 e^{-kx}$ (or $y(x) = c_1 \cosh kx + c_2 \sinh kx$). Substitution of the boundary condition $y(0) = 0$ yields $y(0) = c_1 + c_2 = 0$, so $c_2 = -c_1$. Since $y(p) = 0$ indicates that $y(p) = c_1 e^{kp} + c_2 e^{-kp} = 0$, substitution gives us the equation $y(p) = c_1 e^{kp} - c_1 e^{-kp} = c_1(e^{kp} - e^{-kp}) = 0$. Notice that $e^{kp} - e^{-kp} = 0$ only if $e^{kp} = e^{-kp}$, and this can occur only when $k = 0$ or $p = 0$. If $k = 0$, then $\lambda = -k^2 = -0^2 = 0$, contradicting the assumption that $\lambda < 0$. We also assumed that $p > 0$, so $e^{kp} - e^{-kp} \neq 0$. Hence, $y(p) = c_1(e^{kp} - e^{-kp}) = 0$ implies that $c_1 = 0$, so $c_2 = -c_1 = 0$ as well. Therefore, $\lambda < 0$ leads to the trivial solution $y(x) = 0$, so there are no negative eigenvalues.

Case III: ($\lambda > 0$)
To represent λ as a positive value, we let $\lambda = k^2 > 0$. Then we have the characteristic equation $m^2 + k^2 = 0$ with complex conjugate roots $m_{1,2} = \pm ki$. Thus, a general solution is $y(x) = c_1 \cos kx + c_2 \sin kx$. Because $y(0) = c_1 \cos k(0) + c_2 \sin k(0) = c_1$, the boundary condition $y(0) = 0$ indicates that $c_1 = 0$. Hence, $y(x) = c_2 \sin kx$. Application of $y(p) = 0$ yields $y(p) = c_2 \sin kp = 0$, so either $c_2 = 0$ or $\sin kp = 0$. Selecting $c_2 = 0$ leads to the trivial solution that we want to avoid, so we determine the values of k that satisfy $\sin kp = 0$. Since the sine of integer multiples of π is zero, $\sin kp = 0$ if $kp = n\pi$, $n = 1, 2, \ldots$. Solving for k, we have $k = \dfrac{n\pi}{p}$, so the eigenvalues are

$$\lambda = \lambda_n = \left(\frac{n\pi}{p}\right)^2, n = 1, 2, \ldots.$$

Notice that the subscript n is used to indicate that the parameter depends on the value of n. For each eigenvalue, the corresponding eigenfunction is obtained by substitution into $y(x) = c_2 \sin kx$. Because c_2 is arbitrary, let $c_2 = 1$.

Therefore, the eigenvalue $\lambda = \lambda_n = \left(\dfrac{n\pi}{p}\right)^2$, $n = 1, 2, \ldots$ has corresponding eigenfunction

$$y(x) = y_n(x) = \sin\left(\frac{n\pi x}{p}\right), n = 1, 2, \ldots.$$

We will find the eigenvalues and eigenfunctions in the previous example quite useful in future sections. The following eigenvalue problem will be useful as well.

EXAMPLE: Solve $y'' + \lambda y = 0$, $0 < x < p$, subject to $y'(0) = 0$ and $y'(p) = 0$.

SOLUTION: Notice that the only difference between this problem and that in the previous example is in the boundary conditions. Again, the characteristic equation is $m^2 + \lambda = 0$, so we must consider the three cases $\lambda = 0$, $\lambda < 0$, and $\lambda > 0$. Note that the general solution in each case is the same as that obtained in the previous example. However, the final results may differ due to the boundary conditions.

Case I: ($\lambda = 0$)
Since $y(x) = c_1 x + c_2$, $y'(x) = c_1$. Therefore, $y'(0) = c_1 = 0$, so $y(x) = c_2$. Notice that this constant function satisfies $y'(p) = 0$ for all values of c_2. Hence, if we let $c_2 = 1$, then $\lambda = 0$ is an eigenvalue with corresponding eigenfunction $y(x) = y_0(x) = 1$.

Case II: ($\lambda < 0$)
If $\lambda = -k^2 < 0$, then $y(x) = c_1 e^{kx} + c_2 e^{-kx}$ and $y'(x) = c_1 k e^{kx} - c_2 k e^{-kx}$. Then $y'(0) = c_1 k - c_2 k = 0$, so $c_1 = c_2$. Therefore, $y'(p) = c_1 k e^{kp} - c_1 k e^{-kp} = 0$, and this is not possible unless $c_1 = 0$ because $k \neq 0$ and $p \neq 0$. Thus, $c_1 = c_2 = 0$, so $y(x) = 0$. Because we have the trivial solution, there are no negative eigenvalues.

Case III: ($\lambda > 0$)
By letting $\lambda = k^2 > 0$, $y(x) = c_1 \cos kx + c_2 \sin kx$ and $y'(x) = -c_1 k \sin kx + c_2 k \cos kx$. Hence, $y'(0) = c_2 k = 0$, so $c_2 = 0$. Consequently, $y'(p) = -c_1 k \sin kp = 0$ that is satisfied if $kp = n\pi$, $n = 1, 2, \ldots$. Therefore, the eigenvalues are

$$\lambda = \lambda_n = \left(\frac{n\pi}{p}\right)^2, n = 1, 2, \ldots.$$

Note that we found $c_2 = 0$ in $y(x) = c_1 \cos kx + c_2 \sin kx$, so the corresponding eigenfunctions are

$$y(x) = \cos \frac{n\pi x}{p}, n = 1, 2, \ldots$$

if we let $c_1 = 1$.

Summarizing our results, this problem has the following eigenvalues and corresponding eigenfunctions.

$$\lambda_n = \begin{cases} 0, n = 0 \\ \left(\dfrac{n\pi}{p}\right)^2, n = 1, 2, \ldots \end{cases}$$

and

$$y_n(x) = \begin{cases} 1, n = 0 \\ \cos \dfrac{n\pi x}{p}, n = 1, 2, \ldots . \end{cases}$$

☰ Sturm-Liouville Problems

Because of the importance of eigenvalue problems, we express these equations in the general form

$$a_2(x)y''(x) + a_1(x)y'(x) + [a_0(x) + \lambda]\, y(x) = 0, a < x < b,$$

where the boundary conditions at the endpoints $x = a$ and $x = b$ can be written as $\alpha_1 y(a) + \alpha_2 y'(a) = \alpha_3$ and $\beta_1 y(b) + \beta_2 y'(b) = 0$ for the constants α_1, α_2, α_3, β_1, β_2, and β_3, where at least one of α_1, α_2 and one of β_1, β_2 is not zero. This equation can be rewritten by letting

$$p(x) = e^{\int \frac{a_1(x)}{a_2(x)} dx}, q(x) = \frac{a_0(x)}{a_2(x)}p(x), \text{ and } s(x) = \frac{p(x)}{a_2(x)}.$$

By making this change, we obtain the equivalent equation

$$\frac{d}{dx}\left(p(x)\frac{dy}{dx}\right) + (q(x) + \lambda s(x))y = 0.$$

This is called a **Sturm-Liouville equation**, and with boundary conditions, it is called the **Sturm-Liouville problem**. This particular from of the equation is known as **self-adjoint** form. This form is of particular interest because of the relationship of the function $s(x)$ and the solutions of the problem.

EXAMPLE: Place the equation $x^2 y'' + 2xy' + \lambda y = 0$ in self-adjoint form.

SOLUTION: In this case, $a_2(x) = x^2$, $a_1(x) = 2x$, and $a_0(x) = 0$. Hence,

$$p(x) = e^{\int \frac{a_1(x)}{a_2(x)} dx} = e^{\int \frac{2x}{x^2} dx} = e^{2\ln x} = x^2, q(x) = \frac{a_0(x)}{a_2(x)}p(x) = 0, and s(x) = \frac{p(x)}{a_2(x)} = \frac{x^2}{x^2} = 1,$$

so the self-adjoint form of the equation is $\dfrac{d}{dx}\left(x^2\dfrac{dy}{dx}\right) + \lambda y = 0$. We see that our result is correct by differentiating.

Solutions of Sturm-Liouville problems have several interesting properties, two of which are included in the following theorem.

. . . .

THEOREM   *Linear Independence and Orthogonality of Eigenfunctions*

If $y_m(x)$ and $y_n(x)$ are eigenfunctions of the regular Sturm-Liousville problems, where $m \neq n$, then $y_m(x)$ and $y_n(x)$ are **linearly independent** and the **orthogonality condition** $\int_a^b s(x)y_m(x)y_n(x)\,dx = 0$ $(m \neq n)$ holds.

. . . .

EXAMPLE: Consider the eigenvalue problem $y'' + \lambda y = 0$ subject to $y(0) = 0$ and $y(p) = 0$, which we solved in a previous example. Verify that the eigenfunctions $y_1(x) = \sin\dfrac{\pi x}{p}$ and $y_2(x) = \sin\dfrac{2\pi x}{p}$ are linearly independent. Also, verify the orthogonality condition.

SOLUTION: We can verify that $y_1(x) = \sin\dfrac{\pi x}{p}$ and $y_2(x) = \sin\dfrac{2\pi x}{p}$ are linearly independent by finding the Wronskian,

$$W(y_1, y_2) = \begin{vmatrix} \sin\dfrac{\pi x}{p} & \sin\dfrac{2\pi x}{p} \\ \dfrac{\pi}{p}\cos\dfrac{\pi x}{p} & \dfrac{2\pi}{p}\cos\dfrac{2\pi x}{p} \end{vmatrix}.$$

After defining $y_1(x) = \sin\dfrac{\pi x}{p}$ and $y_2(x) = \sin\dfrac{2\pi x}{p}$ and loading the linalg package, we use det and wronskian, which are both contained in the linalg package, to find the Wronskian.

```
> y[1]:=x->sin(Pi*x/p):
  y[2]:=x->sin(2*Pi*x/p):
  with(linalg):
  step_1:=det(Wronskian([y[1](x),y[2](x)],x));
```

$$step\_1 := -\frac{\pi\left(-2\sin\left(\frac{\pi x}{p}\right)\cos\left(\frac{2\pi x}{p}\right) + \sin\left(\frac{2\pi x}{p}\right)\cos\left(\frac{\pi x}{p}\right)\right)}{p}$$

We then simplify the result obtained in step_1 using combine together with the trig option.

```
> combine(step_1,trig);
```

$$\frac{1}{2} \frac{\pi \sin\left(\frac{3\pi x}{p}\right)}{p} - \frac{3}{2} \frac{\pi \sin\left(\frac{\pi x}{p}\right)}{p}$$

Because $W(y_1, y_2)$ is not identically zero, the two functions are linearly independent.

The equation $y'' + \lambda y = 0$ is in self-adjoint form with $s(x) = 1$. Hence, the orthogonality condition is

$$\int_0^p (1)y_1(x)y_2(x)\,dx = \int_0^p (1)y_1(x)y_2(x)\,dx = \int_0^p \sin\frac{\pi x}{p} \sin\frac{2\pi x}{p}\,dx$$

$$= \int_0^p 2\sin^2\frac{\pi x}{p} \cos\frac{\pi x}{p}\,dx = \frac{2p}{3\pi}\left[\sin^3\frac{\pi x}{p}\right]_0^p = 0,$$

which is verified with $\texttt{int}$.

```
> int(y[1](x)*y[2](x),x=0..p);
```

0

11.2 FOURIER SINE SERIES AND COSINE SERIES

☰ *Fourier Sine Series*

Recall the eigenvalue problem $\begin{cases} y'' + \lambda y = 0 \\ y(0) = 0, y(p) = 0 \end{cases}$ that was solved in Section 11.1. The eigenvalues of this problem are $\lambda = \lambda_n = \left(\frac{n\pi}{p}\right)^2, n = 1, 2, \ldots$ with corresponding eigenfunctions $\phi_n(x) = \sin\frac{n\pi x}{p}, n = 1, 2, \ldots$. Therefore, for some functions f, we can find coefficients c_n so that

$$f(x) = \sum_{n=1}^{\infty} c_n \sin\frac{n\pi x}{p}.$$

A series of this form is called a **Fourier sine series**. In order to make use of these series, we must determine the coefficients c_n. We accomplish this by taking advantage of the orthogonality properties of eigenfunctions. Because the differential equation is in self-adjoint form, we have that $s(x) = 1$. Therefore, the orthogonality condition is $\int_0^{\pi} \sin\frac{n\pi x}{p} \sin\frac{m\pi x}{p}\,dx = 0 (m \neq n)$. In order to use this condition, multiply both sides of $f(x) = \sum_{n=1}^{\infty} c_n \sin \dfrac{n\pi x}{p}$ by the eigenfunction $\sin \dfrac{m\pi x}{p}$ and $s(x) = 1$. Then integrate the results from $x = 0$ to $x = p$ (because the boundary conditions of the corresponding eigenvalue problem are given at these two values of x). This yields

$$\int_0^p f(x) \sin \frac{m\pi x}{p} \, dx = \int_0^p \sum_{n=1}^{\infty} c_n \sin \frac{n\pi x}{p} \sin \frac{m\pi x}{p} \, dx.$$

Assuming that term-by-term integration is allowed on the right-hand side of the equation, we have

$$\int_0^p f(x) \sin \frac{m\pi x}{p} \, dx = \sum_{n=1}^{\infty} \int_0^p c_n \sin \frac{n\pi x}{p} \sin \frac{m\pi x}{p} \, dx = \sum_{n=1}^{\infty} c_n \int_0^p \sin \frac{n\pi x}{p} \sin \frac{m\pi x}{p} \, dx.$$

Recall that the eigenfunctions $\phi_n(x) = \sin \dfrac{n\pi x}{p}$, $n = 1, 2, \ldots$ are orthogonal, so

$\displaystyle\int_0^p \sin \frac{n\pi x}{p} \sin \frac{m\pi x}{p} \, dx = 0$ if $m \neq n$. In contrast, if $m = n$,

$$\int_0^p \sin \frac{n\pi x}{p} \sin \frac{m\pi x}{p} \, dx = \int_0^p \sin^2 \frac{n\pi x}{p} \, dx = \int_0^p \frac{1}{2}\left(1 - \cos \frac{2n\pi x}{p}\right) dx = \frac{1}{2}\left[x - \frac{p}{2\pi n} \sin \frac{2n\pi x}{p}\right]_0^p = \frac{p}{2}.$$

Therefore, each term in the sum $\displaystyle\sum_{n=1}^{\infty} c_n \int_0^p \sin \frac{n\pi x}{p} \sin \frac{m\pi x}{p} \, dx$ equals zero except if $m = n$. Hence,

$\displaystyle\int_0^p f(x) \sin \frac{n\pi x}{p} \, dx = c_n \left(\frac{p}{2}\right)$, so the Fourier sine series coefficients are given by

$$c_n = \frac{2}{p} \int_0^p f(x) \sin \frac{n\pi x}{p} \, dx,$$

where we assume that f is integrable on $(0, p)$.

EXAMPLE: Find the Fourier sine series for $f(x) = x$, $0 \leq x \leq \pi$.

SOLUTION: In this case, $p = \pi$, so

$$c_n = \frac{2}{\pi} \int_0^{\pi} f(x) \sin \frac{n\pi x}{\pi} \, dx = \frac{2}{\pi} \int_0^{\pi} x \sin nx \, dx,$$

which is computed with `int`.

```
> n:='n':
  2/Pi*int(x*sin(n*x),x=0..Pi);
```

$$-2\frac{-\sin(n\pi) + 2n\cos(n\pi)\pi}{\pi n^2}$$

Notice $\cos n\pi = (-1)^n$, because $\cos \pi = -1$, $\cos 2\pi = 1$, $\cos 3\pi = -1, \ldots$, and $\sin n\pi = 0$.
Hence, $c_n = -\dfrac{2(-1)^n}{n} = \dfrac{2(-1)^{n+1}}{n}$, and the Fourier sine series is

$$f(x) = \sum_{n=1}^{\infty} c_n \sin \frac{n\pi x}{\pi} = \sum_{n=1}^{\infty} \frac{2(-1)^{n+1}}{n} \sin nx$$

$$= 2\sin x - \sin 2x + \frac{2}{3} \sin 3x - \frac{1}{2} \sin 4x + \cdots$$

Notice that we can use a finite number of terms of the series to obtain a (trigonometric) polynomial. We begin by defining the function **c** that computes the nth coefficient c_n and then
computing the first five coefficients with **array** and **seq**. Note that the option **remember** is
included in the command so that Maple "remembers" the values of c_n computed and thus avoids
recomputing previously computed values. In this particular case, we could replace

```
2/Pi*int(x*sin(n*x),x=0..Pi)
```

by

```
2*(-1)^(n+1)/n
```

because we know the exact value of c_n. However, in cases when the coefficients can be computed
exactly but the computation is difficult to implement, you can use **int**, as follows. If numerical
approximations of the coefficients are desired, use **evalf** and **Int** or `evalf/int` after
loading the command by entering **readlib(`evalf/int`)**.

```
> c:='c':n:='n':
  c:=proc(n) option remember;
    2/Pi*int(x*sin(n*x),x=0..Pi)
    end:
  array([seq([n,c(n)],n=1..5)]);
```

$$\begin{bmatrix} 1 & 2 \\ 2 & -1 \\ 3 & \dfrac{2}{3} \\ 4 & -\dfrac{1}{2} \\ 5 & \dfrac{2}{5} \end{bmatrix}$$

Let **approx(k)** denote the kth partial sum of $\displaystyle\sum_{n=1}^{\infty} c_n \sin nx$. Note that

$$\mathbf{approx(k)} = \sum_{n=1}^{k} c_n \sin nx = \sum_{n=1}^{k-1} c_n \sin nx + c_k \sin kx$$

$$= \mathbf{approx(k-1)} + c_k \sin kx$$

Thus, to calculate the kth partial sum of $\displaystyle\sum_{n=1}^{\infty} c_n \sin nx$, we need only add $c_k \sin kx$ to the $(k-1)$st partial sum: we need not recompute all k terms of the k the partial sum if we know the $(k-1)$st partial sum. Using this observation, we define the recursively defined function **approx** to return the kth partial sum of the series

```
> approx:=proc(k) option remember;
    approx(k-1)+c(k)*sin(k*x)
    end:
  approx(1):=c(1)*sin(x):
```

and then use **approx** to compute the third, sixth, ninth, and twelfth partial sums.

```
> k_vals:=3,6,9,12:
  array([seq([k,approx(k)],k=k_vals)]);
```

$$\left[3, 2\sin(x) - \sin(2x) + \frac{2}{3}\sin(3x) \right]$$

$$\left[6, 2\sin(x) - \sin(2x) + \frac{2}{3}\sin(3x) - \frac{1}{2}\sin(4x) + \frac{2}{5}\sin(5x) - \frac{1}{3}\sin(6x) \right]$$

$$\left[9, 2\sin(x) - \sin(2x) + \frac{2}{3}\sin(3x) - \frac{1}{2}\sin(4x) + \frac{2}{5}\sin(5x) - \frac{1}{3}\sin(6x) \right.$$
$$\left. + \frac{2}{7}\sin(7x) - \frac{1}{4}\sin(8x) + \frac{2}{9}\sin(9x) \right]$$

$$\left[12, 2\sin(x) - \sin(2x) + \frac{2}{3}\sin(3x) - \frac{1}{2}\sin(4x) + \frac{2}{5}\sin(5x) - \frac{1}{3}\sin(6x) \right.$$
$$\left. + \frac{2}{7}\sin(7x) - \frac{1}{4}\sin(8x) + \frac{2}{9}\sin(9x) - \frac{1}{5}\sin(10x) + \frac{2}{11}\sin(11x) - \frac{1}{6}\sin(12x) \right]$$

We then use a **for** loop to graph x and the kth partial sum on the interval $[0, \pi]$ for $k = 3, 6, 9,$ and 12. Similarly, we graph the absolute value of the difference of x and the kth partial sum. The results are shown side by side.

```
> for k in k_vals do plot({x,approx(k)},x=0..Pi) od;
  for k in k_vals do plot(abs(x-approx(k)),x=0..Pi) od;
```

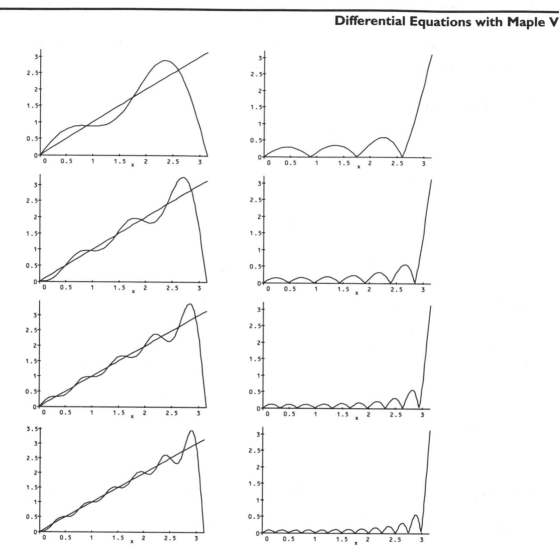

As we increase the number of terms used in approximating f, we improve the accuracy. Notice from the graphs that none of the polynomials attains the value of $f(\pi) = \pi$ at $x = \pi$. This is due to the fact that at $x = \pi$, each of the polynomials yields a value of 0. Hence, our approximation can be reliable only on the interval $0 \leq x < \pi$. In general, however, we are assured of accuracy (at points of continuity of f) only on the open interval.

EXAMPLE: Find the Fourier sine series for $f(x) = \begin{cases} 1, 0 \le x < 1 \\ -1, 1 \le x \le 2 \end{cases}$.

SOLUTION: Because f is defined on $0 \le x \le 2$, $p = 2$. Hence,

$$c_n = \frac{2}{2} \int_0^2 f(x) \sin \frac{n\pi x}{2} \, dx = \int_0^1 \sin \frac{n\pi x}{2} \, dx + \int_1^2 (-1) \sin \frac{n\pi x}{2} \, dx$$

$$= \left[-\frac{2}{n\pi} \cos \frac{n\pi x}{2} \right]_0^1 + \left[\frac{2}{n\pi} \cos \frac{n\pi x}{2} \right]_1^2$$

$$= \left(-\frac{2}{n\pi} \right) \left(\cos \frac{n\pi}{2} - 1 \right) + \left(\frac{2}{n\pi} \right) \left(\cos n\pi - \cos \frac{n\pi}{2} \right)$$

$$= \left(\frac{2}{n\pi} \right) \left(-2 \cos \frac{n\pi}{2} + \cos n\pi + 1 \right).$$

Defining c in the same manner as in the previous example, we confirm the result we obtained.

```
> c:='c':n:='n':
  c:=proc(n) option remember;
     int(sin(n*Pi*x/2),x=0..1)-int(sin(n*Pi*x/2),x=1..2)
     end:
  c(n);
```

$$-4 \frac{\cos\left(\frac{1}{2} n\pi\right)}{n\pi} + 2\frac{1}{n\pi} + 2\frac{\cos(n\pi)}{n\pi}$$

Calculating a few of the c_ns, we find $c_1 = \frac{2}{\pi}(0) = 0$, $c_2 = \frac{1}{\pi}(4) = \frac{4}{\pi}$, $c_3 = \frac{2}{3\pi}(0) = 0$, $c_4 = \frac{2}{4\pi}(0) = 0, \ldots$.

```
> array([seq([n,c(n)],n=1..10)]),
    array([seq([n,c(n)],n=11..20)]);
```

$$\begin{bmatrix} 1 & 0 \\ 2 & 4\dfrac{1}{\pi} \\ 3 & 0 \\ 4 & 0 \\ 5 & 0 \\ 6 & \dfrac{4}{3}\dfrac{1}{\pi} \\ 7 & 0 \\ 8 & 0 \\ 9 & 0 \\ 10 & \dfrac{4}{5}\dfrac{1}{\pi} \end{bmatrix} \quad \begin{bmatrix} 11 & 0 \\ 12 & 0 \\ 13 & 0 \\ 14 & \dfrac{4}{7}\dfrac{1}{\pi} \\ 15 & 0 \\ 16 & 0 \\ 17 & 0 \\ 18 & \dfrac{4}{9}\dfrac{1}{\pi} \\ 19 & 0 \\ 20 & 0 \end{bmatrix}$$

As we can see, most of the coefficients are zero. In fact, only those c_ns where n is an odd multiple of 2 yield a nonzero value. For example, $c_6 = c_{2(3)} = \dfrac{2}{6\pi}(4) = \dfrac{4}{3\pi}$, $c_{10} = c_{2(5)} = \dfrac{2}{10\pi}(4) = \dfrac{4}{5\pi}, \ldots,$ $c_{2(2n-1)} = \dfrac{4}{(2n-1)\pi}$, $n = 1, 2, \ldots,$ so we have the series

$$\begin{aligned} f(x) &= \sum_{n=1}^{\infty} \frac{4}{(2n-1)\pi} \sin \frac{2(2n-1)\pi x}{2} \\ &= \sum_{n=1}^{\infty} \frac{4}{(2n-1)\pi} \sin(2n-1)\pi x \\ &= \frac{4}{\pi} \sin \pi x + \frac{4}{3\pi} \sin 3\pi x + \frac{4}{5\pi} \sin 5\pi x + \cdots \end{aligned}$$

In the same manner as in the previous example, we define **approx(k)** to be the kth partial sum of the series $\displaystyle\sum_{n=1}^{\infty} c_n \sin \frac{n\pi x}{2}$ and then compute **approx(k)** for various values of k.

```
> approx:='approx':
  approx:=proc(k) option remember;
    approx(k-1)+c(k)*sin(k*Pi*x/2)
    end:
  approx(1):=0:
```

```
> k_vals:=10,18,26,34:
  array([seq([k,approx(k)],k=k_vals)]);
```

$$\left[10, 4\frac{\sin(\pi x)}{\pi} + \frac{4}{3}\frac{\sin(3\pi x)}{\pi} + \frac{4}{5}\frac{\sin(5\pi x)}{\pi}\right]$$

$$\left[18, 4\frac{\sin(\pi x)}{\pi} + \frac{4}{3}\frac{\sin(3\pi x)}{\pi} + \frac{4}{5}\frac{\sin(5\pi x)}{\pi} + \frac{4}{7}\frac{\sin(7\pi x)}{\pi} + \frac{4}{9}\frac{\sin(9\pi x)}{\pi}\right]$$

$$\left[26, 4\frac{\sin(\pi x)}{\pi} + \frac{4}{3}\frac{\sin(3\pi x)}{\pi} + \frac{4}{5}\frac{\sin(5\pi x)}{\pi} + \frac{4}{7}\frac{\sin(7\pi x)}{\pi} + \frac{4}{9}\frac{\sin(9\pi x)}{\pi}\right.$$
$$\left. + \frac{4}{11}\frac{\sin(11\pi x)}{\pi} + \frac{4}{13}\frac{\sin(13\pi x)}{\pi}\right]$$

$$\left[34, 4\frac{\sin(\pi x)}{\pi} + \frac{4}{3}\frac{\sin(3\pi x)}{\pi} + \frac{4}{5}\frac{\sin(5\pi x)}{\pi} + \frac{4}{7}\frac{\sin(7\pi x)}{\pi} + \frac{4}{9}\frac{\sin(9\pi x)}{\pi}\right.$$
$$\left. + \frac{4}{11}\frac{\sin(11\pi x)}{\pi} + \frac{4}{13}\frac{\sin(13\pi x)}{\pi} + \frac{4}{15}\frac{\sin(13\pi x)}{\pi} + \frac{4}{17}\frac{\sin(17\pi x)}{\pi}\right]$$

On the interval $[0, 2]$, note that $f(x) = \begin{cases} 1, 0 \le x < 1 \\ -1, 1 \le x \le 2 \end{cases} = 1 - 2\,\text{Heaviside}(x-1)$, where **Heaviside** represents the unit step function. Thus, to compare $f(x)$ to the partial sums we computed, we first define f by entering

```
> f:='f':
  f:=x->1-2*Heaviside(x-1):
```

and then generate the graphs along with the graphs of the absolute values of their differences by entering

```
>  for k in k_vals do plot({f(x),approx(k)},x=0..2) od;
   for k in k_vals do plot(abs(f(x)-approx(k)),x=0..2) od;
```

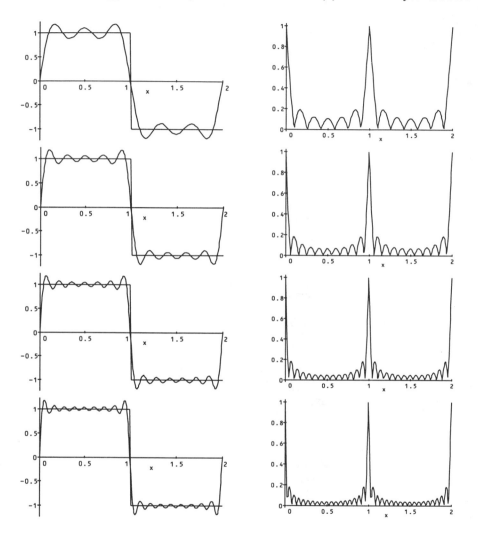

Notice that with a large number of terms the approximation is quite good at values of x, where f is continuous. The behavior of the series near points of discontinuity in that the approximation overshoots the function is called the **Gibbs phenomenon**. The approximation continues to "miss" the function even though more and more terms from the series are used!

☰ Fourier Cosine Series

Another important eigenvalue problem that has useful eigenfunctions is

$$\begin{cases} y'' + \lambda y = 0 \\ y'(0) = 0, y'(p) = 0 \end{cases}'$$

which has eigenvalues and eigenfunctions given by

$$\lambda_n = \begin{cases} 0, n = 0 \\ \left(\dfrac{n\pi}{p}\right)^2, n = 1, 2, \dots \end{cases}$$

and

$$y_n(x) = \begin{cases} 1, n = 0 \\ \cos \dfrac{n\pi x}{p}, n = 1, 2, \dots. \end{cases}$$

Therefore, for many functions $f(x)$, we can find a series expansion of the form

$$f(x) = \frac{1}{2}a_0 + \sum_{n=1}^{\infty} a_n \cos \frac{n\pi x}{p}.$$

We call this expansion a **Fourier cosine series**, where the first term (associated with $\lambda_0 = 0$), the constant $\frac{1}{2}a_0$, is written in this form for convenience in finding the formula for the coefficients $a_n, n = 0, 1, 2, \dots$. We find these coefficients in a manner similar to that followed to find the coefficients in the Fourier sine series. Notice that in this case, the orthogonality condition is

$$\int_0^p \cos \frac{n\pi x}{p} \cos \frac{m\pi x}{p} dx = 0 \ (m \neq n).$$

We use this condition by multiplying both sides of the series expansion by $\cos \dfrac{m\pi x}{p}$ and integrating from $x = 0$ to $x = p$. This yields

$$\int_0^p f(x) \cos \frac{m\pi x}{p} dx \int_0^p \frac{1}{2}a_0 \cos \frac{m\pi x}{p} dx \int_0^p \sum_{n=1}^{\infty} a_n \cos \frac{n\pi x}{p} \cos \frac{m\pi x}{p} dx.$$

Assuming that term-by-term integration is allowed,

$$\int_0^p f(x) \cos \frac{m\pi x}{p} dx = \int_0^p \frac{1}{2}a_0 \cos \frac{m\pi x}{p} dx + \sum_{n=1}^{\infty} \int_0^p a_n \cos \frac{n\pi x}{p} \cos \frac{m\pi x}{p} dx.$$

If $m = 0$, then this expression reduces to

$$\int_0^p f(x)\,dx = \int_0^p \frac{1}{2}a_0\,dx + \sum_{n=1}^{\infty}\int_0^p a_n\cos\frac{n\pi x}{p}\,dx,$$

where $\int_0^p \cos\dfrac{n\pi x}{p}\,dx = 0$ and $\int_0^p \frac{1}{2}a_0\,dx = \dfrac{p}{2}a_0$. Therefore, $\int_0^p f(x)\,dx = \dfrac{p}{2}a_0$, so

$$a_0 = \frac{2}{p}\int_0^p f(x)\,dx.$$

If $m > 0$, then we note that, by the orthogonality property, $\int_0^p a_n\cos\dfrac{n\pi x}{p}\cos\dfrac{m\pi x}{p}\,dx = 0$ if $m \ne n$.

We also note that $\int_0^p \frac{1}{2}a_0\cos\dfrac{m\pi x}{p}\,dx = 0$ and $\int_0^p \cos^2\dfrac{n\pi x}{p}\,dx = \dfrac{p}{2}$. Hence, $\int_0^p f(x)\cos\dfrac{n\pi x}{p}\,dx = 0 + a_n\left(\dfrac{p}{2}\right)$. Solving for a_n, we have

$$a_n = \frac{2}{p}\int_0^p f(x)\cos\frac{n\pi x}{p}\,dx,\, n = 1, 2, \ldots.$$

Notice that this formula also works for $n = 0$ because $\cos\dfrac{(0)\pi x}{p} = \cos 0 = 1$.

EXAMPLE: Find the Fourier cosine series for $f(x) = x, 0 \le x \le \pi$.

SOLUTION: In this case, $p = \pi$. Hence,

$$a_0 = \frac{2}{\pi}\int_0^{\pi} x\,dx = \frac{2}{\pi}\left[\frac{x^2}{2}\right]_0^{\pi} = \pi$$

and

$$a_n = \frac{2}{\pi}\int_0^{\pi} x\cos\frac{n\pi x}{\pi}\,dx = \frac{2}{\pi}\int_0^{\pi} x\cos nx\,dx,$$

which is defined and computed now with int.

```
> a:='a':n:='n':
  a:=proc(n) option remember;
    2/Pi*int(x*cos(n*x),x=0..Pi)
    end:
  simplify(a(n));latex("");
```

$$2\frac{\cos(n\pi) + 2n\sin(n\pi)\pi - 1}{\pi n^2}$$

Calculating a few of the a_ns, we see that if n is even and positive, $a_n = 0$.

```
>   array([seq([n,a(n)],n=1..10)]),
    array([seq([n,a(n)],n=11..20)]);
```

$$\begin{bmatrix} 1 & -4\dfrac{1}{\pi} \\ 2 & 0 \\ 3 & -\dfrac{4}{9}\dfrac{1}{\pi} \\ 4 & 0 \\ 5 & -\dfrac{4}{25}\dfrac{1}{\pi} \\ 6 & 0 \\ 7 & -\dfrac{4}{49}\dfrac{1}{\pi} \\ 8 & 0 \\ 9 & -\dfrac{4}{81}\dfrac{1}{\pi} \\ 10 & 0 \end{bmatrix}, \quad \begin{bmatrix} 11 & -\dfrac{4}{121}\dfrac{1}{\pi} \\ 12 & 0 \\ 13 & -\dfrac{4}{169}\dfrac{1}{\pi} \\ 14 & 0 \\ 15 & -\dfrac{4}{225}\dfrac{1}{\pi} \\ 16 & 0 \\ 17 & -\dfrac{4}{289}\dfrac{1}{\pi} \\ 18 & 0 \\ 19 & -\dfrac{4}{361}\dfrac{1}{\pi} \\ 20 & 0 \end{bmatrix}$$

Notice that for even values of n, $[(-1)^n - 1] = 0$. Therefore, $a_n = 0$ if n is even. Conversely, if n is odd, $[(-1)^n - 1] = -2$. Hence, $a_1 = -\dfrac{4}{\pi}$, $a_3 = -\dfrac{4}{9\pi}$, $a_5 = -\dfrac{4}{25\pi}, \ldots, a_{2n-1} = -\dfrac{4}{(2n - 1)^2\pi}$, so the Fourier cosine series is

$$f(x) = \frac{1}{2}(\pi) - \sum_{n=1}^{\infty} \frac{1}{n^2\pi}\cos\frac{2n\pi x}{\pi} = \frac{\pi}{2} - \frac{4}{\pi}\sum_{n=1}^{\infty}\frac{1}{(2n-1)^2}\cos(2n-1)x.$$

As in the previous two examples, we plot the function with several terms of the series along with the absolute value of the difference of the function and the partial sum. Compare these results to those obtained when approximating this function with a sine series. Which series yields the better approximation with the fewer number of terms?

```
>  approx:='approx':
   approx:=proc(k) option remember;
     approx(k-1)+a(k)*cos(k*x)
     end:
   approx(0):=Pi/2:

>  k_vals:=5,11,17,23:
   for k in k_vals do plot({x,approx(k)},x=0..Pi) od;
   for k in k_vals do plot(abs(x-approx(k)),x=0..Pi) od;
```

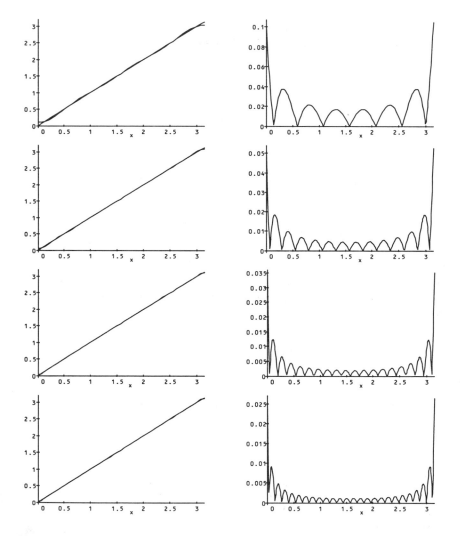

11.3 FOURIER SERIES

Consider the eigenvalue problem

$$\begin{cases} y'' + \lambda y = 0, -p < x < p \\ y(-p) = y(p), y'(-p) = y'(p) \end{cases},$$

which has eigenvalues

$$\lambda_n = \begin{cases} 0, n = 0 \\ \left(\dfrac{n\pi}{p}\right)^2, n = 1, 2, \ldots \end{cases}$$

and eigenfunctions

$$y_n(x) = \begin{cases} 1, n = 0 \\ a_n \cos \dfrac{n\pi x}{p} + b_n \sin \dfrac{n\pi x}{p}, n = 1, 2, \ldots \end{cases},$$

so we can consider a series made up of these functions. Hence, we write

$$f(x) = \frac{1}{2}a_0 + \sum_{n=1}^{\infty} \left(a_n \cos \frac{n\pi x}{p} + b_n \sin \frac{n\pi x}{p} \right),$$

which is called a **Fourier series**. As was the case with Fourier sine and Fourier cosine series, we must determine the coefficients a_0, $a_n(n = 1, 2, \ldots)$, and $b_n(n = 1, 2, \ldots)$. Because we use a method similar to those used to find the coefficients in Section 11.2, we give the value of several integrals in the following table:

$\displaystyle\int_{-p}^{p} \cos \frac{n\pi x}{p}\, dx = 0$	$\displaystyle\int_{-p}^{p} \sin \frac{n\pi x}{p}\, dx = 0$	$\displaystyle\int_{-p}^{p} \cos \frac{m\pi x}{p} \sin \frac{n\pi x}{p}\, dx = 0$
$\displaystyle\int_{-p}^{p} \cos \frac{m\pi x}{p} \cos \frac{n\pi x}{p}\, dx = \begin{cases} 0, m \neq n \\ p, m = n \end{cases}$		$\displaystyle\int_{-p}^{p} \sin \frac{m\pi x}{p} \sin \frac{n\pi x}{p}\, dx = \begin{cases} 0, m \neq n \\ p, m = n \end{cases}$

We begin by finding a_0 and a_n $(n = 1, 2, \ldots)$. Multiplying both sides of

$$f(x) = \frac{1}{2}a_0 + \sum_{n=1}^{\infty} \left(a_n \cos \frac{n\pi x}{p} + b_n \sin \frac{n\pi x}{p} \right)$$

by $\cos \dfrac{m\pi x}{p}$ and integrating from $x = -p$ to $x = p$ (because of the boundary conditions) yields

$$\int_{-p}^{p} f(x) \cos \frac{m\pi x}{p}\, dx = \int_{-p}^{p} \frac{1}{2} a_0 \cos \frac{m\pi x}{p}\, dx + \int_{-p}^{p} \sum_{n=1}^{\infty} \left(a_n \cos \frac{n\pi x}{p} \cos \frac{m\pi x}{p} + b_n \sin \frac{n\pi x}{p} \cos \frac{m\pi x}{p} \right) dx$$

$$= \int_{-p}^{p} \frac{1}{2} a_0 \cos \frac{m\pi x}{a}\, dx + \sum_{n=1}^{\infty} \left(\int_{-p}^{p} a_n \cos \frac{n\pi x}{p} \cos \frac{m\pi x}{p}\, dx \right.$$

$$\left. + \int_{-p}^{p} a_n \sin \frac{n\pi x}{p} \cos \frac{m\pi x}{p}\, dx \right).$$

If $m = 0$, we notice that all of the integrals that we are summing have the value zero. Thus, this expression simplifies to

$$\int_{-p}^{p} f(x)\, dx = \int_{-p}^{p} \frac{1}{2} a_0\, dx$$

$$\int_{-p}^{p} f(x)\, dx = \frac{1}{2} a_0 [2p]$$

$$a_0 = \frac{1}{p} \int_{-p}^{p} f(x)\, dx$$

If $m \neq 0$, then only one of the integrals on the right-hand side of the expression yields a value other than zero, and this occurs with $\displaystyle\int_{-p}^{p} \cos \frac{m\pi x}{p} \cos \frac{n\pi x}{p}\, dx = \begin{cases} 0, m \neq n \\ p, m = n \end{cases}$ if $m = n$. Hence,

$$\int_{-p}^{p} f(x) \cos \frac{n\pi x}{p}\, dx = p \cdot a_n$$

$$a_n = \frac{1}{p} \int_{-a}^{a} f(x) \cos \frac{n\pi x}{p}\, dx, n = 1, 2, \ldots$$

We find b_n $(n = 1, 2, \ldots)$ by multiplying the series by $\sin \dfrac{m\pi x}{p}$ and integrating from $x = -p$ to $x = p$. This yields

$$\int_{-p}^{p} f(x) \sin \frac{m\pi x}{p}\, dx = \int_{-p}^{p} \frac{1}{2} a_0 \sin \frac{m\pi x}{p}\, dx + \int_{-p}^{p} \sum_{n=1}^{\infty} \left(a_n \cos \frac{n\pi x}{p} \sin \frac{m\pi x}{p} + b_n \sin \frac{n\pi x}{p} \sin \frac{m\pi x}{p} \right) dx$$

$$= \int_{-p}^{p} \frac{1}{2} a_0 \sin \frac{m\pi x}{p}\, dx + \sum_{n=1}^{\infty} \left(\int_{-p}^{p} a_n \cos \frac{n\pi x}{p} \sin \frac{m\pi x}{p}\, dx \right.$$

$$\left. + \int_{-p}^{p} a_n \sin \frac{n\pi x}{p} \sin \frac{m\pi x}{p}\, dx \right).$$

Again, we note that only one of the integrals on the right-hand side is not zero. In this case, we use

$$\int_{-p}^{p} \sin \frac{m\pi x}{p} \sin \frac{n\pi x}{p} \, dx = \begin{cases} 0, m \neq n \\ p, m = n \end{cases} \text{ to obtain}$$

$$\int_{-p}^{p} f(x) \sin \frac{n\pi x}{p} \, dx = p \cdot b_n$$

$$b_n = \frac{1}{p} \int_{-p}^{p} f(x) \sin \frac{n\pi x}{p} \, dx, n = 1, 2, \ldots.$$

DEFINITION

Fourier Series

Suppose that f is defined on $-p < x < p$. Then the **Fourier series** for f is

$$f(x) = \frac{1}{2}a_0 + \sum_{n=1}^{\infty} \left(a_n \cos \frac{n\pi x}{p} + b_n \sin \frac{n\pi x}{p} \right),$$

where

$$a_0 = \frac{1}{p} \int_{-p}^{p} f(x) \, dx,$$

$$a_n = \frac{1}{p} \int_{-p}^{p} f(x) \cos \frac{n\pi x}{p} \, dx, n = 1, 2, \ldots,$$

and

$$b_n = \frac{1}{p} \int_{-p}^{p} f(x) \sin \frac{n\pi x}{p} \, dx, n = 1, 2, \ldots.$$

The following theorem tells us that the Fourier series for any function converges to the function except at points of discontinuity.

· ·

THEOREM   *Convergence of Fourier Series*

Suppose that f and f' are piecewise continuous functions on $-p < x < p$. Then the Fourier series for f on $-p < x < p$ converges to $f(x)$ at every x where f is continuous. At $x = x_0$ where f is discontinuous, the Fourier series converges to the average

$$\frac{f(x_0^+) + f(x_0^-)}{2},$$

where $f(x_0^+) = \lim_{x \to x_0^+} f(x)$ and $f(x_0^-) = \lim_{x \to x_0^-} f(x)$.

· ·

EXAMPLE: Find the Fourier series of $f(x) = \begin{cases} 1, -2 \le x < 0 \\ 2, 0 \le x < 2 \end{cases}$, where $f(x + 4) = f(x)$.

SOLUTION: In this case, $p = 2$. Hence,

$$a_0 = \frac{1}{2} \int_{-2}^{2} f(x)\, dx = \frac{1}{2} \int_{-2}^{0} (1)\, dx + \frac{1}{2} \int_{0}^{2} (2)\, dx = \frac{1}{2} [x]_{-2}^{0} + \frac{1}{2} \cdot 2[x]_{0}^{2} = 1 + 2 = 3,$$

while

$$a_n = \frac{1}{2} \int_{-2}^{2} f(x) \cos \frac{n\pi x}{2}\, dx = \frac{1}{2} \int_{-2}^{0} (1) \cos \frac{n\pi x}{2}\, dx + \frac{1}{2} \int_{0}^{2} (2) \cos \frac{n\pi x}{2}\, dx$$

and

$$b_n = \frac{1}{2} \int_{-2}^{2} f(x) \sin \frac{n\pi x}{2}\, dx = \frac{1}{2} \int_{-2}^{0} (1) \sin \frac{n\pi x}{2}\, dx + \frac{1}{2} \int_{0}^{2} (2) \sin \frac{n\pi x}{2}\, dx$$

are computed with `int`.

After defining **a**, we compute **a(n)**. Thus, for $n > 0$, $a_n = 0$ because $\sin n\pi = 0$.

```
> a:='a':n:='n':
  a:=proc(n) option remember;
    1/2*(int(cos(n*Pi*x/2),x=-2..0)+
    int(2*cos(n*Pi*x/2),x=0..2));
  end:
  simplify(a(n));
```

$$3\frac{\sin(n\pi)}{n\pi}$$

Similarly, we define b and compute b(n).

```
> b:='b':
  b:=proc(n) option remember;
    1/2*(int(sin(n*Pi*x/2),x=-2..0)+
     int(2*sin(n*Pi*x/2),x=0..2));
    end:
  simplify(b(n));
```

$$-\frac{-1 + \cos(n\pi)}{n\pi}$$

Several of the b_ns are computed using **seq** and **array**. Note that the even terms are zero.

```
>  array([seq([n,b(n)],n=1..10)]);
```

$$
\begin{bmatrix}
1 & 2\dfrac{1}{\pi} \\
2 & 0 \\
3 & \dfrac{2}{3}\dfrac{1}{\pi} \\
4 & 0 \\
5 & \dfrac{2}{5}\dfrac{1}{\pi} \\
6 & 0 \\
7 & \dfrac{2}{7}\dfrac{1}{\pi} \\
8 & 0 \\
9 & \dfrac{2}{9}\dfrac{1}{\pi} \\
10 & 0
\end{bmatrix}
$$

Therefore,

$$
f(x) = \frac{3}{2} + \sum_{n=1}^{\infty} (1 - (-1)^n)\frac{1}{n\pi} \sin \frac{n\pi x}{2}
$$

$$
= \frac{3}{2} + \frac{2}{\pi} \sin \frac{\pi x}{2} + \frac{2}{3\pi} \sin \frac{3\pi x}{2} + \frac{2}{5\pi} \sin \frac{5\pi x}{2} + \cdots.
$$

We graph $f(x)$ with several terms of the series in the same way as we did in Section 11.2.

Note that for $x > 0$,

```
f(x) = 2-Heaviside(x-2)+Heaviside(x-4)-Heaviside(x-6)+
       Heaviside(x-8)- ··· ,
```

where `Heaviside` represents the unit step function.
    Thus, entering

```
> f:=x->2-Heaviside(x-2)+Heaviside(x-4)-Heaviside(x-6)+
    Heaviside(x-8):
  plot(f(x),x=0..8,0..3);
```

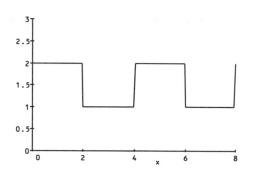

defines f (the definition is valid on the interval $[0,8]$) and then graphs f on the interval $[0,8]$.
Note that the graph corresponds to the graph of two periods of f.
    In the same manner as in Section 11.2, we define `approx(k)` to be the kth partial sum of the
Fourier series. We then compute the fifth, ninth, and thirteenth partial sums.

```
> approx:=proc(k) option remember;
    approx(k-1)+b(k)*sin(k*Pi*x/2)
    end:
  approx(0):=3/2:
  k_vals:=5,9,13:
  array([seq([k,approx(k)],k=k_vals)]);
```

$$\left[\begin{array}{l} 5, \dfrac{3}{2}+2\dfrac{\sin\left(\frac{1}{2}\pi x\right)}{\pi}+\dfrac{2}{3}\dfrac{\sin\left(\frac{3}{2}\pi x\right)}{\pi}+\dfrac{2}{5}\dfrac{\sin\left(\frac{5}{2}\pi x\right)}{\pi} \\[2ex] 9, \dfrac{3}{2}+2\dfrac{\sin\left(\frac{1}{2}\pi x\right)}{\pi}+\dfrac{2}{3}\dfrac{\sin\left(\frac{3}{2}\pi x\right)}{\pi}+\dfrac{2}{5}\dfrac{\sin\left(\frac{5}{2}\pi x\right)}{\pi}+\dfrac{2}{7}\dfrac{\sin\left(\frac{7}{2}\pi x\right)}{\pi}+\dfrac{2}{9}\dfrac{\sin\left(\frac{9}{2}\pi x\right)}{\pi} \\[2ex] 13, \dfrac{3}{2}+2\dfrac{\sin\left(\frac{1}{2}\pi x\right)}{\pi}+\dfrac{2}{3}\dfrac{\sin\left(\frac{3}{2}\pi x\right)}{\pi}+\dfrac{2}{5}\dfrac{\sin\left(\frac{5}{2}\pi x\right)}{\pi}+\dfrac{2}{7}\dfrac{\sin\left(\frac{7}{2}\pi x\right)}{\pi}+\dfrac{2}{9}\dfrac{\sin\left(\frac{9}{2}\pi x\right)}{\pi} \\[2ex] \qquad +\dfrac{2}{11}\dfrac{\sin\left(\frac{11}{2}\pi x\right)}{\pi}+\dfrac{2}{13}\dfrac{\sin\left(\frac{13}{2}\pi x\right)}{\pi} \end{array}\right]$$

We graph $f(x)$ and the kth partial sum for $k = 5, 9,$ and 13 on the interval $[0, 8]$ using the following **for** loop.

```
> for k in k_vals do plot({f(x),approx(k)},x=0..8,0..2.5) od;
```

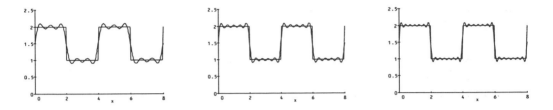

We graph the error by graphing the absolute value of the difference of $f(x)$ and the kth partial sum using the following **for** loop.

```
> for k in k_vals do plot(abs(f(x)-approx(k)),x=0..8) od;
```

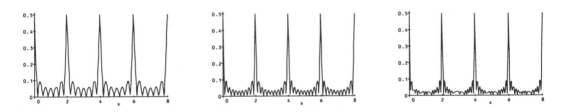

If we extend f over more periods, then the approximation by the Fourier series carries over to these intervals.

EXAMPLE: Find the Fourier series of $f(x) = \begin{cases} 0, & -1 \le x < 0 \\ \sin \pi x, & 0 \le x \le 1 \end{cases}$, where $f(x + 2) = f(x)$.

SOLUTION: In this case, $p = 1$, so

$$a_0 = \frac{1}{1} \int_{-1}^{1} f(x)\,dx = \int_{-1}^{0} (0)\,dx + \int_{0}^{1} \sin \pi x\,dx = -\left[\frac{\cos \pi x}{\pi}\right]_0^1 = -\frac{1}{\pi}(\cos \pi - \cos 0) = \frac{2}{\pi}$$

and

$$a_n = \frac{1}{1} \int_{-1}^{1} f(x) \cos n\pi x\,dx = \int_{-1}^{0} (0) \cos n\pi x\,dx + \int_{0}^{1} \sin \pi x \cos n\pi x\,dx = \int_{0}^{1} \sin \pi x \cos n\pi x\,dx.$$

The value of this integral depends on the value of n. If $n = 1$, we have

```
> int(sin(Pi*x)*cos(Pi*x),x=0..1);
    0
```

$a_1 = 0$. If $n \neq 1$, then

```
> an:=int(sin(Pi*x)*cos(n*Pi*x),x=0..1);
```

$$an := -\frac{\cos(n\pi)}{\pi(1+n)(-1+n)} - \frac{1}{\pi(1+n)(-1+n)}$$

and computing several of the coefficients results in

```
> a:='a':n:='n':
  a:=proc(n) option remember;
     if n > 1 then an else 0 fi;
     end:
  array([seq([n,a(n)],n=1..10)]);
```

$$\begin{bmatrix} 1 & 0 \\ 2 & -\dfrac{2}{3}\dfrac{1}{\pi} \\ 3 & 0 \\ 4 & -\dfrac{2}{15}\dfrac{1}{\pi} \\ 5 & 0 \\ 6 & -\dfrac{2}{35}\dfrac{1}{\pi} \\ 7 & 0 \\ 8 & -\dfrac{2}{63}\dfrac{1}{\pi} \\ 9 & 0 \\ 10 & -\dfrac{2}{99}\dfrac{1}{\pi} \end{bmatrix}$$

Notice that if n is odd, then both $(1 - n)$ and $(1 + n)$ are even. Hence, $\cos(1 - n)\pi x = \cos(1 + n)\pi x = 1$, so

$$a_n = -\frac{1}{2}\left\{\left[\frac{1}{(1-n)\pi} + \frac{1}{(1+n)\pi}\right] - \left[\frac{1}{(1-n)\pi} + \frac{1}{(1+n)\pi}\right]\right\} = 0.$$

If n is even, $(1-n)$ and $(1+n)$ are odd. Therefore, $\cos(1-n)\pi x = \cos(1+n)\pi x = -1$, so

$$a_n = -\frac{1}{2}\left\{\left[\frac{-1}{(1-n)\pi} + \frac{-1}{(1+n)\pi}\right] - \left[\frac{1}{(1-n)\pi} + \frac{1}{(1+n)\pi}\right]\right\}$$

$$= \frac{1}{(1-n)\pi} + \frac{1}{(1+n)\pi} = \frac{2}{(1-n)(1+n)\pi} = -\frac{2}{(n-1)(1+n)\pi}$$

if n is even. Putting this information together, we can write the coefficients as

$$a_{2n} = -\frac{2}{(2n-1)(1+2n)\pi}, n = 1, 2, \dots.$$

Similarly,

$$b_n = \frac{1}{1}\int_{-1}^{1} f(x)\sin n\pi x\, dx = \int_0^1 \sin \pi x \sin n\pi x\, dx,$$

so if $n = 1$, then

```
> int(sin(Pi*x)^2,x=0..1);
```

$$\frac{1}{2}$$

$b_1 = \frac{1}{2}$. If $n \ne 1$, then

```
> bn:=int(sin(Pi*x)*sin(n*Pi*x),x=0..1);
```

$$bn := -\frac{\sin(n\pi)}{\pi(1+n)(-1+n)}$$

Hence, $b_n = 0, n = 2, 3, \dots.$
Therefore, we write the Fourier series as

$$f(x) = \frac{1}{\pi} + \frac{1}{2}\sin \pi x - \frac{2}{\pi}\sum_{n=1}^{\infty} \frac{1}{(2n-1)(1+2n)}\cos 2n\pi x.$$

As in the previous example, we graph f along with the kth partial sum of the Fourier series for various values of k. First, we note that in terms of the unit step function, **Heaviside**, we have

$$f(x) = \sin(\text{Pi} * x)(1 - \text{Heaviside}(x - 1) + \text{Heaviside}(x - 2) - \text{Heaviside}(x - 3) + \cdots), x > 0.$$

For example, on the interval $[0, 4]$,

$$f(x) = \sin(\text{Pi} * x)(1 - \text{Heaviside}(x - 1) + \text{Heaviside}(x - 2)-$$
$$\text{Heaviside}(x - 3) + \text{Heaviside}(x - 4)).$$

Thus, entering

```
> f:=x->sin(Pi*x)*(1-Heaviside(x-1)+Heaviside(x-2)-
  Heaviside(x-3)+Heaviside(x-4)):
  plot(f(x),x=0..4,-.5..1.5);
```

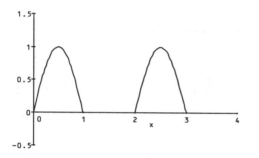

graphs f on the interval $[0, 4]$. Note that the result corresponds to two periods of f.

In the same manner as in the previous examples, we define `approx(k)` to be the kth partial sum of the Fourier series.

```
> approx:=proc(k) option remember;
    approx(k-1)+a(k)*cos(k*Pi*x)
    end:
  approx(0):=1/Pi+1/2*sin(Pi*x):
```

We graph f along with the second, sixth, and tenth partial sums of the series.

```
> k_vals:=2,6,10:
  for k in k_vals do
    plot({f(x),approx(k)},x=0..4,-.5..1.5) od;
```

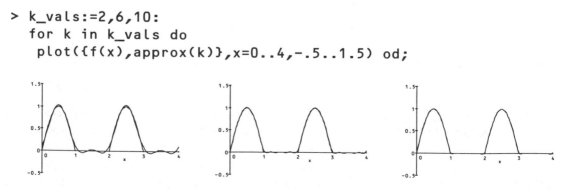

The corresponding errors are graphed as well.

```
> for k in k_vals do
    plot(abs(f(x)-approx(k)),x=0..4) od;
```

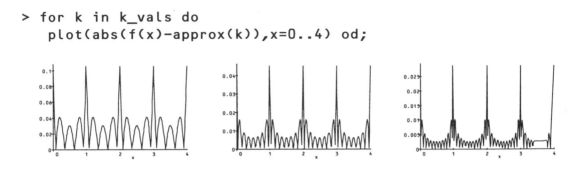

11.4 GENERALIZED FOURIER SERIES: BESSEL-FOURIER SERIES

In addition to the trigonometric eigenfunctions that were used to form the Fourier series in Sections 11.2 and 11.3, the eigenfunctions of other eigenvalue problems can be used to form what we call **generalized Fourier series**. We will find that these series will assist in solving problems in applied mathematics that involve physical phenomena that cannot be modeled with trigonometric functions.

Recall **Bessel's equation of order zero**

$$x^2 y'' + xy' + \lambda^2 x^2 y = 0$$

with linearly independent solutions $J_0(\lambda x)$ and $Y_0(\lambda x)$. If we require that the solutions of this differential equations satisfy the boundary conditions $|y(0)| < \infty$ (meaning that the solution is bounded at $x = 0$) and $y(p) = 0$, then we can find the eigenvalues of the boundary value problem

$$\begin{cases} x^2 y'' + xy' + \lambda^2 x^2 y = 0, 0 < x < p \\ |y(0)| < \infty, y(p) = 0 \end{cases}$$

A general solution to Bessel's equation is $y = c_1 J_0(\lambda x) + c_2 Y_0(\lambda x)$. Because $|y(0)| < \infty$, we must choose $c_2 = 0$ because $\lim_{x \to 0^+} Y_0(\lambda x) = -\infty$. Hence, $y(p) = c_1 J_0(\lambda p) = 0$. Just as we did with the eigenvalue problems solved earlier in Section 11.1, we can't choose $c_1 = 0$, so we must select λ so that $J_0(\lambda p) = 0$. Unfortunately, the values of x, where $J_0(\lambda p) = 0$, are not as easily expressed as they are with trigonometric functions. From our study of Bessel functions in Section 6.5, we know that this function intersects the x-axis in infinitely many places. In this case, we let α_n represent the nth zero of the Bessel function of order zero, J_0, where $n = 1, 2, \ldots$. To approximate the α_ns, we begin by graphing $J_0(x)$ with **plot**.

```
> plot(BesselJ(0,x),x=0..40);
```

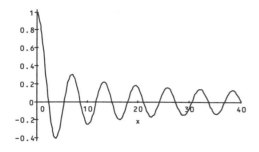

The command `fsolve(equation,x,a..b)` attempts to locate an approximation of the solution to `equation`, which represents an equation in `x`, in the interval `(a,b)`. For example, from the graph we see that the first zero of the Bessel function of order zero occurs in the interval $(2,3)$. Thus, entering

```
> fsolve(BesselJ(0,x)=0,x,2..3);
```

    2.404825558

returns an approximation of the solution to $J_0(x) = 0$ in the interval $(2,3)$. We interpret the result to mean $\alpha_1 \approx 2.4048$. To approximate the first 10 zeros, we see from the graph that the second zero occurs in the interval $(5,6)$, the third in $(8,9)$, the fourth in $(11,12)$, the fifth in $(14,15)$, the sixth in $(18,19)$, the seventh in $(21,22)$, the eighth in $(24,25)$, the ninth in $(27,28)$, and the tenth in $(30,31)$. These 10 ordered pairs are entered in the list `guesses` using `..` to indicate that they represent intervals.

```
> guesses:=[2..3,5..6,8..9,11..12,14..15,18..19,
    21..22,24..25,27..28,30..31];
```

Next we define the function `approx`. Given `guess`, where `guess` is of the form `a..b`, `approx(guess)` attempts to find a numerical solution to $J_0(x) = 0$ in the interval `guess`.

```
> approx:=proc(guess)
    fsolve(BesselJ(0,x)=0,x,guess)
    end:
```

We then use `map` to apply the function `approx` to the list `guesses`, naming the resulting list `alpha`.

```
> alpha:=map(approx,guesses);
```

   $\alpha :=$ [2.404825558, 5.520078110, 8.653727913, 11.79153444,
        14.93091771, 18.07106397, 21.21163663, 24.35247153,
        27.49347913, 30.63460647]

Thus, entering

```
> alpha[5];
```
14.93091771

returns an approximation of the fifth zero of $J_0(x)$: $\alpha_5 \approx 14.9309$.

Therefore, in trying to find the eigenvalues, we must solve $J_0(\lambda p) = 0$. From our definition of α_n, this equation is satisfied when $\lambda p = \alpha_n$, $n = 1, 2, \ldots$. Hence, the eigenvalues are

$$\lambda = \lambda_n = \frac{\alpha_n}{p}, n = 1, 2, \ldots,$$

and the corresponding eigenfunctions are

$$y(x) = y_n(x) = J_0(\lambda_n x) = J_0\left(\frac{\alpha_n x}{p}\right), n = 1, 2, \ldots.$$

As with the trigonometric eigenfunctions that we found in Section 11.2 and 11.3 , $J_0\left(\frac{\alpha_n x}{p}\right)$ can be used to build an eigenfunction series expansion of the form

$$f(x) = \sum_{n=1}^{\infty} c_n J_0\left(\frac{\alpha_n x}{p}\right),$$

where we use the orthogonality properties of $J_0\left(\frac{\alpha_n x}{p}\right)$ to find the coefficients c_n.

We determine the orthogonality condition by placing Bessel's equation of order zero in the self-adjoint form

$$\frac{d}{dx}[xy'] + \lambda^2 xy = 0.$$

Because $s(x) = x$, the orthogonality condition is

$$\int_0^p x J_0\left(\frac{\alpha_n x}{p}\right) J_0\left(\frac{\alpha_m x}{p}\right) dx = 0, n \neq m.$$

Multiplying $f(x) = \sum_{n=1}^{\infty} c_n J_0 \left(\frac{\alpha_n x}{p} \right)$ by $x J_0 \left(\frac{\alpha_m x}{p} \right)$ and integrating from $x = 0$ to $x = p$ yield

$$\int_0^p x f(x) J_0 \left(\frac{\alpha_m x}{p} \right) dx = \int_0^p \sum_{n=1}^{\infty} c_n x J_0 \left(\frac{\alpha_n x}{p} \right) J_0 \left(\frac{\alpha_m x}{p} \right) dx$$

$$= \sum_{n=1}^{\infty} c_n \int_0^p x J_0 \left(\frac{\alpha_n x}{p} \right) J_0 \left(\frac{\alpha_m x}{p} \right) dx.$$

However, by the orthogonality condition, each of the integrals on the left-hand side of the equation equals zero except when $m = n$. Therefore,

$$c_n = \frac{\int_0^p x f(x) J_0 \left(\frac{\alpha_n x}{p} \right) dx}{\int_0^p x \left[J_0 \left(\frac{\alpha_n x}{p} \right) \right]^2 dx}, n = 1, 2, \ldots.$$

The value of the integral in the denominator can be found through the use of several of the identities associated with the Bessel functions. Since $\lambda = \lambda_n = \frac{\alpha_n}{p}, n = 1, 2, \ldots$, the function $J_0 \left(\frac{\alpha_n x}{p} \right) = J_0(\lambda_n x)$ satisfies Bessel's equation of order zero given by

$$\frac{d}{dx} \left[x \frac{d}{dx} J_0(\lambda_n x) \right] + \lambda_n^2 x J_0(\lambda_n x) = 0.$$

Multiplying by the factor $2x \frac{d}{dx} J_0(\lambda_n x)$, we can write the expression as

$$\frac{d}{dx} \left[x \frac{d}{dx} J_0(\lambda_n x) \right]^2 + \lambda_n^2 x^2 \frac{d}{dx} [J_0(\lambda_n x)]^2 = 0.$$

Integrating this expression from $x = 0$ to $x = p$, we have

$$2\lambda_n^2 \int_0^p x [J_0(\lambda_n x)]^2 dx = \lambda_n^2 p^2 [J_0'(\lambda_n p)]^2 + \lambda_n^2 p^2 [J_0(\lambda_n p)]^2.$$

Since $\lambda_n p = \frac{\alpha_n}{p} p = \alpha_n$, we make the following substitutions:

$$2\lambda_n^2 \int_0^p x [J_0(\lambda_n x)]^2 dx = \lambda_n^2 p^2 [J_0'(\alpha_n)]^2 + \lambda_n^2 p^2 [J_0(\alpha_n)]^2.$$

Then $J_0(\alpha_n) = 0$ because α_n is the nth zero of J_0. Also, with $n = 0$, the identity $\dfrac{d}{dx}[x^{-n}J_n(x)] = -x^{-n}J_{n+1}(x)$ indicates that $J_0'(\alpha_n) = -J_1(\alpha_n)$. Therefore,

$$2\lambda_n^2 \int_0^p x[J_0(\lambda_n x)]^2 dx = \lambda_n^2 p^2[-J_1(\alpha_n)]^2 + \lambda_n^2 p^2(0)$$

$$\int_0^p x[J_0(\lambda_n x)]^2 dx = \frac{p^2}{2}[J_1(\alpha_n)]^2,$$

where the value of $J_1(\alpha_n)$ can be approximated using **Bessel**. Therefore, the series coefficients are found with

$$c_n = \frac{2}{p^2[J_1(\alpha_n)]^2} \int_0^p xf(x)J_0\left(\frac{\alpha_n x}{p}\right) dx, n = 1, 2, \ldots.$$

As was the case with Fourier series, we can make a statement about the convergence of the Bessel-Fourier series.

· ·

THEOREM   *Convergence of Bessel-Fourier Series*

Suppose that f and f' are piecewise continuous functions on $0 < x < p$. Then the Bessel-Fourier series for f on $-p < x < p$ converges to $f(x)$ at every x where f is continuous. At $x = x_0$ where f is discontinuous, the Bessel-Fourier series converges to the average

$$\frac{f(x_0^+) + f(x_0^-)}{2}.$$

· ·

EXAMPLE: Find the Bessel-Fourier series for $f(x) = 1 - x^2$ on $0 < x < 1$.

SOLUTION: In this case, $p = 1$, so

$$c_n = \frac{2}{[J_1(\alpha_n)]^2} \int_0^1 x(1 - x^2)J_0(\alpha_n x) dx = \frac{2}{[J_1(\alpha_n)]^2}\left\{\int_0^1 xJ_0(\alpha_n x) dx - \int_0^1 x^3 J_0(\alpha_n x) dx\right\}.$$

Using the formula, $\dfrac{d}{dx}[x^n J_n(x)] = -x^n J_{n-1}(x)$ with $n = 1$,

$$\int_0^1 x J_0(\alpha_n x)\, dx = \left[\frac{1}{\alpha_n} x J_1(\alpha_n x)\right]_0^1 = \frac{1}{\alpha_n} J_1(\alpha_n).$$

Note that the factor $\dfrac{1}{\alpha_n}$ is due to the chain rule for differentiating the argument of $J_1(\alpha_n x)$. We use integration by parts with $u = x^2$ and $dv = x J_0(\alpha_n x)$ to evaluate $\displaystyle\int_0^1 x^3 J_0(\alpha_n x)\, dx$. As in the first integral, we obtain $v = -\dfrac{1}{\alpha_n} J_1(\alpha_n x)$. Then, since $du = 2x\, dx$, we have

$$\int_0^1 x^3 J_0(\alpha_n x)\, dx = \left[\frac{1}{\alpha_n} x^3 J_1(\alpha_n x)\right]_0^1 - \frac{2}{\alpha_n}\int_0^1 x^2 J_1(\alpha_n x)\, dx$$

$$= \frac{1}{\alpha_n} J_1(\alpha_n) - \frac{2}{\alpha_n}\left[\frac{1}{\alpha_n} x^2 J_2(\alpha_n x)\right]_0^1 = \frac{1}{\alpha_n} J_1(\alpha_n) - \frac{2}{\alpha_n^2} J_2(\alpha_n).$$

Thus, the coefficients are

$$c_n = \frac{2}{[J_1(\alpha_n)]^2}\int_0^1 x(1 - x^2) J_0(\alpha_n x)\, dx = \frac{2}{[J_1(\alpha_n)]^2}\left\{\int_0^1 x J_0(\alpha_n x)\, dx - \int_0^1 x^3 J_0(\alpha_n x)\, dx\right\}$$

$$= \frac{2}{[J_1(\alpha_n)]^2}\left[\frac{1}{\alpha_n} J_1(\alpha_n) - \left(\frac{1}{\alpha_n} J_1(\alpha_n) - \frac{2}{\alpha_n^2} J_2(\alpha_n)\right)\right] = \frac{4 J_2(\alpha_n)}{\alpha_n^2 [J_1(\alpha_n)]^2}, \quad n = 1, 2, \ldots,$$

so that the Bessel-Fourier series is

$$f(x) = \sum_{n=1}^{\infty} \frac{4 J_2(\alpha_n)}{\alpha_n^2 [J_1(\alpha_n)]^2} J_0(\alpha_n x).$$

To graph f along with several terms of the series, we first define the function **c** and then compute **c(n)** for $n = 1, 2, 3, 4,$ and 5.

```
> readlib(`evalf/int`):
  c:=proc(n) option remember;
    2/BesselJ(1,alpha[n])^2*
      `evalf/int`(x*(1-x^2)*BesselJ(0,alpha[n]*x),x=0..1)
    end:
  array([seq([n,c(n)],n=1..5)]);
```

$$
\begin{bmatrix}
1 & 1.108022261 \\
2 & -0.1397775052 \\
3 & 0.04547647070 \\
4 & -0.02099090194 \\
5 & 0.01163624312
\end{bmatrix}
$$

Then, in the same manner as in Sections 11.2 and 11.3, we define approx(k) to be the kth partial sum of the series.

```
> approx:=proc(k) option remember;
    approx(k-1)+c(k)*BesselJ(0,alpha[k]*x)
    end:
  approx(1):=c(1)*BesselJ(0,alpha[1]*x):
```

We graph f along with several terms of the series. Notice that the polynomial with two terms yields an accurate approximation of f. W' 'n using four or six terms, the graphs are practically indistinguishable.

```
> f:=x->1-x^2:
  k_vals:=2,4,6:
  for k in k_vals do plot({f(x),approx(k)},x=0..1) od;
```

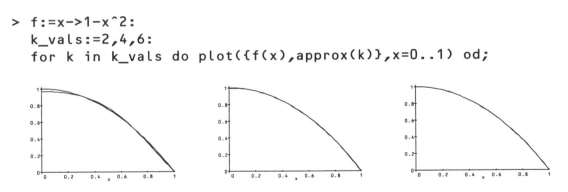

We verify that the error is "small" by graphing the absolute value of the difference between $f(x)$ and the kth partial sum for $k = 2, 4,$ and 6.

```
> for k in k_vals do plot(abs(f(x)-approx(k)),x=0..1) od;
```

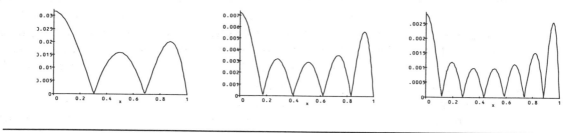

Constructing a Table of Zeros of Bessel Functions

EXAMPLE: Since the zeros of the Bessel functions play an important role in the generalized Fourier series involving Bessel functions (See Section 12.5), use Maple to find the first eight zeros of the Bessel functions of the first kind, $Jn(x)$, of order $n = 0, 1, 2, \dots, 6$.

SOLUTION: The **Bessel function of the first kind of order** n, $Jn(x)$, is represented by BesselJ(n,x). We graph the Bessel functions of the first kind of order n for $n = 0, 1,$ $\dots, 8$ on the interval $[0, 40]$:

```
> for i from 0 to 8 do
    plot(BesselJ(i,x),x=0..40,numpoints=100) od;
```

The results are displayed as follows:

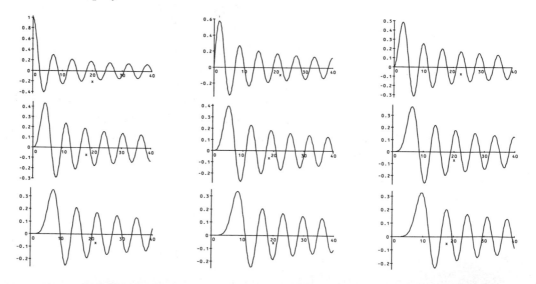

To approximate the zeros of the Bessel functions we will use the command `fsolve`. Recall that `fsolve(equation,x,a..b)` attempts to locate an approximation of the solution to `equation`, which represents an equation in x, on the interval (a, b). We use these graphs to approximate (open) intervals containing the zeros. For example, we see that the first zero of the Bessel function of order zero occurs in the interval $(2, 3)$, the second in $(5, 6)$, the third in $(8, 9)$, the fourth in $(11, 12)$, the fifth in $(14, 15)$, the sixth in $(18, 19)$, the seventh in $(21, 22)$, and the eighth in $(24, 25)$. Hence, these initial intervals are entered in the array `guesses`. Note that `guesses[1]` corresponds to a list of eight intervals containing the first eight zeros of the Bessel function of order zero. Hence, `guesses[i]` corresponds to a list of eight intervals containing the first eight zeros of the Bessel function of order $i - 1$; `guesses[i,j]` corresponds to an interval containing the jth zero of the Bessel function of order $i - 1$.

```
> guesses:=array(
    [[2..3,5..6,8..9,11..12,14..15,
    18..19,21..22,24..25],
    [3..4,7..8,10..11,13..14,16..17,
    19..20,22..23,25..26],
    [5..6,8..9,11..12,14..15,17..18,
    21..22,24..25,27..28],
    [6..7,9..10,13..14,16..17,19..20,
    22..23,25..26,28..29],
    [7..8,11..12,14..15,17..18,20..21,
    24..25,27..28,30..31],
    [8..9,12..13,15..16,18..19,22..23,
    25..26,28..29,31..32],
    [9..10,13..14,17..18,20..21,23..24,
    26..27,30..31,33..34]]):
```

The function `alpha` uses `fsolve` and the intervals in `guesses` to approximate the zeros of the Bessel functions. In the two following `for` loops, `i` corresponds to the Bessel function of order `i` and `j` represents the jth zero. For example, for $i = 0$, the first eight zeros of the Bessel function of order zero are computed by using `fsolve` to approximate the zero on each of the intervals in the first list in the array `guesses`. This is carried out for $i = 0$ to $i = 6$ to yield a table of zeros of the Bessel functions of order zero to order six. In the resulting table, each zero corresponds to an ordered pair (i, j) that represents the jth zero of the Bessel function of order i.

```
> i:='i':j:='j':
  alpha:=table():
  for i from 0 to 6 do
   for j to 8 do
    alpha[i,j]:=
    fsolve(BesselJ(i,x)=0,x,guesses[i+1,j])
    od od:
```

```
> ALPHA:=array(1..7,1..8):
```

```
> for i from 1 to 7 do
   for j from 1 to 8 do
    ALPHA[i,j]:=evalf(alpha[i-1,j],5) od od:
  print(ALPHA);
```

[2.4048, 5.5201, 8.6537, 11.792, 14.931, 18.071, 21.212, 24.352]
[3.8317, 7.0156, 10.173, 13.324, 16.471, 19.616, 22.760, 25.904]
[5.1356, 8.4172, 11.620, 14.796, 17.960, 21.117, 24.270, 27.421]
[6.3802, 9.7610, 13.015, 16.223, 19.409, 22.583, 25.748, 28.908]
[7.5883, 11.065, 14.373, 17.616, 20.827, 24.019, 27.199, 30.371]
[8.7715, 12.339, 15.700, 18.980, 22.218, 25.430, 28.627, 31.812]
[9.9361, 13.589, 17.004, 20.321, 23.586, 26.820, 30.034, 33.233]

We then save this table of numbers for later use in a text editor and name the resulting file **BesselTable** in the Maple folder. In this particular case, a word processor was used to manipulate the list so that the result is an array of numbers. The array of numbers is displayed as follows. Note that the first row corresponds to the zeros of the Bessel function of order zero, the second row corresponds to the zeros of the Bessel function of order one, and so on. An alternative approach would be to save a Maple scratchpad as `BesselTable`.

In order to use this file, we first load the `readdata` command by entering `readlib(readdata)` and then load the `BesselTable` file, naming the resulting array `Alpha_Array`.

```
> readlib(readdata):
  Alpha_Array:=readdata(`BesselTable`,float,8);
```

$Alpha_Array := $ [[2.404800000000000, 5.520100000000000, 8.653700000000001,
11.79200000000000, 14.93100000000000, 18.07100000000000,
21.21200000000000, 24.35200000000000], [3.831700000000000,
7.015600000000000, 10.17300000000000, 13.32400000000000,
16.47100000000000, 19.61600000000000, 22.76000000000000,
25.90400000000000], [5.135600000000000, 8.417199999999999,
11.62000000000000, 14.79600000000000, 17.96000000000000,
21.11700000000000, 24.27000000000000, 27.42100000000000],
[6.380200000000000, 9.760999999999999, 13.01500000000000,
16.22300000000000, 19.40900000000000, 22.58300000000000,
25.74800000000000, 28.90800000000000], [7.588300000000000,
11.06500000000000, 14.37300000000000, 17.61600000000000,
20.82700000000000, 24.01900000000000, 27.19900000000000,
30.37100000000000], [8.771500000000000, 12.33900000000000,
15.70000000000000, 18.98000000000000, 22.21800000000000,
25.43000000000000, 28.62700000000000, 31.81200000000000],
[9.936100000000000, 13.58900000000000, 17.00400000000000,
20.32100000000000, 23.58600000000000, 26.82000000000000,
30.03400000000000, 33.23300000000000]]

We then define the function alpha(i,j) that represents the *j*th zero of the Bessel function of order *i*.

```
> alpha:=(i,j)->Alpha_Array[i+1][j]:
```

Partial Differential Equations

12

In the previous chapters, we have seen that many physical and mathematical situations are described by ordinary differential equations. We have also seen that others are described by partial differential equations. In this chapter, we investigate one way to solve some partial differential equations, the method of separation of variables, which was mostly developed by Fourier in his study of the heat equation.

12.1 INTRODUCTION TO PARTIAL DIFFERENTIAL EQUATIONS AND SEPARATION OF VARIABLES

We begin our study of partial differential equations with an introduction of some of the terminology associated with the topic. A **linear second-order partial differential equation** in the two independent variables x and y has the form

$$Au_{xx} + Bu_{xy} + Cu_{yy} + Du_x + Eu_y + Fu = G(x, y),$$

where A, B, C, D, E, and F are constants and the solution is $u(x, y)$. If $G(x, y) = 0$ for all x and y, then we say that the equation is **homogeneous**. Otherwise, the equation is **nonhomogeneous**.

EXAMPLE: Classify the following partial differential equations: (a) $u_{xx} + u_{yy} = u$ and (b) $u_{xx} + uu_x = x$.

SOLUTION: (a) This equation satisfies the form of the linear second-order partial differential equation with $A = C = 1$, $F = -1$, and $B = D = E = 0$. Because $G(x, y) = 0$, the equation is homogeneous. (b) This equation is nonlinear because the coefficient of u_x is not a constant. It is also nonhomogeneous because $G(x, y) = x$.

A method that can be used to solve linear partial differential equations is called **separation of variables** (or the **product method**). The goal of the method of separation of variables is to transform the partial differential equation into a system of ordinary differential equations by separating the variables. Suppose that the function $u(x, y)$ is a solution of a partial differential equation in the independent variables x and y. In separating variables, we assume that u can be written as the product of a function of x and a function of y. Hence,

$$u(x, y) = X(x)Y(y),$$

and we substitute this product into the partial differential equation to determine $X(x)$ and $Y(y)$. Of course, to substitute into the differential equation, we must be able to differentiate this product. However, this is accomplished by following the differentiation rules of multivariate calculus. For example, using the following shorthand notation for convenience, we have

$$u_x = X'Y, u_{xx} = X''Y, u_{xy} = X'Y', u_y = XY', \text{ and } u_{yy} = XY'',$$

where X' represents $\dfrac{d}{dx}X(x)$ and Y' represents $\dfrac{d}{dy}Y(y)$. After these substitutions are made, if the equation is separable, then we can obtain an ordinary differential equation for X and an ordinary differential equation for Y. These two equations are then solved to find $X(x)$ and $Y(y)$.

EXAMPLE: Use separation of variables to solve $xu_x = u_y$.

SOLUTION: If $u(x, y) = X(x)Y(y)$, then $u_x = X'Y$ and $u_y = XY'$. The equation then becomes

$$xX'Y = XY',$$

which can be written as the separated equation

$$\frac{xX'}{X} = \frac{Y'}{Y}.$$

Notice that the left-hand side of the equation is a function of x, while the right-hand side is a function of y. Hence, the only way that this situation can be true is for $\dfrac{xX'}{X}$ and $\dfrac{Y'}{Y}$ both to be constant. Therefore,

$$\frac{xX'}{X} = \frac{Y'}{Y} = k,$$

so we obtain the ordinary differential equations $xX' - kX = 0$ and $Y' - kY = 0$. We find x with `dsolve`.

> `step_1:=dsolve(x*diff(X(x),x)-k*X(x)=0,X(x));`

$step\_1 := X(x) = x^k \_C1$

Similarly, we find Y.

> `step_2:=dsolve(diff(Y(y),y)-k*Y(y)=0,Y(y));`

$step\_2 := Y(y) = e^{ky} \_C1$

Therefore, the solution is $u(x, y) = X(x)Y(y) = (C_1 x^k)(C_2 e^{ky}) = C_3 x^k e^{ky}$, where k and C_3 are arbitrary constants. (Notice that the first equation is a Cauchy-Euler equation, so we could have used the techniques covered in Section 6.1 to solve it. Similarly, we could have used an auxiliary equation to solve the second equation.)

12.2 THE ONE-DIMENSIONAL HEAT EQUATION

One of the more important differential equations is the **heat equation**:

$$u_t = c^2 u_{xx}.$$

In one spatial dimension, the solution of the heat equation represents the temperature (at any position x and any time t) in a thin rod or wire of length p. Since the rate at which heat flows through the rod depends on the material that makes up the rod, the constant c^2 that is related to the thermal diffusivity of the material is included in the heat equation. Several different situations can be considered when determining the temperature in the rod. The ends of the wire can be held at a constant temperature, the ends may be insulated, or there can be a combination of these situations.

☰ The Heat Equation with Homogeneous Boundary Conditions

The first problem that we investigate is the situation in which the temperature at the ends of the rod is constantly kept at zero and the initial temperature distribution in the rod is represented as the function $f(x)$. Hence, the fixed-end zero temperature is given in the boundary conditions (ii) while the initial heat distribution is given in (iii). Because the temperature is zero at the endpoints, we say that the problem has homogeneous boundary conditions. We call problems of this type **initial boundary value problems**, because they include initial as well as boundary conditions:

$$\begin{cases} u_t = c^2 u_{xx}, 0 < x < p, t > 0 \\ u(0, t) = 0, u(p, t) = 0, t > 0 . \\ u(x, 0) = f(x), 0 < x < p \end{cases}$$

We solve this problem through separation of variables by assuming that

$$u(x, t) = X(x)T(t).$$

Substitution into the differential equation yields

$$\frac{T'}{c^2 T} = \frac{X''}{X} = -\lambda^2,$$

where $-\lambda^2$ is the separation constant. Separating the variables, we have the two equations $T' + c^2\lambda^2 T = 0$ and $X'' + \lambda^2 X = 0$. Now that we have successfully separated the variables, we turn our attention to the homogeneous boundary conditions given. In terms of the functions $X(x)$ and $T(t)$, these boundary conditions become $u(0, t) = X(0)T(t) = 0$ and $u(p, t) = X(p)T(t) = 0$.

In each case, we must avoid setting $T(t) = 0$ for all t, because if this were the case, our solution would be the trivial solution $u(x, t) = X(x)T(t) = 0$. Therefore, we have the boundary conditions $X(0) = 0$ and $X(p) = 0$, so we solve the eigenvalue problem

$$\begin{cases} X'' + \lambda^2 X = 0 \\ X(0) = 0, X(p) = 0 \end{cases}.$$

The eigenvalues of this problem are

$$\lambda = \lambda_n = \left(\frac{n\pi}{p}\right)^2, n = 1, 2, \ldots.$$

with corresponding eigenfunctions

$$X(x) = X_n(x) = \sin\frac{n\pi x}{p}, n = 1, 2, \ldots.$$

Similarly, we solve $T' + c^2\lambda^2 T = 0$, which is a first-order equation with characteristic equation $m + c^2\lambda^2 = 0$. Since $m = -c^2\lambda^2$, a general solution is

$$T(t) = Ae^{-c^2\lambda^2 t},$$

where A is an arbitrary constant and $\lambda = \lambda_n = \left(\frac{n\pi}{p}\right)^2, n = 1, 2, \ldots$. Because $X(x)$ and $T(t)$ both depend on n, the solution $u(x, t) = X(x)T(t)$ does as well. Hence,

$$u_n(x, t) = X_n(x)T_n(t) = c_n \sin\frac{n\pi x}{p}e^{-c^2\lambda^2 t},$$

where we have replaced the constant A by one that depends on n. To find the value of c_n, we must apply the initial condition $u(x, 0) = f(x)$. Because

$$u_n(x, 0) = c_n \sin \frac{n\pi x}{p} e^{-c^2\lambda^2(0)} = c_n \sin \frac{n\pi x}{p}$$

is satisfied only by functions of the form $\sin \frac{\pi x}{p}, \sin \frac{2\pi x}{p}, \sin \frac{3\pi x}{p}, \ldots$, we use the principle of superposition to state that

$$u(x, t) = \sum_{n=1}^{\infty} c_n \sin \frac{n\pi x}{p} e^{-c^2\lambda^2 t}$$

is also a solution of the problem. Then, when we apply the initial condition $u(x, 0) = f(x)$, we find that

$$u(x, 0) = \sum_{n=1}^{\infty} c_n \sin \frac{n\pi x}{p} e^{-c^2\lambda^2(0)} = \sum_{n=1}^{\infty} c_n \sin \frac{n\pi x}{p} = f(x).$$

Therefore, c_n represents the Fourier sine series coefficients for $f(x)$ that are given by

$$c_n = \frac{2}{p} \int_0^p f(x) \sin \frac{n\pi x}{p} dx, n = 1, 2, \ldots.$$

EXAMPLE: Solve

$$\begin{cases} u_t = u_{xx}, 0 < x < 1, t > 0 \\ u(0, t) = 0, u(1, t) = 0, t > 0 \\ u(x, 0) = 50, 0 < x < 1 \end{cases}$$

SOLUTION: In this case, $c = 1$, $p = 1$, and $f(x) = 50$. Hence,

$$u(x, t) = \sum_{n=1}^{\infty} c_n \sin n\pi x e^{-\lambda^2 t},$$

where $c_n = \dfrac{2}{1} \displaystyle\int_0^1 50 \sin n\pi x \, dx, n = 1, 2, \ldots$, and $\lambda = \lambda_n = n\pi$.

We use `int` to compute c_n.

```
> 2*int(50*sin(n*Pi*x),x=0..1);
```

$$-100 \frac{\cos(n\pi)}{n\pi} + 100 \frac{1}{n\pi}$$

Simplifying this result gives us

$$c_n = \frac{2}{1}\int_0^1 50\sin n\pi x\,dx = -\frac{100}{n\pi}(\cos n\pi - 1) = -\frac{100}{n\pi}[(-1)^n - 1].$$

Therefore, because $c_n = 0$ if n is even, we write u as

$$u(x, t) = \sum_{n=1}^{\infty} \frac{200}{(2n-1)\pi} \sin(2n-1)\pi x e^{-(2n-1)^2\pi^2 t}.$$

To sketch the graph of u at various times, we begin by defining `approx(k)` that corresponds to the kth partial sum of $u(x, t)$.

```
> approx:='approx':t:='t':x:='x':
  approx:=proc(k) option remember;
   approx(k-1)+200/((2*k-1)*Pi)*sin((2*k-1)*Pi*x)*
    exp(-(2*k-1)^2*Pi^2*t)
   end:
  approx(1):=200/Pi*sin(Pi*x)*exp(-Pi^2*t):
```

For example, entering

```
> approx(10);
```

$$200\frac{\sin(\pi x)e^{-\pi^2 t}}{\pi} + \frac{200}{3}\frac{\sin(3\pi x)e^{-9\pi^2 t}}{\pi} + 40\frac{\sin(5\pi x)e^{-25\pi^2 t}}{\pi}$$

$$+ \frac{200}{7}\frac{\sin(7\pi x)e^{-49\pi^2 t}}{\pi} + \frac{200}{9}\frac{\sin(9\pi x)e^{-81\pi^2 t}}{\pi}$$

$$+ \frac{200}{11}\frac{\sin(11\pi x)e^{-121\pi^2 t}}{\pi} + \frac{200}{13}\frac{\sin(13\pi x)e^{-169\pi^2 t}}{\pi}$$

$$+ \frac{40}{3}\frac{\sin(15\pi x)e^{-225\pi^2 t}}{\pi} + \frac{200}{17}\frac{\sin(17\pi x)e^{-289\pi^2 t}}{\pi}$$

$$+ \frac{200}{19}\frac{\sin(19\pi x)e^{-361\pi^2 t}}{\pi}$$

returns the tenth partial sum of the series.

We then use the following `for` loop to graph the tenth partial sum for $t = 0, 0.1, 0.2, 0.3, 0.4$, and 0.5. Notice what happens to the temperature as t increases. Because of the zero fixed-end boundary conditions, the temperature eventually becomes zero at every point in the wire.

```
> for t from 0 to 0.5 by 0.1 do
    plot(approx(5),x=0..1,0..60) od;
```

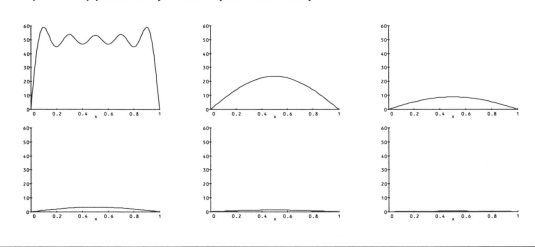

≡ Nonhomogeneous Boundary Conditions

The ability to apply the method of separation of variables depends on the presence of homogeneous boundary conditions, as we saw in the previous problem. However, with the heat equation, the temperature at the endpoints may not be held constantly at zero. Instead, the temperature at the left-hand endpoint is T_0, and at the right-hand endpoint it is T_1. Mathematically, we state these **nonhomogeneous boundary conditions** as $u(0, t) = T_0$ and $u(p, t) = T_1$, so we would be faced with solving the problem

$$\begin{cases} u_t = c^2 u_{xx}, 0 < x < p, t > 0 \\ u(0, t) = T_0, u(p, t) = T_1, t > 0 \\ u(x, 0) = f(x), 0 < x < p \end{cases}.$$

In this case, we must modify the problem to introduce homogeneous boundary conditions to the problem. We do this by using the physical observance that as $t \to \infty$, the temperature in the wire does not depend on t. Hence,

$$\lim_{t \to \infty} u(x, t) = S(x),$$

where we call $S(x)$ the **steady-state temperature**. Therefore, we let

$$u(x, t) = v(x, t) + S(x),$$

where $v(x, t)$ is called the **transient temperature**. We use these two functions to obtain two problems that we can solve. In order to substitute $u(x, t)$ into the heat equation $u_t = c^2 u_{xx}$, we calculate the derivatives $u_t(x, t) = v_t(x, t) + 0$ and $u_{xx}(x, t) = v_{xx}(x, t) + S''(x)$. Hence,

$$u_t = c^2 u_{xx}$$

and

$$v_t = c^2 v_{xx} + c^2 S'',$$

so we have the two equations $v_t = c^2 v_{xx}$ and $S'' = 0$. We then consider the boundary conditions. Because $u(0, t) = v(0, t) + S(0) = T_0$ and $u(p, t) = v(p, t) + S(p) = T_1$, we can choose the boundary conditions for S to be the nonhomogeneous conditions $S(0) = T_0$ and $S(p) = T_1$ and the boundary conditions for $v(x, t)$ to be the homogeneous conditions $v(0, t) = 0$ and $v(p, t) = 0$. Of course, we have failed to include the initial temperature. Applying this condition, we have $u(x, 0) = v(x, 0) + S(x) = f(x)$, so the initial condition for v is $v(x, 0) = f(x) - S(x)$.

Therefore, we have two problems, one for v with homogeneous boundary conditions and one for S with nonhomogeneous boundary conditions. These two problems are stated as follows:

$$\begin{cases} S'' = 0, 0 < x < p \\ S(0) = T_0, S(p) = T_1 \end{cases}$$

and

$$\begin{cases} v_t = c^2 v_{xx}, 0 < x < p, t > 0 \\ v(0, t) = 0, v(p, t) = 0, t > 0 \\ v(x, 0) = f(x) - S(x), 0 < x < p \end{cases}$$

Because S is needed in the determination of v, we begin by finding S. A general solution of $S'' = 0$ is $S(x) = c_1 + c_2 x$. Then $S(0) = c_1 = T_0$ and $S(p) = T_0 + c_2 p = T_1$, so $c_2 = \dfrac{T_1 - T_0}{p}$. Hence,

$$S(x) = T_0 + \left(\frac{T_1 - T_0}{p} \right) x.$$

We are now able to find $v(x, t)$ by solving the heat equation with homogeneous boundary conditions for v. Since we solved this problem at the beginning of this section, we will not go through the separation of variables procedure. Instead, we use the formula that we derived there using the initial temperature $f(x) - S(x)$. Therefore,

$$v(x, t) = \sum_{n=1}^{\infty} c_n \sin \frac{n \pi x}{p} e^{-c^2 \lambda^2 t},$$

where $v(x,0) = \sum_{n=1}^{\infty} c_n \sin \frac{n\pi x}{p} = f(x) - S(x)$. Hence, c_n represents the Fourier sine series coefficients for the function $f(x) - S(x)$ given by

$$c_n = \frac{2}{p} \int_0^p (f(x) - S(x)) \sin \frac{n\pi x}{p} dx, n = 1, 2, \ldots.$$

EXAMPLE: Solve

$$\begin{cases} u_t = u_{xx}, 0 < x < 1, t > 0 \\ u(0, t) = 10, u(1, t) = 60, t > 0 \;. \\ u(x, 0) = 10, 0 < x < 1 \end{cases}$$

SOLUTION: In this case, $c = 1$, $p = 1$, $T_0 = 10$, $T_1 = 60$, and $f(x) = 10$. Hence, the steady-state solution is

$$S(x) = T_0 + \left(\frac{T_1 - T_0}{p}\right) x = 10 + \left(\frac{60 - 10}{1}\right) x = 10 + 50x.$$

Then, the initial transient temperature is

$$v(x, 0) = 10 - (10 + 50x) = -50x,$$

so that the series coefficients are given by $c_n = \frac{2}{1} \int_0^1 (-50x) \sin n\pi x\, dx, n = 1, 2, \ldots$, which is computed as follows with `int`.

```
> 2*int(-50*x*sin(n*Pi*x),x=0..1);
```

$$100 \frac{-\sin(n\pi) + n\pi \cos(n\pi)}{n^2 \pi^2}$$

Simplifying, we see that $c_n = \frac{100(-1)^n}{n\pi}, n = 1, 2, \ldots$, so the transient temperature is

$$v(x, t) = \sum_{n=1}^{\infty} c_n \sin \frac{n\pi x}{p} e^{-c^2 \lambda^2 t} = \sum_{n=1}^{\infty} \frac{100(-1)^n}{n\pi} \sin n\pi x e^{-n^2 \pi^2 t}.$$

Therefore,

$$u(x, t) = v(x, t) + S(x) = \sum_{n=1}^{\infty} \frac{100(-1)^n}{n\pi} \sin n\pi x e^{-n^2 \pi^2 t} + 10 + 50x.$$

In the same manner as in the previous example, we define `approx(k)` to be the *k*th partial sum of the series for $u(x, t)$.

```
> approx:='approx':t:='t':x:='x':
  approx:=proc(k) option remember;
   approx(k-1)+100*(-1)^k/(k*Pi)*sin(k*Pi*x)*
    exp(-k^2*Pi^2*t)
   end:
  approx(0):=10+50*x:
```

We then graph the fifteenth partial sum for $t = 0.025$, 0.05, 0.1, 0.2, and 0.4. Notice that as t increases, u approaches the steady-state temperature.

```
> t_vals:=0.025,0.05,0.1,0.2,0.4:
  to_plot:={seq(subs(t=i,approx(15)),i=t_vals)}:
  plot(to_plot,x=0..1);
```

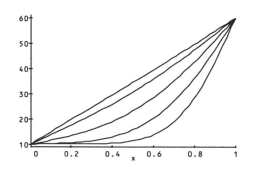

☰ Insulated Boundary

Another important situation concerning the flow of heat in a wire involves insulated ends. In this case, heat is not allowed to escape from the ends of the wire. Mathematically, we express these boundary conditions as $u_x(0, t) = 0$ and $u_x(p, t) = 0$ because the rate at which the heat changes along the x-axis at the endpoints $x = 0$ and $x = p$ is zero. Therefore, if we want to determine the temperature in a wire of length p with insulated ends, we solve the initial-boundary value problem

$$\begin{cases} u_t = c^2 u_{xx}, 0 < x < p, t > 0 \\ u_x(0, t) = 0, u_x(p, t) = 0, t > 0 \, . \\ u(x, 0) = f(x), 0 < x < p \end{cases}$$

Notice that the boundary conditions are homogeneous, so we can use separation of variables to find $u(x,t) = X(x)T(t)$. By following the steps taken in the solution of the problem with homogeneous boundary conditions, we obtain the ordinary differential equations $T' + c^2\lambda^2 T = 0$ and $X'' + \lambda^2 X = 0$. However, when we consider the boundary conditions $u_x(0,t) = X'(0)T(t) = 0$ and $u_x(p,t) = X'(p)T(t) = 0$, we see that to avoid letting $T(t) = 0$ for all t to obtain the trivial solution. Hence, we have $X'(0) = 0$ and $X'(p) = 0$, so we solve the problem

$$\begin{cases} X'' + \lambda^2 X = 0, 0 < x < p \\ X'(0) = 0, X'(p) = 0 \end{cases}$$

to find $X(x)$. The eigenvalues and corresponding eigenfunctions of this problem are

$$\lambda_n = \begin{cases} 0, n = 0 \\ \left(\dfrac{n\pi}{p}\right)^2, n = 1, 2, \ldots \end{cases}$$

and

$$y_n(x) = \begin{cases} 1, n = 0 \\ \cos\dfrac{n\pi x}{p}, n = 1, 2, \ldots \end{cases}.$$

Next, we solve the equation $T' + c^2\lambda^2 T = 0$ for the λ_n given earlier. First, for $\lambda = \lambda_0 = 0$, we have the equation $T' = 0$ that has the solution $T(t) = A_0$, where A_0 is a constant. Therefore, when $\lambda = \lambda_0 = 0$, the solution is the product

$$u_0(x,t) = X_0(x)T_0(t) = a.$$

Then, for $\lambda_n = \dfrac{n\pi x}{p}$, we solve $T' + c^2\lambda^2 T = 0$ that has as a general solution $T(t) = T_n(t) = a_n e^{-\lambda_n^2 t}$. For these eigenvalues, we have the solution

$$u_n(x,t) = X_n(x)T_n(t) = a_n \cos\frac{n\pi x}{p} e^{-\lambda_n^2 t}.$$

Therefore, by the principle of superposition, the solution is

$$u(x,t) = A_0 + \sum_{n=1}^{\infty} a_n \cos\frac{n\pi x}{p} e^{-\lambda_n^2 t}.$$

Application of the initial temperature yields

$$u(x, 0) = A_0 + \sum_{n=1}^{\infty} a_n \cos \frac{n \pi x}{p} = f(x),$$

which is the Fourier cosine series for $f(x)$, where the coefficient A_0 is equivalent to $\frac{1}{2} a_0$ in the original Fourier series given in Section 11.2. Therefore,

$$A_0 = \frac{1}{2} a_0 = \frac{1}{2} \frac{2}{p} \int_0^p f(x) \, dx = \frac{1}{p} \int_0^p f(x) \, dx$$

and

$$a_n = \frac{2}{p} \int_0^p f(x) \cos \frac{n \pi x}{p} \, dx, n = 1, 2, \ldots$$

EXAMPLE: Solve

$$\begin{cases} u_t = u_{xx}, 0 < x < \pi, t > 0 \\ u_x(0, t) = 0, u_x(\pi, t) = 0, t > 0 \, . \\ u(x, 0) = x, 0 < x < \pi \end{cases}$$

SOLUTION: In this case, $p = \pi$ and $c = 1$. Since we need the Fourier cosine series coefficients for $f(x) = x$, we refer back to Section 11.2 where we found that

$$A_0 = \frac{1}{2} a_0 = \frac{1}{\pi} \int_0^\pi x \, dx = \frac{1}{\pi} \left[\frac{x^2}{2} \right]_0^\pi = \frac{\pi}{2}$$

and

$$a_n = \frac{2}{\pi} \int_0^\pi x \cos \frac{n \pi x}{\pi} \, dx = \frac{2}{\pi n^2} [(-1)^n - 1], n = 1, 2, \ldots.$$

Therefore, the solution is

$$u(x, t) = \frac{\pi}{2} - \sum_{n=1}^{\infty} \frac{4}{(2n - 1)^2 \pi} \cos(2n - 1) x e^{-(2n-1)^2 \pi^2 t},$$

where we have used the fact that $a_n = 0$ if n is even.

As in the previous two examples, we define `approx(k)` to be the kth partial sum of the series for $u(x, t)$.

```
> approx:='approx':t:='t':x:='x':
  approx:=proc(k) option remember;
   approx(k-1)-4/((2*k-1)^2*Pi)*cos((2*k-1)*x)*
    exp(-(2*k-1)^2*Pi^2*t)
   end:
  approx(0):=Pi/2:
```

We then graph the fifteenth partial sum for various values of t.

```
> t_vals:=0.025,0.05,0.1,0.4:
  to_plot:={seq(subs(t=i,approx(15)),i=t_vals)}:
  plot(to_plot,x=0..Pi);
```

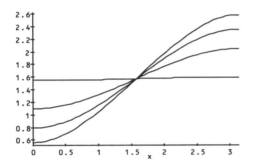

Notice that the temperature eventually becomes $A_0 = \dfrac{\pi}{2}$ throughout the wire. Temperatures to the left of $x = \dfrac{\pi}{2}$ increase while those to the right decrease.

12.3 THE ONE-DIMENSIONAL WAVE EQUATION

The one-dimensional **wave equation** is important in solving an interesting problem. Suppose that we pluck a string (i.e., a guitar or violin string) of length p and constant mass density that is fixed at each end. The question that we need to answer is, What is the position of the string at a particular instance of time? We answer this question by modeling the physical situation with a differential equation—namely, the wave equation in one spatial variable. We will not go through this derivation as we did with the heat equation, but we point out that it is based on determining the forces that act on a small segment of the string and applying Newton's Second Law of Motion. The partial differential equation that is found is

$$c^2 u_{xx} = u_{tt}$$

and is called the (one-dimensional) **wave equation**. In this equation $c^2 = \dfrac{T}{\rho}$, where T is the tension of the string and ρ is the mass of the string per unit length. The solution $u(x, t)$ represents the height of the string above the x-axis at time t. To determine u we must describe the boundary and initial conditions that model the physical situation. At the ends of the string, the height above the x-axis is zero, so we use the homogeneous conditions $u(0, t) = 0$ and $u(p, t) = 0$ for $t > 0$. We also must describe the initial position as well as the initial velocity of the string. This is done with $u(x, 0) = f(x)$ and $u_t(x, 0) = g(x)$ for $0 < x < p$.

Therefore, we determine the height (or displacement) of the string with the initial boundary value problem

$$\begin{cases} c^2 u_{xx} = u_{tt}, 0 < x < p, t > 0 \\ u(0, t) = 0, u(p, t) = 0, t > 0 \\ u(x, 0) = f(x), u_t(x, 0) = g(x), 0 < x < p \end{cases}$$

(Notice that the wave equation requires two initial conditions while the heat equation needed only one. This is due to the fact that there is a second derivative with respect to t, while there is only one derivative with respect to t in the heat equation.)

This problem is solved through separation of variables by assuming that $u(x, t) = X(x)T(t)$. Substitution into the wave equation yields

$$c^2 X''T = XT''$$

and

$$\frac{X''}{X} = \frac{T''}{c^2 T} = -\lambda^2,$$

so we obtain the two second-order ordinary differential equations $X'' + \lambda^2 X = 0$ and $T'' + c^2\lambda^2 T = 0$.

At this point, we solve the equation that involves the homogeneous boundary conditions. As was the case with the heat equation, the boundary conditions in terms of $u(x, t) = X(x)T(t)$ are $u(0, t) = X(0)T(t) = 0$ and $u(p, t) = X(p)T(t) = 0$, so we have $X(0) = 0$ and $X(p) = 0$. Therefore, we determine $X(x)$ by solving the eigenvalue problem

$$\begin{cases} X'' + \lambda^2 X = 0 \\ X(0) = 0, X(p) = 0 \end{cases}.$$

The eigenvalues of this problem are

$$\lambda = \lambda_n = \left(\frac{n\pi}{p}\right)^2, n = 1, 2, \ldots.$$

with corresponding eigenfunctions

$$X(x) = X_n(x) = \sin\frac{n\pi x}{p}, n = 1, 2, \ldots.$$

Next, we solve the equation $T'' + c^2\lambda^2 T = 0$ using the eigenvalues given here. From our experience with second-order equations, a general solution is

$$T_n(t) = a_n \cos c\lambda_n t + b_n \sin c\lambda_n t = a_n \cos\frac{cn\pi t}{p} + b_n \sin\frac{cn\pi t}{p},$$

where the coefficients a_n and b_n must be determined. Putting this information together, we obtain

$$u_n(x, t) = \left(a_n \cos\frac{cn\pi t}{p} + b_n \sin\frac{cn\pi t}{p}\right)\sin\frac{n\pi x}{p},$$

so by the principle of superposition, we have

$$u(x, t) = \sum_{n=1}^{\infty}\left(a_n \cos\frac{cn\pi t}{p} + b_n \sin\frac{cn\pi t}{p}\right)\sin\frac{n\pi x}{p}.$$

Applying the initial position yields

$$u(x, 0) = \sum_{n=1}^{\infty} a_n \sin\frac{n\pi x}{p} = f(x),$$

so a_n is the Fourier sine series coefficient for $f(x)$ that is given by

$$a_n = \frac{2}{p}\int_0^p f(x)\sin\frac{n\pi x}{p}\,dx, n = 1, 2, \ldots.$$

To determine b_n, we must use the initial velocity. Therefore, we compute

$$u_t(x, t) = \sum_{n=1}^{\infty}\left(-a_n\frac{cn\pi}{p}\sin\frac{cn\pi t}{p} + b_n\frac{cn\pi}{p}\cos\frac{cn\pi}{p}\right)\sin\frac{n\pi x}{p}.$$

Then

$$u_t(x, 0) = \sum_{n=1}^{\infty} b_n \frac{cn\pi}{p} \sin \frac{n\pi x}{p} = g(x),$$

so $b_n \dfrac{cn\pi}{p}$ represents the Fourier sine series coefficient for $g(x)$, and this means that

$$b_n = \frac{p}{cn\pi} \frac{2}{p} \int_0^p g(x) \sin \frac{n\pi x}{p} \, dx = \frac{2}{cn\pi} \int_0^p g(x) \sin \frac{n\pi x}{p} \, dx, n = 1, 2, \ldots.$$

EXAMPLE: Solve the wave equation with $c = 1$ and $p = 1$ subject to the indicated conditions:

$$\begin{cases} \dfrac{\partial^2 u}{\partial x^2} = \dfrac{\partial^2 u}{\partial t^2}, 0 < x < 1, t > 0 \\ u(0, t) = 0, u(1, t) = 0, t \geq 0 \\ u(x, 0) = \sin(\pi x), \left. \dfrac{\partial u}{\partial t} \right|_{t=0} = 3x + 1, 0 \leq x \leq 1 \end{cases}$$

SOLUTION: The appropriate parameters and initial condition functions are entered as follows:

```
> f:=x->sin(Pi*x):
  g:=x->3*x+1:
```

Next, the functions to determine the coefficients in the series approximation of the solution $u(x, t)$ are defined in **a** and **b**. The use of `evalf/int` in these functions causes the calculations to be performed more quickly in most cases, even though the integrations could be performed exactly, if desired. We then use **array** and **seq** to calculate the first 10 coefficients.

```
> readlib(`evalf/int`):
  a:=proc(n) option remember;
   evalf(2*`evalf/int`(f(x)*sin(n*Pi*x),x=0..1))
   end:
  b:=proc(n) option remember;
   evalf(2/(n*Pi)*`evalf/int`(g(x)*sin(n*Pi*x),x=0..1))
   end:
```

```
> array([seq([n,a(n),b(n)],n=1..10)]);
```

$$
\begin{bmatrix}
1 & 1.000000000 & 1.013211836 \\
2 & -.3146904858 \times 10^{-16} & -.1519817755 \\
3 & .2960594732 \times 10^{-15} & .1125790929 \\
4 & .2591358472 \times 10^{-16} & -.03799544385 \\
5 & -.2467162276 \times 10^{-16} & .04052847345 \\
6 & -.1586682824 \times 10^{-15} & -.01688686394 \\
7 & -.4852085812 \times 10^{-15} & .02067779258 \\
8 & .199504843 \times 10^{-15} & -.009498860961 \\
9 & -.1171902082 \times 10^{-15} & .01250878810 \\
10 & .5913456454 \times 10^{-15} & -.006079271017
\end{bmatrix}
$$

The function **u** defined next computes the nth term in the series expansion. Hence, **approx** determines the approximation of order k by summing the first k terms of the expansion as illustrated with **approx(2)**.

```
> u:=n->(a(n)*cos(n*Pi*t)+b(n)*sin(n*Pi*t))*sin(n*Pi*x):
  approx:=proc(k) option remember;
   approx(k-1)+u(k)
   end:
  approx(1):=u(1):
```

```
> approx(2);
```

$$
(1.000000000 \cos(\pi t) + 1.013211836 \sin(\pi t)) \sin(\pi x)
$$
$$
+ (-3.146904858 \times 10^{-16} \cos(2\pi t) - .1519817755 \sin(2\pi t)) \sin(2\pi x)
$$

To view the motion of the string, we use **approx(5)**, the fifth partial sum of the series, to approximate the motion of the spring together with **animate**, which is contained in the **plots** package, to sketch the graph on the interval $[0, 1]$ for 20 equally spaced values of t between 0 and 2. Several frames from the resulting animation are displayed.

```
> with(plots):
  animate(approx(5),x=0..1,t=0..2,frames=20);
```

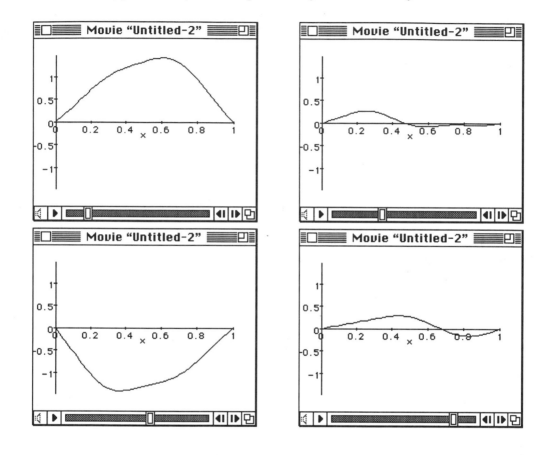

☰ *D'Alembert's Solution*

An interesting version of the wave equation is to consider a string of infinite length. Therefore, the boundary conditions are no longer of importance. Instead, we simply work with the wave equation with the initial position and velocity functions. To solve the problem

$$\begin{cases} c^2 u_{xx} = u_{tt}, \ -\infty < x < \infty, t > 0 \\ u(x,0) = f(x), u_t(x,0) = g(x) \end{cases}$$

we use the change of variables $r = x + ct$ and $s = x - ct$. Using the chain rule, we compute the derivatives u_{xx} and u_{tt} in terms of the variables r and s:

$$u_x = u_r r_x + u_s s_x = u_r + u_s,$$
$$u_{xx} = (u_r + u_s)_r r_x + (u_r + u_s)_s s_x = u_{rr} + 2u_{rs} + u_{ss},$$
$$u_t = u_r r_t + u_s s_t = cu_r - cu_s = c(u_r - u_s),$$

and

$$u_{tt} = c[(u_r - u_s)_r r_t + (u_r - u_s)_s s_t] = c^2[u_{rr} - 2u_{rs} + u_{ss}].$$

Substitution into the wave equation yields

$$c^2 u_{xx} = u_{tt}$$
$$c^2[u_{rr} + 2u_{rs} + u_{ss}] = c^2[u_{rr} - 2u_{rs} + u_{ss}]$$
$$4c^2 u_{rs} = 0$$
$$u_{rs} = 0.$$

The partial differential equation $u_{rs} = 0$ can be solved by first integrating with respect to s to obtain

$$u_r = f(r),$$

where $f(r)$ is an arbitrary function of r. Then, integrating with respect to s, we have

$$u(r, s) = F(r) + G(s),$$

where F is an anti-derivative of f and G is an arbitrary function of s. Returning to our original variables then gives us

$$u(x, t) = F(x + ct) + G(x + ct).$$

The functions F and G are determined by the initial conditions that indicate that

$$u(x, 0) = F(x) + G(x) = f(x)$$

and

$$u_t(x, 0) = cF'(x) - cG'(x) = g(x).$$

We can rewrite the second equation by integrating to obtain

$$F(x) - G'(x) = \frac{g(x)}{c}$$

$$F(x) - G(x) = \frac{1}{c} \int_0^x g(v) \, dv.$$

Therefore, we solve the system of equations

$$F(x) + G(x) = f(x)$$

$$F(x) - G(x) = \frac{1}{c} \int_0^x g(v) \, dv.$$

Adding these equations yields

$$F(x) = \frac{1}{2} \left[f(x) + \frac{1}{c} \int_0^x g(v) \, dv \right],$$

and subtracting gives us

$$G(x) = \frac{1}{2} \left[f(x) - \frac{1}{c} \int_0^x g(v) \, dv \right].$$

Therefore,

$$F(x + ct) = \frac{1}{2} \left[f(x + ct) + \frac{1}{c} \int_0^{x+ct} g(v) \, dv \right]$$

and

$$G(x - ct) = \frac{1}{2} \left[f(x - ct) - \frac{1}{c} \int_0^{x-ct} g(v) \, dv \right],$$

so the solution is

$$u(x, t) = \frac{1}{2} [f(x + ct) + f(x - ct)] + \frac{1}{2c} \int_{x-ct}^{x+ct} g(v) \, dv.$$

EXAMPLE: Solve $\begin{cases} u_{xx} = u_{tt}, -\infty < x < \infty, t > 0 \\ u(x,0) = \dfrac{2}{1+x^2}, u_t(x,0) = 0 \end{cases}$.

SOLUTION: Since $c = 1$, $f(x) = \dfrac{1}{1+x^2}$, and $g(x) = 0$, we have the solution

$$u(x,t) = \frac{1}{2}[f(x+t) + f(x-t)] = \frac{1}{2}\left[\frac{2}{1+(x+t)^2} + \frac{2}{1+(x-t)^2}\right]$$

$$= \frac{1}{1+(x+t)^2} + \frac{1}{1+(x-t)^2}.$$

After we define $u(x,t)$, we illustrate the motion of the string of infinite length using **animate**. Several frames from the resulting animation are shown now.

```
> u:=(x,t)->1/(1+(x+t)^2)+1/(1+(x-t)^2):
  with(plots):
  animate(u(x,t),x=-20..20,t=0..15,frames=60);
```

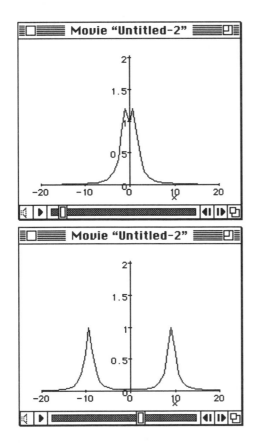

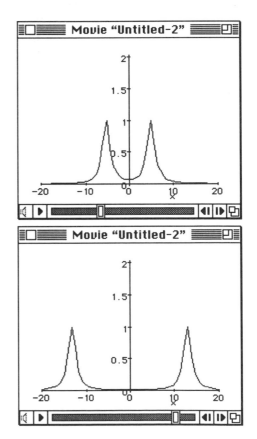

D'Alembert's solution is sometimes referred to as the **traveling wave solution** due to the behavior of its graph. The waves appear to move in opposite directions along the x-axis as t increases, as we can see in the animation.

12.4 PROBLEMS IN TWO DIMENSIONS: LAPLACE'S EQUATION

Laplace's equation, often called the **potential equation**, is given by

$$u_{xx} + u_{yy} = 0$$

in rectangular coordinates and is one of the most useful partial differential equations in that it arises in many fields of study. These include fluid flows as well as electrostatic and gravitational potential. Because the potential does not depend on time, no initial condition is required. We are faced with solving a pure boundary value problem when working with Laplace's equation. The boundary conditions can be stated in different forms. If the value of the solution is given around the boundary of the region, then the boundary value problem is called a **Dirichlet problem**, whereas if the normal derivative of the solution is given around the boundary, the problem is known as a **Neumann problem**. We now investigate the solutions to Laplace's equation in a rectangular region by, first, stating the general form of the Dirichlet problem:

$$\begin{cases} u_{xx} + u_{yy} = 0, 0 < x < a, 0 < y < b \\ u(x,0) = f_1(x), u(x,b) = f_2(x), 0 < x < a \\ u(0,y) = g_1(y), u(a,y) = g_2(y), 0 < y < b \end{cases}.$$

This boundary value problem is solved through separation of variables. We begin by considering the problem

$$\begin{cases} u_{xx} + u_{yy} = 0, 0 < x < a, 0 < y < b \\ u(x,0) = 0, u(x,b) = f(x), 0 < x < a \\ u(0,y) = 0, u(a,y) = 0, 0 < y < b \end{cases}.$$

In this case, we assume that

$$u(x,y) = X(x)Y(y),$$

so substitution into Laplace's equation yields

$$X''Y + XY'' = 0$$

$$\frac{X''}{X} = -\frac{Y''}{Y} = -\lambda,$$

where $-\lambda$ is the separation constant. Therefore, we have the ordinary differential equations $X'' + \lambda X = 0$ and $Y'' - \lambda Y = 0$. Notice that the boundary conditions along the lines $x = 0$ and $x = a$ are homogeneous. In fact, because $u(0, y) = X(0)Y(y) = 0$ and $u(a, y) = X(a)Y(y) = 0$, we have $X(0) = 0$ and $X(a) = 0$. Therefore, we first solve the eigenvalue problem

$$\begin{cases} X'' + \lambda X = 0 \\ X(0) = 0, X(a) = 0 \end{cases}.$$

The eigenvalues and corresponding eigenfunctions are $\lambda_n = \left(\dfrac{n\pi}{a}\right)^2, n = 1, 2, \ldots,$ and $X_n(x) = \sin\dfrac{n\pi x}{a}, n = 1, 2, \ldots$. We then solve the equation $Y'' - \lambda^2 Y = 0$ using these eigenvalues. From our experience with second-order equations, we know that $Y_n(y) = a_n e^{\lambda_n y} + b_n e^{-\lambda_n y}$, which can be written in terms of the hyperbolic trigonometric functions as

$$Y(y) = Y_n(y) = A_n \cosh \lambda_n y + B_n \sinh \lambda_n y = A_n \cosh\frac{n\pi y}{a} + B_n \sinh\frac{n\pi y}{a}.$$

Then, using the boundary condition $u(x, 0) = X(x)Y(0) = 0$, which indicates that $Y(0) = 0$, we have

$$Y(0) = Y_n(0) = A_n \cosh 0 + B_n \sinh 0 = A_n = 0,$$

so $A_n = 0$ for all n. Therefore, $Y_n(y) = B_n \sinh \lambda y$, and the solution is

$$u_n(x, y) = B_n \sinh\frac{n\pi y}{a} \sin\frac{n\pi x}{a},$$

so by the principle of superposition,

$$u(x, y) = \sum_{n=1}^{\infty} B_n \sinh\frac{n\pi y}{a} \sin\frac{n\pi x}{a},$$

where the coefficients are determined with the boundary condition $u(x, b) = f(x)$. Substitution into the solution yields

$$u(x, b) = \sum_{n=1}^{\infty} B_n \sinh\frac{n\pi b}{a} \sin\frac{n\pi x}{a} = f(x),$$

where $B_n \sinh\dfrac{n\pi b}{a}$ represents the Fourier sine series coefficients given by

$$B_n \sinh\frac{n\pi b}{a} = \frac{2}{a} \int_0^a f(x) \sin\frac{n\pi x}{a} dx$$

so

$$B_n = \frac{2}{a \sinh \frac{n\pi b}{a}} \int_0^a f(x) \sin \frac{n\pi x}{a} \, dx.$$

EXAMPLE: Solve

$$\begin{cases} u_{xx} + u_{yy} = 0, 0 < x < 1, 0 < y < 2 \\ u(x,0) = 0, u(x,2) = x(1-x), 0 < x < 1. \\ u(0,y) = 0, u(1,y) = 0, 0 < y < 2 \end{cases}$$

SOLUTION: In this case, $a = 1$, $b = 2$, and $f(x) = x(1-x)$. Therefore, $B_n = \dfrac{2}{\sinh 2n\pi} \displaystyle\int_0^1 x(1-x) \sin n\pi x \, dx$, which we now compute with **int**.

```
> B:=n->2/sinh(2*n*Pi)*int(x*(1-sin(x))*sin(n*Pi*x),x=0..1):
  simplify(B(n));
```

$$\begin{aligned}
\Big(& 2\sin(n\pi) - 2n^3\pi^3\cos(n\pi - 1) - 4\sin(n\pi)n^2\pi^2 \\
& - n^4\pi^4\sin(n\pi + 1) + 2\sin(n\pi)n^4\pi^4 - n^5\pi^5\sin(n\pi - 1) \\
& + 4n^3\pi^3\cos(n\pi) + \sin(n\pi - 1)n^3\pi^3 - 2n\pi\cos(n\pi) \\
& + \sin(n\pi - 1)n^2\pi^2 + n^4\pi^4\cos(n\pi + 1) - \cos(n\pi - 1)n^2\pi^2 \\
& - 2n^3\pi^3\cos(n\pi + 1) - 2n^5\pi^5\cos(n\pi) + n^5\pi^5\sin(n\pi + 1) \\
& + \sin(n\pi + 1)n^2\pi^2 - \sin(n\pi + 1)n^3\pi^3 - n^4\pi^4\cos(n\pi - 1) \\
& + \cos(n\pi + 1)n^2\pi^2 - n^4\pi^4\sin(n\pi - 1) + 4n^3\pi^3 \Big) \Big/ \Big(\pi^2 n^2 \sinh(2n\pi)(n^4\pi^4 - 2n^2\pi^2 + 1) \Big)
\end{aligned}$$

Simplifying this expression results in

$$B_n = \frac{2}{\sinh 2n\pi} \int_0^1 x(1-x)\sin n\pi x \, dx = \frac{1}{\sinh 2n\pi} \left(-\frac{4\cos n\pi}{n^3\pi^3} + \frac{4}{n^3\pi^3} \right)$$

$$= \frac{4}{n^3\pi^3\sinh 2n\pi}(1 - (-1)^n), n = 1, 2, \ldots,$$

so the solution is

$$u(x,y) = \sum_{n=1}^{\infty} B_n \sinh n\pi y \sin n\pi x = \sum_{n=1}^{\infty} \frac{8\sinh(2n-1)\pi y \sin(2n-1)\pi x}{(2n-1)^3\pi^3 \sinh 2(2n-1)\pi}.$$

In this case, we may prefer to use approximations of the coefficients. Taking advantage of `` `evalf/int` ``, we redefine B using option remember and then compute approximations of $B_1, B_2, \ldots, B_7$.

```
> readlib(`evalf/int`):
  B:=proc(n) option remember;
   evalf(2/sinh(2*n*Pi)*
    `evalf/int`(x*(1-sin(x))*sin(n*Pi*x),x=0..1))
   end:
  array([seq([n,B(n)],n=1..7)]);
```

$$
\begin{bmatrix}
1 & 0.0010702974 \times 10 \\
2 & -.2489757042 \times 10^{-6} \\
3 & .5097376515 \times 10^{-9} \\
4 & -.5701025419 \times 10^{-12} \\
5 & .9698140300 \times 10^{-15} \\
6 & -.1381966414 \times 10^{-17} \\
7 & .2350796443 \times 10^{-20}
\end{bmatrix}
$$

After defining approx(k) to be the kth partial sum of the series, we plot $u(x, y)$ using the first five terms of the series solution with plot3d together with the axes=BOXED option, which instructs Maple to place a box around the resulting graph.

```
> approx:=proc(k) option remember;
   approx(k-1)+B(k)*sinh(k*Pi*y)*sin(k*Pi*x)
   end:
  approx(1):=B(1)*sinh(Pi*y)*sin(Pi*x):
  plot3d(approx(5),x=0..1,y=0..2,axes=BOXED);
```

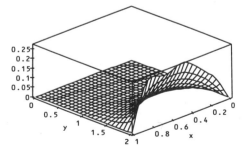

We notice that the value of u decreases to zero away from the boundary $y = 2$.

Any version of Laplace's equation on a rectangular region can be solved through separation of variables as long as we have a pair of homogeneous boundary conditions in the same variable.

EXAMPLE: Solve

$$\begin{cases} u_{xx} + u_{yy} = 0, 0 < x < \pi, 0 < y < 1 \\ u(x,0) = 0, u(x,1) = 0, 0 < x < \pi \\ u(0,y) = \sin 2\pi y, u(\pi, y) = 4, 0 < y < 1 \end{cases}.$$

SOLUTION: As we did in the previous problem, we assume that $u(x,y) = X(x)Y(y)$. Notice that this problem differs from the previous problem in that the homogeneous boundary conditions are in terms of the variable y. Hence, when we separate variables, we use a different constant of separation. This yields

$$X''Y + XY'' = 0$$

so

$$\frac{X''}{X} = -\frac{Y''}{Y} = \lambda,$$

so we have the ordinary differential equations $X'' - \lambda X = 0$ and $Y'' + \lambda Y = 0$. Therefore, with the homogeneous boundary conditions $u(x,0) = X(x)Y(0) = 0$ and $u(x,1) = X(x)Y(1) = 0$, we have $Y(0) = 0$ and $Y(1) = 0$. The eigenvalue problem

$$\begin{cases} Y'' + \lambda Y = 0 \\ Y(0) = 0, Y(1) = 0 \end{cases}$$

has the eigenvalues $\lambda_n = \left(\frac{n\pi}{1}\right)^2 = n^2\pi^2, n = 1, 2, \ldots$, and $Y_n(y) = \sin n\pi y, n = 1, 2, \ldots$. We then solve the equation $X'' - \lambda X = 0$ using these eigenvalues to obtain $X_n(x) = a_n e^{n\pi x} + b_n e^{-n\pi x}$, which can be written in terms of the hyperbolic trigonometric functions as

$$X(x) = X_n(x) = A_n \cosh n\pi x + B_n \sinh n\pi x.$$

Now, because the boundary conditions on the boundaries $x = 0$ and $x = \pi$ are nonhomogeneous, we use the principle of superposition to obtain

$$u(x,y) = \sum_{n=1}^{\infty} (A_n \cosh n\pi x + B_n \sinh n\pi x) \sin n\pi y.$$

Therefore,

$$u(0, y) = \sum_{n=1}^{\infty} A_n \sin n\pi y = \sin 2\pi y,$$

so $A_2 = 1$ and $A_n = 0$ for $n \neq 2$. Similarly,

$$u(\pi, y) = A_2 \cosh 2\pi^2 + \sum_{n=1}^{\infty} B_n \sinh n\pi^2 \sin n\pi y = 4,$$

which indicates that $\sum_{n=1}^{\infty} B_n \sinh n\pi^2 \sin n\pi y = 4 - \cosh 2\pi^2$. Then $B_n \sinh n\pi^2$ are the Fourier sine series coefficients for the constant function $4 - \cosh 2\pi^2$ that are obtained with `int`

```
> 2*int((4-cosh(2*Pi^2))*sin(n*Pi*y),y=0..1);
```

$$2\frac{\cos(n\pi)(-4 + \cosh(2\pi^2))}{n\pi} - 2\frac{-4 + \cosh(2\pi^2)}{n\pi}$$

and simplified to yield

$$B_n \sinh n\pi^2 = \frac{2}{1} \int_0^1 (4 - \cosh 2\pi^2) \sin n\pi y \, dy = \frac{-2(4 - \cosh 2\pi^2)}{n\pi} [(-1)^n - 1], n = 1, 2, \ldots.$$

From this formula, we see that $B_n = 0$ if n is even. Therefore, we express these coefficients as

$$B_{2n-1} = \frac{4(4 - \cosh 2\pi^2)}{(2n - 1)\pi}, n = 1, 2, \ldots,$$

so that the solution is

$$u(x, y) = \cosh 2\pi x + \sum_{n=1}^{\infty} \frac{4(4 - \cosh 2\pi^2)}{(2n - 1)\pi} \sin(2n - 1)\pi y.$$

As with the previous example, we may prefer to deal with approximations of the coefficients instead of exact values. We now define **B(n)** to be the approximate value of the nth coefficient B_n and **approx(k)** to be the kth partial sum of the series $u(x, y) = \cosh 2\pi x + \sum_{n=1}^{\infty} B_n \sin n\pi y$, and then we use **plot3d** to graph the fifth partial sum.

```
> readlib(`evalf/int`):
  B:=proc(n) option remember;
   evalf(2/sinh(n*Pi^2)*
    `evalf/int`((4-cosh(2*Pi^2))*sin(n*Pi*y),y=0..1))
   end:
  approx:=proc(k) option remember;
   approx(k-1)+B(k)*sin(k*Pi*y)
   end:
  approx(0):=cosh(2*Pi*x):

> plot3d(approx(5),x=0..Pi,y=0..1,axes=BOXED);
```

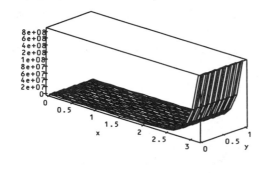

12.5 TWO-DIMENSIONAL PROBLEMS IN A CIRCULAR REGION

In some situations, the region on which we solve a boundary value problem or an initial boundary value problem is not rectangular in shape. For example, we usually do not have rectangular-shaped drumheads or square heating elements on top of the stove. Instead, these objects are typically circular in shape, so we find the use of polar coordinates convenient. In this section, we discuss problems of this type by presenting two important problems solved on circular regions: Laplace's equation, which is related to the steady-state temperature, and the wave equation, which is used to find the displacement of a drumhead.

≡ *Laplace's Equation in a Circular Region*

With the change of variables

$$\begin{cases} x = r \cos \theta \\ y = r \sin \theta \end{cases},$$

we transform Laplace's equation in rectangular coordinates, $u_{xx} + u_{yy} = 0$, to polar coordinates

$$u_{rr} + \frac{1}{r}u_r + \frac{1}{r^2}u_{\theta\theta} = 0, 0 < r < \rho, -\pi < \theta < \pi.$$

Recall that for the solution of Laplace's equation in a rectangular region, we had to specify a boundary condition on each of the four boundaries of the rectangle. However, in the case of a circle, there are not four sides, so we must alter the boundary conditions. Because in polar coordinates the points (r, π) and $(r, -\pi)$ are equivalent for the same value of r, we want our solution and its derivative with respect to θ to match at these points. Therefore, two of the boundary conditions are $u(r, -\pi) = u(r, -\pi)$ and $u_\theta(r, -\pi) = u_\theta(r, -\pi)$ for $0 < r < \rho$. Also, we want our solution to be bounded at $r = 0$, so another boundary condition is $|u(0, \theta)| < \infty$ for $-\pi < \theta < \pi$. Finally, we can specify the value of the solution around the boundary of the circle. This is given by $u(\rho, \theta) = f(\theta)$ for $-\pi < \theta < \pi$.

Therefore, we solve the following boundary value problem to solve Laplace's equation (the Dirichlet problem) in a circular region of radius ρ:

$$\begin{cases} u_{rr} + \dfrac{1}{r}u_r + \dfrac{1}{r^2}u_{\theta\theta} = 0, 0 < r < \rho, -\pi < \theta < \pi \\ u(r, -\pi) = u(r, -\pi), u_\theta(r, -\pi) = u_\theta(r, -\pi), 0 < r < \rho \\ |u(0, \theta)| < \infty, u(\rho, \theta) = f(\theta), -\pi < \theta < \pi \end{cases}.$$

Using separation of variables, we assume that $u(r, \theta) = R(r)H(\theta)$. Then substitution into Laplace's equation yields

$$R''H + \frac{1}{r}R'H + RH'' = 0$$

$$R''H + \frac{1}{r}R'H = -RH''$$

$$\frac{rR'' + R'}{rR} = -\frac{H''}{H} = \lambda.$$

Therefore, we have the ordinary differential equations $H'' + \lambda H = 0$ and $r^2R'' + rR' - \lambda R = 0$. Notice that the boundary conditions become

$$\begin{array}{ccc} u(r, -\pi) = u(r, -\pi) & & u_\theta(r, -\pi) = u_\theta(r, -\pi) \\ R(r)H(-\pi) = R(r)H(\pi) & \text{and} & R(r)H'(-\pi) = R(r)H'(\pi) \\ H(-\pi) = H(\pi) & & H'(-\pi) = H'(\pi). \end{array}$$

This means that we begin by solving the eigenvalue problem

$$\begin{cases} H'' + \lambda H = 0 \\ H(-\pi) = H(\pi), H'(-\pi) = H'(\pi) \end{cases}.$$

The eigenvalues and corresponding eigenfunctions of this problem are

$$\lambda_n = \begin{cases} 0, n = 0 \\ n^2, n = 1, 2, \ldots \end{cases}$$

and

$$H_n(\theta) = \begin{cases} 1, n = 0 \\ a_n \cos n\theta + b_n \sin n\theta, n = 1, 2, \ldots \end{cases}$$

Because $r^2R'' + rR' - \lambda^2 R = 0$ is a Cauchy-Euler equation, we assume that $R(r) = r^m$:

$$m(m - 1)r^2 r^{m-2} + mr r^{m-1} - \lambda r^m = 0$$
$$r^m[m(m - 1) + m - \lambda] = 0.$$

Therefore,

$$m^2 - \lambda^2 = 0$$
$$m = \pm\lambda.$$

If $\lambda_0 = 0$, then $R_0(r) = c_1 + c_2 \ln r$. However, because we require that the solution be bounded near $r = 0$ and $\lim_{r \to 0^+} \ln r = -\infty$, we must choose $c_2 = 0$. Therefore, $R_0(r) = c_1$. In contrast, if $\lambda_n = n^2$, $n = 1, 2, \ldots$, then $R_n(r) = c_3 r^n + c_4 r^{-n}$. Similarly, since $\lim_{r \to 0^+} r^{-n} = \infty$, we must let $c_4 = 0$, so $R_n(r) = c_3 r^n$. By the principle of superposition, we have the solution

$$u(r, \theta) = A_0 + \sum_{n=1}^{\infty} r^n (A_n \cos n\theta + B_n \sin n\theta),$$

where $A_0 = c_1$, $A_n = c_3 a_n$, and $B_n = c_3 b_n$. We find these coefficients by applying the boundary condition $u(\rho, \theta) = f(\theta)$. This yields

$$u(\rho, \theta) = A_0 + \sum_{n=1}^{\infty} \rho^n (A_n \cos n\theta + B_n \sin n\theta) = f(\theta),$$

so A_0, A_n, and B_n are related to the Fourier series coefficients in the following way:

$$A_0 = \frac{1}{2\pi} \int_{-\pi}^{\pi} f(\theta) \, d\theta,$$

$$A_n = \frac{1}{\pi\rho^n} \int_{-\pi}^{\pi} f(\theta) \cos n\theta \, d\theta \qquad (n = 1, 2, \ldots),$$

and

$$B_n = \frac{1}{\pi \rho^n} \int_{-\pi}^{\pi} f(\theta) \sin n\theta \, d\theta.$$

EXAMPLE: Solve

$$
\begin{cases}
u_{rr} + \dfrac{1}{r} u_r + \dfrac{1}{r^2} u_{\theta\theta} = 0, 0 < r < 2, -\pi < \theta < \pi \\[2mm]
u(r, -\pi) = u(r, -\pi), u_\theta(r, -\pi) = u_\theta(r, -\pi), 0 < r < 2 \\[2mm]
|u(0, \theta)| < \infty, u(2, \theta) = 0, -\pi < \theta < \pi
\end{cases}
$$

SOLUTION: Notice that $f(\theta) = \theta$ is an odd function on $-\pi < \theta < \pi$. Therefore, $A_0 = 0$ and $A_n = 0$ for $n = 1, 2, \dots$. Then, using **int**, we obtain

> **1/(Pi*2^n)*int(theta*sin(n*theta),theta=-Pi..Pi);**

$$-2 \frac{-\sin(n\pi) + n\cos(n\pi)\pi}{\pi \, 2^n n^2}$$

and simplifying yields

$$B_n = \frac{1}{\pi \, 2^n} \int_{-\pi}^{\pi} \theta \sin n\theta \, d\theta = -\left(\frac{1}{\pi \, 2^n}\right) \frac{2\pi \cos n\pi}{n} = -\frac{1}{2^{n-1}n}(-1)^n = \frac{1}{2^{n-1}n}(-1)^{n+1}, n = 1, 2, \dots,$$

so the solution is

$$u(r, \theta) = \sum_{n=1}^{\infty} r^n \frac{1}{2^{n-1}n}(-1)^{n+1} \sin n\theta, n = 1, 2, \dots.$$

In the same manner as in the previous sections, we define **approx(k)** to be the kth partial sum of the series and then use **plot3d** to graph the fifth partial sum.

```
> approx:=proc(k) option remember;
    approx(k-1)+r^k*(-1)^(k+1)/(2^(k-1)*k)*sin(k*theta)
    end:
  approx(1):=r*sin(theta):

> plot3d([r*cos(theta),r*sin(theta),approx(9)],
    r=0..2,theta=-Pi..Pi,axes=BOXED);
```

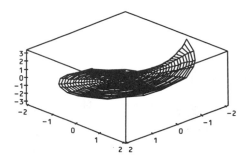

☰ The Wave Equation in a Circular Region

One of the more interesting problems involving two spatial dimensions (x and y) is the wave equation,

$$c^2(u_{xx} + u_{yy}) = u_{tt}.$$

This is due to the fact that the solution to this problem represents something with which we are all familiar, the displacement of a drumhead. Since most drumheads are circular in shape, we investigate the solution of the wave equation in a circular region. Therefore, we transform the wave equation into polar coordinates. Then the wave equation in polar coordinates is

$$c^2\left(u_{rr} + \frac{1}{r}u_r + \frac{1}{r^2}u_{\theta\theta}\right) = u_{tt}.$$

If we assume that the height of the drumhead above the xy-plane at time t is the same at equal distances from the origin, then we say that the solution u is **radially symmetric**. (In other words, the value of u does not depend on the angle θ.) Therefore, $u_{\theta\theta} = 0$, so the wave equation can be expressed in terms of r and t only as

$$c^2\left(u_{rr} + \frac{1}{r}u_r\right) = u_{tt}.$$

Of course, to find $u(r, t)$ we need the appropriate boundary and initial conditions. Because the circular boundary of the drumhead must be fixed so that it doesn't move, we say that $u(\rho, t) = 0$ for $t > 0$. Then, as we had in Laplace's equation on a circular region, we require that the solution u be bounded near the origin, so we have the condition $|u(0, t)| < \infty$ for $t > 0$. The initial position and initial velocity functions are given as functions of r as $u(r, 0) = f(r)$ and $u_t(r, 0) = g(r)$ for $0 < r < \rho$. Therefore, the initial boundary value problem to find the displacement u of a circular drumhead (of radius ρ) is given by

$$\begin{cases} c^2 \left(u_{rr} + \dfrac{1}{r} u_r \right) = u_{tt} \\ u(\rho, t) = 0, |u(0, t)| < \infty, t > 0 \\ u(r, 0) = f(r), u_t(r, 0) = g(r), 0 < r < \rho \end{cases}$$

As with other problems, we are able to use separation of variables to find u by assuming that $u(r, t) = R(r)T(t)$. Substitution into the wave equation yields

$$c^2 \left(R''T + \frac{1}{r} R'T \right) = RT''$$

so

$$\frac{rR'' + R'}{rR} = \frac{T''}{c^2 T} = -k^2,$$

where $-k^2$ is the separation constant. Separating the variables, we have the ordinary differential equations $r^2 R'' + rR' + k^2 r^2 R = 0$ and $T'' + c^2 k^2 T = 0$. We recognize the equation $r^2 R'' + rR' + k^2 r^2 R = 0$ as Bessel's equation of order zero that has solution

$$R(r) = c_1 J_0(kr) + c_2 Y_0(kr),$$

where J_0 and Y_0 are the Bessel functions of order zero of the first and second kinds, respectively. In terms of R, we express the boundary condition $|u(0, t)| < \infty$ as

$$|R(0)| < \infty.$$

Therefore, because $\lim\limits_{r \to 0^+} Y_0(kr) = -\infty$, we must choose $c_2 = 0$. Then, applying the other boundary condition $R(\rho) = 0$, we have

$$R(\rho) = c_1 J_0(k\rho) = 0,$$

so to avoid the trivial solution with $c_1 = 0$, we have

$$k\rho = \alpha_n,$$

where α_n is the nth zero of J_0, which we discussed in Section 11.4. Because k depends on n, we write

$$k_n = \frac{\alpha_n}{\rho}.$$

The solution of $T'' + c^2 k^2 T = 0$ is

$$T_n(t) = A_n \cos ck_n t + B_n \sin ck_n t,$$

so with the principle of superposition, we have

$$u(r, t) = \sum_{n=1}^{\infty} (A_n \cos ck_n t + B_n \sin ck_n t) J_0(k_n r),$$

where the coefficients A_n and B_n are found through application of the initial position and velocity functions. With $u(r, 0) = \sum_{n=1}^{\infty} A_n J_0(k_n r) = f(r)$ and the orthogonality conditions of the Bessel functions, we find that

$$A_n = \frac{\int_0^\rho f(r) J_0(k_n r)\, dr}{\int_0^\rho r[J_0(k_n r)]^2\, dr} = \frac{2}{[J_1(\alpha_n)]^2} \int_0^\rho f(r) J_0(k_n r)\, dr, n = 1, 2, \ldots.$$

Similarly, because

$$u_t(r, t) = \sum_{n=1}^{\infty} (-ck_n A_n \sin ck_n t + ck_n B_n \cos ck_n t) J_0(k_n r),$$

we have $u_t(r, 0) = \sum_{n=1}^{\infty} ck_n B_n J_0(k_n r) = g(r)$. Therefore,

$$B_n = \frac{\int_0^\rho g(r) J_0(k_n r)\, dr}{ck_n \int_0^\rho r[J_0(k_n r)]^2\, dr} = \frac{2}{ck_n [J_1(\alpha_n)]^2} \int_0^\rho g(r) J_0(k_n r)\, dr, n = 1, 2, \ldots.$$

EXAMPLE: Solve

$$
\begin{cases}
c^2 \left(u_{rr} + \dfrac{1}{r} u_r \right) = u_{tt} \\[4pt]
u(1, t) = 0, |u(0, t)| < \infty, t > 0 \\[4pt]
u(r, 0) = r(r - 1), u_t(r, 0) = \sin \pi r, 0 < r < 1
\end{cases}
$$

SOLUTION: In this case, $\rho = 1$, $f(r) = r(r - 1)$, and $g(r) = \sin \pi r$. To calculate the coefficients, we will need to have approximations of the zeros of the Bessel functions, so we read in the table of zeros that we compiled earlier in **BesselTable** and call this table **Alpha_Array** (see Section 11.4).

```
> readlib(readdata):
  Alpha_Array:=readdata(`BesselTable`,float,8):
```

We then define the function **alpha(i,j)** that represents the jth zero of the Bessel function of order i.

```
> alpha:=(i,j)->Alpha_Array[i+1][j]:
```

To estimate the coefficients, we will use `evalf/int`, which we load as follows, and then define the constants **R** and **c** and the functions **f**, **g**, **k**, and **lambda**.

```
> a:='a':b:='b':
  readlib(`evalf/int`):
  R:=1:
  c:=1:
  f:=r->r*(r-1):
  g:=r->sin(Pi*r):
  k:=m->alpha(0,m)/R:
  lambda:=m->c/R*alpha(0,m):
```

The formulas for the coefficients A_m and B_m, which were derived earlier, are then defined in **a** and **b** so that an approximate solution may be determined. Note that we use `evalf/int` to avoid the difficulties in integration associated with the presence of the Bessel function of order zero.

```
> a:=proc(n) option remember;
  2/BesselJ(1,alpha(0,n))^2*
    `evalf/int`(r*f(r)*BesselJ(0,k(n)*r),r=0..R)
  end:
  b:=proc(n) option remember;
  2/(c*alpha(0,n)*BesselJ(1,alpha(0,n))^2)*
    `evalf/int`(r*g(r)*BesselJ(0,k(n)*r),r=0..R)
  end:
```

We now compute the first six values of **a(n)** and **b(n)**. Because **a** and **b** are defined using the option `remember`, Maple remembers these values for later use.

```
> array([seq([n,a(n),b(n)],n=1..6)]);latex(");
```

$$
\begin{bmatrix}
1 & -.3235010276 & .5211819702 \\
2 & .2084692034 & -.1457773395 \\
3 & .007640292448 & -.01342290349 \\
4 & .03838004574 & -.008330225220 \\
5 & .005341000924 & -.002504216150 \\
6 & .01503575900 & -.002082788164
\end{bmatrix}
$$

The mth term of the series solution is defined in **u**. Then an approximate solution is obtained in **uapprox** by summing the first six terms of u given before. The form of this approximation is printed as output to show that it is of the correct form.

```
> u:=(m,r,t)->(a(m)*cos(lambda(m)*t)+
  b(m)*sin(lambda(m)*t))*BesselJ(0,k(m)*r):
  uapprox:=(r,t)->sum('u(m,r,t)','m'=1..6):
  uapprox(r,t);
```

$(-.3235010276 \cos(2.404800000000000t) + .5211819702 \sin(2.404800000000000t))$
$\text{BesselJ}(0, 2.404800000000000r) + (.2084692034 \cos(5.520100000000000t)$
$\qquad - .1457773395 \sin(5.520100000000000t))\text{BesselJ}(0, 5.520100000000000r)$
$\qquad + (.007640292448 \cos(8.653700000000001t)$
$\qquad - .01342290349 \sin(8.653700000000001t))\text{BesselJ}(0, 8.653700000000001r)$
$\qquad + (.03838004574 \cos(11.79200000000000t)$
$\qquad - .008330225220 \sin(11.79200000000000t))\text{BesselJ}(0, 11.79200000000000r)$
$\qquad + (.005341000924 \cos(14.93100000000000t)$
$\qquad - .002504216150 \sin(14.93100000000000t))\text{BesselJ}(0, 14.93100000000000r)$
$\qquad + (.01503575900 \cos(18.07100000000000t)$
$\qquad - .002082788164 \sin(18.07100000000000t))\text{BesselJ}(0, 18.07100000000000r)$

Without the assistance of the computer, we would be forced to stop at this point. Fortunately, we can use Maple to produce the graphics associated with this function. Since this function is independent of the angular coordinate θ, we can plot this function over the interval $[0, 1]$ to yield a side view of half of the drumhead. This is accomplished with **animate**, which is contained in the **plots** package, by plotting **uapprox** for 30 equally spaced values of t between 0 and 2. Several frames from the resulting animation are shown here.

```
> with(plots):
  animate(uapprox(r,t),r=0..1,t=0..2,fr$A_m$es=30);
```

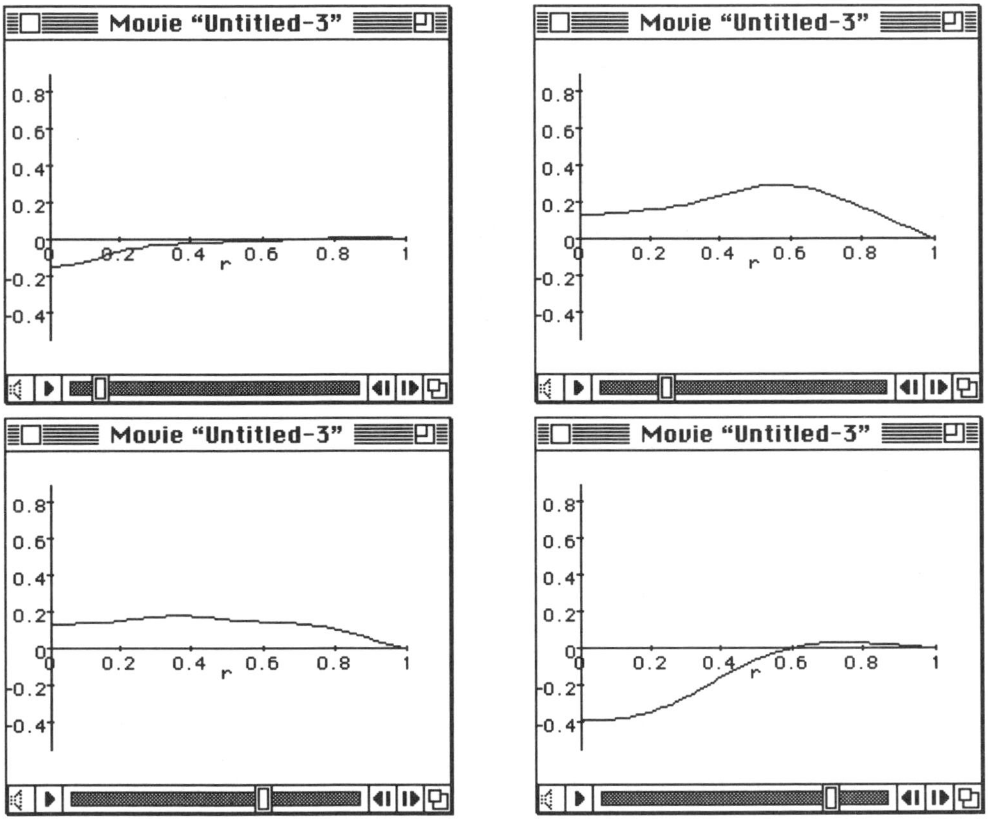

The problem that depends on the angle θ is more complicated to solve. Due to the presence of $u_{\theta\theta}$ we must include two more boundary conditions to solve the initial boundary value problem. So that the solution is a smooth function, we require the "artificial" boundary conditions $u(r, \pi, t) = u(r, -\pi, t)$ and $u_{\theta}(r, \pi, t) = u_{\theta}(r, -\pi, t)$ for $0 < r < \rho$ and $t > 0$.

Therefore, we solve the problem

$$
\begin{cases}
c^2 \left(u_{rr} + \dfrac{1}{r} u_r + \dfrac{1}{r^2} u_{\theta\theta} \right) = u_{tt} \\[2mm]
u(\rho, \theta, t) = 0, |u(0, \theta, t)| < \infty, -\pi < \theta < \pi, t > 0 \\[2mm]
u(r, \pi, t) = u(r, -\pi, t), u_\theta(r, \pi, t) = u_\theta(r, -\pi, t), 0 < r < \rho, t > 0 \\[2mm]
u(r, \theta, 0) = f(r, \theta), u_t(r, \theta, 0) = g(r, \theta), 0 < r < \rho, -\pi < \theta < \pi
\end{cases},
$$

to describe the displacement of a drumhead that is not radially symmetric. Assuming that $u(r, \theta, t) = R(r)H(\theta)T(t)$ and using separation of variables yields

$$
c^2 \left(R''HT + \frac{1}{r}R'HT + \frac{1}{r^2}RH''T \right) = RHT''.
$$

Then division by RHT gives us

$$
c^2 \left(\frac{R''}{R} + \frac{1}{r}\frac{R'}{R} + \frac{1}{r^2}\frac{H''}{H} \right) = \frac{T''}{T} = -\lambda^2,
$$

where $-\lambda^2$ is the constant of separation. Separating variables, we obtain

$$
T'' + \lambda^2 c^2 T = 0
$$

and

$$
\frac{R''}{R} + \frac{1}{r}\frac{R'}{R} + \frac{1}{r^2}\frac{H''}{H} = -\lambda^2.
$$

Separating variables in the second equation yields

$$
r^2\frac{R''}{R} + r\frac{R'}{R} + r^2\lambda^2 = \frac{-H''}{H} = \mu^2,
$$

where μ^2 is the constant of separation. Therefore, we have two more ordinary differential equations, $H'' + \mu^2 H = 0$ and $r^2 R'' + rR' + (r^2\lambda^2 - \mu^2)R = 0$. The boundary conditions in terms of H and R become $R(\rho) = 0, |R(0)| < \infty, H(-\pi) = H(\pi)$, and $H'(-\pi) = H'(\pi)$. Recall that the problem

$$
\begin{cases}
H'' + \mu^2 H = 0 \\
H(-\pi) = H(\pi), H'(-\pi) = H'(\pi)
\end{cases}
$$

has solutions

$$H_n(x) = \begin{cases} 1, n = 0 \\ a_n \cos n\theta + b_n \sin n\theta \end{cases}$$

that correspond to the eigenvalues $\mu_n^2 = \begin{cases} 0, n = 0 \\ n^2, n = 1, 2, \ldots \end{cases}$

The corresponding solutions of $r^2 R'' + rR' + (r^2\lambda^2 - \mu^2)R = 0$, which we recognize as Bessel's equation of order n, are $R_n(r) = c_n J_n(\lambda r) + d_n Y_n(\lambda r)$. Since $|R(0)| < \infty$, $d_n = 0$. Then, since $R(\rho) = 0$, $J_n(\lambda\rho) = 0$, so $\lambda_{mn} = \dfrac{\alpha_{mn}}{\rho}$, where α_{mn} denotes the nth zero of the Bessel function $J_m(x)$. Then the solution is

$$u(r, \theta, t) = \sum_n a_{0n} J_0(\lambda_{0n}r) \cos(\lambda_{0n}ct)$$

$$+ \sum_{m,n} a_{mn} J_m(\lambda_{mn}r) \cos(m\theta) \cos(\lambda_{mn}ct)$$

$$+ \sum_{m,n} b_{mn} J_m(\lambda_{mn}r) \sin(m\theta) \cos(\lambda_{mn}ct) + \sum_n A_{0n} J_0(\lambda_{0n}r) \sin(\lambda_{0n}ct)$$

$$+ \sum_{m,n} A_{mn} J_m(\lambda_{mn}r) \cos(m\theta) \sin(\lambda_{mn}ct)$$

$$+ \sum_{m,n} B_{mn} J_m(\lambda_{mn}r) \sin(m\theta) \sin(\lambda_{mn}ct),$$

where

$$a_{0n} = \frac{\int_0^{2\pi} \int_0^\rho f(r, \theta) J_0(\lambda_{0n}r) r \, dr \, d\theta}{2\pi \int_0^\rho [J_0(\lambda_{0n}r)]^2 r \, dr},$$

$$a_{mn} = \frac{\int_0^{2\pi} \int_0^\rho f(r, \theta) J_m(\lambda_{mn}r) \cos(m\theta) r \, dr \, d\theta}{\pi \int_0^\rho [J_m(\lambda_{mn}r)]^2 r \, dr},$$

$$b_{mn} = \frac{\int_0^{2\pi} \int_0^\rho f(r, \theta) J_m(\lambda_{mn}r) \sin(m\theta) r \, dr \, d\theta}{\pi \int_0^\rho [J_m(\lambda_{mn}r)]^2 r \, dr},$$

$$A_{0n} = \frac{\int_0^{2\pi} \int_0^\rho g(r, \theta) J_0(\lambda_{0n}r) r \, dr \, d\theta}{2\pi\lambda_{0n}c \int_0^\rho [J_0(\lambda_{0n}r)]^2 r \, dr},$$

$$A_{mn} = \frac{\int_0^{2\pi} \int_0^\rho g(r, \theta) J_m(\lambda_{mn}r) \cos(m\theta) r \, dr \, d\theta}{\pi\lambda_{mn}c \int_0^\rho [J_m(\lambda_{mn}r)]^2 r \, dr},$$

and

$$B_{mn} = \frac{\int_0^{2\pi} \int_0^\rho g(r, \theta) J_m(\lambda_{mn}r) \sin(m\theta) r \, dr \, d\theta}{\pi\lambda_{mn}c \int_0^\rho [J_m(\lambda_{mn}r)]^2 r \, dr}.$$

EXAMPLE: Solve

$$\begin{cases} 10^2\left(u_{rr} + \dfrac{1}{r}u_r + \dfrac{1}{r^2}u_{\theta\theta}\right) = u_{tt} \\[2mm] u(1, \theta, t) = 0, |u(0, \theta, t)| < \infty, -\pi < \theta < \pi, t > 0 \\[2mm] u(r, \pi, t) = u(r, -\pi, t), u_\theta(r, \pi, t) = u_\theta(r, -\pi, t), 0 < r < 1, t > 0 \\[2mm] u(r, \theta, 0) = \cos\dfrac{\pi}{2r}\sin\theta, u_t(r, \theta, 0) = (r-1)\cos\dfrac{\pi\theta}{2}, 0 < r < 1, -\pi < \theta < \pi \end{cases}$$

SOLUTION: As in the previous example, we read in the table of zeros, which was found earlier and saved as `BesselTable`, with `readdata` and called `Alpha_Array`. Then a function `alpha` is defined so that these zeros of the Bessel functions can be obtained more easily from the list.

```
> readlib(readdata):
  Alpha_Array:=readdata(`BesselTable`,float,8);
  alpha:=(i,j)->Alpha_Array[i+1][j]:
```

The appropriate parameter values as well as the initial condition functions are defined as follows. Notice that the functions describing the initial position and velocity are defined as the product of functions. This enables the subsequent calculations to be carried out in the following manner:

```
> readlib(`evalf/int`):
  f:='f':f1:='f1':f2:='f2':
  a:='a':g1:='g1':g2:='g2':A:='A':c:='c':
  g:='g':capa:='capa':capb:='capb':
  b:='b':
  c:=10:
  A:=1:
  f1:=r->r*sin(2*Pi*r):
  f2:=theta->sin(2*theta+cos(4*theta)):
  f:=proc(r,theta) option remember;
   f1(r)*f2(theta)
   end:
  g1:=r->r*(r-1):
  g2:=theta->cos(2*theta):
  g:=proc(r,theta) option remember;
   g1(r)*g2(theta)
   end:
```

The coefficients a_{0n} are determined with the function a0.

```
> a0:=proc(n) option remember;
  `evalf/int`(f1(r)*
  BesselJ(0,alpha(0,n)*r)*r,r=0..A)*
  `evalf/int`(f2(t),t=0..2*Pi)/
  (2*Pi*`evalf/int`(r*
  BesselJ(0,alpha(0,n)*r)^2,r=0..A))
  end:
```

The first five coefficients are computed with the following for loop and then displayed using op. As in the previous example, the remember option is used with proc so that the coefficient values are "remembered" by Maple.

```
> for n from 1 to 5 do a0(n) od:
  op(4,op(a0));
```

table([

$$5 = -.3655277067 \ 10^{-13} \frac{1}{\pi}$$

$$1 = -.1136145667 \ 10^{-12} \frac{1}{\pi}$$

$$2 = .4942812760 \ 10^{-12} \frac{1}{\pi}$$

$$3 = -.4439121267 \ 10^{-12} \frac{1}{\pi}$$

$$4 = .8747891165 \ 10^{-13} \frac{1}{\pi}$$

])

Since the denominators of the integral formulas to find a_{mn} and b_{mn} are the same, the function bjmn that computes this value is defined here. A table of nine values of this coefficient is then determined and displayed.

```
> bjmn:=proc(m,n) option remember;
  `evalf/int`(r*BesselJ(m,alpha(m,n)*r)^2,
  r=0..A)
  end:
```

```
> for n from 1 to 3 do
   for m from 1 to 3 do bjmn(m,n) od od;
   op(4,op(bjmn));
```
table([

       $(1,1) = .08110781816$

       $(2,1) = .05768792844$

       $(3,1) = .04448295755$

       $(1,2) = .04503456132$

       $(2,2) = .03682464114$

       $(3,2) = .03110451636$

       $(1,3) = .03117913218$

       $(2,3) = .02701416557$

       $(3,3) = .02382360434$

])

We also note that in evaluating the integrals in the numerators a_{mn} and b_{mn}, we must compute $\int_0^A f_1(r)[J_m(\alpha_{mn}r)]^2\,dr$. This integral is defined in **fbjmn** and the corresponding values found for $n = 1, 2, 3$, and $m = 1, 2, 3$.

```
> fbjmn:=proc(m,n) option remember;
   `evalf/int`(f1(r)*
   BesselJ(m,alpha(m,n)*r)^2,
   r=0..A)
   end:
```

```
> for n from 1 to 3 do
   for m from 1 to 3 do fbjmn(m,n) od od;
```

```
> op(4,op(fbjmn));
```
table([

       $(1,1) = -.01567088416$

       $(2,1) = -.02670827240$

       $(3,1) = -.02773427236$

       $(1,2) = .001339693423$

       $(2,2) = -.0006193412555$

       $(3,2) = -.003681231888$

       $(1,3) = .0007431918596$

       $(2,3) = .0008130001095$

       $(3,3) = -.0001060502972$

])

The formula to compute a_{mn} is then defined and uses the information calculated in `fbjmn` and `bjmn`. As in the previous calculation, the coefficient values for $n = 1, 2, 3$ and $m = 1, 2, 3$ are determined (but in this case they are not displayed).

```
> a:=proc(m,n) option remember;
    fbjmn(m,n)*
      `evalf/int`(f2(t)*cos(m*t),
     t=0..2*Pi)/
     (Pi*bjmn(m,n))
    end:
  for n from 1 to 3 do
    for m from 1 to 3 do a(m,n) od od;
```

A similar formula is then defined for the computation of b_{mn}. Values are computed, but not displayed, for $n = 1, 2, 3$ and $m = 1, 2, 3$.

```
> b:=proc(m,n) option remember;
    (fbjmn(m,n)*
      `evalf/int`(f2(t)*sin(m*t),
     t=0..2*Pi))/
     (Pi*bjmn(m,n))
    end:
  for n from 1 to 3 do
    for m from 1 to 3 do b(m,n) od od;
```

The formula for A_{0n} is defined now, and the first six values are determined.

```
> capa0:=proc(n) option remember;
    (`evalf/int`(g1(r)*
    BesselJ(0,alpha(0,n)*r)*r,
    r=0..A)*
      `evalf/int`(g2(t),t=0..2*Pi))/
    (2*Pi*c*alpha(0,n)*
      `evalf/int`(r*
    BesselJ(0,alpha(0,n)*r)^2,
    r=0..A))
    end:
  for n from 1 to 6 do capa0(n) od:
```

As with a_{mn} and b_{mn}, the similarities between a_{mn} and b_{mn} are used to determine the coefficient values. Now, we define `gbjmn` to determine the integral $\int_0^A g_1(r) J_m(\alpha_{mn}r) r \, dr$, which must be computed in each case. The values for $m = 1, 2, 3$ and $n = 1, 2, 3$ are then found but not displayed.

```
> gbjmn:=proc(m,n) option remember;
   `evalf/int`(g1(r)*
   BesselJ(m,alpha[m,n]*r)*r,
   r=0..A)
   end:
  for n from 1 to 3 do
   for m from 1 to 3 do gbjmn(m,n) od od;
```

We next define the formula to determine A_{mn}. Notice that this formula depends on `gbjmn` and `bjmn` that were used to calculate a_{mn} and b_{mn}, respectively. The corresponding values for $m = 1, 2, 3$ and $n = 1, 2, 3$ are then found but not displayed.

```
> capa:=proc(m,n) option remember;
   (gbjmn(m,n)*`evalf/int`(
   g2(t)*cos(m*t),t=0..2*Pi))/
   (Pi*alpha[m,n]*c*bjmn(m,n))
   end:
  for n from 1 to 3 do
   for m from 1 to 3 do capa(m,n) od od;
```

A similar formula is defined to compute B_{mn}. This formula also involves `gbjmn` and `bjmn`.

```
> capb:=proc(m,n) option remember;
   (gbjmn(m,n)*`evalf/int`(
   g2(t)*sin(m*t),t=0..2*Pi))/
   (Pi*alpha[m,n]*c*bjmn(m,n))
   end:
  for n from 1 to 3 do
   for m from 1 to 3 do capb(m,n) od od;
```

The approximate solution can now be constructed with the coefficients we computed. We do this by defining the functions `term1`, `term2`,..., `term6`, which represent the terms corresponding to each component of the series solution $u(r, \theta, t)$ given earlier.

```
> term1:='term1':term2:='term2':
  term3:='term3':term4:='term4':
  term5:='term5':term6:='term6':
  term1:=(r,t,n)->
  a0(n)*BesselJ(0,alpha(0,n)*r)*
   cos(alpha(0,n)*c*t):
  term2:=(r,t,th,m,n)->
  a(m,n)*BesselJ(m,alpha(m,n)*r)*cos(m*th)*
   cos(alpha(m,n)*c*t):
  term3:=(r,t,th,m,n)->
  b(m,n)*BesselJ(m,alpha(m,n)*r)*sin(m*th)*
   cos(alpha(m,n)*c*t):
  term4:=(r,t,n)->
  capa0(n)*BesselJ(0,alpha(0,n)*r)*
   sin(alpha(0,n)*c*t):
  term5:=(r,t,th,m,n)->
  capa(m,n)*BesselJ(m,alpha(m,n)*r)*
   cos(m*th)*sin(alpha(m,n)*c*t):
  term6:=(r,t,th,m,n)->
  capb(m,n)*BesselJ(m,alpha(m,n)*r)*
   sin(m*th)*sin(alpha(m,n)*c*t):
```

The approximate solution is then given by summing over each of these terms. The first term is summed from $n = 1$ to $n = 5$ because these coefficients, a_{0n}, were calculated earlier for these values of n. In all other cases, we sum over $n = 1$ to $n = 3$ and $m = 1$ to $m = 3$. We call this approximate solution u; it depends on the variables r, t, and th.

```
u:=proc(r,t,th)
 sum('term1(r,t,n)','n'=1..5)+
  sum('sum('term2(r,t,th,m,n)','n'=1..3)',
  'm'=1..3)+
  sum('sum('term3(r,t,th,m,n)','n'=1..3)',
  'm'=1..3)+
  sum('term4(r,t,n)','n'=1..3)+
  sum('sum('term5(r,t,th,m,n)','n'=1..3)',
  'm'=1..3)+
  sum('sum('term6(r,t,th,m,n)','n'=1..3)',
  'm'=1..3)
 end:
```

The position of the waves on the circular region can be viewed with Maple through the use of polar coordinates (in the *xy*-plane) and plot3d. We plot the position for $t = 0$ to $t = 1$, using increments of one-eighth with the following for loop. A sequence of nine three-dimensional plots is the result.

```
> for t from 0 by 1/8 to 1 do
    plot3d([r*cos(th),r*sin(th),u(r,t,th)],r=0..1,
     th=-Pi..Pi) od;
```

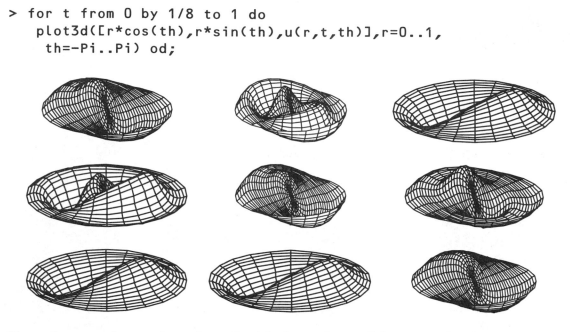

These three graphs can be animated with the animate3d command, which, like the animate command, is contained in the plots package. Be forewarned: generating nine graphs like these and animating the results require a large amount of memory on your computer.

Appendix:
Getting Help from
Maple V

A NOTE REGARDING DIFFERENT VERSIONS OF MAPLE

Many new functions and features have been added to Maple V with Release 2. Particular examples of features available with Release 2 but not Release 1 include standard mathematical notation for output, contour and implicit plots, animation of graphics, and the interactive help browser. We encourage users of Release 1 to update to Release 2 as soon as they can. All examples in *Differential Equations with Maple V* were completed with Release 2. In most cases, the same results will be obtained if you are using Release 1. Occasionally, however, particular features of Release 2 are taken advantage of in an example that are not available with Release 1. If you are using an earlier or later version of Maple, your results may not appear in a form identical to those found in this book: some commands found in Release 2 are not available in earlier versions of Maple, and in later versions some commands will certainly be changed, new commands added, and obsolete commands removed. If you are using Release 2, enter the command `?updates,v5.2` at the prompt to see a summary of the new features of Maple V Release 2.

In this book, we assume that Maple V has been correctly installed on the computer you are using. If you need to install Maple V on your computer, please refer to the documentation that came with the Maple V software package.

GETTING STARTED WITH MAPLE V

After the Maple V program has been properly installed, a user can access Maple V. We now briefly describe methods of starting Maple V on several platforms.

Macintosh: In the same manner that folders are opened, the Maple application can be started by selecting the Maple icon, going to File, and selecting Open or by simply clicking twice on the Maple icon. Of course, opening an existing Maple document also opens the Maple program.

Windows: To start Maple V for Windows, double-click on the Maple V for Windows application icon. This operation opens Maple V for Windows with a blank worksheet.

DOS: To run Maple, type the command MAPLE at the DOS prompt. After a few seconds, the screen will clear, the Maple logo will appear briefly, and a prompt will appear at the bottom of the screen above the status line, and you will be in the Maple session. You can now begin typing Maple commands.

UNIX: To start Maple V, type xmaple at the UNIX prompt. A new window will open that contains Maple V Release 2. You can now begin typing Maple commands.

If you are using a text-based interface (like UNIX), Maple V is started with the operating system command maple or xmaple. If you are using a windows interface (like Macintosh, Windows, or NeXT), Maple V is started by selecting the Maple V icon and double-clicking or selecting the Maple V icon and selecting Open from the File menu. The untitled worksheet that appears if you have not started Maple V by opening an existing document looks like this:

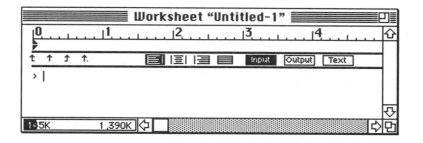

Once Maple V has been started, computations can be carried out immediately. Maple V commands are typed to the right of the prompt (>), a semicolon is placed at the end of the command, and then commands are evaluated by pressing **ENTER**. On some computer systems, commands are evaluated by pressing **SHIFT-RETURN**.

PLEASE DO NOT TYPE THE PROMPT WHEN ENTERING COMMANDS.

When a colon is placed at the end of the command, the resulting output is **not** displayed. Note that pressing **ENTER** evaluates commands and pressing **RETURN** yields a new line. Output is displayed below input. We illustrate some of the typical steps involved in working with Maple V in the following calculations. In each case, we type the command, place a semicolon at the end of the command, and press **ENTER**. Maple V evaluates the command, displays the result, and inserts a new prompt. For example, entering

> `evalf(Pi,50);`

3.1415926535897932384626433832795028841971693993751

returns a 50-digit approximation of π.

The next calculation can then be typed and entered in the same manner as the first. For example, entering

> `solve(x^3-2*x+1=0);`

$$1, -\frac{1}{2} + \frac{1}{2}\sqrt{5}, -\frac{1}{2} - \frac{1}{2}\sqrt{5}$$

solves the equation $x^3 - 2x + 1 = 0$ for x. Subsequent calculations are entered in the same way. For example, entering

> `plot({sin(x),2*cos(2*x)},x=0..3*Pi);`

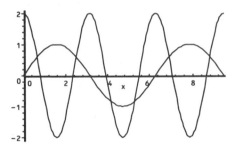

graphs the functions $\sin x$ and $2\cos 2x$ on the interval $[0, 3\pi]$. Similarly, entering

> `plot3d(sin(x+cos(y)),x=0..4*Pi,y=0..4*Pi);`

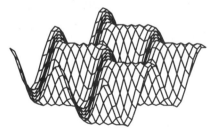

graphs the function $\sin(x + \cos y)$ on the rectangle $[0, 4\pi] \times [0, 4\pi]$.

Be sure to include a semicolon (`;`) or colon (`:`) at the end of each command. However, if you do forget to include a semicolon, you do not need to retype your command. After Maple returns a new prompt, type a semicolon and press **ENTER**.

Maple displays mathematical output using standard mathematical notation. In fact, Maple offers the user flexibility in how output is displayed. Different output formats may be selected by choosing FORMAT and then OUTPUT from the Maple menu, as follows:

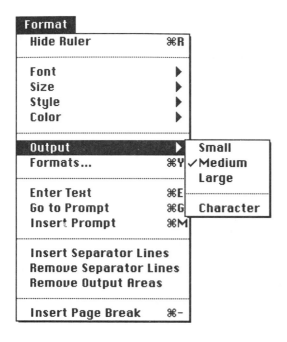

In addition, you can change your prompt and evaluate input using the RETURN key by selecting SETTINGS from the Maple menu and then selecting PREFERENCES....

This book includes real input and output from Maple V Release 2. Appearances of input and output may vary, depending on the version of Maple used, the fonts used to display input and ouput, the quality of the monitor, and the resolution and type of printer used to print the Maple worksheet: the results displayed on your computer may not be physically identical to those shown here.

Maple V sessions are terminated by entering `quit`, `done`, or `stop`. On platforms with a windows interface (like Macintosh, Windows, and NeXT), Maple V sessions are ended by selecting QUIT from the FILE menu, or by using the keyboard shortcut COMMAND Q, as with other applications. They can be saved by selecting SAVE(COMMAND S) from the FILE menu.

On these platforms, input and text regions in notebook interfaces can be edited. Editing input can create a worksheet in which the mathematical output does not make sense in the sequence it appears. It is also possible simply to go into a notebook and alter input without doing any recalculation. This also creates misleading worksheets. Hence, common sense and caution should be used when editing the input regions of notebooks. Recalculating all commands in the notebook will clarify any confusion.

For the Maple V user to take full advantage of the capabilities of this software, an understanding of its syntax is imperative. Although all of the rules of Maple V syntax are far too numerous to list here, knowledge of the following five rules equips the beginner with the necessary tools to start using the Maple V program with little trouble.

1. The arguments of functions are given in parentheses (**. . .**).
2. A semicolon (**;**) or colon (**:**) must be included at the end of each command.
3. Multiplication is represented by a *****.
4. Powers are denoted by a **^**.
5. If you get no response or an incorrect response, you may have entered or executed the command incorrectly. In some cases, the amount of memory allocated to Maple V can cause a crash; like people, Maple V is not perfect and some errors can occur.

GETTING HELP FROM MAPLE V

Becoming competent with Maple can take a serious investment of time. We hope the messages that result from syntax errors are viewed lightheartedly. Ideally, instead of becoming frustrated, beginning Maple users will find it challenging and fun to locate the source of errors. Maple's error messages frequently indicate where the error(s) has (have) occurred. In this process, users will naturally become more proficient with Maple.

One way to obtain information about commands and functions is with the commands **?** and **help**. In general, **?f** and **help(f)** give information on the Maple function **f**. This information appears in a separate window. We demonstrate these commands as follows. A typical help window includes a detailed explanation of the syntax of the command, several examples illustrating typical results obtained with the command, and a list of related commands.

EXAMPLE: Use **?** to obtain information about the command **plot**.

SOLUTION: Notice that when we use **?** to obtain help, we do not need to include a semicolon or colon at the end of the command. The same results results are obtained by entering **help(plot)**.

> ?plot

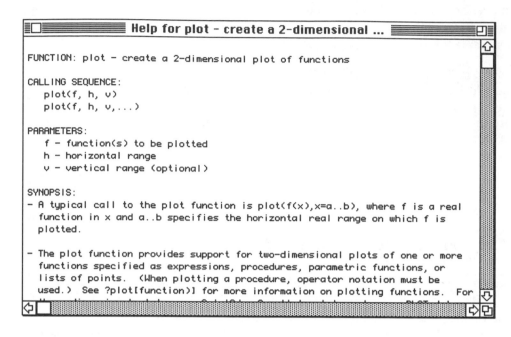

☰ *Additional Ways of Obtaining Help from Maple V*

On some platforms with a windows interface (like Macintosh, Windows, and NeXT), additional help features are accessed from the Maple menu. The Maple menu offers other ways of obtaining help. For example, if a command is selected, and then HELP FOR SELECTION is selected from the WINDOWS menu, Maple displays the help window for that command. This is illustrated with the `dsolve` command. We begin by typing and selecting the `dsolve` command.

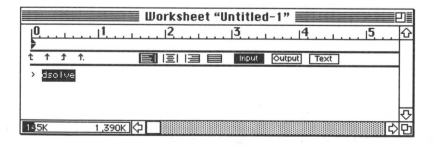

We then select HELP FOR SELECTION from the WINDOWS menu. Notice that in this case the word SELECTION is replaced by the word DSOLVE.

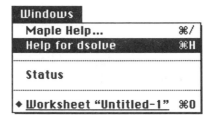

A portion of the resulting help window is displayed here. Note that we would obtain the same help window by entering `?dsolve`.

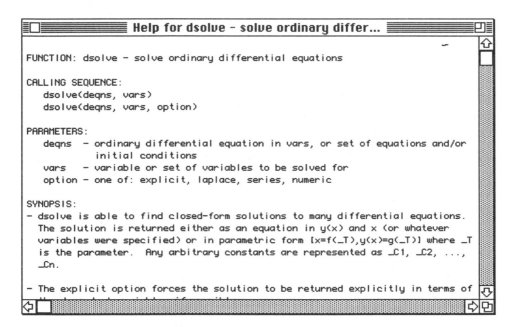

Assistance can also be obtained by selecting MAPLE HELP from the WINDOWS menu.

This yields the following help window:

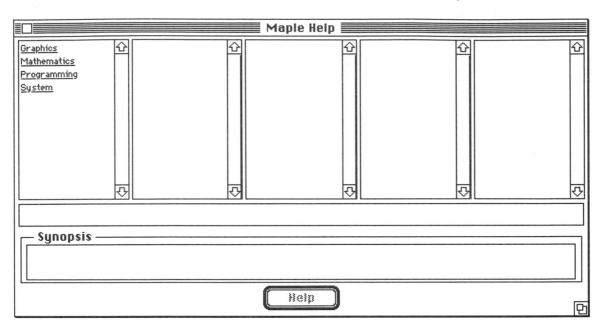

Any of the four topics can be selected to yield a variety of subtopics. In the following window, we show the result of selecting **Mathematics**. Notice that each underlined entry contains subtopics. To illustrate, we then select **Calculus** from the subtopics associated with the **Mathematics** category. We then select **Differential Equations** followed by **DEtools**, resulting in the following:

Next, we select DEplot.

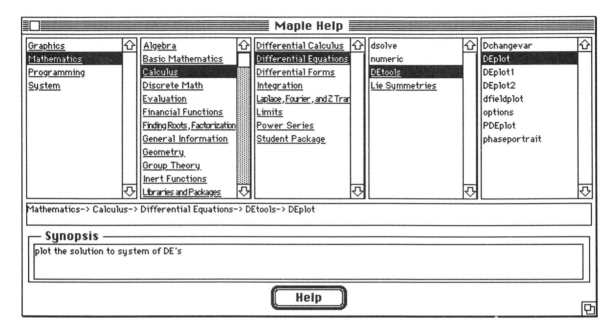

When we select HELP, the help window for the DEplot command is displayed as follows. The same help window would have been displayed if we had entered ?DEplot instead.

```
========= Help for DEtools[DEplot] - plot the sol... =========

FUNCTION: DEtools[DEplot] - plot the solution to system of DE's

CALLING SEQUENCES:
    DEplot(diffeq, vars, trange, inits, <options>)
    DEplot(diffeq, vars, trange, xrange, yrange, <options>)

PARAMETERS:
    diffeq    - either a sytem of n first order Differential Equations, or one
                nth order Differential Equation
    vars      - the names of the variables
    trange    - range of the independent variable
    inits     - initial conditions
    xrange    - range of the first dependent variable to be plotted
    yrange    - range of the second dependent variable to be plotted
    <options> - described below

SYNOPSIS:
- Given a sytem of first order differential equations of the form
    x' = f1(t,x,y), y' = f2(t,x,y), or a higher order equation of the form
    diff(y(x),x$n) = f(x,y)  and a set of initial conditions, DEplot produces a
```

THE MAPLE V TUTORIAL

For Release 2 users, an additional help facility is available by entering the command `tutorial()`. This beginning-level tutorial offers a quick way to get started with Maple V Release 2. We show the results of entering `tutorial()` at the input prompt here. **Don't forget to include a semicolon at the end of the command!**

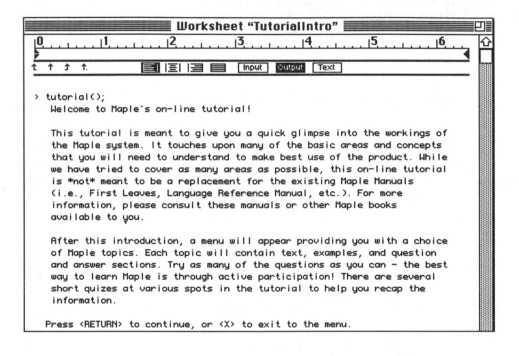

At this point, we press **RETURN** to continue, and Maple provides additional introductory information about the tutorial.

```
The programming of this tutorial was done entirely within Maple V
Release 2, taking advantage of the new I/O capabilities of that
release. As such, the tutorial will not run on any earlier versions of
Maple. The Maple source files that created this program are available
to you, with the intention that any user can thereby customize the
tutorial to suit his/her specific needs. For more information, read the
comments that appear at the beginning of the file "tutorial.src" which
resides in your Maple tutorial directory/folder.

This tutorial was written specifically to run on *all* Maple V Release
2 platforms. While it may not take full advantage of many of the
interface capabilities of some platforms, it is intended that future
releases of Maple will contain improved versions of this program. If
you have any questions or suggestions, please contact a Maple Support
Representative near you.

Now we begin!
Press <RETURN> to continue, or <X> to exit to the menu.
```

Pressing **RETURN** yields a list of the available chapters of the tutorial. Note that entering `tutorial(n)`, where *n* is an integer between 1 and 14, instead of `tutorial()` goes directly to chapter *n* of the tutorial, skipping over the introductory information.

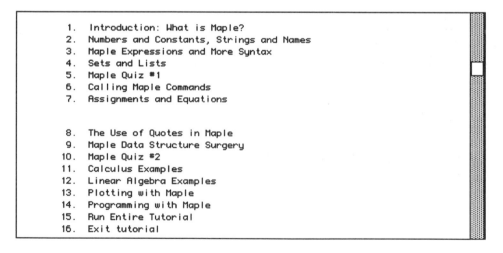

```
 1.   Introduction: What is Maple?
 2.   Numbers and Constants, Strings and Names
 3.   Maple Expressions and More Syntax
 4.   Sets and Lists
 5.   Maple Quiz #1
 6.   Calling Maple Commands
 7.   Assignments and Equations


 8.   The Use of Quotes in Maple
 9.   Maple Data Structure Surgery
10.   Maple Quiz #2
11.   Calculus Examples
12.   Linear Algebra Examples
13.   Plotting with Maple
14.   Programming with Maple
15.   Run Entire Tutorial
16.   Exit tutorial
```

At this point we are asked to make a selection. For example, we choose Chapter 3, "Maple Expressions and More Syntax."

```
  Please make a selection.
> 3
```

```
   MAPLE EXPRESSIONS AND SYNTAX

   Expressions are a very important structure in Maple. Most Maple objects
   are, at one level or another, made up entirely of these expressions.
   At the most basic level, an expression consists of a single value or
   unknown. Conversely, Maple expressions can consist of thousands upon
   thousands of values and unknowns strung together with the use of
   various arithmetic operators.

   Maple's arithmetic operators include:

      +              addition
      -              subtraction
      *              multiplication
      /              division
      ^              exponentiation
      !              factorial
      abs()          absolute value
      iqou()         integer quotient
      irem()         integer remainder
```

Note that we are frequently given the option to exit the tutorial. In this case, we continue and Maple illustrates various examples.

```
  Press <RETURN> to continue, or <X> to exit to the menu.
  The following are some examples of simple Maple expressions.

> a+b+c;
```

$$a + b + c$$

```
> 3*x^3-4*x^2+x-7;
```

$$3\,x^3 - 4\,x^2 + x - 7$$

```
> x^2/25+y^2/36;
```

$$\frac{1}{25}\,x^2 + \frac{1}{36}\,y^2$$

```
  As you can see, Maple echoes these expressions in a "pretty" form, the
  quality of which depends upon the capabilities of your monitor.
```

When Maple is finished with its demonstration, we are again given the option to exit the tutorial. We continue and Maple provides a brief introduction to the order of operations.

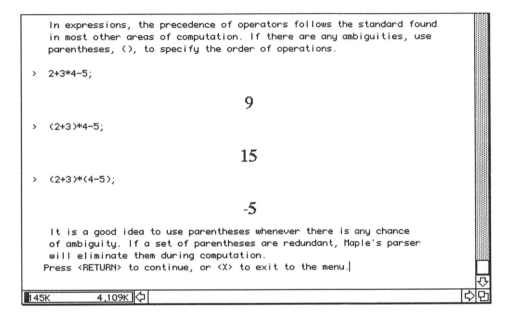

At this point, we exit the tutorial.

LOADING MISCELLANEOUS LIBRARY FUNCTIONS

Maple V's modularity, which gives Maple a great deal of flexibility, helps to minimize Maple's memory requirements. Nevertheless, although Maple contains many built-in functions that are loaded immediately when called, some other functions are contained in packages that must be loaded separately. Other functions, miscellaneous library functions, must also be loaded separately. A help window listing the miscellaneous library functions is obtained by entering `?index[libmisc]`.

Miscellaneous library commands must be loaded with the command `readlib(command)` before using them during a Maple session.

A typical example of a command that must be loaded with `readlib` is `invlaplace`, which is used to compute the inverse Laplace transform of functions. For example, entering

```
> invlaplace(1/(s^3*(s^2+9)),s,t);
```

$$\text{invlaplace}\left(\frac{1}{s^3(s^2+9)}, s, t\right)$$

does not compute the inverse Laplace transform of $\dfrac{1}{s^3(s^2+9)}$ because we have not loaded the invlaplace command. However, when we load the invlaplace command by entering

```
> readlib(laplace):
```

and then re-execute the command

```
> invlaplace(1/(s^3*(s^2+9)),s,t);
```

$$\frac{1}{18}t^2 - \frac{1}{81} + \frac{1}{81}\cos(3t)$$

Maple computes the inverse Laplace transform of $\dfrac{1}{s^3(s^2+9)}$.

LOADING PACKAGES

In addition to the standard library functions and miscellaneous library functions, a tremendous number of additional commands are available in various packages. All the commands in a particular package are loaded by entering with(packagename). Alternatively, commands contained in a package can be entered in their long form: packagename[command](arguments). We now use Maple's help facility to view a list of the available packages. Entering the command

```
> ?packages
```

displays the following window that lists the available packages.

```
═□═════════════ Help for Index of descriptions for pack...  ═════╣╚╣
                                                                           ⇧
HELP FOR: Index of descriptions for packages of library functions

SYNOPSIS:
- The following packages are available:

            approx:      Numerical Approximation
            combinat:    combinatorial functions
            DEtools:     Differential Equation Tools
            difforms:    differential forms
            Gauss:       create domains of computation
            GaussInt:    Gaussian Integers
            geom3d:      three-dimensional Euclidean geometry
            geometry:    two-dimensional Euclidean geometry
            grobner:     Grobner bases
            group:       permutation and finitely-presented groups
            liesymm:     Lie symmetries
            linalg:      linear algebra
            logic:       Boolean logic
            networks:    Graph Networks
            np:          Newman-Penrose formalism
            numtheory:   number theory
            orthopoly:   orthogonal polynomials
            padic:       Padic Numbers
            plots:       graphics package
            powseries:   formal power series                      ⇩
⇦█▒▒▒▒▒▒▒▒▒▒▒▒▒▒▒▒▒▒▒▒▒▒▒▒▒▒▒▒▒▒▒▒▒▒▒▒▒▒▒▒▒▒▒▒▒▒▒▒▒▒▒▒▒▒▒▒▒⇨ ⬚
```

Information about particular packages is obtained with `?packagename`. For example, with the command `?DEtools` we obtain information about the `DEtools` package, including a list of the commands contained in the package.

```
> ?DEtools
```

```
┌─────────────────────────────────────────────────────────────┐
│ ▤□▤▤▤▤▤▤▤▤▤▤ Help for Differential Equations Tools p... ▤▤▤▤▤ ▣▤▤ │
│                                                          ⇧    │
│  HELP FOR: Differential Equations Tools package DEtools      │
│                                                              │
│  CALLING SEQUENCE:                                           │
│     <function>(args)                                         │
│     DEtools[<function>](args)                                │
│                                                              │
│  SYNOPSIS:                                                   │
│  - To use a DEtools function, either define that function alone using the com- │
│    mand with(DEtools, <function>), or define all DEtools functions using the   │
│    command with(DEtools).  Alternatively, invoke the function using the long   │
│    form readlib(DEtools)[<function>].  This long form notation is necessary    │
│    whenever there is a conflict between a package function name and another    │
│    function used in the same session.                         │
│                                                              │
│  - The functions available are:                              │
│                                                              │
│        DEplot DEplot1 DEplot2 Dchangevar PDEplot  dfieldplot  phaseportrait │
│                                                              │
│  - For more information on a particular function see DEtools[<function>]. │
│                                                          ⇩    │
│ ◁▭▤▤▤▤▤▤▤▤▤▤▤▤▤▤▤▤▤▤▤▤▤▤▤▤▤▤▤▤▤▤▤▤▤▤▤▤▤▤▤▤▤▤▤▤▤▤▤▤▤▤ ▷▣ │
└─────────────────────────────────────────────────────────────┘
```

A typical example of a package is the linear algebra package, called `linalg`. For example, to compute the determinant of the matrix $\mathbf{A} = \begin{pmatrix} 10 & -6 & -9 \\ 6 & -5 & -7 \\ -10 & 9 & 12 \end{pmatrix}$, we must first define the matrix and then select a Maple command, or define a new one, that computes the determinant of a square matrix.

In this case, we first define $\mathbf{A}$ and then use the command `linalg[det]` to compute the determinant of $\mathbf{A}$. Note that the command `det` is contained in the package `linalg`.

```
> A:=array([[10,-6,-9],[6,-5,-7],[-10,9,12]]);
```

$$A := \begin{bmatrix} 10 & -6 & -9 \\ 6 & -5 & -7 \\ -10 & 9 & 12 \end{bmatrix}$$

```
> linalg[det](A);
```

$$6$$

However, we can use the `with` command to load a package. After a package has been loaded, subsequent calculations involving commands contained in the package can be entered directly.

We now use `with` to load the package `linalg`. The commands contained in the `linalg` package are displayed after the package is loaded. If a colon were used at the end of the command instead of a semicolon, then the package commands would be loaded but not listed.

```
> with(linalg);
```

[*BlockDiagonal, GramSchmidt, JordanBlock, Wronskian, add,*
    addcol, addrow, adj, adjoint, angle, augment, backsub,
    band, basis, bezout, blockmatrix, charmat, charpoly, col, coldim,
    colspace, colspan, companion, concat, cond, copyinto,
    crossprod, curl, definite, delcols, delrows, det, diag, diverge,
    dotprod, eigenvals, eigenvects, entermatrix, equal,
    exponential, extend, ffgausselim, fibonacci, frobenius,
    gausselim, gaussjord, genmatrix, grad, hadamard, hermite,
    hessian, hilbert, htranspose, ihermite, indexfunc, innerprod,
    intbasis, inverse, ismith, iszero, jacobian, jordan, kernel,
    laplacian, leastsqrs, linsolve, matrix, minor, minpoly, mulcol,
    mulrow, multiply, norm, normalize, nullspace, orthog,
    permanent, pivot, potential, randmatrix, randvector, range,
    rank, ratform, row, rowdim, rowspace, rowspan, rref,
    scalarmul, singularvals, smith, stack, submatrix, subvector,
    sumbasis, swapcol, swaprow, sylvester, toeplitz, trace,
    transpose, vandermonde, vecpotent, vectdim, vector]

We can then compute the determinant of **A** using the command `det` instead of the long form `linalg[det]`.

```
> det(A);
```

6

Glossary

Symbols

+

The symbol **+** is used to represent addition. Thus, entering **a+b** returns the sum of **a** and **b**.

−

The symbol **−** is used to represent subtraction. Thus, entering **a−b** returns the result of subtracting **b** from **a**.

*

The symbol ***** represents multiplication. Thus, entering **a*b** returns the product of **a** and **b**. Be careful when entering expressions. For example, it is common to write *ab* to represent the product of *a* and *b*, but with Maple be sure you enter **a*b**.

/

The symbol **/** is used to represent division. Thus, entering **a/b** returns the quotient of dividing **a** by **b**.

^

The symbol **^** represents exponentiation. Thus, entering **a^b** returns **a** raised to the **b**th power.

**

The symbol ****** represents exponentiation. Thus, entering **a**b** returns **a** raised to the **b**th power.

->

The symbol **->**, which we will call an **arrow**, is a functional operator. The symbol **->** is obtained by typing a minus sign (−) followed by a greater than sign (>). For example, the expression **x->x^3** represents the function that cubes its argument. Use **:=** to assign function names.

:=

:= represents the assignment operator. Thus, entering **lefthandside:=righthandside** assigns **lefthandside** the value **righthandside**. The command **name:='name'** clears all prior definitions, if any, of **name**. Use **:=** to define functions and name objects.

$

$ is used to produce a sequence of elements. For example, **expr $i=m..n** evaluates **expr** from **i=m** to **i=n**. Also, **expr$n** creates a sequence of length **n** of the indicated expression. The ellipsis symbol **..** (obtained by typing two periods) is used to indicate the range of a list.

@

@ represents the composition operator. If f and g are functions, the composition $(f \circ g)(x) = f(g(x))$ can be computed with both **f(g(x))** and **(f@g)(x)**.

@@

The symbol **@@** represents the repeated composition operator. Thus, if f is a function, **(f@@n)(x)** computes the composition $(\underbrace{f \circ f \circ \cdots \circ f}_{n\text{-times}})(x)$.

&*

&* represents noncommutative multiplication. Thus, if **A** and **B** are matrices, **evalm(A &* B)** computes the matrix product **AB**, provided that the product is defined.

=

Use **=** to separate the left- and right-hand sides of equations.

> and >=

The symbols **>** and **>=** represent greater than and greater than or equal to symbols, respectively. Use these symbols to define inequalities in the same manner that you use **=** to define equations.

< and <=
The symbols **<** and **<=** represent less than and less than or equal to symbols, respectively. Use these symbols to define inequalities in the same manner that you use = to define equations.

"
The symbol **"** represents the last expression. Similarly, **""** represents the second to last expression, and **"""** represents the third to last expression. Often, you can use **"** to refer to results that have not been named.

%
The symbol **%** is used as a label, often appearing in lengthy results.

A

abs
abs(val) computes the absolute value of the real or complex number **val**.

alias
alias(name=expression) allows the user to denote **name** to be an abbreviation of **expression**. The **alias** command is particularly useful when dealing with complicated expressions repeatedly.

allvalues
allvalues(expr) evaluates all of the possible values that involve **RootOf**. It is especially useful in evaluating expressions that result from **solve**.

arccos
arccos(x) represents the inverse cosine function.

arccosh
arccosh(x) represents the inverse hyperbolic cosine function.

arccot
arccot(x) represents the inverse cotangent function.

arccoth
`arccoth(x)` represents the inverse hyperbolic cotangent function.

arccsc
`arccsc(x)` represents the inverse cosecant function.

arccsch
`arccsch(x)` represents the inverse hyperbolic cosecant function.

arcsec
`arcsec(x)` represents the inverse secant function.

arcsech
`arcsech(x)` represents the inverse hyperbolic secant function.

arcsin
`arcsin(x)` represents the inverse sine function.

arcsinh
`arcsinh(x)` represents the inverse hyperbolic sine function.

arctan
`arctan(x)` represents the inverse tangent function.

arctanh
`arctanh(x)` represents the inverse hyperbolic tangent function.

array
`array(indexf,irange,vals)` creates an array using the index function `indexf`, the integer ranges listed in `irange`, and the list of initial values indicated in `vals`. Note that the arguments of array are all optional. Frequently, we use `array` or `linalg[matrix]` to define matrices.

assign
`assign(a,b)` or `assign(a=b)` makes the assignment `a:=b`. This command can be used with a list or set of equations, especially with results returned by commands such as `solve` and `dsolve`.

assume
The `assume` facility allows you to make assumptions about functions or variables. For detailed information about the `assume` facility, enter the command `?assume`. The syntax of this command is generally `assume(name,property)`, where `name` is the name of a function or variable and `property` is a property like continuous, real, or integer. `assume(name,property)` assigns the property `property` to `name`, `additionally(name,property)` adds the additional `property` to `name` (properties assigned to `name` with `assume` are retained), and `is(name,property)` determines if `name` has the property `property`. Clear all properties given to `name` using the `assume` facility with `unassign`: enter `name:='name'`.

B

BesselJ
`BesselJ(v,x)` represents the Bessel function of the first kind, $J_v(x)$.

BesselY
`BesselY(v,x)` represents the Bessel function of the second kind, $Y_v(x)$.

C

Ci
`Ci(x)` represents the cosine integral function, Ci(x), which is given by

$$\mathtt{Ci(x)} = \gamma + \ln(ix) - \frac{1}{2}i\pi \int_0^x \frac{\cos t - 1}{t}$$

$$= \gamma + \ln x + \int_0^x \frac{\cos t - 1}{t}, \ x \text{ real.}$$

collect
`collect(expr,x)` collects all of the coefficients of `x` that are raised to the same power in the expression `expr`.

combine
combine(exprlist,namelist) applies the transformations that combine the expressions indicated in exprlist according to the rules associated with namelist that may include exp, ln, power, Psi, and trig. Notice that combine tends to perform the inverse transformations of those carried out by expand.

convert
convert(expression,type) is used to convert expression to the type of expression specified in type. For example, convert(list,set) converts the list list to a set; convert(fraction,parfrac) returns the partial fraction decomposition of fraction.

cos
cos(x) represents the cosine function.

cosh
cosh(x) represents the hyperbolic cosine function.

cot
cot(x) represents the cotangent function.

D

D
D represents the differential operator. If f is a function of one variable, then D(f) computes f'; D(f)(x) returns $f'(x)$. If f is a function of more than one variable, then D[i](f) calculates $\dfrac{\partial f}{\partial x_i}$, where x_i represents the ith argument of f and D[i,j](f) computes $\dfrac{\partial^2 f}{\partial x_i \partial x_j}$.

diff
diff(expr,varlist) computes the partial derivative of expr with respect to the variables in varlist. For example, if f(x) is a function of one variable, then diff(f(x),x) computes f'(x). Also, for higher-order derivatives, the abbreviation var$n may be used in varlist to indicate differentiation n times with respect to var.

Dirac
Dirac(t) represents the Dirac delta function, while **Dirac(n,t)** represents the nth derivative of this function.

dsolve
dsolve(deqs,vars) solves the differential equation or set of differential equations using any indicated initial conditions in **deqs** for the variables given in **vars**. If the third argument **numeric** is used, then a numeric solution is obtained, which may be graphed using the **odeplot** command contained in the **plots** package. Other options include **explicit**, **laplace**, and **series**.

E

Ei
Ei(n,x) represents the exponential integral that is given by the formula

$$Ei(n,x) = \int_1^\infty \frac{e^{-xt}}{t^n}\, dt,$$

where the real part of x must be greater than zero. For real values of x, **Ei(x)** returns the principal value of $-\int_{-x}^\infty \frac{e^{-t}}{t}\, dt$.

eval
eval is used to evaluate an expression fully or at one level. **eval(expr)** fully evaluates the indicated expression, while **eval(expr,n)** yields an n-level evaluation.

evalf
evalf(expr) evaluates expressions that involve constants such as **E**, **gamma**, and **Pi** and functions such as **exp**, **erf**, **cos**, **cosh**, **arccos**, and **GAMMA** to floating-point expressions. If the second argument is used as in **evalf(expr,n)**, then the result is calculated with n-digit floating-point arithmetic.

evalm
evalm(expr) evaluates expressions that involve matrices.

exp
exp(x) represents the exponential function e^x, the inverse of the natural logarithm function $\ln x$.

expand

`expand(expr)` distributes products over sums in expressions involving polynomials. When the expression contains a quotient of polynomials, then the expansion is carried out in the numerator only. The expression may include the standard trigonometric functions, the hyperbolic trigonometric functions, the exponential function, special functions such as the Bessel functions, as well as standard Maple commands such as `limit, min`, and `det`.

F

factor

`factor(expr)` factors the algebraic expression `expr` that involves integer, rational, or algebraic numbers as coefficients.

for

The command

```
for name from start by increment to
        finish do commands od
```

executes the commands listed in `commands` for `name=start` to `name=finish` using a step size of `increment`. If `by increment` is omitted, the step size is 1. Similarly, if `from start` is omitted, the commands are executed from `name=1` to `name=finish`. In Release 2, the clauses `by, from, to`, and `while` can appear in any order.

fsolve

`fsolve(eqs,vars,opts)` uses floating-point arithmetic to solve the equation (or set of equations) in `eqs` in terms of the variables `vars` using the indicated options in `opts`. Note that `opts` may include `complex` to include complex floating-point numbers, `fulldigits` to restrict `fsolve` from reducing the setting of `Digits, interval` to indicate a particular interval over which to solve, and `maxsols` to indicate the number of least roots to find.

G

GAMMA

`GAMMA(x)` represents the gamma function, where

$$\text{GAMMA}(x) = \Gamma(x) = \int_0^x e^{-t} t^{x-1} \, dt.$$

H

help (?)

help(t) or **?t** yields information concerning the command **t**. Maple V Release 2 help facilities are discussed in more detail in the Appendix.

Heaviside

Heaviside(t) repesents the unit step function $f(t) = \begin{cases} 1, t \geq 0 \\ 0, t < 0 \end{cases}$ that also satisfies the relationship **diff(Heaviside(t),t)=Dirac(t)**.

I

I

I represents the imaginary number $I = \sqrt{-1}$.

if

if statements are conditional operators that direct the flow of the execution of the procedure.

int

int represents Maple's integration function for computing definite as well as indefinite integrals. **int(f(x),x)** yields the integral of **f(x)** with respect to the variable **x**, $\int f(x)\,dx$, when possible. If the closed form of this integral cannot be found, then the command is returned unevaluated. **int(f(x),x=a..b)** attempts to determine the definite integral $\int_a^b f(x)\,dx$. If this command is returned unevaluated, then **evalf** can be used so that numerical integration with the Newton-Cotes method is used. Alternatively, you may load the command `evalf/int` by entering **readlib(`evalf/int`)** and then use `evalf/int` instead of **int** to perform numerical integration immediately.

invlaplace

invlaplace(expr,s,t) computes the inverse Laplace transform of the expression **expr** (which is given in the variable **s**) to yield an expression in terms of the variable **t**. **expr** may involve polynomials, sums of rational functions of polynomials, as well as expressions that are Laplace transforms of functions involving exponentials and Bessel functions with linear arguments. If the result involves a convolution integral, then the variable U_ appears in the output of the command. **readlib(laplace)** must be entered before using this function.

isolate
isolate(eq,expr) isolates the expression expr in the equation eq. Note that if eq is not entered in the form of an equation, then the entry is assumed to be of this form. A third argument of this command may be included as in isolate(eq,expr,n) in order to specify that the maximum number of transformation steps used to isolate the expression should be the positive integer n. readlib(isolate) must be entered before this command is used.

L

laplace
laplace(expr,t,s) gives the Laplace transform (in the variables) of the expression expr of the variable t. The Laplace transform of $f(t)$ is defined to be $L\{f\} = \int_0^\infty e^{-st} f(t)\, dt$. This command recognizes derivatives and integrals as well as expressions that include the exponential function, the trigonometric functions sin(x) and cos(x) with linear arguments, the hyperbolic trigonometric functions cosh(x) and sinh(x) with linear arguments, and the Bessel functions BesselJ(x) and BesselY(x) with linear arguments. In addition, the method of Laplace transforms can be used to solve some differential equations by using the dsolve command together with the laplace option.

lhs
lhs(expr) yields the left-hand side of the expression expr that may be an equation, an inequality, a relation, or a range. This command gives the same result as op(1,expr).

limit
limit(f,x=x0) computes the limit of the algebraic expression f as the variable x approaches x0, where x0 may be infinity or -infinity. This command may be entered as limit(f, x=x0,d) so that the direction d, which includes left, right, real, or complex, may be stated. You should exercise caution when interpreting the results returned by executing a limit command.

ln
ln(x) represents the natural logarithm function.

M

map
map(f,expr) applies f to the operands of expr. If expr is a table or an array, then f is applied to each member of expr.

match

`match(expr=mpatt,var,'sym')` yields a value of `true` if Maple can match the expression `expr` with the pattern given in `mpatt`, where `var` indicates the main variable and the match is assigned the name `sym`. Hence, if the match is found, then the relationship `subs(sym, mpatt)=expr` is satisfied.

minus

`A minus B` determines the set difference of the two sets A and B. This is defined as the intersection of A and the complement of B.

N

nops

`nops(expr)` gives the number of operands or components of the expression `expr`. Use `op` to extract operands of an expression.

O

op

`op(j,expr)` extracts the jth component of the expression `expr`. This command may be entered as `op(i..j,expr)` so that the ith through jth components are given. If `op(expr)` is entered, then all components of `expr` are given. Generally, `expr[j]` and `op(j,expr)` both return the jth part of the expression; both `expr[i..j]` and `op(i..j,expr)` return the ith through jth parts of the expression. Use `op` to display abbreviated portions of long lists.

Order

`Order` is a variable representing the truncation order of results returned by `series`. The default value of `Order` is 6. Thus, `series(f(x),x=a)` returns the terms of the power series for $f(x)$ about $x = a$ to order 6.

P

plot

`plot(f,x=a..b,y=c..d,opts)` plots the two-dimensional graph of the function `f` over the horizontal range `a..b` and the vertical range `c..d` (which is optional) using the indicated options in `opts`. The options of `plot` include (with default settings in parentheses) `coords (polar)`, `numpoints (49)`, `resolution (200)`, `xtickmarks`, `style (LINE)`. `coords=polar` indicates that the plot be generated in polar coordinates, `numpoints=n` specifies the minimum number of points to be generated, `resolution=n` sets the horizontal display resolution of the device (in pixels), `xtickmarks` (and/or `ytickmarks`)=n indicates the number of tick marks to be placed along the x-axis (y-axis), `style=s` states the interpolation style (`POINT`, `LINE`) that should be used to create the plot, and `title=s` states the character string that should be used as the title of the graph. A multiple plot is produced with `plot({f1,f2,...},x=a..b,y=c..d,opts)`. Generally, you should use `plot` to graph functions of a single variable, lists of two-dimensional points, and parametric functions in two dimensions.

plot3d

`plot3d(f,x=a..b,y=c..d,opts)` generates the three-dimensional plot of the function `f` of two variables `x` and `y` using the indicated options listed in `opts`. A multiple plot is generated with `plot3d({f,g,...},x=a..b,y=c..d,opts)`. The options of `plot3d` are given as follows. Note that information in parentheses is given in the form (other possible settings; default setting) `axes` (BOXED, FRAME, NORMAL; NONE); `coords=c`(cylindrical, spherical; cartesian); `grid=[m,n]` (m and n any real numbers; equally spaced); `labels=[x,y,z]` (x, y, and z are any real numbers; no label); `numpoints=n` (n any positive integer; 625); `orientation=[theta,phi]` (angles in degrees; [45,45]); `projection=r` (r any real number on 0..1; 1); `scaling=s` (CONSTRAINED; UNCONSTRAINED); `shading=s` (XYZ, XY, Z; device dependent); `style=s` (PATCH, POINT, WIREFRAME; HIDDEN); `title=t` (t any character string; no title); `view=zmin..zmax` (or `[xmin..xmax,ymin..ymax,zmin..zmax]`; entire surface). The `shading` option indicates how the surface is colored; the `view` option indicates the minimum and maximum coordinates of the surface that should be displayed; the `orientation` option indicates the point in spherical coordinates from which the surface is viewed; the `projection` option indicates the perspective from which the surface is viewed, with 1 indicating an orthogonal projection and 0 a wide-angle view; and the `scaling` option indicates if the surface should be scaled in order to fit the screen where scaling occurs with the `CONSTRAINED` setting. Generally, you should use `plot3d` to graph functions of two variables or parametric curves in three dimensions.

R

readlib

`readlib` is used to read library files. The command `readlib(f)` executes

```
read ".lname."/`.f.`.m`;.
```

The command `readlib(f,f1,f2,...)` causes all of the indicated files to be read as in `read f1; read f2; ...`. Note that placing a colon at the end of a `readlib` command eliminates the procedure output from being displayed. Generally, use `readlib` to load miscellaneous library functions.

rhs
rhs(expr) yields the right-hand side of the equation, inequality, relation, or range given in expr. Note that the result of this command is the same as that of op(2,expr).

RootOf
RootOf is used to represent the roots of an equation of one variable. For example, RootOf(expr), where expr is a univariate expression or an expression of _Z, represents the roots of the expression expr. The variable may be entered in the command as RootOf(expr,x). If expr is not an equation, then expr=0 is implied. If expr is irreducible, then RootOf represents an algebraic extension field. The commands diff, evalf, series, and simplify can all be applied to expressions involving RootOf.

S

seq
seq is used to create a sequence. seq(f(k),k=i..j) creates the sequence f(i),f(i+1), ...,f(j-1),f(j), where i and j need not be integers.

series
series(f(x),x=a,n) computes the series expansion of f about x=a up to order n, where n is a nonnegative integer. Note that if the third argument is omitted, then the order is taken to be the setting of Order, which has default setting 6.

Si
Si(x) represents the sine integral function that is given by $Si(x) = \int_0^x \frac{\sin t}{t} dt.$

simplify
simplify(expr) applies simplification rules to simplify the expression expr.

sin
sin(x) represents the sine function.

sinh
sinh(x) represents the hyperbolic sine function.

solve

solve(eqs,vars) determines the solution of the single equation or system of equations given in eqs in terms of the variable vars. The result of this command is a set of equations that indicate the solution. However, if no solution is found, then NULL is given.

sqrt

sqrt(x) calculates the square root of integers or polynomials x that are perfect squares. evalf must be used in other cases to obtain the square root.

subs

subs(s1,s2,...,expr) returns the expression that is generated by substituting s1,s2,... into the expression expr. Since subs does not cause evaluation of the expression, eval must be used to evaluate the resulting expression. If eval is unable to evaluate the resulting expression exactly, try using evalf.

sum

sum(f,k) computes the indefinite sum of $f(k)$ with respect to k. Hence, the formula g is determined that satisfies the equation $g(k + 1) - g(k) = f(k)$. Also, sum(f,k=i..j) computes the sum $\sum_{k=1}^{j} f(k)$. Finally, sum(f,k=a) calculates the definite sum over the roots of the polynomial a, where the roots of a must be in terms of RootOf.

T

table

table(infunc,init) generates the table of values using the indexing function infunc and the initial values given in the list init. Note that the values in init do not have to be integers. Note that the table command may be used with only the first argument or only the second.

tan

tan(x) represents the tangent function.

taylor

taylor(f,x=a,n) computes the Taylor series expansion of the function f about the value x=a up to order n. Note that the third argument is optional. Note that a series object s may be converted to a polynomial with the command convert(s,polynom).

trunc

`trunc(x)` truncates the numerical argument `x` to an integer. If `x` is not an integer, then the expression is returned in an unevaluated form.

tutorial

The command `tutorial()` begins the on-line introductory Maple V Release 2 tutorial. The fourteen chapters of the tutorial are as follows:

1. Introduction: What Is Maple?
2. Numbers and Constants, Strings and Names
3. Maple Expressions and More Syntax
4. Sets and Lists
5. Maple Quiz #1
6. Calling Maple Commands
7. Assignments and Equations
8. The Use of Quotes in Maple
9. Maple Data Structure Surgery
10. Maple Quiz #2
11. Calculus Examples
12. Linear Algebra Examples
13. Plotting with Maple
14. Programming with Maple

If you enter `tutorial(n)`, where *n* is an integer between 1 and 14, instead of `tutorial()`, Maple skips the introductory information about the tutorial and starts at chapter *n*. See the Appendix for further details about the Maple `tutorial`.

U

union

`A union B` yields the union of sets `A` and `B`.

updates

The command `?updates,version` (where `version` is usually like `v5.2`, `v5`, or `v4.3`) returns a help window indicating the new features available for the `version` release of Maple.

W

W

`W(x)` represents the omega function that satisfies the relationship $\omega(x)e^{\omega(x)} = x$ that is analytic at $x = 0$. `W(k,x)`, where k is an integer, can be used to obtain other branches of the omega function.

with

`with(pack,func1,func2,...)` defines the function names `func1`, `func2`,... from the Maple package `pack`. Similarly, the command `with(pack)` defines all of the functions that are contained in the package.

DEtools

Load all the commands contained in the `DEtools` package by entering `with(DEtools)`. A help window describing the options available with the various commands is obtained by entering `?DEtools[options]`.

DEplot

The command

```
DEplot(differentialequation,variables,
     to..t1,set_of_conditions)
```

graphs the solutions to the nth-order differential equation or system of differential equations `differentialequation` that satisfy the conditions specified by `set_of_conditions`. Unlike `DEplot1`, which handles a single first-order equation, or `DEplot2`, which handles a system of two equations, `DEplot` can be used to graph solutions of systems involving more than two equations.

DEplot1

The command

```
DEplot1(differentialequation,y(x),
        x=x0..x1,y=y0..y1)
```

graphs the direction field for the first-order differential equation `differentialequation` on the rectangle $[x0,x1] \times [y0,y1]$. The command

```
DEplot1(differentialequation,y(x),
  x=x0..x1,y=y0..y1,set_of_conditions)
```

graphs the direction field along with the solutions specified in `set_of_conditions`. If the option `arrows=NONE` is included in the command, the associated direction field is not displayed.

Unlike `dsolve` together with the `numeric` option, `DEplot1` does not return a numerical solution that can be evaluated for particular numbers. Thus, you should use `DEplot1` to graph solutions to an equation when a formula for the solution is not desired. In other cases, use `dsolve` followed by `plot` (when an exact solution is needed and can be computed), or use `dsolve` together with the `numeric` option followed by the `odeplot` command, contained in the `plots` package.

DEplot2

The command

```
DEplot2([eq1,eq2],[t,x,y],t=t0..t1,
        initialconditions)
```

graphs the solutions to the system of differential equations $\begin{cases} x'(t) = \text{eq1} \\ y'(t) = \text{eq2} \end{cases}$ that satisfy the conditions specified in `initialconditions` for `t0<=t<=t1`. If the system can be written in the form

$$\begin{pmatrix} x' \\ y' \end{pmatrix} = \begin{pmatrix} a & b \\ c & d \end{pmatrix} \begin{pmatrix} x \\ y \end{pmatrix},$$

then you can replace `[eq1,eq2]` by the matrix $\begin{pmatrix} a & b \\ c & d \end{pmatrix}$ (see `dfieldplot`). If the system is autonomous and the command is entered as

```
DEplot2([eq1,eq2],[x,y],t=t0..t1,
        x=x0..x1,y0..y1)
```

or

```
DEplot2([eq1,eq2],[x,y],t=t0..t1,
    initialconditions,x=x0..x1,y=y0..y1),
```

then the direction field for the equation is graphed along with any solutions specified by `initialconditions` on the rectangle $[x0,x1] \times [y0,y1]$.

linalg

Load all the commands contained in the `linalg` package by entering `with(linalg)`.

adj

`adj(mat)` or `adjoint(mat)` computes the adjoint of the matrix `mat`.

augment
`augment(matrixlist)` yields the matrix that results when the rows of the matrices in `matrixlist` are joined.

charpoly
The command `charpoly(mat,lbda)` calculates the characteristic polynomial, the determinant of the characteristic matrix, `lbda*mat-Id`, where `Id` represents the identity matrix of the same dimensions of the square matrix `mat`.

det
The command `det(mat)` or `det(mat,sparse)` computes the determinant of the square matrix `mat`.

dotprod
`dotprod(u,v)` calculates the dot product of the n-dimensional vectors using the formula

$$\texttt{dotprod(u,v)} = u \bullet v = \sum_{i-1}^{n} u_i v_i,$$

if $u = \langle u_1, u_2, \ldots, u_n \rangle$ and $v = \langle v_1, v_2, \ldots, v_n \rangle$.

eigenvals
`eigenvals(mat)` lists the eigenvalues `lbda` of the matrix `mat` by evaluating `det(lbda*I-mat)`, where `I` is the identity matrix of the same size as `mat`. If the characteristic polynomial associated with `mat` is of degree greater than 4, then `eigenvals(mat, 'implicit')` should be used, in which case the eigenvalues are given in terms of the `RootOf` command. Also, the generalized eigenvalue problem is solved with `eigenvals(mat1,mat2)`, which solves `det(lbda*mat2-mat1)`. The built-in command `Eigenvals` is the inert form of the command `eigenvals`. `Eigenvals` is a built-in function and need not be loaded before using. Approximations of eigenvalues and the corresponding eigenvectors of a given matrix can be approximated with `Eigenvals`. When the command is entered in the form `Eigenvals(mat, 'vecs')`, then the eigenvectors of the matrix mat are given in the array `vecs`.

eigenvects
`eigenvects(mat)` computes the exact value of the eigenvectors of the square matrix `mat`. The entries of the output list of this command have the form `[eigenval[i],mult[i], {vec[1,i],..., vec[m,i]}]`, where `eigenval[i]` represents the eigenvalue, `mult[i]` the multiplicity of the eigenvalue, and `{vec[1,i],..., vec[m,i]}` the set of linearly inde-

pendent eigenvectors associated with eigenval[i]. Eigenvectors are determined by finding and factoring the characteristic polynomial associated with mat. If nonlinear factors are encountered in this factoring process, then results are given in terms of the RootOf command.

inverse

The command inverse(mat) yields the inverse of the square matrix mat. Note that an error message results if mat is singular. The method by which this matrix is found depends on the size of mat. If mat is a 4 x 4 or smaller matrix, then Cramer's rule is used. Otherwise, Gauss-Jordan elimination is used to transform mat into the corresponding identity matrix.

jacobian

jacobian(flist,vlist) computes the Jacobian matrix of the vector-valued function flist in terms of the variables indicated in vlist. Hence, the outcome is a matrix in which the element in the (i, j)th position is given by $\frac{\partial f_i}{\partial x_j}$, where f_i represents the ith element of flist and x_j the jth element of vlist.

linsolve

The command linsolve(A,b) yields the vector solution that satisfies $Ax = b$ if a unique solution exists. Otherwise, if no solution exists, then NULL is given, and if infinitely many solutions exist, then the solutions are given in terms of the parameters t1, t2,.... This command can be used to solve matrix equations as well. Thus, linsolve(A,B) solves the matrix equation $AX = B$.

matrix

matrix(m,n,vlist) generates the m $\times$ n rectangular array using the m*n values given in vlist. (The first n values in vlist are placed in the first row of the matrix, the next n values in the second row, etc.) matrix(veclist) creates a matrix with the first element of veclist assigned to the first row, the second element to the second row, etc. Also, matrix(m,n) generates an m $\times$ n matrix in which the elements are not specified. A function func can be used as the third argument of matrix in order that element values be determined according to func(m,n).

Wronskian

The command Wronskian(listoffunctions,variable) determines the Wronskian matrix for the list of functions listoffunctions with respect to the variable variable.

orthopoly

Load the commands contained in the orthopoly package by entering with(orthopoly).

P

`P(n,a,b,x)` or `orthopoly[P](n,a,b,x)` evaluates the nth Jacobi polynomial with the parameters a and b at x where n is a nonnegative integer, a and b are rational numbers greater than −1 or nonrational algebraic expressions, and x is an algebraic expression. Similarly, `P(n,x)` or `orthopoly[P](n,x)` determines `P(n,0,0,x)`, the nth Legendre polynomial $P_n(x)$. These polynomials satisfy the orthogonality condition

$$\int_{-1}^{1} (1-x)^a (1-x)^b P(m,a,b,x) P(n,a,b,x)\, dx = 0$$

for $m \neq n$. Also, $P_n(x)$ satisfies the differential equation $\left(1 - x^2\right) \dfrac{d^2y}{dx^2} - 2x\dfrac{dy}{dx} + n(n+1)y = 0.$

plots

Load all the commands contained in the `plots` package by entering `with(plots)`.

animate

The command

```
animate(f(x,t),x=x0..x1,t=t0..t1,
              options
```

generates an animation of f by graphing $f(x, t)$ on the interval $[x_0, x_1]$ for 16 equally spaced values of t between t_0 and t_1. `animate` has the same options as `plot` in addition to the option `frames=n`, where n is the number of graphs to generate. Generally, larger values of n result in a smoother animation.

animate3d

The command

```
animate3d(f(x,y,t),x=x0..x1,y=y0..y1,
          t=t0..t1,options)
```

generates an animation of f by graphing $f(x, y, t)$ for $x_0 \le x \le x_1$ and $y_0 \le y \le y_1$ for eight equally spaced values of t between t_0 and t_1. `animate3d` has the same options as `plot` in addition to the option `frames=n`, where n is the number of graphs to generate. Generally, larger values of n result in a smoother animation.

contourplot

The **level curves** of the function $f(x, y)$ are curves in the xy-plane that satisfy the equation $f(x, y) = c$, where c is a constant. Maple graphs several of the level curves of the function $f(x, y)$ on the rectangle

$[x_0, x_1] \times [y_0, y_1]$ with the command

```
counterplot(f(x,y),x=x0..x1,y=y0..y1,
                    options).
```

The result of entering a `contourplot` command is a three-dimensional graphics object. Thus, `contourplot` has the same options as `plot3d`.

display
The command

```
display(plotlist,options)
```

displays the previously created two-dimensional graphics that are listed in `plotlist`. `plotlist` must be either a set (enclosed in braces `{}`) or a list (enclosed in brackets `[]`). The options given in `options` include `r=c..d` to specify the dimensions of the plotting window or `view=[x0..x1, y0..y1]` or `view=[x=x0..x1,y=y0..y1]` to specify the horizontal and vertical bounds on the viewing window. `title=name` may also be used as an option. In addition to displaying several graphs together, `display` can also be used to animate lists of two-dimensional graphics when the option `insequence=true` is included.

display3d
The command

```
display3d(plotlist,options)
```

displays the previously created three-dimensional graphics that are listed in `plotlist`. Hence, `display3d` may include graphics produced by `cylinderplot`, `matrixplot`, `plot3d`, `pointplot`, `spacecurve`, `sphereplot`, and `tubeplot`. All options of `plot3d` can be used in the option list of `display3d` except `coords`. As with the command `display`, `display3d` can be used to animate lists of graphics when the option `insequence=true` is included in the `display3d` command.

fieldplot
The command

```
fieldplot([f(x,y),g(x,y)],x=x0..x1,
                    y=y0..y1)
```

graphs the vector field $\langle f(x, y), g(x, y) \rangle$ on the rectangle $[x_0, x_1] \times [y_0, y_1]$. In addition to the options `grid=[m,n]` and `arrows=LINE, THIN, SLIM,` or `THICK` (the default is `THIN`), `fieldplot` has the same options as `plot`.

fieldplot3d
The command

```
fieldplot3d([f(x,y,z),g(x,y,z),
        h(x,y,z)],x=x0..x1,y=y0..y1,z=z0..z1)
```

graphs the vector field $\langle f(x, y, z), g(x, y, z), h(x, y, z) \rangle$ on the parallelepiped $[x_0, x_1] \times [y_0, y_1] \times [z_0, z_1]$. In addition to the options `grid=[m,n,p]` and `arrows=LINE, THIN, SLIM,` or `THICK` (the default is `THIN`), `fieldplot3d` has the same options as `plot3d`.

implicitplot
The command

```
implicitplot(eq,x=x0..x1,y=y0..y1,
                options)
```

graphs **eq** in the rectangle $[x_0, x_1] \times [y_0, y_1]$.

odeplot
The command

```
odeplot(procedure,[x,y(x)],x0..x1,
                options)
```

graphs the function $y(x)$ on the interval $[x_0, x_1]$, where $y(x)$ is specified in the procedure **procedure**, returned by using **dsolve** together with the **numeric** option.

spacecurve
The command

```
spacecurve(cplist,options)
```

yields a three-dimensional plot of the curve or curve(s) given by the components listed in **cplist**. The options that may be listed in **options** include all of those used with **plot3d** except the one to specify the grid size.

powseries

Load all the commands contained in the **powseries** package by entering **with(powseries)**.

powsolve
The command **powsolve(differentialequation)** uses power series methods to solve the linear differential equation or system of differential equations specified in **differentialequation**.

tpsform

The command

    tpsform(p,variable,n)

displays the terms of the power series **p** in terms of the variable **variable** up to, but not including, order n. The 0-term in the result indicates the omitted higher-order terms of the power series.

student

Load all the commands contained in the **student** package by entering **with(student)**.

completesquare

The command **completesquare(expr)** completes the square by writing **expr** as the sum of perfect squares and a remainder. If more than one variable is present in **expr**, then **completesquare(expr,var)** should be used to indicate the variable with which to complete the square in **var**.

equate

The command **equate(A,B)** returns the equation or set of equations obtained by equating components of A and B.

Selected References

Abell, M., and Braselton, J., *Maple V by Example*, AP PROFESSIONAL, 1994.

Abell, M., and Braselton, J., *The Maple V Handbook*, AP PROFESSIONAL, 1994.

Auer, J.W., *Linear Algebra with Applications*, Prentice-Hall, Canada, 1991.

Auer, J.W., *Maple Solutions Manual for Linear Algebra with Applications*, Prentice-Hall, Canada, 1991.

Bauldry, W.C., and Fielder, J.R., *Calculus Laboratories with Maple*, Brooks/Cole, 1991.

Burbulla, C.C.M., and Dodson, C.T.J., *Self-Tutor for Computer Calculus Using Maple*, Prentice-Hall, Canada.

Char, B.W., Geddes, K.O., Gonnet, G.H., Leong, B.L., Monagan, M.B., and Watt, S.M., *First Leaves: A Tutorial Introduction to Maple V*, Springer-Verlag, 1992.

Char, B.W., Geddes, K.O., Gonnet, G.H., Leong, B.L., Monagan, M.B., and Watt, S.M., *Maple V Library Reference Manual*, Springer-Verlag, 1991.

Char, B.W., Geddes, K.O., Gonnet, G.H., Leong, B.L., Monagan, M.B., and Watt, S.M., *Maple V Language Reference Manual*, Springer-Verlag, 1991.

Cheung/Harer, *A Guide to Multivariable Calculus with Maple V*. New York: John Wiley & Sons.

Cox, D., Little, J., and O'Shea, D., *Ideals, Varieties and Algorithms—An Introduction to Computational Algebraic Geometry and Commutative Algebra*, Springer-Verlag.

Devitt, J.S., *Calculus with Maple V*, Brooks/Cole.

Ellis, W., and Lodi, E., *Maple for the Calculus Student: A Tutorial*, Brooks/Cole, 1989.

Ellis, W., Lodi, E., Johnson, and Schwalbe, *The Maple V Flight Manual*, Brooks/Cole, 1992.

Fattahi, A., *Maple V Calculus Labs*, Brooks/Cole, 1992.

Geddes, K.O., Czapor, and Labahn, *Algorithms for Computer Algebra*, Kluwer Academic Publishers, 1992.

Geddes, K., Marshman, McGee, Ponzo, and Char, *Maple Calculus Workbook: Problems and Solutions* (Available from Waterloo Maple Software).

Gloggengieber, H. *Maple V Software fur Mathematiker*, Markt & Technik.

Harper, D., Wooff, C., and Hodgkinson, D., *A Guide to Computer Algebra Systems*, John Wiley & Sons.

Harris, K., *Discovering Calculus with Maple*, John Wiley & Sons.

Heck, A., *Introduction to Maple—A Computer Algebra System*, Springer-Verlag, 1992.

Holmes, M.H., Ecker, J.G., Boyce, W.E., and Siegmann, W.L., *Exploring Calculus with Maple*. Reading, Mass.: Addison-Wesley Publishing Company.

Johnson, E., *Linear Algebra with Maple V*, Brooks/Cole.

Kamerich, E., *A Guide to Maple*, Springer-Verlag.

Kreyszig, H.E., *Maple Manual to Accompany Advanced Engineering Mathematics*, 7th ed. New York: John Wiley & Sons.

Redfern, D., *The Maple Handbook*, Springer-Verlag.

Some of the Maple texts in progress include the following:

Blachman, N., *Maple V Quick Reference Guide*, Brooks/Cole Publishing Co.

Char, B.W., Geddes, K.O., Gonnet, G.H., Leong, B.L., Monagan, M.B., and Watt, S.M., *Maple V for the PC*, Brooks/Cole Publishing Co.

Char, B.W., Geddes, K.O., Gonnet, G.H., Leong, B.L., Monagan, M.B., and Watt, S.M., *Maple V for the Macintosh*, Brooks/Cole Publishing Co.

For information, including purchasing information, about Maple contact:

Waterloo Maple Software
160 Columbia Street West
Waterloo, Ontario, Canada
N2L 3L3

Phone: (519) 747-2373
Sales: 1-800-877-6583
Fax: (519) 747-5284
E-mail: info@maplesoft.on.ca

In Europe, contact:

Waterloo Maple Software
Tiergartenstraße 17
W-6900 Heidelberg 1
Germany

Phone: +49-6221-487 180
Fax: +49-6221-487 184

Index

QUICK REFERENCE

≣ Frequently Used Abbreviations

+	addition	!	factorial
->	arrow	>	greater than
:=	assignment	>=	greater than or equal to
@	composition	%	label
.	concatenation or decimal	"	latest expression
		<	less than
&*	noncommutative multiplication	<=	less than or equal to
		*	multiplication
/	division	<>	not equal
..	ellipses	@@	repeated composition
=	equal	$	sequence operator
**	exponentiation	-	subtraction
^	exponentiation		

≣ Frequently Used Commands

Topic	Example	Command
Compute a **determinant**	Compute $\begin{vmatrix} 4 & -3 \\ 2 & 1 \end{vmatrix}$.	`with(linalg):` `det([[4,-3],[2,1]]);`
Compute a **limit**	Calculate $\lim\limits_{x \to 2} \dfrac{x^2 + x - 6}{x^2 + 2x - 8}$.	`limit((x^2+x-6)/` `(x^2+2*x-8),x=2);`
Compute a **power series**	Compute the first five terms of the power series for $\sin x$ about $x = 0$.	`series(sin(x),x=0,5);`
Compute the **partial fraction decomposition** of an expression	Find the partial fraction decomposition of $\dfrac{x}{x^2 - 3x - 4}$.	`convert(` `x/(x^2-3*x-4),parfrac);`
Define a **function**	Define $f(x) = \dfrac{x^2}{x^2 + 1}$.	`f:=x->x^2/(x^2+1);`
Define a **matrix**	Define $A = \begin{pmatrix} 4 & 3 \\ 5 & 0 \end{pmatrix}$.	`A:=array([[4,3],[5,0]]);`
Differentiate an expression	Compute $\dfrac{d}{dx}(x \sin x)$.	`diff(x*sin(x),x);`

(continued on other side)

Display an expression as a single **fraction**	Write $1 + \dfrac{1}{x}$ as a single fraction.	`simplify(1+1/x);`
Factor a polynomial	Factor $5x^2 - 8x - 4$.	`factor(5*x^2-8*x-4);`

Find the **eigenvalues** and corresponding **eigenvectors** of a matrix	Find the eigenvalues and corresponding eigenvectors of $\begin{pmatrix} -17 & -15 \\ 20 & 18 \end{pmatrix}$.	`with(linalg):`
		`eigenvects(`
		`[[-17,-15],[20,18]]);`

Find the **inverse** of a matrix	Calculate $\begin{pmatrix} -4 & 3 \\ 4 & -4 \end{pmatrix}^{-1}$.	`with(linalg):` `inverse([[-4,3],[4,-4]]);`
Generate a random number	Generate a random integer between 0 and 10.	`rand() mod 11;`
Graph a function	Graph $\dfrac{x}{x^2+1}$ on the interval $[-6,6]$.	`plot(x/(x^2+1),x=-6..6);`
Graph a **function of two variables** in three dimensions	Graph $\sin x \cos y$ for $0 \le x \le 4\pi$ and $0 \le y \le 2\pi$.	`plot3d(sin(x)*cos(y),` `  x=0..4*Pi,y=0..2*Pi);`
Graph parametric equations	Graph $\begin{cases} x = \cos t \\ y = 4 \sin t \end{cases}$ for $0 \le t \le 2\pi$.	`plot([cos(t),4*sin(t),` `  t=0..2*Pi]);`
Graph several functions	Graph $\sin x^2$ and $\sin^2 x$ on the interval $[0, 2\pi]$.	`plot({sin(x^2),sin(x)^2},` `  x=0..2*Pi);`
Integrate an expression	Compute $\int x \sin x \, dx$.	`int(x*sin(x),x);`
Multiply an algebraic expression	Compute $(5x + 2)(x - 2)$.	`expand((5*x+2)*(x-2));`

Multiply together two matrices	Compute AB if $A = \begin{pmatrix} 4 & 2 \\ 7 & 4 \end{pmatrix}$ and $B = \begin{pmatrix} 1 & 9 \\ 6 & 6 \end{pmatrix}$	`A:=array([[4,2],[7,4]]);`
		`B:=array([[1,9],[6,6]]);`
		`evalm(A &* B);`

Reduce a fraction to lowest terms	Reduce $\dfrac{x-1}{x^2-1}$ to lowest terms.	`simplify((x-1)/(x^2-1));`
Solve a **differential equation**	Solve $y' = 1 + y$.	`dsolve(diff(y(x),x)` `=1+y(x),y(x));`
Solve a **system of equations**	Solve $\begin{cases} 2x - y = 7 \\ 4x + 2y = 2 \end{cases}$.	`solve({2*x-` `  y=7,4*x+2*y=2});`
Solve an **equation**	Solve $x^2 - 4x - 5 = 0$.	`solve(x^2-4*x-5=0);`